Trace Elements in the Terrestrial Environment

D.C. Adriano

Trace Elements
in the Terrestrial Environment

With 99 illustrations

Springer-Verlag New York Berlin Heidelberg Tokyo

D.C. Adriano
Senior Scientist
Biogeochemical Ecology Division
Savannah River Ecology Laboratory
University of Georgia Institute of Ecology
Savannah River Plant
Aiken, South Carolina 29801, USA

Library of Congress Cataloging in Publication Data
Adriano, D.C.
    Trace elements in the terrestrial environment.
    Includes bibliographies and index.
    1. Trace elements—Environmental aspects. 2. Trace elements in nutrition.
3. Biogeochemical cycles.
I. Title.
QH545.T7A37   1986   574.19'214   85-9914

Typeset by David E. Seham Associates, Inc., Metuchen, New Jersey.
Printed and bound by R.R. Donnelley & Sons Co., Harrisonburg, Virginia.
Printed in the United States of America.

9 8 7 6 5 4 3 2 1

ISBN 0-387-96158-5 Springer-Verlag New York Berlin Heidelberg Tokyo
ISBN 3-540-96158-5 Springer-Verlag Berlin Heidelberg New York Tokyo

In gratitude to

Larry S. Murphy and Parker F. Pratt
for introducing me to and strengthening my background in agronomy and soil science,

Albert L. Page
for invigorating my interest in trace element research, and

Eugene P. Odum
for founding our laboratory and inspiring us to enhance environmental quality and conservation.

Preface

I intend to fill, with this book, a need that has long been felt by students and professionals in many areas of agricultural, biological, natural, and environmental sciences—the need for a comprehensive reference book on many important aspects of trace elements in the "land" environment. This book is different from other books on trace elements (also commonly referred to as heavy metals) in that each chapter focuses on a particular element, which in turn is discussed in terms of its importance in our economy, its natural occurrence, its fate and behavior in the soil-plant system, its requirement by and detriment to plants, its health limits in drinking water and food, and its origin in the environment. Because of long-distance transport to pristine areas of cadmium, lead, copper, and zinc in relatively large quantities, these elements have an extra section on natural ecosystems. A blend of pictorial and tabular data are provided to enhance understanding of the relevant information being conveyed. Since individual chapters are independent of one another, they are arranged alphabetically. However, readers with weak backgrounds in soil science are advised to start with the chapter on zinc, since soil terminology is discussed in more detail here. Sections on sorption, forms and speciation, complexation, and transformations become more technical as soil physical–(bio)chemical phenomena are discussed. The less important "environmental" trace elements are discussed together in the "Other Trace Elements" chapter.

This book could also serve as complementary reading material in soil chemistry, soil fertility and plant nutrition, biogeochemistry, and related courses. Since complete coverage of the immense and growing literature on trace elements in geologic material, soils, plants, natural ecosystems, and drinking water and food is impossible, it is hoped that the most pertinent ones are included.

This book is an outgrowth of my interest in trace elements research. My early interest in trace elements was kindled by the then poorly understood "macro-micronutrient" interactions in the nutrition of agronomic crops when I was a graduate student in the 1960s under Drs. L.S. Murphy, (the late) R. Ellis, Jr., and G.M. Paulsen at Kansas State University. It was temporarily sidetracked by my other interest in environmental chemistry of pollutants, which blossomed during my postdoctoral research with Dr. P.F. Pratt at the University of California at

Riverside, and during my stay at Michigan State University with Drs. B.G. Ellis and A.E. Erickson. Since 1975, I have been conducting research in trace elements (in the form of environmental wastes) that include radionuclides (i.e., primarily the actinides) from utilization of nuclear energy. In 1978–79, I spent my sabbatical researching heavy metals in the form of major solid wastes with Dr. A.L. Page at the University of California, Riverside. It was during my sabbatical that the basic nucleus of the book was born.

The book is geared to answer some of the most commonly asked questions about trace elements—when? what? where? and why? It is hoped, therefore, that students and specialists find this book a handy source of information on the more important aspects of trace elements in our ever-changing "land" environment.

D.C. Adriano
Aiken, South Carolina

Acknowledgments

A book of this nature could have not been written without the cooperation and assistance of many people. Foremost was the support of my family: my wife Zena, for her continuous encouragement and help in assembling reading and writing material; and my children Aileen, Mary Ann, Larry, and Cindy, for their help in collecting, assembling, and archiving the literature. More importantly, their incredible understanding and patience enabled me to devote so many long weeks to this book. I am also very grateful to my colleagues at the laboratory, particularly our director, Dr. Michael H. Smith, for their cooperation and encouragement throughout the preparation of the manuscript. I am also indebted to the following individuals: Mrs. Jean Coleman for preparing the illustrations; Mrs. Jean Mobley for ordering the hundreds of articles; and Mmes. Nancy Barber, Cathy Houck, and Debbie Perks for typing parts of the manuscript; Dr. D. Mokma of Michigan State University for his suggestions in the glossary of soil terms; Mrs. Peggy Burkman for preparing the scientific names for plants; to my "inner" colleagues Brenda Rosier, Bob Lide, and especially Bill Burkman, for their assistance in proofreading the galley and page proofs.

It is with great pleasure to acknowledge the superb cooperation and constructive criticisms of the following reviewers: Dr. A.L. Page, University of California (Riverside)—who served as reviewer general; Dr. F.C. Boswell, University of Georgia—chapter on manganese; Dr. E.E. Cary, USDA-ARS, Cornell University—chapter on selenium; Dr. M. Chino, Tokyo University—chapter on cadmium; Dr. F. D'Itri, Michigan State University—chapter on mercury; Dr. B.G. Ellis, Michigan State University—chapter on chromium; Dr. U.C. Gupta, Agriculture Canada—chapter on boron; Dr. T.C. Hutchinson, University of Toronto, Canada—chapter on nickel; Dr. W.M. Jarrell, University of California (Riverside)—chapter on molybdenum; Dr. D.R. Keeney, University of Wisconsin—chapter on arsenic; Dr. T.J. Logan, Ohio State University—introductory chapter; Dr. D.C. Martens, Virginia Polytechnic Institute and State University—chapter on zinc; Dr. J.J. Mortvedt, Tennessee Valley Authority—chapter on copper; Dr. P.J. Peterson, Monitoring and Assessment Research Center, University of London—chapter on other trace elements; and Dr. D.J. Swaine, CSIRO, Australia—chapter on lead.

To these reviewers, particularly Dr. A.L. Page, who reviewed all chapters, my sincere gratitude, although errors and oversights that may still exist in the text are fully my responsibility.

Finally, I am indebted to my colleagues all over the world for their cooperation by sending reprints, answering inquiries, and supplying material for inclusion in the book.

Contents

1
Introduction

The ever-increasing production and demand for some elements in developed and developing countries (see Appendix Table 1.1)* suggest the mounting probability of their dispersal and contact with the environment. An element may be dispersed from the time its ore is being mined to the time it becomes usable as a finished product or ingredient of a product. In addition, increasing demands for fertilizers in high-production agriculture may enhance this probability. Land disposal techniques that seem promising for agricultural wastes and other solid wastes may also increase the metal burden of the soil. Trace element research has been intense during the last three or so decades, highlighted by exploration into Itai-itai disease and Minamata disease in Japan. At stake are the integrity and quality of land resources in the United States: 182.1×10^6 hectares (ha) of cropland, 404.7×10^6 ha of forest and range, and 202.4×10^6 ha of nonagricultural land (USDA, 1969). At similar risk are the world's land resources.

It has recently been estimated that about 0.5, 20, 240, 250, and 310 million tonnes of Cd, Ni, Pb, Zn, and Cu have been mined and ultimately deposited in the biosphere (Nriagu, 1984). Other estimates place the amounts of anthropogenic As, Cd, Pb, Cu, and Zn that are currently being disseminated annually via the atmosphere to distant ecosystems at about 22,000, 7,300, 400,000, 56,000, and 214,000 tonnes, respectively. Nriagu (1984) indicated that, in many instances, the inputs from anthropogenic sources exceed the contributions from natural sources by several fold. Thus, it has become evident that human activities have altered the global cycles of trace elements.

Current universal interests in trace element studies are being spurred by our needs to (1) increase food, fiber, and energy production; (2) determine trace element requirements and tolerance by organisms, including relationships to animal and human health and disease; (3) evaluate the potential biomagnification and biotoxicity of trace elements; (4) understand

---

*This appendix is provided on pages 22–41.

trace element cycling in nature, including their biogeochemistry; (5) assess
trace element enrichment in the environment by recycling wastes; (6) dis-
cover additional ore deposits; and (7) comply with stringent state and federal regulations on releases of effluents (both aqueous and gaseous) to the
environment.

The significance of trace elements in the environment was underlined
during the Conference on Human Environment organized by the Royal
Swedish Academy of Sciences. In this conference, held in Rättvik, Sweden
in 1982, the following environmental research and management priorities
for the 1980s were identified by carefully selected scientists from various
countries around the world (Munro, 1983):

RESEARCH PRIORITIES FOR THE 1980S:

Depletion of tropical forests
Reduction of biological diversity
Cryptic spread of mutant genes
Droughts and floods
CO_2 buildup and climate change
Impact of hazardous substances on ecosystems and man
Loss of productive land due to salinization
Impact of urbanization
Meeting current and future energy needs

MANAGEMENT PRIORITIES FOR THE 1980S:

Management of hazardous chemicals, processes, and wastes
Depletion of tropical forests
Desertification due to overgrazing
Control of pathogens from human wastes and their aquatic vectors
River basin management
Population growth and urbanization
Acid deposition
Species loss
Protection of the marine environment
Fuelwood crises

1. Definitions and Functions of Trace Elements

The trace elements that have been studied most extensively in soils are
those that are essential for the nutrition of higher plants: B, Cu, Co, Fe,
Mn, Mo, and Zn. Similarly, those extensively studied in plants and foodstuff because of their essentiality for animal nutrition are: Cu, Co, Fe,
Mn, Mo, Zn, Cr, F, Ni, Se, Sn, and V.

The term "trace element" is rather loosely used in the literature and has differing meanings in various scientific disciplines. Often it designates a group of elements that occur in natural systems in minute concentration. Sometimes it is defined as those elements used by organisms in small quantities but believed essential to their nutrition. However, it broadly encompasses elements including those with no known physiological functions. Earth scientists generally view "trace elements" as those other than the eight abundant rock-forming elements found in the biosphere (O, Si, Al, Fe, Ca, Na, K, and Mg). It is a general consensus that an element is considered "trace" in natural materials (i.e., lithosphere) when present at levels of less than 0.1%. In biochemical and biomedical research, trace elements are considered to be those that are ordinarily present in plant or animal tissue in concentrations comprising less than 0.01% of the organism. In food nutrition, "trace element" may be defined as an element which is of common occurrence but whose concentration rarely exceeds 20 parts per million (ppm) in the foodstuffs as consumed. It should be noted that some of the "nutritive" trace elements (for example, Mn and Zn) may often exceed this concentration.

In this book, "trace elements" refers to elements that occur in natural and perturbed systems in small amounts and that, when present in sufficient concentrations, are toxic to living organisms. Other terms that have been used and, for all practical purposes, considered synonymous for "trace elements" are: "trace metals," "heavy metals," "micronutrients," "microelements," "minor elements," and "trace inorganics." The use of "micronutrients" usually is restricted to those elements required by higher plants (Zn, Mn, Cu, Fe, Mo, and B). The term "heavy metals" usually refers to elements having densities greater than 5.0. In this book, the trace elements to be considered are: arsenic (As), silver (Ag), boron (B), barium (Ba), beryllium (Be), cadmium (Cd), cobalt (Co), chromium (Cr), copper (Cu), fluorine (F), mercury (Hg), manganese (Mn), molybdenum (Mo), nickel (Ni), lead (Pb), antimony (Sb), selenium (Se), tin (Sn), titanium (Ti), thallium (Tl), vanadium (V), and zinc (Zn). Iron (Fe) and aluminum (Al) are not included because of their abundance in environmental matrices, especially in the earth's crust.

In the plant kingdom, especially with higher plants, an element is considered essential if it meets the following criteria (Hewitt and Smith, 1974):

1. Omission of the element must directly cause abnormal growth or failure to complete the life cycle, or premature senescence and death;
2. The effect must be specific, and no other element can be substituted in its place; and
3. The effect must be direct on some aspect of growth or metabolism. Indirect or secondary beneficial effects of an element, such as reversal of the inhibitory effect on some other element, do not qualify an element as essential.

In animal nutrition, the criteria for essentiality are somewhat similar

(Underwood, 1977): (1) consistent significant growth response to dietary supplements of the element and this element alone; (2) development of the deficiency state on diets otherwise adequate and satisfactory; and (3) correlation of the deficiency state with the occurrence of subnormal levels of the element in the blood or tissues of animals exhibiting the response. The essentiality, benefits, and potential toxicity of trace elements are shown in Table 1.1. For higher plants, B, Cu, Mn, Mo, and Zn are essential; Co is required for N fixation in bacteria and algae; Se and V are essential for *E. coli* and alga *Scenedesmus obliquus* (Hewitt and Smith, 1974). In animals, Cu, Co, Zn, Mn, Se, Cr, V, Sn, Ni, F, As, and Mo are essential (Underwood, 1977).

TABLE 1.1. Essentiality and effects of trace elements on plant and animal nutrition in terrestrial environment.[a]

Element	Essential or beneficial to		Potential toxicity to		Comments
	Plants	Animals	Plants	Animals	
Ag	No	No		Yes	Interacts with Cu and Se
As	No	Yes	Yes	Yes	Phytotoxic before animal toxicity; may be carcinogenic
B	Yes	No	Yes		Narrow margin, especially in plants
Ba	No	Possible			Insoluble; relatively nontoxic
Be	No	No	Yes	Yes	Speciation important; carcinogenic
Bi	No	No	Yes	Yes	Relatively nontoxic
Cd	No	No	Yes	Yes	Narrow margin; enriched in food chain; carcinogenic; Itai-itai disease
Co	Yes	Yes	Yes	Yes	Relatively nontoxic; high enrichment factor; carcinogenic
Cr	No	Yes	Yes		Speciation important; Cr^{6+} very toxic; otherwise relatively nontoxic; carcinogenic
Cu	Yes	Yes	Yes		Easily complexed in soils; narrow margin for plants
F	No	Yes	Yes		Accumulative toxicity for plants and animals
Hg	No	No		Yes	Enriched in food chain; aquatic accumulation; Minamata disease
Mn	Yes	Yes	<pH 5		Wide margin; toxic in acid soils; among the least toxic
Mo	Yes	Yes		5–20 ppm	High enrichment in plants; narrow margin for animals
Ni	No	Yes	Yes	Yes	Very mobile in plants; relatively nontoxic; carcinogenic
Pb	No	No	Yes	Yes	Aerial dispersion and primarily surface deposited; cumulative poison
Sb	No	No		Yes	Insoluble; relatively nontoxic

TABLE 1.1. *Continued*

Element	Essential or beneficial to		Potential toxicity to		Comments
	Plants	Animals	Plants	Animals	
Se	Yes	Yes	Yes	4 ppm	Narrow margin for animals; interacts with other trace metals
Sn	No	Yes		Yes	Relatively nontoxic; very low uptake by plants
Ti	No	Possible			Insoluble; relatively nontoxic; possibly carcinogenic
Tl	No	No		Yes	Very mobile in plants
V	Yes	Yes	Yes	Yes	Narrow margin and highly toxic in animals; high enrichment factor; carcinogenic
W	No	No			Very mobile in plants; very rare and insoluble
Zn	Yes	Yes			Wide margin; easily complexed in soils; maybe lacking in some diets; relatively nontoxic

[a] Extracted from Allaway (1968); Hewitt and Smith (1974); Loehr et al (1979); Zingaro (1979); Wood and Goldberg (1977); Luckey and Venugopal (1977); Underwood (1975; 1977); Van Hook and Shults (1976); Miller and Neathery (1977).

2. Biogeochemical Cycles of Trace Elements

2.1 In Agro-Ecosystems

In agricultural ecosystems, most trace elements are cycled as visualized in Figure 1.1. In this conceptualization, the relative importances of the various transfer pathways vary considerably, depending on the element, plant species, soil type, location, terrain, management practices, and others. There are basically two components: plant and soil. The equilibrium between soluble, available, and unavailable fractions of elements is also schematically shown by Hodgson (1963). This equilibrium indicates that the soil solution is the focal "crossroad" of trace element transformation. Practically speaking, there are two major routes for input of trace elements into agro-ecosystems: aerial (e.g., aerosols, particulate matter, resuspended and airborne dusts, etc.), and land (fertilizers, pesticides, solid wastes, other soil amendments, etc.). The output pathways can be represented primarily by losses through plant tissue removal for food, feedstuff, and fiber, and by leaching and erosion. Because of almost constant changes in outputs and inputs, agro-ecosystems are practically in nonequilibrium state with respect to elemental cycling. Depending on several factors, an agro-ecosystem often is either enriched or deficient in one or more (trace) elements. Very seldom can they sustain maximum productivity for several years without supplemental addition of some "micronutrients."

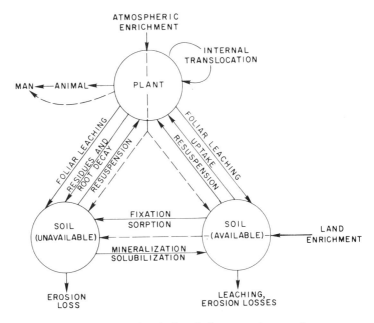

FIGURE 1.1. Generalized biogeochemical cycle for trace elements in agro-ecosystems.

2.2 In Forest Ecosystems

A conceptualized model for trace element cycling in forest ecosystems (Fig. 1.2) is similar to that for agro-ecosystems in many respects. Again, the significance of various transfer pathways is dependent on several factors, foremost of which are the element, soil type, and forest type. But unlike agro-ecosystems, there is practically only one input pathway in established forest ecosystems, and that is atmospheric input. Erosion seldom occurs in aged forests; however, leaching can still occur. Because of the presence of an understory and forest floor, there are more transfer pathways than in agro-ecosystems. In contrast to agro-ecosystems, aged and established forests are usually in a state of equilibrium with respect to elemental cycling.

The soil–plant system serves as an effective barrier against toxicity for some trace elements (e.g., As, Be, F, Ni, Zn, Cr, Cu, Pb, V, etc.) in that plant growth will be greatly depressed or cease before these elements will be taken up from the soil and accumulated in concentrations that would be dangerous for animals. The soil–plant system exerts an effective buffering action on the environmental cycling of trace elements. The amount of each trace element listed in Table 1.1 that is present in the plant root zone in the soil is a hundred to a hundred thousand times the amount likely to be removed by any one crop (Allaway, 1968).

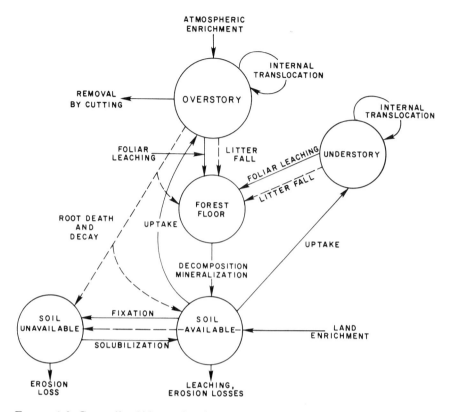

FIGURE 1.2. Generalized biogeochemical cycle for trace elements in forest ecosystems. *Source:* Modified from Van Hook et al, 1980, with permission of Wiley Interscience, New York.

3. Sources of Trace Elements

There are several sources of trace elements in the environment, both natural and manmade: soil parent material (rocks), commercial fertilizers, liming materials, sewage sludges, animal wastes, pesticides, irrigation waters, coal combustion residues, metal-smelting industries, auto emissions, and others. With the exception of the parent material, all are anthropogenic in nature. The first seven sources are the primary input sources in agro-ecosystems. The rest may impact on natural ecosystems, as well as on urban and rural areas. The projected annual production rates for different types of solid and semisolid wastes in the United States are indicated in Figure 1.3.

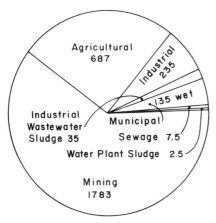

FIGURE 1.3. Estimated annual production rates for different types of solid and semisolid wastes in the USA, 1970–1974 (dry weight in 10^6 tonnes/year). *Source:* Lippman and Schlesinger, 1979, with permission.

3.1 Soil Parent Material

This subject matter has been reviewed by Cannon (1978) and Mitchell (1964) and will not be discussed at length here. Free of human interference, the trace element content of the soil is largely dependent on that of the rocks from which the soil parent material was derived and on the process of weathering to which the soil-forming materials have been subjected. The more aged and older the soil, the less may be the influence of parent rocks. The extreme variable nature of trace element concentration in various soil-forming rocks is demonstrated in Appendix Table 1.2.

3.2 Commercial Fertilizers

A micronutrient survey in the 1960s revealed that soils are deficient in plant-available forms of one or more of the six microelements (Zn, Mn, Fe, Cu, Mo, and B) and that applications are recommended in every state of the United States (Sparr, 1970). The approximate quantities of micronutrients sold for agricultural use in the United States in 1968, in thousands of tonnes, are: Zn, 14.5; Mn, 10.6; Cu, 2.4; B, 2.5; and Mo, 0.8. Although micronutrient requirement by crop varied considerably from state to state, more vegetable crops needed these elements than did field, forage, or fruit and nut crops. The form, composition, and rates of these micronutrients commonly used in agriculture are given by Murphy and Walsh (1972).

   In addition to these micronutrient carriers, other commercial fertilizers contain small amounts of trace elements (Appendix Table 1.3). Phosphatic fertilizers contain varying amounts of Zn, Cd, and other trace elements that originated in phosphate rock. Differences in trace element content of phosphate rocks mined in various areas are caused by impurities co-precipitated with the phosphates at the time of deposition (Mortvedt and Giordano, 1977). In general, phosphate rock from the western United

States contains higher concentrations of most trace elements than phosphate rock from eastern phosphate deposits. Mortvedt and Giordano (1977) calculated that application of 500 kg/ha of diammonium phosphate (20-48-0) for 100 years using North Carolina phosphate rock contributed to the soil the following amounts, in kg/ha: Zn, 14.25; Cu, 0.05; Cd, 1.50; Cr, 9.75; Ni, 1.90; and Pb, 0.24.

3.3 Pesticides and Lime

Pesticides (herbicides, insecticides, fungicides, rodenticides, etc.) are widely used in high-production agriculture for the control of insects and diseases in fruit, vegetables, and other crops. They can be applied by spraying, dusting, or soil application. Sometimes, seed treatment is used. Typical pesticides in use in Canada for several decades are shown in Appendix Table 1.4. The amounts of elements applied vary according to the type of pesticide used, ranging as low as 0.002 kg/ha Hg from methyl mercurials to as high as 0.5 kg/ha of As and 2.3 kg/ha of Pb from lead arsenate per application (Frank et al, 1976). Such contributions can total several kilograms of certain elements in a given fruit orchard.

Lime is also widely used in agriculture to correct soil acidity and optimize nutrient uptake. The most commonly used liming material is agricultural limestone, which is a mixture of calcite and dolomite. The trace elements (Appendix Table 1.5) originate from the rock from which the product is derived. The amount of lime applied depends on the degree of acidity and on other chemical properties of the soil (buffering capacity).

In the southern United States, nearly 10×10^6 tonnes of agricultural limestone are spread annually on acid soils (Pearson and Adams, 1967). The minimum rate of lime is generally 2.2 tonnes/ha, except in special cases where "overliming" injury is likely to occur, or where exact pH control is necessary. Based on the chemical composition of lime as given in Appendix Table 1.5, application of 4.5 tonnes/ha would add, per hectare, an average of 0.140 kg of Zn, 1.48 kg of Mn, and 0.012 kg of Cu.

3.4 Sewage Sludges

Land disposal of municipal sewage sludge and effluents is an old practice that recently has been attracting the public eye because of the disposal problems associated with increasing amounts of these wastes produced by urban and industrial activities. There were about 7×10^6 tonnes of municipal sewage sludge produced in the United States in 1980 (NAS, 1977), and this amount was projected to about double by 1990. With the concern over disposal of sludge in the ocean and the high cost of sludge incineration, land application and landfilling are becoming more popular. The primary objective of waste disposal is to get rid of wastes in the most environmentally acceptable and economical manner. Land disposal is often more economical than other treatment alternatives. However, with disposal as the goal, it is more economical to concentrate the wastes on small land areas and at locations of short distance from treatment facilities. This will increase the pollutant burden of the soil, and numerous environmental consequences may ensue.

Sewage effluents have been applied to land in Europe and Australia for nearly a hundred years, and at several sites in the United States, applications have been in progress for up to 50 years (Carlson and Menzies, 1971). A good example of how treated sewage effluents can be used on the land is the system used by Sopper and associates at Pennsylvania State University (Kardos, 1970). The system involves using the water and nutrients in sewage effluents on forest and cultivated crops. Recycling nutrients to the land, restoring ground water, preventing stream pollution, and eliminating the need to add commercial fertilizers to cropland are some of the benefits derived from this system. This particular system holds promise for solving waste disposal problems in small villages and cities. The potential for recycling sewage effluent in cropland is demonstrated at a sewage farm in Braunschweig, Federal Republic of Germany, where even after operation since 1895, the trace element burden of the soil is still within the tolerable limits with regard to their plant compatibility as proposed by the Federal German Board of Environmental Protection (El-Bassam et al, 1979). The nondeleterious enrichment of the soil with trace elements from this long-term application is shown in Appendix Table 1.6.

Sewage sludges and effluents contain a wide range of plant nutrients and metals. At the present time, recommendations for sludge application rates on land are based on the fertilizer value (nitrogen and phosphorus) and on the concentration of trace elements present in sludge. The elements of primary concern are Cd, Zn, Cu, Pb, and Ni, which, when applied to soils in excessive amounts, may depress plant yields or degrade the quality of food or fiber produced. The concentrations of trace elements in soils in areas treated with sewage sludge can be expected to increase with increasing concentrations of these elements in the sludge (Fig. 1.4). The variable nature of trace element concentration in sewage sludge has been reviewed by Page (1974) and Sommers (1977). Guidelines for land application of sewage sludge have been formulated in Ohio (Miller et al, 1979), Wisconsin (Keeney et al, 1975), and other states. Appraisals of the potential hazards of trace elements in sewage sludge to plants and/or animals from land application of sewage sludge have been conducted (CAST, 1976, 1980).

The variable nature of trace element concentration in sewage sludges is demonstrated in Table 1.2 and in Appendix Tables 1.7 and 1.8. The concentration is largely dependent on the types and amounts of urban and industrial discharges into the sewage treatment systems and to the amount added in the conveyance and treatment systems. The data in Table 1.2 are typical of sludge from communities without excessive industrial waste

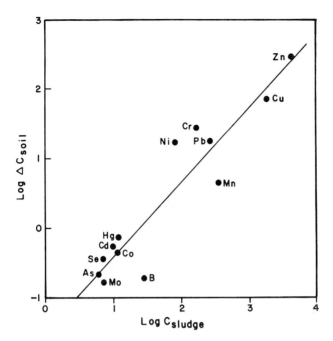

FIGURE 1.4. Relationship between concentration of trace elements in sewage sludge (C_{sludge}) and the increase of trace elements in the soil (ΔC_{soil}) from using sewage sludge. *Source:* Andersson and Nilsson, 1972, with permission of the authors and the publisher.

TABLE 1.2. Concentrations of potentially harmful trace elements in domestic digested sewage sludges (ppm, dry wt. basis).[a,b]

Element	Mean	Range
Cadmium	<15	0.6–3,400
Copper	<1,000	250–17,000
Chromium	<1,000	50–30,000
Mercury	<10	<1–100
Nickel	<200	20–8,000
Lead	<1,000	100–10,000
Zinc	<2,000	500–50,000
Cadmium/zinc	<1%	0.1–110 %

[a] *Source:* Adapted from Stewart and Chaney, 1975, with permission of the authors.
[b] Typical of sludge from USA communities without excessive industrial waste inputs or with adequate abatement.

inputs, such as those of small cities in Iowa (Appendix Table 1.8). As urbanization and industrialization progress, the concentrations of most trace elements increase (Appendix Tables 1.7 and 1.8). As a general rule, sludges should be analyzed for nutrients and trace elements before application on land to aid in calculating the fertilizer and metal loads for a particular application rate.

3.5 Animal Wastes

Concentration of elements in animal wastes is variable. The variability is related to the classes of livestock. Within a class, additional variation is associated with age of the animal, type of ration, housing type, and even waste management practice. The trace elements in animal wastes originate from the rations and whatever dietary supplement is added. For example, in the production of milk and pork, Cu and Co may be added to the diet as 1% copper sulfate and as 1% cobalt iodized salts. Some swine and poultry are fed diets containing up to 250 ppm Cu and 100–200 ppm Zn in the absence of antibiotics. The manure from these animals is then 10 to 40 times higher in Cu and 4 to 10 times higher in Zn than normal. Application of these manures at N fertilization rate would add 3 to 6 kg Cu/ha/year (Baker, 1974). Typical trace element composition of animal wastes is shown in Appendix Table 1.9. Unlike sewage sludge, however, animal wastes are rarely high in potentially harmful trace elements. But there are also land disposal problems associated with these abundant wastes, produced at a daily rate of about 3.4×10^6 tonnes in the United States. Foremost problems are likely to be excess N and salts (Adriano et al, 1971, 1973) and nutrient imbalance in plants (causing grass tetany in cattle with the use of poultry manure).

3.6 Coal Residues

Use of coal combustion to generate electrical power may increase greatly during the next few decades as US dependence on foreign oil is being deemphasized. World production of bituminous and lignite coals was estimated in 1975 at 3.4×10^9 tonnes, 19% of which was produced in the United States (US Bureau of Mines, 1977). Approximately 73% of the 556×10^6 tonnes of coal produced in the United States in the late 1970s was consumed by the electric utilities. Current total ash production by electric utilities is estimated at about 70×10^6 tonnes yearly; 84% of this amount is fly ash (Adriano et al, 1980). Of the total ash produced in the United States, only about 20% is being used for cement making, concrete mixing, ceramics, and other products. Since there is no consistent trend of effective utilization, the accumulating fly ash becomes a continuing waste disposal problem. Excellent reviews on coal residues and its characterization, potential for utilization, and potential hazards to plants and/

or animals have been conducted recently (Adriano et al, 1980; Page et al, 1979; Terman, 1978).

The concentration of trace elements in coal residues is extremely variable and depends on the composition of the parent coal, conditions during coal combustion, efficiency of emission control devices, storage and handling of the by-product, and climate. Typical concentrations of trace elements that can be expected from a given coal fly ash are shown in Appendix Table 1.10. The flows and discharges of trace elements from US combustion of coal are shown in Appendix Table 1.11. When compared to elemental flows due to natural processes, such as weathering, some values are quite substantial. Particularly notable are the amounts of As, Cd, Mo, Se, and Zn.

Trace elements in coal residues that can biomagnify and therefore, would be critical in the food chain are As, B, Mo, Se, and V (Adriano et al, 1980).

3.7 Municipal Refuse

Disposal of municipal refuse, which consists primarily of paper, glass, ceramics, metals, and food wastes, is a problem of increasing importance. At present land-filling is the most widely used and least expensive method of disposal in the United States. Composting of refuse shows potential for horticultural use for home lawns and gardens. But this is only of limited market potential. In the United States it was estimated that about 1% of the domestic refuse might be used as compost for intensive, luxury-type agriculture (Hart, 1968). Boron may be a problem trace element in refuse composts (Purves and Mackenzie, 1974). Potential phytotoxicity also exists for Zn, Cu, and Ni from soil application of refuse.

3.8 Waste Waters

Reviews on potential hazards from trace elements in waste waters applied to land have been conducted (Bouwer and Chaney, 1975; Knezek, 1972; Leeper, 1972). The same elements listed under sewage sludge (Cd, Zn, Cu, Ni, and Pb) are important potential hazards if present in waste waters. In addition, B is considered as potentially hazardous in any irrigation water if present at greater than 0.75 ppm (Appendix Table 1.12). This concentration might be phytotoxic already to some forest tree species (Neary et al, 1975).

3.9 Mining and Smelting

The metal industry can be an important point source of trace elements in the environment from (1) the mining and milling operations with problems of grinding, concentrating, and transporting ores, and disposal of tails along with mine and mill waste water, and (2) the smelter-refinery process with problems of concentrate, haulage, storage, sintering, refining, atmospheric discharges, and blowing dust. The proportion of trace elements released into the environment depends on the ores being processed. In the lead industry, Pb, Cu, Zn, and Cd can be released in substantial amounts (Wixson et al, 1973; Ragaini et al, 1977). In the case of Ni and Cu smelting, Co, Zn, Pb, and Mn, as well as Ni and Cu, can also enrich the surrounding environment significantly (Hutchinson and Whitby, 1974). Similarly, in smelting zinc ores, sizable releases of Zn, Cd, Cu, and Pb can occur (Jordan, 1975). The impacts of atmospheric discharges from smelters can be detected within several kilometers from the point of release (Appendix Table 1.13).

In Great Britain, Davies and co-workers (Davies and Roberts, 1978; Davies and Ginnever, 1979) found widespread contamination by Pb, Cd, and Zn in some mining areas.

3.10 Auto Emission

The greatest single source of air contamination in the United States is automobile exhaust. Roadside soils and vegetation have been shown to be contaminated with various trace elements primarily from this emission (Lagerwerff and Specht, 1970; Motto et al, 1970). It includes Pb, Zn, Cd, Cu, and Ni, the more important being Pb from fuels and Zn from tires. Depending on the location and traffic intensity, contaminated zones can extend up to several hundred meters from the road.

Kloke et al (1984) reported that the most important sources of trace element contamination in industrialized countries, in descending order of importance, are: air pollution, river sediments, municipal sewage sludges, municipal waste composts, agricultural chemicals, and industrial wastes.

4. Soil Capacity for Trace Elements

Soil is the ultimate and most important sink of trace elements in the terrestrial environment. Most of the trace element inventory in a particular ecosystem is in the soil. However, the soil has only a finite capacity for holding these elements. Approaching or exceeding this capacity may lead into several environmental consequences, including increased mobility in the soil system. In the soil–plant–animal system, it is best not to exceed the soil capacity that will result in biotoxicity. Geochemically, an element introduced into the soil may end up in one or more of the following forms: (1) dissolved in soil solution, (2) held onto exchange sites of organic solids or inorganic constituents, (3) occluded or fixed into soil minerals, (4) precipitated with other compounds in soils, and (5) incorporated into biological material. The first two forms are the mobile forms and are phytoavailable;

the last three are immobile and sometimes become mobile and phytoavailable with time. The kinetics governing the chemical equilibrium of each trace element between the solution phase and the solid phase of the soil are complicated and not well understood. Metal activity in the soil solution is generally considered to be the result of metal equilibria among clay mineral, organic matter, hydrous oxides of Fe, Mn, and Al, and soluble chelators, with the soil pH strongly affecting these equilibria (Lindsay, 1979). Because of soil heterogeneity and inherent variability in chemical, mineralogical, and physical properties of the soil, it is extremely difficult to predict the fate and behavior of trace elements in soil without understanding the factors described in the sections below.

4.1 pH

In general, the capacity of soil for most trace elements is increased with increasing pH, with the maximum under neutral and slightly alkaline conditions. Exceptions are As, Mo, Se and some valency states of Cr that are commonly more mobile under alkaline or calcareous soil conditions. The relative mobility of some trace elements in soils as influenced by soil pH was summarized by Fuller (1977): In acidic soils (pH 4.2 to 6.6), Cd, Hg, Ni, and Zn are relatively "mobile"; As, Be, and Cr are "moderately mobile"; and Cu, Pb, and Se are "slowly mobile." In neutral to alkaline soils (pH 6.7 to 7.8), As and Cr are relatively mobile; Be, Cd, Hg, and Zn are moderately mobile; and Cu, Pb, and Ni are "slowly mobile." A decrease in plant uptake of B, Co, Cu, Mn, and Zn was observed when soil pH was increased from pH 5 to 8 (Hodgson, 1963). White et al (1979) found that soil pH was an important factor in determining the relative Zn tolerance of several soybean cultivars. In a recent CAST report (CAST, 1980), it was concluded that soil pH is the most critical factor in controlling plant uptake of Cd and Zn from sludge-treated soils. However, it was found that, even at a soil pH of 6.5, the Cd added in many sludges is sufficient to increase the Cd concentrations in most crops.

4.2 Cation Exchange Capacity

The cation exchange capacity (CEC) of soil is largely dependent on the amount and type of clay, organic matter, and iron, manganese, and aluminum oxides. These soil components have different cation exchange properties. In general, the higher the CEC of soil, the greater the amount of metal a soil can accept without potential hazards (Table 1.3). Numerous evidences indicate that CEC can best be viewed as a general, but imperfect, indicator of the soil components (i.e., clays, organic matter, and oxides of iron, aluminum, and manganese) that limit the solubility of metals instead of a specific factor in the availability of these metals. For example, less than 1% of the total Cd and Zn applied to soils in sludge is found in the exchangeable form (Silviera and Sommers, 1977; Latterell et al, 1978).

TABLE 1.3. Guidelines on cumulative metal additions in soil based on cation exchange capacity.[a]

Element	Cumulative metal additions (kg/ha)[b]		
	<5 meq/100 g	5–15 meq/100 g	>15 meq/100 g
Cadmium	5.5	11	22
Nickel	56	112	224
Copper	140	280	560
Zinc	280	560	1,120
Lead	560	1,120	2,240

[a] Source: Logan and Chaney, 1983, with permission of the authors.
[b] Proposed in 1976 by the North Central (USA) Regional Research Committee on Land Application of Sewage Sludges (NC-118). The Ni levels were later raised to be the same as for Cu.

4.3 Organic Matter

Some trace elements (Co, Cu, Mn, Ni, Pb, Zn, and others) exhibit rather high affinities for soil organic matter, otherwise known as humus. Somewhat stable soluble and insoluble complexes between the metals and soil organic matter may form. These complexes result from the binding of the metals through the carboxyl and phenolic functional groups in the organic matter. Thus, indigenous soil organic matter and that added in animal manures, sewage sludges, composts, peat, and plant residues bind trace elements in soils. Organic matter has both the cation exchange property and the chelating ability. Therefore, organic matter is sometimes viewed as a source of soluble complexing agents for trace elements. However, this is contradicted by frequent observation of micronutrient deficiencies, especially Cu, of crops grown on organic soils (muck soils). In addition, it is now established that trace elements in sludge are taken up less readily than are trace elements added to sludge as inorganic salts or trace elements directly added to soil in the same concentration as present in sludge (Kirkham, 1977). Haghiri (1974) indicated that the retaining power of organic matter for Cd was predominantly through its CEC property rather than its chelating ability.

4.4 Amount and Type of Clay

The amount of clay in relation to the amount of silt and sand determines the soil texture, which in turn influences the CEC of soils. In general, the higher the clay content, the higher the CEC. For example: clay loam soils may have CEC of 4 to 58 meq/100 g, whereas sandy loams may only have 2.5 to 17 meq/100 g. The 2:1 type clay minerals usually have higher CEC than 1:1 type minerals. For example, montmorillonite (2:1 type) may have 80 to 120 meq/100 g compared to only 3 to 15 meq/100 g for kaolinite (1:1 type). Dr. Fuller and his group at the University of Arizona showed that the soil capacity for trace metals is regulated by soil properties such as texture, clay, surface area, and iron oxide content. Of the trace elements present in cationic form, Cu and Pb are highly immobile in most soils, while Hg is relatively more mobile (Fig. 1.5.) Of the elements in anionic form, soil capacity is decreased in the order Se, V, As, and Cr (Fig. 1.5, bottom half). Korte et al (1976) found that soil capacity for elements in the cationic form is best correlated with the surface area (amount of clay), while those present in anionic form are more strongly correlated with the free iron oxides in the soil.

4.5 Oxides of Iron, Manganese, and Aluminum

Jenne (1968) and Gadde and Laitinen (1974) showed these oxides to be important in sorbing and occluding various trace elements. Jenne (1968) proposed that the hydrous oxides of Fe and Mn provide the principal control on the fixation of trace elements in soils and freshwater sediments.

SOIL PROPERTY				CAPACITY						
TEXTURE	S.A. (cm²/g)	%Fe₂O₃	%Clay	Cu	Pb	Be	Zn	Cd	Ni	Hg
Clay	67	23	52							
Silty Clay	120	5.6	29		HIGH CAPACITY					
Clay	128	2.5	40							
Clay	122	3.7	46							
Sandy Loam	38	1.7	11			MODERATE				
Clay	51	17	61			CAPACITY				
Silty Clay Loam	62	4	31							
Sand	9	1.8	5				LOW			
Sandy Loam	20	1.8	15				CAPACITY			
Loamy Sand	8	0.6	4							

TEXTURE	S.A. (cm²/g)	%Fe₂O₃	%Clay	$SeO_3^=$	VO_3^-	$AsO_4^=$	$Cr_2O_7^=$
Clay	67	23	52				
Silty Clay	121	5.6	50	HIGH CAPACITY			
Clay	51	17	61				
Silty Clay	62	4	31		MODERATE		
Clay	122	3.7	46		CAPACITY		
Clay Loam (Ca)	128	2.5	40				
Sand	9	1.8	5				
Sandy Loam	38	1.7	11				
Loamy Sand	8	0.6	4	LOW CAPACITY			
Sandy Loam	20	1.8	15				

FIGURE 1.5. Relative mobility of trace elements, in both cationic and anionic forms, through soils. The soils used came from the seven most prominent soil orders. *Source:* Adapted from Korte et al (1976).

4.6 Redox Potential

The water content of soils influences their capacity for trace elements through biological or chemical oxidation–reduction reactions. In oxidized soils, redox potential may range from about $+400$ to $+700$ mV. In sediments and flooded soils, redox potential may range from around -400 mV (strongly reduced) to $+700$ mV (well oxidized) (Gambrell and Patrick, 1978). Under reducing conditions, sulfides of elements such as Cd, Zn, Ni, Co, Cu, Pb, and Sn can form. The sulfides of these elements are quite insoluble, so that their mobility and phytoavailability are considerably less than would be expected under well-oxidized soils. Exceptions are Mn and Fe in that they are more soluble under reducing than in oxidizing conditions. The data of Bingham et al (1976) on trace element concentrations in soil solution extracted from sludge-treated soil indicate the reduced solubility of Cd, Cu, and Zn and increased solubility of Mn and Fe under reducing conditions.

5. Plant Capacity for Trace Elements

As with soils, the capacity of plants to accumulate trace elements is limited. Excess of trace elements in soils, even among the essential micronutrients, could result in phytotoxicity. In addition to the various soil factors that can influence phytoavailability of trace elements, the following factors can also affect the ability of plants to accumulate trace elements.

5.1 Plant Species

Crops differ widely in their sensitivity to excess trace elements (Table 1.4). Some crops, such as kale, turnips, and the beet family are very sensitive to metals. Other vegetable crops are more tolerant. Corn, soybeans, and small grains are more tolerant, and most grasses are even more tolerant. Some crops, such as tomato, rice, squash, and cabbage could tolerate high soil Cd concentration (>100 ppm) before showing adverse effects on yield. Others (like spinach, lettuce, curly cress, and soybeans), however, are fairly sensitive to Cd added to soils (Bingham, 1979). In general, leafy vegetables are the greatest accumulators of soil Cd, whereas the edible portions of squash, tomato, and radish tend to have low Cd levels (CAST, 1980). Crops differ considerably not only in their general sensitivity to trace elements, but also in their relative sensitivity to the individual trace elements. Generally, at about pH 5.5 to 6.5, Cu could be twice as toxic as Zn, and Ni four times as toxic as Zn.

TABLE 1.4. Relative sensitivity of various crop species to sludge-borne heavy metals.[a,b]

Very sensitive[c]	Sensitive[d]	Tolerant[e]	Very tolerant[f]
Chard	Mustard	Cauliflower	Corn
Lettuce	Kale	Cucumber	Sudangrass
Redbeet	Spinach	Zucchini squash	Smooth bromegrass
Carrot	Broccoli	Flatpea	"Merlin" red fescue
Turnip	Radish	Oat	
Peanut	Tomato	Orchardgrass	
Ladino clover	Marigold	Japanese bromegrass	
Alsike clover	Zigzag, Red, Kura	Switchgrass	
Crownvetch	and crimson clover	Redtop	
"Arc" alfalfa	Alfalfa	Buffelgrass	
White sweetclover	Korean lespedeza	Tall fescue	
Yellow sweetclover	Sericea lespedeza	Red fescue	
Weeping lovegrass	Blue lupin	Kentucky bluegrass	
Lehman lovegrass	Birdsfoot trefoil		
Deertongue	Hairy vetch		
	Soybean		
	Snapbean		
	Timothy		
	Colonial bentgrass		
	Perennial ryegrass		
	Creeping bentgrass		

[a] Source: Logan and Chaney, 1983, with permission of the authors.
[b] Sassafras sandy loam amended with a highly stabilized and leached digested sludge containing 5,300 mg Zn, 2,400 mg Cu, 320 mg Ni, 390 mg Mn, and 23 mg Cd/kg dry sludge. At 5% sludge, maximum cumulative recommended applications of Zn and Cu are made.
[c] Injured at 10% of a high-metal sludge at pH 6.5 and at pH 5.5.
[d] Injured at 10% of a high-metal sludge at pH 5.5, but not at pH 6.5.
[e] Injured at 25% high-metal sludge at pH 5.5, but not at pH 6.5, and not at 10% sludge at pH 5.5 or 6.5.
[f] Not injured even at 25% sludge, pH 5.5.

5.2 Plant Cultivars

Differences in uptake of trace elements among cultivars have long been established. Evidence exists for a genetic basis of differential translocation of elements within plants (Epstein and Jefferies, 1964). Investigators recently found differential uptake and translocation of Cd and Zn among lettuce and corn cultivars (Giordano et al, 1979; Hinesly et al, 1978); and differential Cd uptake among soybean cultivars (Boggess et al, 1978).

5.3 Plant Parts and Age

It is a common observation that trace elements are not uniformly distributed among plant tissues. In general, the seeds (or grain) contain lower concentrations of most trace elements than do the vegetative tissues. Thus, grain crops are very likely to contribute smaller amounts of trace elements to the diet than do leafy vegetables, subterranean crops, forages, and pastures. Dr. Page and his co-workers at the University of California, Riverside found that Cd and Zn concentrations in parts of corn grown on sludge-treated soils follow the order: leaf > stem > husk > kernel. In trees, concentrations of trace elements generally follow the pattern: roots > foliage > branch > bole (Van Hook et al, 1977). Garland et al (1981) suggested that the distribution of elements in various vegetative tissues (exclusive of stems) is a characteristic of xylem transport and that the ultimate concentration of an element in a specific tissue is related to the flux of transpirational water lost through evapotranspiration and the duration of this process for specific tissues.

Age also has an effect on trace element concentration in plants. Van Loon (1973) showed that spring shoots of quack grass had higher trace element contents (Cd, Cr, Cu, Hg, Mn, Ni, Pb, and Zn) than later season tissues. Boswell's (1975) data indicate a tendency for trace element concentrations (Cd, Cr, Cu, Mn, Pb, and Zn) in fescue leaves to decrease during the growing season.

5.4 Ion Interactions

Phosphorus–micronutrient interactions in crops have been intensively studied during the last two decades. This physiological phenomenon has been reviewed recently (Murphy et al, 1981; Olsen, 1972) and it was shown that severe nutrient imbalance in crops could result in yield reduction. Observed common P-micronutrient interactions include P-Zn, P-Fe, P-Cu, P-Mn, P-Mo, and P-B. In addition to macronutrient-micronutrient interactions, interactions among trace elements have also been observed (Allaway, 1977b).

5.5 Management Practices

Some practices geared to maximize agricultural production may affect the ability of plants to take up elements. In irrigated agriculture, land often has to be leveled, and removal of topsoil exposes usually nutrient-poor subsoils. Consequently, it is not uncommon to observe micronutrient deficiencies under this situation.

The rate and method of fertilizer application, as well as the source of fertilizers may also affect elemental uptake by plants. Banding of micronutrient fertilizers along the plant row may result in more elevated uptake than when the fertilizers are broadcast. Organic sources, such as chelated micronutrients, are generally more effective sources of micronutrients than the inorganic sources.

Disposal or recycling of solid wastes such as animal manures, composts, and sewage sludges on agricultural lands would tend to increase the elemental burden of the soil and may consequently result in increased phytoavailability of trace elements.

5.6 Soil and Climatic Conditions

Under abnormal soil and climatic conditions, micronutrient deficiencies may occur that normally would not develop under more favorable growing conditions. Crops growing in cold, wet soils may be unable to obtain sufficient micronutrients for normal growth because of restricted root growth under such conditions. Other factors that can affect elemental uptake by limiting root growth are excessive rainfall or irrigation, as well as drought.

6. Trace Elements in the Food Chain

Other than occupational exposure and inhalation, the primary route of entry of trace elements into humans is via ingestion of foods of plant and animal origin (Fig. 1.6). Thus, it is important to understand the soil–plant–animal interrelationships. Plants can be considered as intermediate reservoirs through which trace elements from primary sources are transferred to other organisms. Plant growth and quality largely depend upon the soil environment and whatever sources of elements are present (Appendix Table 1.14). Elemental uptake can occur via the root pathway, or through foliar and stem pathway, or both. In turn, the growth and performance of animals largely depend on the quantity and quality of plants or foodstuff they are feeding on. The sources and pathways of trace elements to animals are listed in Appendix Table 1.15. In addition to the sources listed for terrestrial animals, some livestock, such as cattle and sheep, ingest soil particles at a rate equal to 2% to 14% of their diet (dry weight). On an annual basis, this amounts to about 45 kg of soil for sheep and 10 times as much for cattle (Sillanpaa, 1972). Also ingested are whatever foreign materials are present on foliage surfaces, such as sewage sludge or manure applied on pastures. In studying trace element interrelationships in natural systems between soil, vegetation, and animals (represented by rock squirrels), Sharma and Shupe (1977) found that, in general, there was a greater correlation for the means of certain elements (Cu, Mo, Se, and Zn) between soil and vegetation than between vegetation and animals, or between soil and animals.

Excellent reviews on the food chain or soil–plant–animal aspects of trace elements have been recently conducted (Allaway, 1977a; Nicholas and Egan, 1975; Underwood, 1977). Because of extreme variability in trace element concentrations in various food classes and differences among countries in the selective proportion that these food classes are consumed (Appendix Table 1.16), dietary intake of trace elements among peoples of the world is expected to vary.

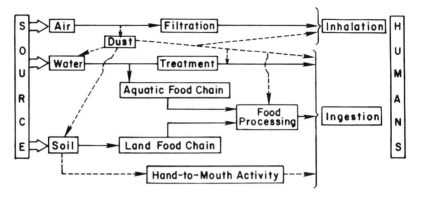

FIGURE 1.6. Pathways of trace elements to humans (cross-hatching denotes path of fine airborne fallout). *Source:* Nriagu (1984), with permission of the author.

Appendix

A series of tables (Tables 1.1 to 1.18) is appended to this chapter rather than to the end of the book to be more convenient to the readers.

However, scientific names for plants and an abbreviated glossary of soil terms are provided in the backmatter of this book.

APPENDIX TABLE 1.1. Trace element production and use.[a]

Element	Production, 1979[b]										USA Consumption[c] (1976)	Manmade Air pollution in USA 1 year (1976)[c]
	World	USA	Canada	Australia	Brazil	China	Germany[d]	Japan	USSR	UK		
Antimony	79.381	0.722	3.300	2.300	0.280	17.000	—	—	9.000	—	34.581	0.25
Arsenic	36.2	NA[e]	—	—	—	—	0.400	NA	8.5	—	4.196	9.76
Beryllium	3.082	NA	—	—	0.800	—	—	—	2.000	—	0.519	0.45
Boron	2.900	1.590	—	—	—	30	—	—	220	—	110.23	10.5
Cadmium	18.280	1.715	1.480	0.800	—	0.120	1.190	2.590	2.850	0.410	5.886	0.4
Chromium (chromite)	10.498	—	—	—	330	—	—	10	2,600	—	284.44	16.46
Cobalt	31.437	0.464	1.522	1.700	—	—	0.365	2.866	2,000	0.500	8.287	0.75
Copper	7.606	1.444	644	234	5	150	16.9	60	885	NA	2,206.60	200.0
Lead	3.513	526	316	415	24	120	38	47	525	2	1,466.47	234.5
Manganese (ore)	26.963	—	—	1,836	1,875	1,650	—	100	10,500	—	—	—
Mercury (by flasks)	198.845[f]	29,343	—	2	—	20,000	2,500	—	61,000	—	2.241	20.0
Molybdenum	113.5	71.9	12.4	—	—	2.2	—	0.15	11.2	—	—	—
Nickel	776.0	15	145	81	4.4	11	3	—	168	—	160.35	4.885
Selenium	1.72	0.293	0.564	—	0.5	—	—	0.500	—	—	0.486	0.50
Silver (10^6 troy Oz.)	344	38	38	25	8	2	2.4	8.7	46	—	—	—
Tin	256.0	NA	0.362	11.4	—	25	1.6	0.6	35	2.8	61.934	30.0
Vanadium	41.4	5.5	—	—	—	7.2	—	—	11.0	—	4.65	20.0
Zinc	5.978	267	1,149	530	80	120	97	243	770	NA	—	—

[a] 10^3 tonnes, unless otherwise stated.
Sources: [b] US Bureau of Mines (1980); [c] Jenkins (1981).
[d] If both available, for two Germanys.
[e] Data not available.
[f] Equivalent to approx. 8.429×10^2 tonnes.

APPENDIX TABLE 1.2. Concentrations (ppm) of trace elements in various soil-forming rocks and other natural materials.[a,b]

Element	Ultramafic igneous	Basaltic igneous	Granitic igneous	Shales and clays	Black shales	Deep-sea clays	Limestones	Sandstones
Arsenic	0.3–16 / 3.0	0.2–10 / 2.0	0.2–13.8 / 2.0	— / 10	—	— / 13	0.1–8.1 / 1.7	0.6–9.7 / 2
Barium	0.2–40 / 1	20–400 / 300	300–1,800 / 700	460–1,700 / 700	70–1,000 / 300	— / 2,300	— / 10	— / 20
Beryllium	—	1.0	2–3	3	—	2.6	—	—
Cadmium	0–0.2 / 0.05	0.006–0.6 / 0.2	0.003–0.18 / 0.15	0–11 / 1.4	<0.3–8.4 / 1.0	0.1–1 / 0.5	— / 0.05	— / 0.05
Chromium	1,000–3,400 / 1,800	40–600 / 220	2–90 / 20	30–590 / 120	26–1,000 / 100	— / 90	— / 10	— / 35
Cobalt	90–270 / 150	24–90 / 50	1–15 / 5	5–25 / 20	7–100 / 10	— / 74	— / 0.1	— / 0.3
Copper	2–100 / 15	30–160 / 90	4–30 / 15	18–120 / 50	20–200 / 70	— / 250	— / 4	— / 2
Fluorine	—	20–1,060 / 360	20–2,700 / 870	10–7,600 / 800	—	— / 1,300	0–1,200 / 220	10–880 / 180
Iron	94,000	86,500	14,000–30,000	47,200	20,000	65,000	3,800	9,800
Lead	— / 1	2–18 / 6	6–30 / 18	16–50 / 20	7–150 / 30	— / 80	— / 9	<1–31 / 12
Mercury	0.004–0.5 / 0.1?	0.002–0.5 / 0.05	0.005–0.4 / 0.06	0.005–0.51 / 0.09	0.03–2.8 / 0.5	0.02–2.0 / 0.4	0.01–0.22 / 0.04	0.001–0.3 / 0.05
Molybdenum	0.3	0.9–7 / 1.5	1–6 / 1.4	— / 2.5	1–300 / 10	— / 27	— / 0.4	— / 0.2
Nickel	270–3,600 / 2,000	45–410 / 140	2–20 / 8	20–250 / 68	10–500 / 50	— / 225	— / 20	— / 2
Selenium	0.05	0.05	0.05	0.6	—	0.17	0.08	0.05
Vanadium	17–300 / 40	50–360 / 250	9–90 / 60	30–200 / 130	50–1,000 / 150	— / 120	— / 20	— / 20
Zinc	— / 40	48–240 / 110	5–140 / 40	18–180 / 90	34–1,500 / 100?	— / 165	— / 20	2–41 / 16

[a] Source: Cannon, 1978, with permission of the author.
[b] The upper values are the usually reported range, the lower values the averages.

APPENDIX TABLE 1.3. Concentrations (ppm) of trace elements in fertilizers.

Fertilizer	Zn	Cu	Mn	B	Mo	Co	Cr	Ni	Pb	V	Cd
Diammonium phosphate (20-48-0)[a]:											
Reagent grade	1.0	1.6	0.6	—	—	—	0.2	1.1	0.5	0.3	0.9
Idaho phosphate rock	715	2.7	195	—	—	—	485	64	4.4	1,600	50
North Carolina phosphate rock	285	1.0	93	—	—	—	195	38	4.7	90	30
Rock phosphate[b]	187	32	975	72	555	109	184	—	962	—	—
Single superphosphate (0-16-0)[b]	165	15	890	132	335	77	87	—	488	—	—
Triple superphosphate (0-45-0)[b]	418	49	75	212	270	47	392	—	238	—	—
Diammonium phosphate[b]	112	7.2	307	396	75	16	80	—	195	—	—
Fluid fertilizer (10-15-0)[a]:											
Idaho phosphate rock	673	1.1	125	—	—	—	344	8.0	9.0	1,150	44
North Carolina phosphate rock	500	1.4	25	—	—	—	175	35	5.2	52	17
Urea (45-0)[b]	4.0	0.6	0.5	1.0	5.3	6.6	6.3	—	—	—	—
Calcium ammonium nitrate (25-0-0)[b]	7.6	2.8	25	9.0	56	24	8.5	—	116	—	—
Ammonium sulfate (21-0-0)[b]	11	0.8	3.5	—	6.0	22	4.0	—	—	—	—
Muriate of potash (0-0-60)[b]	10	3.1	3.5	16	26	51	13	—	117	—	—
N-P-K mixture (12-12-12)[b]	88	18	132	61	200	—	116	—	444	—	—
Superphosphate from apatite[c]	—	—	—	—	10	—	20	5	—	5	—

Sources: [a] Mortvedt and Giordano (1977); [b] Arora et al (1975); [c] Ermolenko (1972)

APPENDIX TABLE 1.4. Pesticides containing metals recommended in Ontario, Canada from period 1892–1975.[a]

Chemical	Metal composition of product	Period of recommendation	Crops
Insecticides			
Copper aceto-arsenite (Paris Green)	2.3% As, 39% Cu	1895–1920	Apples & cherries
		1895–1957	Vegetable & small fruit (foliar & bait)
Calcium arsenate	0.8–26% As	1910–1953	Fruit & vegetables
Lead arsenate	4.2–9.1% As	1910–1975	Apples
	11–26% Pb	1910–1971	Cherries
		1910–1956	Peaches
		1910–1955	Vegetables
Mercuric chloride	6% Hg	1932–1954	Cruciferous crops
Zinc sulfate	20–30% Zn	1939–1955	Peaches
Fungicides			
Copper sulfate–calcium salts	4–6% Cu	1892–1975	Fruit & vegetables
Fixed copper salts	2–56% Cu	1940–1975	Fruit & vegetables
		1948–1975	Fruit
Maneb	1–17% Mn	1947–1975	Fruit & vegetables
Mancozeb	16% Mn, 2% Zn	1966–1975	Fruit & vegetables
Methyl & phenyl mercuric salts	0.6–6% Hg	1932–1972	Seed treatment
Phenyl mercuric acetate	6% Hg	1954–1973	Apples
Zineb & ziram	1–18% Zn	1947–1975	Vegetables
		1957–1975	Fruit
Topkiller			
Calcium arsenite	30% As	1930–1972	Vegetables
Sodium arsenite	26% As	1920–1972	Vegetables

[a] *Source:* Frank et al, 1976, with permission of the authors.

APPENDIX TABLE 1.5. Chemical analysis of
agricultural limestones from the USA.[a]

Element	Concentration (No.of Samples = 194)	
	Average	Range
	%	%
Calcium	30.21	17.98–39.76
Magnesium	4.92	0.04–12.86
Silicon	2.36	0.03–15.53
Aluminum	0.45	0.01–2.15
Iron	0.43	0.01–3.11
Potassium	0.23	<0.0001–1.80
Sulfur	0.11	<0.01–1.35
Sodium	0.03	<0.001–0.15
	ppm	ppm
Manganese	330	20–3,000
Fluorine	230	<10–1,410
Phosphorus	210	10–5,660
Zinc	31	<1–425
Vanadium	11	<1–106
Boron	4	<1–21
Copper	2.7	<0.3–89
Molybdenum	1.1	<0.1–92
Cobalt	<1	<1–6

[a] Source: Adapted from Chichilo and Whittaker (1961).

APPENDIX TABLE 1.6. Concentrations (ppm) of trace elements at several soil depths in soil from a sewage farm in Braunschweig, Federal Republic of Germany, in operation since 1895.[a]

Element	Treated				Untreated			
	0–24 cm	24–44 cm	44–90 cm	90–158 cm	0–28 cm	28–74 cm	74–104 cm	104–134 cm
Ag	5.1	<1	<1	1	<1	<1	<1	<1
As	8.7	7.7	7.7	6.8	6.2	5.0	5.5	6.5
Ba	571	283	274	252	285	234	242	274
Cd	2.70	0.276	0.047	0.033	0.084	0.011	0.012	0.011
Co	3.6	2.3	2.4	2.1	2.0	1.8	1.7	2.5
Cr	202	52	40	49	38	22	17	26
Cs	1.6	1.2	1.1	1.1	1.3	0.8	0.9	1.1
Cu	68	9.7	2.7	5.0	7.2	1.6	1.6	2.0
Hg	2.15	<0.1	<0.1	0.1	0.11	<0.1	<0.1	<0.1
Ni	32	16	6.6	22	10.5	7.8	7.4	10.8
Pb	69	5.6	7.8	5.3	4.3	4.9	4.9	4.9
Sb	2.3	0.6	0.4	0.5	0.6	0.2	0.3	0.2
Se	1.6	0.95	1.1	0.40	0.5	<0.4	<0.4	<0.4
Zn	393	58	17	4	16	<4	<4	<4

[a] Source: Adapted from El-Bassam et al, 1979, with permission of the authors.

APPENDIX TABLE 1.7. Concentrations (ppm) of trace elements in sewage sludges.

Element	USA[a] Mean	USA[a] Range	UK[b] Mean	UK[b] Range	Sweden[c] Mean	Sweden[c] Range	Canada[d] Mean	New Zealand[e] Mean
Ag	—	—	32	5–150	—	—	—	—
As	14.3	3–30	—	—	—	—	—	—
B	37.0	22–90	70	15–1,000	—	—	—	480
Ba	621	272–1,066	1,700	150–4,000	—	—	1,950	580
Be	<8.5	—	5	1–30	—	—	—	—
Bi	16.8	<1–56	34	<12–100	—	—	—	—
Cd	104	7–444	<200	<60–1,500	13	2–171	38	4.5
Co	9.6	4–18	24	2–260	15	2–113	19	21
Cr	1,441	169–14,000	980	40–8,800	872	20–40,615	1,960	850
Cu	1,346	458–2,890	970	200–8,000	791	52–3,300	1,600	720
F	167	370–739	—	—	—	—	—	—
Hg	8.6	3–18	—	—	6.0	<0.1–55	—	—
Mn	194	32–527	500	150–2,500	517	73–3,861	2,660	610
Mo	14.3	1–40	7	2–30	—	—	13	8
Ni	235	36–562	510	20–5,300	121	16–2,120	380	350
Pb	1,832	136–7,627	820	120–3,000	281	52–2,914	1,700	610
Sb	10.6	2–44	—	—	—	—	—	—
Se	3.1	1–5	—	—	—	—	—	—
Sn	216	111–492	160	40–700	—	—	—	80
Ti	2,331	1,080–4,580	2,000	<1,000–4,500	—	—	—	4,700
V	40.6	15–92	75	20–400	—	—	15	80
W	20.2	1–100	—	—	—	—	—	—
Zn	2,132	560–6,890	4,100	700–49,000	2,055	705–14,700	6,140	700

Sources: [a] Furr et al (1976); includes Atlanta, Chicago, Denver, Houston, Los Angeles, Miami, Milwaukee, Philadelphia, San Francisco, Seattle, Washington, DC, and five cities in New York.
[b] Berrow and Webber (1972); includes 42 samples from different locations in England and Wales.
[c] As summarized by Page (1974) from Berggren and Oden (1972); from 93 treatment plants.
[d] Oliver and Cosgrove (1975); from 10 sites in southern Ontario, Canada.
[e] Wells and Whitton (1977).

APPENDIX TABLE 1.8. Concentrations (ppm) of trace elements in sewage sludges in Iowa cities of varying population.[a]

Element	Group I[b]		Group II		Group III		Group IV		Group V	
	Mean	Range	Mean	Range	Mean	Range	Mean	Range	Mean	Range
Ag	14	6–63	17	6–25	10	1–25	26	6–51	15	1–46
As	132	0–188	141	18–188	163	96–188	152	0–188	163	125–188
B	96	24–150	260	50–625	178	21–900	75	11–163	141	33–350
Be	<0.25	—	<0.25	—	<0.25	—	<0.25	—	<0.25	—
Cd	19	9–31	28	13–50	20	8–50	22	10–38	31	13–54
Co	17	8–26	22	14–28	23	9–50	21	10–38	27	13–50
Cr	45	25–100	6,079(250)[c]	31–45,875	2,018(369)	25–12,000	249	25–863	289	138–600
Cu	363	138–763	442	125–875	235	93–438	282	100–550	853	125–5,125
Hg	3.0	0.1–13.4	3.0	0.3–5.5	1.6	0.1–5.6	1.5	0.4–3.1	0.7	0.1–2.6
Mn	564	88–3,500	1,038	116–3,250	286	163–600	411	225–750	424	213–725
Mo	11	<1–13	12	<1–28	9	<1–14	17	<1–73	12	<1–21
Ni	24	6–63	4,880(38)	25–38,750	250	10–1,425	40	4–100	67	14–225
Pb	223	138–500	207	106–450	269	100–725	498	103–2,275	356	250–625
Se	<25	—	<25	—	<25	—	<25	—	<25	—
Sr	15	<4–83	392(<4)	<4–3,000	31	<4–175	69	<4–438	23	<4–50
V	28	10–56	26	11–50	30	14–63	31	19–53	36	25–50
Zn	1,208	625–2,500	1,755	538–2,250	2,056	300–8,625	1,570	538–4,500	1,557	750–3,250

[a] Source: Extracted from Tabatabai and Frankenberger, 1979, with permission of the authors.

[b] Population sizes ($\times 10^3$): Group I—10 cities, 0.74 to 1.73; Group II—7 cities, 3.47 to 9.27; Group III—7 cities, 10.28 to 22.41; Group IV—9 cities, 26.22 to 46.85; Group V—6 cities, 60.35 to 201.40.

[c] Median values in parentheses.

APPENDIX TABLE 1.9. Concentrations (ppm) of trace elements in animal wastes.

Waste	As	B	Ba	Be	Cd	Co	Cr	Cu	Hg	Mn	Mo	Ni	Pb	Sb	Se	Sn	Ti	V	Zn
Feedlot diet[a-1]	0.10	—	18	<0.03	0.05	0.10	0.75	3.0	<0.01	17	<2.5	—	0.36	<0.08	0.19	<0.8	8.1	0.57	20
Cattle manure (low fiber diet)[a-2]	0.88	—	105	<0.03	0.28	1.7	20	24	0.05	117	30	—	2.1	<0.08	0.35	4.7	55	3.2	115
Cattle manure (high fiber diet)[a-3]	2.2	—	305	<0.03	0.24	2.2	31	21	<0.03	161	49	—	3.3	<0.08	0.32	7.4	129	8.0	86
Processed cattle waste pellets[a-4]	0.60	—	70	<0.03	0.14	1.1	5	19	<0.09	100	16	—	3.3	<0.08	0.36	3.7	50	3.0	77
Poultry waste with litter[a-5]	0.57	—	54	<0.03	0.42	2.0	6	31	0.06	166	5.0	—	2.1	<0.08	0.38	2.0	12	3.9	155
Poultry waste without litter[a-6]	0.66	—	57	<0.03	0.58	1.2	5	20	<0.04	242	7.2	—	3.4	0.10	0.66	4.1	27	4.3	158
Poultry waste[b]	—	—	—	—	—	5.0	10	4.4	—	187	42	—	90?	—	—	—	—	—	36
Swine waste[b]	—	—	—	—	—	11.0	14	13	—	168	34	—	168?	—	—	—	—	—	198
Farmyard manure[b]	—	—	—	—	—	11.0	12	2.8	—	69	21	—	120?	—	—	—	—	—	15
Cow manure[c]	4.0	24	268	—	0.8	5.9	56	62	0.2	286	14	29	16	0.5	2.4	3.8	2,800?	43	71

Sources: [a] Capar et al (1978)
[1] Typical feedlot diet = 70% corn, 3% hay, 5% beet pulp, 20% corn silage, and 2% mineral supplement.
[2] From feedlot heifers fed 59% corn, 2% alfalfa hay, 3% molasses, 33% corn silage, and 3% mineral supplement.
[3] From feedlot heifers fed 24% corn, 29% alfalfa hay, 3% molasses, 41% corn silage, and 3% mineral supplement.
[4] A commercial high-protein feedlot animal waste product similar to item 3 pelletized.
[5] Includes wood shavings; used layer ration.
[b] Arora et al (1975); high Pb results may be due to contamination of samples.
[c] Furr et al (1976).

APPENDIX TABLE 1.10. Average concentrations (ppm) of trace elements in coal and fly ashes.[a]

Element	Coal	Fly Ash		
		Bituminous	Sub-bituminous	Lignite
As	15	82	2.3	34
B	50	36	50	500
Ba	150	974	2,277	6,917
Cd	1.3	0.3	0.3	0.3
Co	7	35	6.3	8
Cr	15	172	50	43
Cs	0.4	10	3.1	3.1
Cu	19	132	45	75
F	80	8.8	1.4	20
Hg	0.18	0.1	0.04	0.1
Mn	100	145	309	543
Mo	3	33	8.4	19
Ni	15	11	1.8	13
Pb	16	15	3.1	12
Sb	1.1	2.2	0.8	2.6
Se	4.1	5.7	1.2	4.4
Sr	100	794	3,855	931
V	20	256	73	94
Zn	39	20	15	14
Radioactivity (cpm/g)	—	19.2	12.2	12.9

[a] *Source:* Adapted from Adriano et al (1980).

APPENDIX TABLE 1.11. Estimated annual US discharge of trace elements from coal combustion.[a]

Element	Flows (10³ tonnes/year)			Annual flow as percent of weathering mobilization	Atmospheric enrichment
	Discharged slag and fly ash	Atmospheric discharge	Weathering mobilization		
As	1.72	0.029	8	21	4.9
Ba	22.7	0.13	480	4.7	1.7
Cd	0.16	0.0046	0.6	27	8.3
Co	1	0.0063	7.3	13.7	1.8
Cr	6.2	0.066	90	6.9	3.1
Cs	0.38	0.003	5.4	7	2.3
Cu	2.9	—	28	10	—
Hg	0.005	0.038	0.12	4	260
Mn	11.7	0.06	770	1.5	1.5
Mo	3.1	—	2.7	115	—
Ni	5.6	—	36	15.6	—
Pb	1.7	0.06	13	13	10.8
Sb	0.17	0.0036	5	3.4	6.1
Se	0.66	0.11	0.5	132	42
Sr	8	—	370	2.2	—
Ti	176	1.1	4,500	3.9	1.8
V	9.9	0.09	90	11	2.7
Zn	26.8	0.5	75	36	5.3

[a] Source: Adapted from Klein et al, 1975, with permission of the authors.

APPENDIX TABLE 1.12. Surface and irrigation water quality criteria (ppm) for trace elements.

| | | Irrigation water | | |
| | | Continuous use | | Short-term use |
Element	Surface water FWPCA[a]	FWPCA Any soil	NAS[b] Coarse-textured soil	FWPCA[c] Fine-textured soil
Ag	0.05	—	—	—
As	0.05	1.0	0.1	10
B	1.0	0.75	0.75	2.0
Ba	1.0	—	—	—
Cd	0.01	0.005	0.01	0.05
Co	—	0.2	0.05	10.0
Cr	0.05	5.0	0.1	20.0
Cu	1.0	0.2	0.2	5.0
Pb	0.05	5.0	5.0	20.0
Mn	0.05	2.0	0.2	20.0
Mo	—	0.005	0.01	0.05
Ni	—	0.5	0.2	2.0
Se	0.01	0.05	0.02	0.05
V	—	10.0	0.1	10.0
Zn	5.0	5.0	2.0	10.0

Sources:
[a] US Dept of Interior, Fed Water Pollut Control Admin (1968). Surface water criteria are virtually the same as drinking water standards (US Dept of Health, Education, and Welfare, 1962).
[b] National Academy of Sciences–National Academy of Engineering (1973). Recommended maximum concentrations of trace elements in irrigation waters used for sensitive crops on soils with low capacities to retain these elements in unavailable forms.
[c] For short-term use only on fine-textured soils.

APPENDIX TABLE 1.13. Concentrations (ppm) of trace elements in surface soils impacted by smelters.

Pb smelter[a]

Distance from smelter, km	Cd	Sb	Ag	Pb	Zn	Se	As
0.4	83	155	30	7,900	13,000	4.6	100
1.1	25	5	9.3	3,200	870	0.76	49
2.4	—	32	6.0	1,700	970	—	69
3.2	32	260	31	6,700	1,400	5.1	94
3.7	—	28	2.7	2,000	200	—	24
5.3	18	18	2.8	1,000	940	—	36
8.1	—	20	3.6	300	320	—	53
12.6	—	20	3.7	890	804	—	24
19.0	—	40	10	2,200	3,000	—	37

Cu and Ni smelter[b]

Distance from smelter, km	Cu	Ni	Co	Zn	Ag	Pb	Mn	V
1.1	2,892	5,104	199	96	7.9	82	255	103
1.6	2,416	1,851	80	65	3.5	53	202	63
2.2	2,418	2,337	92	82	7.8	58	174	115
2.9	1,657	1,202	41	50	3.3	48	143	25
7.4	1,371	1,771	46	87	2.9	46	299	165
10.4	287	282	54	72	2.3	28	364	137
13.5	233	271	42	100	4.3	23	602	151
19.3	184	306	24	61	ND[c]	28	264	55
24.1	45	101	18	46	5.5	19	207	33
32.1	46	35	16	55	1.9	26	195	96
38.6	2	39	29	62	ND	28	192	169
49.8	26	35	22	83	1.0	20	168	23

Sources:
[a] In Kellogg, Idaho. Samples (0–2 cm surface soil) were not taken on a transect (Ragaini et al. 1977).
[b] In Sudbury Basin in Ontario, Canada. Samples (top 10 cm soil) were taken along a transect from smelter (Hutchinson and Whitby, 1974).
[c] Not detectable.

APPENDIX TABLE 1.14. Sources and pathways of trace elements to plants.[a]

	Sb	As	Be	B	Cd	Cr	Co	Cu	Pb	Hg	Mo	Ni	Se	Sn	V	Zn	Mn
Uptake by roots																	
A. Soil or groundwater	×	×	×	×	×	×	×	×	×	×	×	×	×	×	×	×	×
B. Fallout to soil from air pollution	×	×	×	×	×	×	×	×	×	×		×	×		×	×	
C. Sewage sludge soil amendments		×			×	×		×	×	×		×	×			×	×
D. Biocides applied to soil and/or seed		×		×	×			×	×	×				×		×	×
E. Surface water contamination		×				×			×						×		
F. Fertilizers		×			×	×		×		×	×	×		×		×	×
G. Industrial pollution		×			×	×		×		×		×		×		×	×
Uptake by leaves and stems																	
A. Pollutant fallout from industrial sources	×	×	×		×	×	×	×	×	×		×	×	×		×	
B. Pollutant fallout from auto emissions									×							×	
C. Biocide applications to plants		×		×	×			×	×	×				×		×	×
D. Pollution fallout from incineration of fossil fuels and refuge	×	×	×	×	×	×	×	×	×	×	×	×	×	×	×	×	×

[a] *Source:* Adapted from Jenkins (1981).

APPENDIX TABLE 1.15. Sources and pathways of trace elements in animals.[a]

	Sb	As	Be	B	Cd	Cr	Co	Cu	Pb	Hg	Mo	Ni	Se	Sn	V	Zn	Mn
Terrestrial																	
A. Breathing contaminated air	x		x	x	x	x	x	x	x	x		x	x	x	x	x	x
B. Eating contaminated plant or animal tissue	x	x	x	x	x	x	x	x	x	x	x	x	x	x	x	x	x
C. Drinking contaminated water	x	x	x	x	x	x	x	x	x	x	x	x	x	x	x	x	x
D. Licking or preening fur or feathers									x								
E. Receiving therapeutic drugs (domestic animals)		x												x			
F. Eating biocides or poison baits		x															
Aquatic																	
A. Metal in water	x	x	x	x		x	x	x	x	x		x	x	x	x	x	x
B. Runoff and fallout		x			x	x		x	x	x		x	x	x	x	x	x
C. Sewage and industrial waste outfalls	x	x		x	x	x	x	x	x	x		x		x		x	x
D. Mine tailings or smelter waste leachate		x			x	x		x	x	x				x		x	x
E. Contaminated plants, animals, or sediment	x	x	x	x	x	x	x	x	x	x	x	x	x	x	x	x	x
F. Biocides or runoff		x		x				x						x	x	x	
G. Lead shot									x								

[a] *Source:* Adapted from Jenkins (1981).

APPENDIX TABLE 1.16. International comparison of dietary intake by man.[a]

Country	Year	Food intake (wet kg/year/person)										
		Cereals	Potatoes and starches	Sugars and sweeteners	Pulses, seeds, and nuts	Vegetables	Fruit	Meat, poultry, and whales	Eggs	Marine products	Milk and milk products	Oils and fats
Sweden	1975	61.2	81.0	42.5	2.4	42.0	81.1	60.9	12.7	23.5	359.5	13.4
F. R. Germany	1975	66.0	90.1	37.8	3.3	68.5	110.9	90.4	17.2	9.0	270.2	19.4
France	1975	76.1	90.8	35.6	4.3	112.0	74.2	99.1	12.9	18.7	320.6	19.1
Netherlands	1975	66.3	76.6	47.9	11.1	80.1	110.8	72.5	11.5	11.7	270.2	25.1
United Kingdom	1975	74.6	90.4	48.6	4.9	60.5	47.1	73.5	13.8	7.9	363.6	15.1
USA	1975	61.8	47.9	50.1	8.1	94.7	72.3	110.1	16.0	6.9	246.1	22.5
Canada	1975	69.6	91.6	44.8	10.0	67.4	80.7	94.1	12.7	5.8	302.4	13.8
Denmark	1975	63.5	66.6	48.4	2.4	46.7	53.0	69.5	11.5	34.8	333.3	17.4
Switzerland	1975	70.1	48.6	39.7	5.1	84.4	121.8	75.3	11.3	4.6	380.0	16.1
New Zealand	1975	76.2	51.8	36.0	3.2	128.0	73.3	113.5	17.1	4.7	369.9	6.7
Italy	1975	131.1	36.7	30.5	8.1	155.7	100.8	65.2	11.3	10.8	189.7	21.9
Japan	1977	117.9	28.5	26.2	10.4	131.9	57.3	28.6	16.2	34.1	57.0	11.9
Argentina	1972–74	98.7	83.6	40.3	3.6	76.4	106.0	98.2	6.9	4.7	82.1	16.8
Brazil	1972–74	90.8	69.0	43.5	26.8	22.0	113.9	30.9	4.0	4.8	49.7	8.2
P. R. China	1972–74	149.0	103.1	5.7	17.1	63.9	8.9	18.3	3.7	6.5	3.5	3.1
Korea	1972–74	209.5	39.1	7.8	8.6	99.7	21.5	7.0	3.8	32.2	2.0	1.9
Pakistan	1972–74	145.2	4.9	54.0	8.0	25.2	23.3	5.3	0.3	1.1	73.3	7.4
India	1972–74	136.0	10.6	24.5	19.5	46.7	23.3	1.5	0.1	2.3	33.8	4.5
Indonesia	1972–74	136.2	74.3	11.5	19.0	34.8	28.2	3.3	0.9	7.2	0.6	3.5

[a] Source: Kitagishi and Yamane (1981).

...of average (A) contents and range in contents reported for elements in soils and other surficial materials.

Element	Bowen (1979)[a]		Shacklette & Boerngen (1984)[b]		Vinogradov (1959)[c]	Rose et al (1979)[d]	Mitchell (1964)[e]
	Median	Range	Average	Range	Average	Average or median	Range
Ag	0.05	0.01–8	—	—	—	—	—
As	6	0.1–40	7.2	<0.1–97	5	7.5 (M)	—
B	20	2–270	33	<20–300	10	29 (M)	—
Ba	500	100–3,000	580	10–5,000	—	300 (M)	400–3,000
Be	0.3	0.01–40	0.92	<1–15	6	0.5–4	<5–5
Bi	0.2	0.1–13	—	—	—	—	—
Cd	0.35	0.01–2	—	—	—	—	—
Co	8	0.05–65	9.1	<3–70	8	10 (M)	<2–80
Cr	70	5–1,500	54	1–2,000	200	6.3 (M)	5–3,000
Cs	4	0.3–20	—	—	—	—	—
Cu	30	2–250	25	<1–700	20	15 (M)	<10–100
F	200	20–700	430	<10–3,700	200	300 (M)	—
Hg	0.06	0.01–0.5	0.09	0.01–4.6	—	0.056 (M)	—
Mn	1,000	20–10,000	550	<2–7,000	850	320 (M)	200–5,000
Mo	1.2	0.1–40	0.97	<3–15	2	2.5 (A)	1–5
Ni	50	2–750	19	<5–700	40	17 (M)	10–800
Pb	35	2–300	19	<10–700	—	17 (M)	<20–80
Sb	1	0.2–10	0.66	<1–8.8	—	2 (A)	—
Se	0.4	0.01–12	0.39	0.1–4.3	0.001	0.3 (M)	—
Sn	4	1–200	1.3	0.1–10	—	10 (A)	—
Sr	250	4–2,000	240	<5–3,000	300	67 (M)	60–700
Ti	5,000	150–25,000	2,900	70–20,000	4,600	—	—
Tl	0.2	0.1–0.8	—	—	—	—	—
V	90	3–500	80	<7–500	100	57 (M)	20–250
W	1.5	0.5–83	—	—	—	—	—
Zn	90	1–900	60	<5–2,900	50	36 (M)	—

[a] Compilation from numerous sources.
[b] For conterminous USA.
[c] Presumably, average from worldwide sampling.
[d] Elements useful in geochemical prospecting.
[e] For Scottish surface soils.

APPENDIX TABLE 1.18. Mean and range of concentrations (in ppm) of selected trace elements in tropical Asian paddy soils.[a]

Country	No. of samples	B Mean	B Range	Cu Mean	Cu Range	Zn Mean	Zn Range	Mo Mean	Mo Range
Bangladesh	53	68	25–104	27	6–48	68	10–110	3.3	0.0–6.0
Burma	60	197	32–744	43	18–89	88	15–166	3.1	1.0–6.0
Cambodia	16	96	41–249	47	15–105	35	0–87	2.3	1.0–4.0
East Malaysia	40	98	44–282	11	0–32	56	0–112	3.3	1.0–8.0
India	71	69	5–140	44	12–138	58	1–118	2.6	1.0–5.0
Indonesia	43	52	5–91	36	13–184	80	2–144	3.2	1.0–5.0
Philippines	54	71	20–138	37	5–195	81	26–169	3.1	0.0–6.0
Sri Lanka	33	77	14–318	50	11–113	43	0–248	2.3	1.0–4.0
Thailand	23	95	34–363	34	0–119	51	0–104	2.7	1.0–4.0
Vietnam	49	90	33–147	23	1–51	82	0–121	2.9	1.0–6.0
West Malaysia	40	132	19–459	18	1–49	40	0–92	2.3	1.0–4.0
Tropical Asia	482	96	5–744	33	0–195	66	0–248	2.9	0.0–8.0

[a] Source: Domingo and Kyuma, 1983, with permission of the authors.

Co		V		Cr		Ni		Sr	
Mean	Range	Mean	Range	Mean	Range	Mean	Range	Mean	Range
58	24–96	109	50–209	133	89–196	22	0–63	66	0–147
51	25–72	153	5–230	224	131–263	82	0–170	70	5–109
61	34–170	121	1–296	127	67–304	14	0–97	13	0–83
48	22–80	146	1–565	115	56–147	2	0–17	23	0–63
57	14–112	177	61–419	144	62–385	22	0–86	86	0–400
88	32–690	245	98–444	93	47–240	3	0–29	215	17–579
72	35–135	221	121–360	130	51–467	20	0–363	292	55–665
65	41–135	133	16–270	108	50–251	7	0–41	170	0–634
38	1–67	142	39–368	111	47–183	12	0–30	53	0–119
42	16–67	181	35–223	137	53–159	20	0–34	87	0–153
28	3–66	144	55–205	103	46–143	1	0–10	19	0–81
56	1–690	166	1–565	136	46–467	22	0–363	108	0–665

References

Adriano, D.C., P. F. Pratt, and S. E. Bishop. 1971. *Soil Sci Soc Am Proc* 35:759–762.

Adriano, D.C., A. C. Chang, P. F. Pratt, and R. Sharpless. 1973. *J Environ Qual* 2:396–399.

Adriano, D.C., A. L. Page, A. A. Elseewi, A. C. Chang, and I. Straughan. 1980. *J Environ Qual* 9:333–344.

Allaway, W. H. 1968. *Adv Agron* 20:235–274.

Allaway, W. H. 1977a. In L. F. Elliot and F. J. Stevenson, eds. *Soils for management of organic wastes and water wastes,* 283–298. Am Soc Agron, Madison, WI.

———1977b. *Geochem Environ* 2:111–115.

Andersson, A., and K. O. Nilsson. 1972. *Ambio* 1:176–179.

Arora, C. L., V. K. Nayyar, and N. S. Randhawa. 1975. *Indian J Agric Sci* 45:80–85.

Baker, D. E. 1974. *Proc Fed Am Soc Exp Biol* 33:1188–1193.

Berggren, B., and S. Oden. 1972. For citation see Page (1974), p. 88.

Berrow, M. L., and J. Webber. 1972. *J Sci Fed Agric* 23:93–100.

Bingham, F. T. 1979. *Environ Health Perspec* 28:39–43.

Bingham, F. T., G. A. Mitchell, R. J. Mahler, and A. L. Page. 1976. In *Proc of Intl Conf on Environmental Sensing and Assessment,* vol 2. Inst Elec and Electron Engrs, Inc, New York.

Boggess, S. F., S. Willavize, and D. E. Koeppe. 1978. *Agron J* 70:756–760.

Boswell, F. C. 1975. *J Environ Qual* 4:267–273.

Bouwer, H., and R. L. Chaney. 1975. *Adv Agron* 26:133–176.

Bowen, H. J. M. 1979. *Environmental chemistry of the elements.* Academic Press, New York. 333 pp.

Cannon, H. L. 1978. *Geochem Environ* 3:17–31.

Capar, S. G., J. T. Tanner, M. H. Friedman, and K. W. Boyer. 1978. *Environ Sci Technol* 7:785–790.

Carlson, C. W., and J. D. Menzies. 1971. *BioScience* 21:561–564.

CAST. *See* Council for Agricultural Science and Technology.

Chichilo, P., and C. W. Whittaker. 1961. *Agron J* 53:139–144.

Council for Agricultural Science and Technology (CAST). 1976. In *Application of sewage sludge to cropland.* EPA-430/9-76-013. US-EPA, Washington, DC. 63 pp.

———1980. In *Effects of sewage sludge on the cadmium and zinc content of crops.* CAST Rep no 83, Ames, Iowa. 77 pp.

Davies, B. E., and L. J. Roberts. 1978. *Water, Air, Soil Pollut* 9:507–518.

Davies, B. E., and R. C. Ginnever. 1979. *J Agric Sci Camb* 93:753–756.

Domingo, L. E., and K. Kyuma. 1983. *Soil Sci Plant Nutr* 29:439–452.

El-Bassam, N., C. Tietjen, and J. Esser. 1979. In *Management and control of heavy metals in the environment,* 521–524. CEP Consultants Ltd, Edinburgh, UK.

Epstein, E., and R. L. Jefferies. 1964. *Ann Rev Plant Physiol* 15:169–184.

Ermolenko, N. F. 1972. In *Trace elements and colloids in soils.* NTIS, Springfield, VA.

Frank, R., K. Ishida, and P. Suda. 1976. *Can J Soil Sci* 56:181–196.

Furr, K. A., A. W. Lawrence, S. S. C. Tong, M. G. Grandolfo, R. A. Hofstader, C. A. Bache, W. H. Gutenmann, and D. J. Lisk. 1976. *Environ Sci Technol* 7:683–687.

Fuller, W. H. 1977. In *Movement of selected metals, asbestos, and cyanide in soil: application to waste disposal problem.* EPA-600/2-77-020. Solid and Hazardous Waste Res Div, US-EPA, Cincinnati, OH. 243 pp.

Gadde, R. R., and H. A. Laitinen. 1974. *Anal Chem* 46:2022–2026.

Gambrell, R. P., and W. H. Patrick, Jr. 1978. In D. D. Hook and R. M. M. Crawford, eds. *Plant life in anaerobic environments.* 375–423. Ann Arbor Sci Publ, Ann Arbor, MI.

Garland, T. R., D. A. Cataldo, and R. E. Wildung. 1981. *J Agric Food Chem* 29:915–920.

Giordano, P. M., D. A. Mays, and A. D. Behel, Jr. 1979. *J Environ Qual* 8:233–236.

Haghiri, F. 1974. *J Environ Qual* 3:180–183.

Hart, S. A. 1968. In US Public Health Serv Publ no 1826. 40 pp.

Hewitt, E. J., and T. A. Smith. 1974. *Plant mineral nutrition.* The English Universities Press Ltd, London. 298 pp.

Hinesly, T. D., D. E. Alexander, E. L. Ziegler, and G. L. Barrett. 1978. *Agron J* 70:425–428.

Hodgson, J. F. 1963. *Adv Agron* 15:119–158.

Hutchinson, T. C., and L. M. Whitby. 1974. *Environ Conserv* 1:123–132.

Jenkins, D. W. 1981. In *Biological monitoring of toxic trace elements.* EPA-600/S3-80-090. US-EPA, Cincinnati, OH. 9 pp.

Jenne, E. A. 1968. *Adv Chem* 73:337–387.

Jordan, M. J. 1975. *Ecology* 56:78–91.

Kardos, L. T. 1970. *Environment* 12(2):10–27.

Keeney, D. R., K. W. Lee, and L. M. Walsh. 1975. *Guidelines for the application of waste-water sludge to agricultural land in Wisconsin.* Tech Bull no 88, Madison, WI. 36 pp.

Kirkham, M. B. 1977. In R. C. Loehr, ed. *Land as a waste management alternative,* 209–247. Ann Arbor Sci Publ, Ann Arbor, MI.

Kitagishi, K., and I. Yamane. 1981. In *Heavy metal pollution in soils of Japan.* Japan Sci Soc Press, Tokyo.

Klein, D. H., A. W. Andren, and N. E. Bolton. 1975. *Water, Air, Soil Pollut* 5:71–77.

Kloke, A., D. R. Sauerbeck, and H. Vetter. 1984. In J. O. Nriagu, ed. *Changing metal cycles and human health* (Dahlem Konferenzen), 113–141. Springer-Verlag, Berlin.

Knezek, B. D. 1972. *Heavy metal reactions in the soil.* Mich State Univ Inst Water Res Tech Rep 30:27–43.

Korte, N. E., J. Skopp, W. H. Fuller, E. E. Niebla, and B. A. Alesii. 1976. *Soil Sci* 122:350–359.

Lagerwerff, J. V., and A. W. Specht. 1970. *Environ Sci Technol* 4:583–588.

Latterell, J. J., R. H. Dowdy, and W. E. Larson. 1978. *J Environ Qual* 7:435–440.

Leeper, G. W. 1972. *Reactions of heavy metals with soil with special regard to their application of sewage wastes.* Dept of Army, Corps of Engrs, Washington, DC. Contract no DACW 73-73-C-0026. 70 pp.

Lindsay, W. L. 1979. *Chemical equilibria in soils*. John Wiley and Sons, New York. 449 pp.

Lippmann, M., and R. B. Schlesinger. 1979. *Chemical contamination in the human environment*, Oxford Univ Press, New York, 456 pp.

Loehr, R. C., W. J. Jewell, J. D. Novak, W. W. Clarkson, and G. S. Friedman. 1979. *Land applications of wastes*, vol. 2. Van Nostrand Reinhold, New York. 431 pp.

Logan, T. J., and R. L. Chaney. 1983. In A. L. Page, T. L. Gleason, J. E. Smith, I. K. Iskandar, and L. E. Sommers, eds. *Utilization of municipal wastewater and sludge on land*. Univ of California, Riverside, CA. 480 pp.

Luckey, T. D., and B. Venugopal. 1977. *Metal toxicity in mammals*. 1. Physiologic and chemical basis for metal toxicity. Plenum Press, New York. 238 pp.

Miller, W. J., and M. W. Neathery. 1977. *BioScience* 27:674–679.

Miller, R. H., R. K. White, T. J. Logan, D. L. Forster, and J. N. Stitzlein. 1979. *Ohio guide for land application of sewage sludge*. Res Bull 1079. Columbus, OH. 16 pp.

Mitchell, R. L. 1964. In F. E. Bear, ed. *Chemistry of the soil*, 320–368. Reinhold Publ Corp, New York.

Mortvedt, J. J., and P. M. Giordano. 1977. In H. Drucker and R. E. Wildung, eds. *Biological implications of metals in the environment*. CONF-750929. NTIS, Springfield, VA.

Motto, H. L., R. H. Daines, D. M. Chilko, and C. K. Motto. 1970. *Environ Sci Technol* 4:231–237.

Munro, R. D. 1983. *Ambio* 12:61–62.

Murphy, L. S., and L. M. Walsh. 1972. In J. J. Mortvedt, P. M. Giordano, and W. L. Lindsay, eds. *Micronutrients in agriculture*, 347–387. Soil Sci Soc Am Inc, Madison, WI.

Murphy, L. S., R. Ellis, Jr., and D. C. Adriano. 1981. *J Plant Nutr* 3:593–613.

National Academy of Sciences–National Academy of Engineering (NAS–NAE). 1973. In *Water quality criteria, 1972*. US Govt Printing Office, Washington, DC.

National Academy of Sciences (NAS). 1977. *Multimedia management of municipal sludge*. NAS–NRC, vol 9, Washington, DC. 202 pp.

Neary, D. G., G. Schneider, and D. P. White. 1975. *Soil Sci Soc Am Proc* 39:981–982.

Nicholas, D. J. D., and A. R. Egan. 1975. In *Trace elements in soil-plant-animal systems*. Academic Press, New York. 417 pp.

Nriagu, J. O., ed. 1984. *Changing metal cycles and human health* (Dahlem Konferenzen). Springer-Verlag, Berlin. 445 pp.

Oliver, B. G., and E. G. Cosgrove. 1975. *Environ Letters* 9:75–90.

Olsen, S. R. 1972. In J. J. Mortvedt, P. M. Giordano, and W. L. Lindsay, eds. *Micronutrients in agriculture*, 243–261. Soil Sci Soc Am Inc, Macison, WI.

Page, A. L. 1974. *Fate and effects of trace elements in sewage sludge when applied to agricultural lands*. EPA 670/2-74-005. US-EPA, Cincinnati, OH. 98 pp.

Page, A. L., A. Elseewi, and I. Straughan. 1979. *Residue Rev* 71:83–120.

Pearson, R. W., and F. Adams. 1967. In *Soil acidity and liming*. Monog no 12. Am Soc Agron, Madison, WI. 274 pp.

Purves, D., and E. J. Mackenzie. 1974. *Plant Soil* 40:231–235.

Ragaini, R. C., H. R. Ralston, and N. Roberts. 1977. *Environ Sci Technol* 8:773–781.

Rose, A. W., H. E. Hawkes, and J. S. Webb. 1979. *Geochemistry in mineral exploration.* Academic Press, London. 658 pp.

Shacklette, H. T., and J. G. Boerngen. 1984. *Element concentrations in soils and other surficial materials of the conterminous United States.* USGS Prof Paper 1270. US Govt Printing Office, Washington, DC.

Sharma, R. P., and J. L. Shupe. 1977. In H. Drucker and R. E. Wildung, eds. *Biological implications of metals in the environment,* 595–608. CONF-750929. NTIS, Springfield, VA.

Sillanpaa, M. 1972. *Trace elements in soils and agriculture.* Soils Bull no 17. FAO of the United Nations, Rome. 67 pp.

Silviera, D. J., and L. E. Sommers. 1977. *J Environ Qual* 6:47–52.

Sommers, L. E. 1977. *J Environ Qual* 6:225–232.

Sparr, M. C. 1970. *Commun Soil Sci Plant Anal* 1:241–262.

Stewart, B. A., and R. L. Chaney. 1975. *Proc Soil Conserv Soc Am* 30:160–166.

Tabatabai, M. A., and W. T. Frankenberger, Jr. 1979. In *Chemical composition of sewage sludges in Iowa,* 934–944. Res Bull 586. Iowa State Univ Agric Home Econ Exp Station, Ames, Iowa.

Terman, G. L. 1978. *Solid wastes from coal-fired power plants; use or disposal on agricultural lands.* Bull Y-129. TVA, Muscle Shoals, AL.

Underwood, E. J. 1975. In D. J. D. Nicholas and A. R. Egan, eds. *Trace elements in soil-plant-animal systems,* 227–241. Academic Press, New York.

Underwood, E. J. 1977. *Trace elements in human and animal nutrition.* Academic Press, New York. 545 pp.

US Bureau of Mines. 1980. In *Minerals yearbook, 1978–1979,* vol 1. *Metals and minerals.* US Dept of Interior, Washington, DC.

US Bureau of Mines. 1977. In *Minerals yearbook, 1975,* vol 1. *Metals, minerals and fuels.* US Dept of Interior, Washington, DC.

US Department of Agriculture (USDA). 1969. *Agricultural statistics 1969.* 429 pp.

US Department of the Interior, Federal Water Pollution Control Administration. 1968. *Water quality criteria.* Report of the Natl Technical Advisory Comm, Washington, DC.

US Public Health Service (US-PHS). 1962. Drinking water standards. US Dept Health, Education, and Welfare, Washington, DC.

Van Hook, R. I., and W. D. Shults. 1976. In *Effects of trace contaminants from coal combustion.* ERDA 77-64. NTIS, Springfield, VA. 77 pp.

Van Hook, R. I., W. F. Harris, and G. S. Henderson. 1977. *Ambio* 6:281–286.

Van Hook, R. I., D. W. Johnson, and B. P. Spalding. 1980. In J. O. Nriagu, ed. *Zinc in the environment. Part I. Ecological cycling,* 419–437. Wiley, New York.

Van Loon, J. C. 1973. In *Heavy metals in agricultural lands receiving chemical sewage sludges,* vol 1. Res Rep 9. Environ Canada, Toronto, Ontario. 37 pp.

Vinogradov, A. P. 1959. *The geochemistry of rare and dispersed chemical elements in soils.* Consultants Bureau, Inc, New York. 209 pp.

Wells, N., and J. S. Whitton. 1977. *N Z J Exp Agric* 5:363–369.

White, M. C., A. M. Decker, and R. L. Chaney. 1979. *Agron J* 71:121–126.

Wixson, B. G., E. Bolter, N. L. Gale, J. C. Jennett, and K. Purushothaman. 1973. In *Cycling and control of metals.* Natl Environ Res Center, Cincinnati, OH.

Wood, J. M., and E. D. Goldberg. 1977. In W. Stumm, ed. *Global chemical cycles and their alterations by man,* 137–153. Berlin Dahlem Konferenzen.

Zingaro, R. A. 1979. *Environ Sci Technol* 13:282–287.

2
Arsenic

1. General Properties of Arsenic

Arsenic (atomic no. 33) is a steel-gray, brittle, crystalline metalloid with
three allotropic forms that are yellow, black, and gray. It tarnishes in air
and when heated is rapidly oxidized to arsenous oxide (As_2O_3) with the
odor of garlic. It belongs to Group V-A, has an atomic weight of 74.922,
and closely resembles phosphorus chemically. Gray As, the ordinary stable
form, has a density of 5.73 g/cm^3, a melting point of 817°C, and sublimes
at 613°C. The more common oxidation states available to As are − III,
0, III, and V. Arsenic compounds compete with their phosphorus analogs
for chemical binding sites. Arsenic bonds covalently with most nonmetals
and metals and forms stable organic compounds in both its trivalent and
pentavalent states. The most important compounds are white As (As_2O_3),
the sulfide, Paris Green [3 $Cu(AsO_2)_2$•$Cu(C_2H_3O_2)_2$], calcium arsenate, and
lead arsenate, the last three being used as agricultural pesticides and poi-
sons.

2. Production and Uses of Arsenic

World production of As in 1973 was about 47,000 tonnes (Table 2.1). Sweden was the leading world supplier, followed by France, southwestern
Africa, and the USSR. In 1979, however, world production declined significantly, with the USSR becoming the leading producer, followed by
France, Sweden, and Mexico.
  Arsenic trioxide, also known as white arsenic (As_2O_3), constitutes 97%
of As produced that enters end-product manufacturing. The other form
needed by end-product manufacturers is the metal form, which is used
as an additive in special lead and copper alloys. The primary commercial
sources of As are copper and lead ores. Arsenic is recovered as a by-
product during the smelting process.
  In the United States, agriculture accounted for about 81% of As use in

TABLE 2.1. Major producers of arsenic trioxide (white arsenic) in the world.[a]

Country	As production, tonnes			
	1970	1973	1976	1979
France	10,224	9,100	7,281	7,280
Germany, FR	371	473	364	364
Japan	886	455	60	NA[b]
Mexico	9,168	4,393	5,516	6,370
Peru	774	1,092	780	1,365
Portugal	190	20	278	218
S. W. Africa, Terr.	4,075	8,173	5,138	2,275
Sweden	16,450	16,562	6,744	6,734
USSR	7,170	7,270	7,462	7,735
USA	W[c]	W[c]	W[c]	W[c]

[a]*Source:* US Bureau of Mines, *Minerals yearbooks,* 1973, 1978–79.
[b]W—withheld to avoid disclosing company confidential data.
[c]NA—not available.

1973, ceramics and glass for 8%, chemicals for 5%, and the balance for other uses (NAS, 1977). By 1981, agriculture accounted for only 46% of As use in the United States (Fig. 2.1). Arsenic trioxide is the raw material for arsenical pesticides including lead arsenate, calcium arsenate, sodium arsenite, and organic arsenicals. These compounds are used in insecticides, herbicides, fungicides, algicides, sheepdips, wood preservatives, and dyestuffs and for the eradication of tapeworm in sheep and cattle.

Arsenic is used primarily for its toxic properties. Inorganic arsenicals have been used in agriculture as a pesticide or plant defoliant for many years (Table 2.2). Calcium arsenate and lead arsenate were the backbone of the insecticide industry from the early 1900s until the advent of the

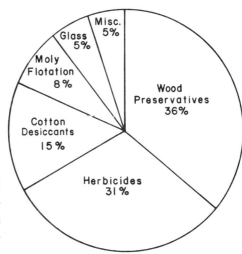

FIGURE 2.1. Estimated markets for arsenic trioxide (As$_2$O$_3$) in the United States in 1981. *Source:* Fitzgerald, 1983, with permission of the author and Reinhold Co., New York.

TABLE 2.2. Names and uses of some important arsenical pesticides.[a]

Pesticide	Application rates and methods	Commercial uses
Arsenic acid	0.0035 m³/ha of the 75% concentrate	Cotton desiccant to facilitate mechanical harvesting
Cacodylic acid	3.4–11.2 kg/ha	Lawn renovation and general weed control in noncrop areas
DSMA (disodium methanearsonate)	Directed after emergence on cotton at 2.52 kg/ha; 2.24–4.26 kg/ha for lawn and ornamental uses	Cotton and noncrop areas; crabgrass
MSMA (monosodium methanearsonate)	Directed after emergence on cotton at 2.52 kg/ha; 2.24–4.26 kg/ha for lawn and ornamental uses	Cotton and noncrop areas; crabgrass
Calcium arsenate	1.68–3.36 kg/ha in 100 gal (0.38 m³) of water or dust at 2.24–28 kg/ha	Cotton insecticide; fruits, vegetables, and potatoes
Lead arsenate	3.36–67.3 kg/ha or 1.2–72 kg/m³ of water	Fruits, vegetables, nuts, turf, and ornamentals
Paris Green	1.12–17.9 kg/ha	Baits and mosquito larvicide
Sodium arsenite	1.12–22.4 kg/ha in dry baits	Baits and a livestock dip; a nonselective herbicide, rodenticide, desiccant, and aquatic weed killer

[a] Source: Modified from NAS (1977).

organic pesticides in the 1940s (Alden, 1983). Since they were less phytotoxic, they replaced the earlier arsenicals, Paris Green and London Purple, which had been in use since the 1860s.

Paris Green (copper acetoarsenite) was successfully used to control the Colorado potato beetle in the eastern United States by about 1867 (Mellanby, 1967). Calcium arsenate has been applied to cotton and tobacco fields for boll weevil, beetle, and other insect control. Lead arsenate has been used for insect control on a variety of fruit trees and was particularly effective in the control of codling moth in apple orchards and horn worm on tobacco. Arsenic trioxide has been widely used as a soil sterilant (Crafts and Rosenfels, 1939) while sodium arsenite has been used for aquatic weed control and as a defoliant for potato prior to tuber harvest (Cunningham et al, 1952).

Recently, organic arsenicals have largely replaced inorganic As compounds as herbicides. The use of sodium arsenite has been limited, and the compound can no longer be used as a defoliant or vine-killer (Walsh and Keeney, 1975). Lead arsenate is now seldom used in orchards, since fruit growers rely primarily on carbamates and organic phosphates for insect control. Calcium arsenate and Paris Green are also seldom used nowadays. In some areas of the United States, however, inorganic arsenicals are still used to control bluegrass on golf greens and fairways.

Since the mid-1970s, use of organic arsenical herbicides—MSMA (monosodium methanearsonate), DSMA (disodium methanearsonate), and cacodylic acid (Table 2.2)—has grown rapidly. MSMA and DSMA have been used as selective herbicides for the postemergence control of crabgrass, Dallisgrass, and other weeds in turf (NAS, 1977). They are also extensively used as selective postemergence herbicides in citrus, cotton, and noncrop areas for the control of Johnsongrass, nutsedge, watergrass, sandbur, foxtail, cocklebur, ragweed, barnyard grass, and pigweed. MSMA is also used for chemical mowing along highway rights-of-way. Cacodylic acid is used for general weed control and is an excellent herbicide for monocotyledonous weeds.

In cotton production, As chemicals currently being used as herbicides and harvest aids are MSMA, DSMA, As acid, and cacodylic acid. Herbicidal control of weeds is obtained with MSMA, DSMA, and cacodylic acid, while As acid and cacodylic acid are used as desiccants and defoliants in cotton (Abernathy, 1983). The use of MSMA and DSMA herbicides now accounts for more than 90% of the As used in agriculture and it is one of the largest volume pesticides (Wauchope and McDowell, 1984). However, because of economic and environmental considerations, glyphosate [N-(phosphonomethyl) glycine] may become a viable alternative to about 46% of the MSMA and DSMA use.

Arsenic is also used as a feed additive, although lesser amounts are used than in pesticides, defoliants, or herbicides (NAS, 1977). Arsenicals are used in poultry feeds to control coccidiosis and to promote chick growth (Calvert, 1975). In a typical year, the following are sold or manufactured in the United States: arsanilic acid—1,360 tonnes; carbarsone—450 tonnes; 3-nitro-4 hydroxyphenylarsonic acid—900 tonnes; and combinations—450 tonnes. They are added to poultry rations at rates of about 100 mg/kg for arsanilic acid and 50 mg/kg for 3-nitro-4-hydroxyphenylarsonic acid. Other uses for As are discussed in detail elsewhere (Dickerson, 1980; NAS, 1977).

3. Natural Occurrence of Arsenic

Arsenic is ubiquitous in nature and is found in detectable concentrations in all soils and nearly all other environmental matrices (Table 2.3, Appendix Table 1.2). The occurrence of As in the continental crust of the earth is generally given as 1.5–2 ppm (NAS, 1977). Arsenic ranks 52nd in crustal abundance, ahead of Mo (Krauskopf, 1979). It is a major constituent of

TABLE 2.3. Commonly observed arsenic concentrations (ppm) in various environmental matrices.

Material	Average concentration	Range
Igneous rocks[a]	1.5	0.2–13.8
Limestone[a]	2.6	0.1–20.1
Sandstone[a]	4.1	0.6–120
Shale[a]	14.5	0.3–500
Petroleum[a]	0.18	<0.003–1.11
Oil[a]	0.01	—
Coal[a,b]	15	0.3–100
Fly ash[b]		
Bituminous	82	—
Sub-bituminous	2.3	—
Lignite	34	—
Sewage sludge[e]	14.3	3–30
Cow manure[f]	1.54	0.88–2.20
Soils (world, normal)[a]	7.2	0.1–55
Forest soils (Norway)[a]	—	0.59–5.70
Common crops[c]	—	0.03–3.50
Ferns[d]	1.3	—
Fungi[d]	—	1.2–2.5
Woody angiosperm[d]	2.0	—
Woody gymnosperm[d]	—	0.2–1.2
Great Lakes sediments[a]	—	0.50–14.00
Ocean sediments[a]	33.7	<0.40–455

Sources: [a]NRCC (1978).
[b]Adriano et al (1980).
[c]Liebig (1965).
[d]Bowen (1979).
[e]Furr et al (1976).
[f]Capar et al (1978).

more than 245 minerals and is found in high concentrations in sulfide deposits: arsenides (27 minerals), sulfides (13 minerals), sulfosalts (65 minerals), and others.

The geologic history of a particular soil determines its native As content. Soils overlying sulfide ore deposits usually contain As at several hundred ppm, the reported average being 126 ppm and the range being from 2 to 8,000 ppm As (NRCC, 1978). A natural background level of 6.3 ppm total As was reported for agricultural soils in Ontario, Canada (Frank et al, 1976). Greaves (1913) found 4 ppm As in western US virgin soils. He attributed soil As as originating from soil-forming rocks.

Arsenic concentrations in igneous and sedimentary rocks are listed in Appendix Table 1.2 and some in Table 2.3. The limited data indicate that shale and clays usually exhibit rather high As concentrations. Coal and its by-product fly ash also contain significant quantities. Thus, combustion of coal for generating power, as well as the disposal of fly ash, may contribute to As input into the environment.

4. Arsenic in Soils

4.1 Total Soil Arsenic

Arsenic levels in uncontaminated, nontreated soils seldom exceed 10 ppm. However, As residues can accumulate to very high levels in agricultural areas where As pesticides or defoliants were repeatedly used (Table 2.4). Thus, it is apparent that agricultural use of arsenicals can cause surface soil accumulation of 600 ppm or more. Survey of soils in the United States indicates that As levels for normal soils ranged from 0.2 to 40 ppm (Olson et al, 1940). More recently, Shacklette and Boerngen (1984) reported an As average of 7.2 ppm (range of < 0.1–97 ppm) for United States surface soils (Appendix Table 1.17).

Selby et al (1974) cited a mean of 8.7 ppm As in 1,140 soil samples collected from 114 counties in Missouri. Thus, it appears that uncontaminated soils seldom contain more than 10 ppm As. The As concentrations of soils of the Russian Plain were estimated to be generally uniform at 1 to 10 ppm (Vinogradov, 1959), with an average of 3.6 ppm. The highest As content is found in the chernozem and gray forest soils, and the lowest, in the northern tundra soils and podzolic soils. Based on analysis of soils from various parts of the world, Berrow and Reaves (1984) reported a mean As content of 10 ppm. In areas near As mineral deposits, soil levels may average 400 to 900 ppm As (NRCC, 1978).

TABLE 2.4. Arsenic levels (total As in ppm) in field soils repeatedly treated with an arsenic pesticide or defoliant and in nontreated soils. [a]

Location	Control soil	Treated soil	Crop
Colorado	1.3–2.3	13–69	orchard
Florida	8	18–28	potato
Idaho	0–10	138–204	orchard
Indiana	2–4	56–250	orchard
Maine	9	10–40	blueberry
Maryland	19–41	21–238	orchard
New Jersey	10	92–270	orchard
New York	3–12	90–625	orchard
North Carolina	4	1–5	tobacco
Nova Scotia	0–7.9	10–124	orchard
Ontario	1.1–8.6	10–121	orchard
Oregon	2.9–14.0	17–439	orchard
	3–32	4–103	orchard
Washington	6–13	106–830	orchard
	8–80	106–2,553	orchard
	4–13	48	orchard
Wisconsin	2.2	6–26	potato

[a] Source: Adapted from Walsh and Keeney, 1975, with permission of the authors and the Am Chem Society.

4.2. Extractable (Available) Arsenic in Soils

As with other trace elements, the soluble (extractable) fraction of As, may give some indication of its phytoavailability and mobility in soils. It is a general consensus that total As in soil does not accurately reflect its phytoavailability. Woolson et al (1971) reported that correlation was better between extractable As and plant growth than between total As and plant growth. The following are some previously used extractants for soil As: distilled water (Deuel and Swoboda, 1972a; Greaves, 1934; Vandecaveye et al, 1936), 1 N and 0.1 N NH_4OAc (Jacobs et al, 1970a; Vandecaveye et al, 1936), Bray P-1 solution (0.03 N NH_4F + 0.025 N HCl) (Jacobs et al, 1970a), mixed acid (0.05 N HCl + 0.025 N H_2SO_4) and 0.5 N $NaHCO_3$ (Woolson et al, 1971), 0.05 M KH_2PO_4 (Woolson et al, 1973), 0.05 N HCl (Reed and Sturgis, 1936), and 0.1 N NH_4NO_3, 0.1 N KNO_3, and hot concentrated HNO_3 (Vandecaveye et al, 1936).

Deuel and Swoboda (1972b) found that in soils under reducing conditions, H_2O-soluble As was well correlated to total As. In addition, H_2O-soluble As increased with increased levels of As added; less than 5% of the added As was H_2O-soluble in samples under room temperature. Jacobs et al (1970a) found that, on the average, about 1% to 5% of the total As was extracted with 1 N NH_4OAc (pH 7.0), while Bray P-1 removed up to 38% of the As in the soil. Johnston and Barnard (1979) rated the efficiency of the basic extractants in removing soil As as follows: 0.5 M NH_4F $\approx$ 0.5 M $NaHCO_3$ < 0.5 M $(NH_4)_2CO_3$ < 0.5 M Na_2CO_3 < 0.1 M NaOH, with the effectiveness increasing with increasing pH. For the acid extractants, the order was: (0.05 N HCl + 0.025 N H_2SO_4) $\approx$ 0.5 N HCl < 0.5 M KH_2PO_4< 0.5 N H_2SO_4. Distilled water, 1 N NH_4Cl, 0.5 M NH_4OAc and 0.5 M NH_4NO_3 did not extract more than 0.1 μg As/g soil even after an 18-hour shaking period. The amounts of As extracted by H_2O, 0.1 N NH_4OAc, 0.1 N NH_4NO_3, 0.1 N $(NH_4)_2SO_4$, and 0.1 N KNO_3 were similar (about 2.8 ppm) and showed no correlation with phytotoxicity (Vandecaveye et al, 1936). In contrast, Deuel and Swoboda (1972a) found that H_2O, 1 N NH_4Cl, and 0.5 N HCl were significantly correlated with plant growth. Jacobs et al (1970a) found that total As, NH_4OAc, and Bray P-1 extractable As were equally effective in predicting reduced yields. Reed and Sturgis (1936) found better correlation between plant growth and 0.05 N HCl-extractable As than with H_2O-soluble As. Both mixed acid (0.025 N H_2SO_4 and 0.05 N HCl) and 0.5 N $NaHCO_3$-extractable As had high and nearly equal correlations (r = 0.81 and 0.82, respectively) with plant growth. The mixed acid method is routinely used to test for available P in weathered acid soils of the eastern United States; the bicarbonate procedure is used to test for available P in alkaline soils in the western states. These available P extractants, along with Bray P-1 solution, may be favorable over the other extractants in determining available (phytotoxic) As because they are less laborious to perform, are routinely used for P analysis, and extract relatively higher As levels, favoring analytical detectability (Jacobs et al, 1970a; Woolson et al, 1971).

4.3 Arsenic in the Soil Profile

Early work by McGeorge (1915) indicated that As in surface soils was fairly immobile. However, more recent work (Frans et al, 1956; Tammes and deLint, 1969; Steevens et al, 1972) have shown that appreciable As can move downward with leaching water, especially in coarse-textured soils. Losses of As applied as sodium arsenite to sandy soils in the Netherlands were directly related to the amounts of As in the soil, with an average half-life in the surface soil of 6.5 ± 0.4 years (Tammes and deLint, 1969). Most of the As lost from the upper 20 cm of soil was found in the 20- to 40-cm depth, although net loss of As from the soil occurred continuously.

The profile distribution of Pb, As, and Zn (Fig. 2.2) in an abandoned apple orchard indicates insignificant leaching of the metals below the 20-cm depth. The site, used for fruit production since the late 1800s, had received large applications of lead arsenate pesticides containing Pb and

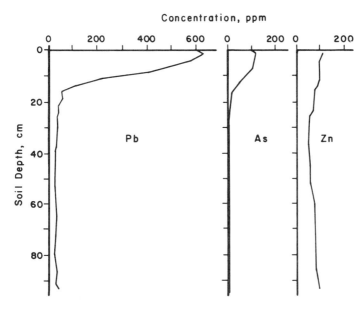

FIGURE 2.2. Arsenic, lead, and zinc distribution with soil depth in an abandoned apple orchard that had received significant applications of lead arsenate pesticides since the late 1800s. *Source:* Veneman et al, 1983, with permission of the authors and the Am Soc of Agronomy, Crop Sci Soc of America, Soil Sci Soc of America.

As, and trace amounts of Zn. Maximum As and Pb concentrations in surface soils were 120 and 870 ppm, respectively.

The profile distribution of As in a Plainfield sand (Table 2.5) that had received varying amounts of As was studied by Steevens et al (1972). These plots were treated with sodium arsenite in 1967 and received about 75 cm of precipitation and 50 cm of irrigation water per year until 1970. Irrigation was stopped after 1970. Phytotoxicity persisted at the 180- and 720-kg/ha As plots from 1967 to 1970. The 720 kg/ha plots were still essentially barren in 1974, seven years after As application. With time, however, significant declines in the amount of total As in the surface soil of all treated plots were observed. In the 720-kg/ha plots, total As in the surface soil had declined to 40% of the original value in only three years, and to 33% by seven years.

Woolson and Isensee (1981) estimated that As, applied at about the recommended rates, will accumulate in the soil gradually and will theoretically reach equilibrium after 25 to 30 years of application. This means that when the amount of As lost equals the amount applied, equilibrium has been attained. These losses in surface soils were attributed in this study largely to leaching that caused increases in As at lower depths. Seven years after As application on the soil surface, As had leached to 83 cm. This study clearly demonstrates that As will leach in sandy soils, that leaching loss is fastest during the first few years, and that As is fairly persistent even in soils of low sorption capacity (Walsh and Keeney, 1975). Similarly, Johnson and Hiltbold (1969) found that As, applied as metha-

TABLE 2.5. Distribution of total arsenic (ppm) in the soil profile of a Plainfield sand after surface applicaton of sodium arsenite. [a]

Year	As applied kg/ha	Depth, cm				
		0–23	23–38	38–53	53–68	68–83
1967	0	3.0	—	—	—	—
	45	11.0	—	—	—	—
	90	23.0	—	—	—	—
	180	73.0	—	—	—	—
	720	250.0	—	—	—	—
1970	0	3.6	1.2	1.8	1.3	1.2
	45	14.1	2.3	1.4	1.9	1.6
	90	27.0	3.4	1.5	1.2	1.4
	180	45.0	4.8	1.9	1.7	1.1
	720	100.0	65.0	8.6	3.3	2.0
1974	0	1.6	1.0	1.3	0.5	0.6
	45	8.1	1.2	1.4	0.7	0.7
	90	14.4	1.8	1.6	1.3	1.3
	180	21.6	2.8	1.5	1.1	1.3
	720	82.8	50.6	19.6	2.6	3.7

[a] *Source:* Walsh and Keeney, 1975, with permission of the authors and the Am Chem Society.

nearsonates to turf over a four-year period on a sandy loam, was highest in concentration in the upper 5 cm of soil and decreased in concentration at the 30-cm depth. Although most of the As residue from MSMA application accumulated in the top 15 cm of soil, Robinson (1975) found that the amount of As in the 15- to 30-cm depth was 2.5 times greater after five years than after three years of application, indicating buildup over time.

4.4 Forms and Sorption of Arsenic in Soils

Because of their chemical similarities, As and P are assumed to react similarly in soils, forming insoluble compounds with Al, Fe, and Ca. Fassbender (1974) found that the fixation capacities for P and As were similar ($r = 0.983$) in 19 acid forest soils in the FRG and that the fixed As and P were primarily bound to Fe and Al. This is expected because both As and P belong to the same chemical group, having comparable dissociation constants for their acids and solubility products for their salts. Consequently, to study the As forms and sorption in soil, the traditional Chang and Jackson (1957) procedure used to fractionate soil P has been adopted. This fractionation scheme can approximate water-soluble or adsorbed As (NH_4Cl-soluble), Al-arsenate (NH_4F-soluble), Fe-arsenate (NaOH-soluble), and Ca-arsenate (H_2SO_4-soluble), as indicated in Table 2.6. The Langmuir adsorption isotherm, commonly used to characterize P sorption in soils, has also been employed for characterizing As sorption by soils and clay minerals (Jacobs et al, 1970b; Huang, 1975). However, the Freundlich isotherm has been recently used to describe arsenite(III) sorption by soils (Elkhatib et al, 1984a,b).

There are now numerous papers elucidating the similar reactions of As and P in soils. Jacobs et al (1970b) found that the sorption of As against extraction by NH_4OAc and Bray P-1 extractants from Wisconsin soils equilibrated with As (80 and 320 μg As/g soil added as Na_2HAsO_4) increased as the sesquioxide content increased. More specifically, the amount of As sorbed increased as the free Fe_2O_3 content of the soil increased. Furthermore, they found that removal of amorphous Fe and Al components by treatment with oxalate eliminated or markedly reduced the As sorption capacity of the soil. In an extensive soil survey encompassing 12 states, Woolson et al (1971) found that most of the As residues in 58 surface soils with a history of inorganic As application was in the Fe-arsenate form. However, when the amount of "reactive" Al or Ca was high and reactive Fe was low, Al- and Ca-arsenate predominated. Water-soluble As was detected in soils from only two states. The distribution of As forms in soils and the influence of soil chemical properties on these forms are shown in Table 2.6.

In another fractionation study of As, added to various soils as DSMA, Fe-arsenate was the predominant form present, accounting for 31% to

TABLE 2.6. Total arsenic and distribution of various forms of arsenic in US soils. [a]

Location and soil	Treatment	pH	Available cations (meq/100 g)			Arsenic[b]						Corn growth reduction, %
			Ca	Fe	Al	Total ppm	Easily soluble, %	Al–As %	Fe–As %	Ca–As %		
Mississippi Memphis sil	Control	5.40	3.2	1,057	6.3	21	0	0	91	9		—
	Treated	5.62	4.0	934	6.6	96	0	10	78	12		-13
Alabama Chesterfield sil	Control	—	—	—	—	8	0	33	67	0		—
	Treated	5.61	3.4	369	4.0	16	0	31	62	7		-28
Idaho Greenleaf sil	Control	7.11	8.6	925	3.3	28	0	0	0	0		—
	Treated	7.01	8.2	906	4.0	170	7	34	46	13		69
Oregon Agate grl	Control	—	—	—	—	10	—	—	—	—		—
	Treated	5.78	8.5	1,155	2.7	67	0	22	56	22		-5
New York Newfane sl	Control	—	—	—	—	4	—	—	—	—		—
	Treated	4.40	1.3	745	4.8	319	0	24	63	13		42
Florida Lakeland fs	Control	—	—	—	—	8	—	—	—	—		—
	Treated	7.14	6.9	35	1.5	28	10	30	0	60		15
Indiana (loess soil)	Control	—	—	—	—	2	—	—	—	—		—
	Treated	4.40	1.4	572	3.6	250	5	26	53	16		67

[a] Source: Extracted from Woolson et al, 1971, with permission of the authors and the Soil Sci Soc of America.
[b] Arsenic forms estimated by a modified Chang and Jackson procedure for P (1957).

54%, with an average of nearly 44% (Akins and Lewis, 1976). Al-arsenate followed in abundance, with over 27% (11% to 41% of added As), followed by Ca-arsenate with 16%, and water-soluble arsenate with 6% and non-extractable fraction, 7%. However, Johnston and Barnard (1979) found that:

$$Fe\text{-}As > > Ca\text{-}As > Al\text{-}As > \text{water-soluble As}$$

The cause for the discrepancy was probably due to incomplete sequential extraction in the latter. Thus, it can be generalized that, in acid soils, As sorption follows the dominance of Fe and Al ions over those of the Ca ions, whereas in alkaline, calcareous soils, the sorption follows the dominance of Ca over the Fe and Al ions.

Analysis of surface soils (0–15 cm) collected in Japan from various agricultural lands polluted with As revealed that arsenate(V) was the dominant form of As in these soils, with lower levels of dimethylarsinate, and detectability of monomethylarsonate in most of the samples (Takamatsu et al, 1982). This study also found seasonal variation of As forms depending on the water status of the paddy field—i.e., under flooded conditions, there is an increase in pH and the amounts of arsenite(III) and dimethylarsinate, whereas under upland conditions pH decreased, but the amounts of arsenate and monomethylarsonate increased.

Woolson et al (1971) pointed out that solubility reactions may play a role in As sorption by soil, with the solubility product constant of $Ca_3(AsO_4)_2$ greater than that of Fe_x and $Al_x(AsO_4)_x$. Because of this increased solubility, the role of Ca in the fixation process is not so pronounced as the role of Fe or Al. For example, Fe-arsenate is extremely insoluble, giving an As concentration in solution of only about $10^{-11}\ M$, compared to about $10^{-5}\ M$ for Ca- or Mg-arsenate (Walsh and Keeney, 1975).

Livesey and Huang (1981) found that the adsorption maxima of soils for arsenate were not related to pH and inorganic C content but were related to ammonium oxalate-extractable Al, and to a lesser extent, to the clay content and ammonium oxalate-extractable Fe. Among the competing anions (i.e., Cl^-, NO_3^-, SO_4^{2-}, and HPO_4^{2-}), the phosphate substantially suppressed As sorption by the soil.

Wauchope (1975) studied the sorption of three As species in comparison with phosphate in aerobic soils. The order of sorption was phosphate < cacodylate < MSMA = arsenate. Similarly, in anaerobic river sediments, the order of sorption was cacodylate < MSMA < arsenate (Holm et al, 1980).

4.5 Transformations of Arsenic in Soils

Arsenic is a labile element present in practically all environmental matrices and can exist in several forms and oxidation states. In strongly reducing environments, elemental As and arsine(-III) can exist, but arsenate(V) is the stable oxidation state in aerobic environments. Under moderately reducing conditions, such as flooded soils, arsenite(III) may be the dominant form (Deuel and Swoboda, 1972b). Arsenite is a common commercial form of As and one of the most toxic As compounds, being 25 times more potent than that of dimethylarsinic acid (Braman and Foreback, 1973). The reduced state of As(III) has been reported to be 4 to 10 times more soluble in soils than the oxidized state(V) (Brenchley, 1914). This was substantiated by Keaton and Kardos (1940) who demonstrated that the V form of As was fixed to a much greater extent than the III form, thereby proposing that the oxidation state of As influences its sorption capacity by soils. Reed and Sturgis (1936) and later Epps and Sturgis (1939) concluded that reduced As compounds are especially toxic to rice grown under flooded conditions. Furthermore, they indicated that arsine (AsH_3) may have been lost in these flooded soils. The $-III$, III, and V oxidation states can form compounds containing the C-As bond and are readily interconverted by microorganisms. In reduced environments, such as sediments, methanogenic bacteria reduced arsenate(V) to arsenite(III) and methylated it to methylarsinic acid(III), or dimethylarsinic (cacodylic) acid(I) (Wood, 1974). These compounds may be further methylated (trimethylarsine, $-III$), or reduced (dimethylarsine, $-III$), and may volatilize to the atmosphere with formation of cacodylic acid via oxidation reactions.

Both oxidative and reductive transformations of methanearsonates could occur in soil. Cheng and Focht (1979) demonstrated that microorganisms in soils amended with inorganic and methylated forms of As could produce volatile arsenicals by a reductive and/or reductive and demethylative pathway. In addition, soil isolates of *Alcaligenes* and *Pseudomonas* were also found to produce arsine from reduction of arsenate and arsenite.

Woolson and Kearney (1973) postulated that cacodylic acid was metabolized by two pathways: an oxidative pathway leading to C-As bond cleavage and a reductive pathway leading to alkyl arsine production. Oxidation of the methyl substituent to CO_2 occurs in association with microbial oxidation of soil organic matter, producing arsenate (Hiltbold, 1975). Dickens and Hiltbold (1967) found amounts of CO_2 evolved from soils receiving methanearsonate ranging from 0.7% to 5.5% of the added compound during a one-month incubation. Similar values (1.7% to 10%) were reported by Von Endt et al (1968) on the oxidation of methyl carbon of MSMA in soils during a three-week period. Woolson and Isensee (1981) found that 14% to 15% of the As applied in soil can be lost through volatilization of alkyl arsines each year.

5. Arsenic in Plants

5.1 Plant Growth and Uptake of Arsenic

There is no evidence that As is essential for plant growth, although stimulation of root growth with small amounts of As in solution culture was reported by Albert and Arndt (1931), Liebig et al (1959), and Stewart and Smith (1922). Liebig's group reported that root growth of lemon plants in solution culture was enhanced by 1 ppm As as arsenate or arsenite. However, at 5 ppm of either form of As, both top and root growth were reduced. In addition, small yield increases have been observed at low levels of As, especially for tolerant crops such as corn, potatoes, rye, and wheat (Jacobs et al, 1970a; Woolson et al, 1971). Stewart (1922) noted that 75 ppm As in calcareous Bench loam caused only slight detrimental effects to the more sensitive plants, but that 25 ppm As in the soil appeared to have stimulated them. Liebig (1965) noted that stimulation does not always occur, is sometimes only temporary, and may result in the reduction of top growth. Two possibilities exist for growth stimulation by As: first, stimulation of plant systems by small amount of As, since other pesticides, like 2,4-D, stimulate plant growth at sublethal dose levels (Woolson et al, 1971); second, displacement of phosphate ions from the soil by arsenate ions, with a resultant increase in phosphate availability (Jacobs et al, 1970b).

Marcus-Wyner and Rains (1982) found that the uptake and translocation of As by cotton plants grown in solution culture were influenced by the source of As. Arsenic, as As_2O_3, was readily taken up by the roots, but was not translocated to the shoots. However, when cacodylic acid was applied, As was translocated to the shoot and reproductive tissues. Arsenic also accumulated in the roots and shoots with MSMA and DSMA as the sources.

5.2 Sensitivity to and Phytotoxicity of Arsenic

Crops have differing degrees of tolerance to soil As (Table 2.7). Members of the bean family, rice, and most of the legumes are fairly sensitive to As. In greenhouse pot experiments, Deuel and Swoboda (1972a) found that the yield-limiting As concentrations in plant tissues were 4.4 ppm in cotton and 1 ppm in soybeans. Jacobs et al (1970a) found that crop tolerance to As was: snap beans < sweet corn < peas < potatoes. In addition, Woolson (1973) found that: green beans < lima beans < spinach < radish < tomato < cabbage. Baker et al (1976) found crop sensitivity to soil As as: snap beans > rice > soybeans > potatoes > cotton. The critical level for barley was 20 ppm As in the leaves and shoots (11–26 ppm range) as determined by sand culture studies (Davis et al, 1978). In rice, the critical level in tops ranges from 20 to 100 ppm As; and in roots, 1,000 ppm (Chino, 1981). Normal leaves from fruit trees contained 0.9 to 1.7 ppm As, but leaves from trees suffering from As excess contained 2.1 to 8.2 ppm (NAS, 1977). Paddy rice is known to be very susceptible to As toxicity as compared to upland rice, since As(III) would be more prevalent under reducing conditions and since As toxicity to paddy rice can be further intensified by the toxic effect of ferrous Fe(II) to the rice plants (Tsutsumi, 1980).

Phytotoxicity symptoms include wilting of new-cycle leaves, followed

TABLE 2.7. Comparative sensitivity to arsenic of various plants. [a]

Tolerant	Moderately tolerant	Low tolerant
	Fruit Crops	
Apples	Cherries	Peaches
Pears	Strawberries	Apricots
Grapes		
Raspberries		
Dewberries		
	Vegetables and Field Crops	
Rye	Beets	Peas
Mint	Corn	Onion
Asparagus	Squash	Cucumber
Cabbage	Turnips	Snap beans
Carrots	Radish	Lima beans
Parsnips		Soybeans
Tomato		Rice
Potato		Spinach
Swiss chard		
Wheat		
Oats		
Cotton		
Peanuts		
Tobacco		
	Forage Crops	
Sudangrass	Crested wheat grass	Alfalfa
Bluegrass	Timothy	Bromegrass
Italian ryegrass		Clover
Kentucky bluegrass		Vetch
Meadow fescue		Other legumes
Red top		

[a] *Sources:* Benson and Reisenauer (1951); Liebig (1965).

by retardation of root and top growth of the plant (Liebig, 1965). It is often accompanied by root discoloration and necrosis of leaf tips and margins. In rice plants, tillering is severely depressed, as in the case of P deficiency (Chino, 1981). These symptoms indicate a restriction in the movement of water into the plant, which may result in death (Woolson et al, 1971).

In applying the GR_{50} (50% reduction in growth) technique, Woolson (1973) found that the GR_{50} values were 76 ppm (8 ppm fresh weight) with unpeeled, washed radish and 10 ppm (1 ppm fresh weight) with spinach, exceeding the tolerance limit of 2.6 ppm of As for vegetables treated with calcium arsenate. The soil for the GR_{50} level for radish contained about 19 ppm available As.

In general, total As in soil does not accurately reflect phytotoxicity. Woolson and co-workers (1971) reported that correlation was better between plant growth and available As than between plant growth and total

As. Likewise, others reported a direct relationship between phytotoxicity and soluble As in soils (Deuel and Swoboda, 1972a: Reed and Sturgis, 1936; Vandecaveye et al, 1936)

In practice, ordinary crop plants do not accumulate enough As to be toxic to man. Instead, growth reductions and crop failure are the main consequences, and only small increases in the total As content of crops are noted in contaminated as compared with noncontaminated soils. Edible portions of crops usually contain less As than the other plant parts. There is very little danger of As residue accumulating to phytotoxic levels under normal application rates.

More recently, Wauchope (1983) reviewed the literature on uptake, translocation, and phytotoxicity of As in plants. He identified some research needs, including the interaction of phosphate with As toxicity and stimulation of plant growth by low levels of As.

6. Factors Affecting the Mobility and Availability of Arsenic

6.1 Soil Factors

A. pH

The effect of soil pH on As mobility as induced by liming is mixed. Since agricultural lime has Ca^{2+} ions as a principal constituent, it may affect the form or solubility of As. Because the solubility product constant of $Ca_3(AsO_4)_2$ is greater than that of Fe_x and $Al_x(AsO_4)_x$, the role of Ca in the fixation process of As is not so pronounced as the role of Fe and Al (Woolson et al, 1971). As previously mentioned, Al-As and Fe-As are the dominant forms of As in acid soils. Therefore, applying lime to acid soils to raise pH is very likely to shift the As form to the Ca-As species. Thus, it is not surprising to find lime as an ineffective soil amendment to alleviate As phytotoxicity on soils containing toxic levels of As (MacPhee et al, 1960). However, Vandecaveye et al (1936) found a positive response in barley with the addition of lime to acid soils. Similarly, Paden and Albert (1930) found improved cotton growth in contaminated soil with lime addition. Apparently in the latter cases, the added lime may have reduced the level of water-soluble As and consequently reduced toxicity.

Wauchope and McDowell (1984) indicated that arsenical (arsenate, MSMA, and cacodylic acid) sorption by selected lake and stream sediments showed some pH dependency, indicative of binding with Ca.

In adsorption studies of $H_2AsO_4^-$ (V) from municipal landfill leachate solutions by various clay minerals, it was found that adsorption of As from solution was maximal at about pH 5 (Frost and Griffin, 1977).

B. Oxides of Iron and Aluminum

Like phosphate, As is strongly adsorbed by amorphous Fe oxide (Jacobs et al, 1970b; Wauchope, 1975; Wauchope and McDowell, 1984; Elkhatib et al, 1984b), but shows less affinity for Al oxides than does phosphate.

C. Soil Texture and Clay Minerals

In general, As mobility and phytotoxicity is greater in sandy than in clayey soils. Woolson (1973) reported that As phytotoxicity to vegetable crops was highest on Lakeland loamy sand and lowest on Hagerstown silty clay loam. Jacobs and Keeney (1970) reported that As was more phytotoxic to corn when grown on a sandy than when grown on the silt loam soil. Similar results were obtained by Reed and Sturgis (1936) with rice, where plants grown on sandy soils were more susceptible to As phytotoxicity than those grown in clayey soils. Others also have noted this inverse relationship between clay content and phytotoxicity increases (Crafts and Rosenfels, 1939). The main reason for this relationship is that both hydrous Fe and Al oxides vary directly with the clay content of the soil. Consequently, the water-soluble fraction of As was highest in soils with low clay content and least in the soil with a high clay content (Akins and Lewis, 1976). In addition, Johnson and Hiltbold (1969) found that about 90% of the soil As, applied repeatedly over a four-year period as either MSMA, MAMA (monoammonium methanearsonate), or DSMA, occurred in the clay fraction. Earlier, Dickens and Hiltbold (1967) found that leaching with 50 cm of water removed 52% of surface-applied DSMA from 24-cm columns of Norfolk loamy sand, while none passed through Decatur clay loam, and 50% of the DSMA remained in the upper 2.5 cm of this soil.

Frost and Griffin (1977) found that montmorillonite adsorbed more As(V) and As(III) than kaolinite. They attributed the higher adsorption by montmorillonite to a higher edge surface area (2.5 times larger than for kaolinite). Another possibility for the higher adsorption by the montmorillonite is that it is known to contain interlayer hydroxy aluminum polymers.

D. Redox Potential

The reduced form of As(III) has been reported to be 4 to 10 times more soluble in soils than the oxidized form (V) (Albert and Arndt, 1931; Brenchley, 1914). Deuel and Swoboda (1972b) found Eh values ranging from 25 to 200 mV and greater soluble As concentrations after adding sugar and water (1:1 soil/water ratio) to samples in an oxygen-free environment. They attributed the higher As solubility to the accompanying reduction of iron from Fe(III) to Fe(II) with subsequent dissolution of ferric arsenate. Arsenate was not reduced to arsenite under the conditions of their experiment. However, they stated that it is possible that flooded soils, such as those in rice paddies, may become reduced to the point that arsenate is reduced to arsenite and the concentration of As in the soil solution becomes very high. The findings of Reed and Sturgis (1936) with rice support this propositon. Using Eh–pH diagrams, Hess and Blanchar (1976) predict that, under waterlogged conditions where 1% dextrose solution was equilibrated with soils, arsenite(III) would be more stable than arsenate(V) at the lower Eh conditions. This could produce arsenite(III) levels phytotoxic to plants. In addition, Ferguson and Gavis (1972) predict that, at high Eh values, arsenic acid species (H_3AsO_4, $H_2AsO_4^-$, $HAsO_4^{2-}$, and AsO_4^{3-}) are stable. Under mildly reducing conditions, arsenious acid species (H_3AsO_3, $H_2AsO_3^-$, and $HAsO_3^{2-}$) become stable. At very low Eh values, arsine (AsH_3) may be formed.

6.2 Management and Fertilization Practices

A. PHOSPHORUS FERTILIZATION

The effects of P on As phytotoxicity are unpredictable. Because of its chemical and physical similarities to As, P competes for As fixation sites in soil and thus may affect As availability. Several nutrient culture studies have demonstrated that the amount of P relative to As (P/As ratio) influences plant growth. Hurd-Karrer (1939) found the phytotoxicity to be a function of P concentration. At P/As ratios of 4:1 or greater, phytotoxicity on wheat was markedly reduced. However, at a ratio of 1:1, stunting occurred at concentrations of 10 ppm As and higher. Others obtained similar results on a variety of crops (Clements and Munson, 1947; Kardos et al, 1941; Rumburg et al, 1960). Tsutsumi (1983) found that the toxicity to rice plants of arsenite(III) was almost independent of added phosphate, whereas arsenate(V) could be antagonistically affected by phosphate.

In soil culture, several investigators noted reduction in phytotoxicity as P levels increased on a wide variety of crops (Benson, 1953; Hurd-Karrer, 1939; Juska and Hanson, 1967; Rumburg et al, 1960). Kardos et al (1941) noted that at P/As ratios above 1.3:1, little As toxicity occurred, whereas below 1:1, toxicity increased. Tsutsumi (1983) noticed that, in rice plants, As was harmless when the P/As ratio was 5:0.4–1.3, but was almost lethal when the ratio was in the range 1:0.8–2.4. Woolson et al (1973) obtained a reduction of As phytotoxicity due to P at ratios of available P/As of 0.7:1 to 42.5:1, but in another instance, no influence at a ratio of 6.8:1 was obtained. Carrow et al (1975) demonstrated that addition of P to soil had only a minimal effect on As toxicity to turfgrass under conditions simulating normal turf management practices. Nevertheless, a common practice when using tricalcium arsenate for control of *Poa annua* L. in intensively managed turfs is to minimize or eliminate P fertilization in order to increase the effectiveness of As. Arsenic rates are often elevated on soils testing high in P. Some investigators noted that additions of P may enhance As phytotoxicity (Jacobs and Keeney, 1970; Schweizer,

1967). Results of Johnson and Hiltbold (1969) indicate considerable differences between As and P distribution among chemical and mineralogical forms in the soil. Arsenic is much more extractable by mild salt solutions, suggesting a greater water solubility and lower extent of adsorption, precipitation, or occlusion. While most of the P is associated with Fe, either as precipitated iron phosphate or occluded with iron oxides, much of the As not readily removed was associated with Al. Thus, the inconsistencies of the effects of P on As availability may be due to which one of the following mechanisms is overriding: (1) differential rate of fixation between As and P by soil (e.g., As was fixed more readily than P in a Hagerstown clay loam [Woolson et al, 1973]); (2) displacement of As ions by P ions on adsorption sites, rendering As more available; (3) antagonism between P and As on plant roots and within the plant system.

B. OTHER PRACTICES

Deep plowing to dilute the As concentration of the surface soil and expose As to more fixation sites appears to be one of the most economical methods of alleviating toxicity (Walsh and Keeney, 1975). Vincent (1944) suggested growing tolerant cover crops such as rye or Sudangrass. If these crops are removed or plowed under, this practice can reduce As phytotoxicity. Leaving the field fallow for a few years is likely to allow As to move into deeper soil horizons by leaching, especially in sandy soils. Another practice that shows promise in alleviating As phytotoxicity, at least in the case of fruit trees, such as peach, is either the soil applications of zinc and iron chelates or foliar applications of ZnEDTA (Batjer and Benson, 1958). Because As shows strong affinity to Fe and Al components of the soil, addition of their salts may reduce As phytotoxicity. Large amounts (5–10 tonnes/ha) of $FeSO_4$ and $Fe_2(SO_4)_3$ occasionally have reduced As toxicity (Kardos et al, 1940; Keaton and Kardos, 1940). Although inconsistent yields of peas and potatoes were obtained, possibly because of low rates of application ($\approx$ 4 tonnes/ha), Steevens et al (1972) obtained lower concentrations of As in the potato tubers by the soil application of $Fe_2(SO_4)_3$ and $Al_2(SO_4)_3$.

6.3 Plant Factors

Numerous investigators (Deuel and Swoboda, 1972a; Baker et al, 1976; Hiltbold, 1975; Johnson and Hiltbold, 1969; Jacobs et al, 1970a; Woolson, 1973) have shown the differential tolerance to As among plant species, already discussed earlier. Arsenic concentrations among plant parts also vary and were found to be below detectable levels (< 0.02 ppm) in corn kernels and shelled peas (Jacobs et al, 1970a). In potatoes, most of the As was found in the peelings and was only slightly above the trace level (< 0.10 ppm) even in potato flesh from plots treated with 720 kg/ha of As.

7. Arsenic in Drinking Water and Food

Standards for maximum allowable As concentrations in drinking water have been established by several agencies. The World Health Organization (WHO, 1958) originally established a permissible limit of 0.20 ppm As but revised it to 0.05 ppm in 1963. The US Public Health Service (US-PHS, 1962) has a recommended maximum concentration of 0.01 ppm and a maximum permissible limit of 0.05 ppm for public drinking water supplies. More recently, the US Environmental Protection Agency (US-EPA, 1976) has set 0.05 ppm as the primary drinking water standard for As. For irrigation of crops, the limit in water is 0.10 ppm. The WHO has also set a 0.05 ppm standard for drinking water (WHO, 1984).

In a recent review, Ferguson and Gavis (1972) concluded that As concentrations in US water supplies are very low, with no potential threat to public health. However, there were some instances where fresh surface waters (rivers and lakes) had As concentrations exceeding the recommended limiting concentration of 0.01 ppm.

Even though As is ubiquitous in the biosphere, most foods contain minute amounts of it, averaging 0.02 ppm, including meats, fish, and poultry (Mahaffey et al, 1975). Russell (1979) reported the As concentrations (ppm, dry weight) for the following food classes: red meat—0.02; poultry—0.02; eggs—0.02; dairy products—0.0033; and aquatic foods—2.6. In their summary, Shacklette et al (1979) had the ranges of As concentrations (ppm, dry weight) for the following fruits and vegetables obtained from retail stores throughout the United States: apple, <0.05–0.20; potato, <0.05–0.05; dry bean, <0.05–0.06; sweet corn, <0.05–0.05; carrot, <0.05–0.08; snap bean, <0.05–0.10; cabbage, <0.05–0.05; head lettuce, <0.05–0.25; tomato, <0.05–0.12; bulb onion, <0.05–0.12; and cucumber, <0.05–0.50. Whole soybeans from different production regions (represented by seven states) had 0.1 ppm or less (fresh weight) As (Wauchope, 1978). The Canadian tolerance limit is 1 ppm for As in fresh vegetables accumulated by absorption from the soil; the US tolerance limit is 2.6 ppm (Woolson, 1973). Arsenic residues from spray applications of arsenicals can be a major source of As in fruits (Bishop and Chisholm, 1966), but the residues can be easily washed off. Fruits, vegetables, cereal products, and even meats and dairy products usually contain less than 0.50 ppm and rarely exceed 1 ppm As on fresh weight basis (Schroeder and Balassa, 1966). However, foods of marine origin, especially the crustaceans, are normally much higher in As.

As shown in Table 2.8, a typical US adult diet supplies about 10 μg As/day, with most of the sources being dairy products and meat, fish, and poultry. However, the WHO reported that the average As intakes for several countries including Canada, the United Kingdom, the United States, and France varied from 25 to 33 μg As/day, with specific values ranging from 7 to 60 μg/day (NAS, 1977). Pfannhauser and Woidich (1980) reported more specific As intakes (mg/person/year) for the following

TABLE 2.8. Estimated dietary intakes of arsenic in various food classes.

	Food intake[a]		As intake[b]	
Class	g food/day	% of total diet	μg As/day	% of total diet
Dairy products	769	22.6	2.34	23.1
Meat, fish and poultry	273	8.0	5.64	55.6
Grain and cereal	417	12.3	1.35	13.7
Potatoes	200	5.9	0.64	6.3
Leafy vegetables	63	1.9	0.0	0.0
Legume vegetables	72	2.1	0.0	0.0
Root vegetables	34	1.0	0.0	0.0
Garden fruits	89	2.6	0.0	0.0
Fruits	222	6.5	0.0	0.0
Oil and fats	51	1.5	0.17	1.7
Sugars and adjuncts	82	2.4	0.0	0.0
Beverages	1,130	33.2	0.0	0.0
Total	3,402	100.0	10.1	

[a] Based on 1972–73 total diet survey by the US Food and Drug Administration (US-FDA) for an adult person (Tanner and Friedman, 1977).
[b] Based on 1973 total diet survey by the US-FDA (Russell, 1979).

countries: Canada (1978)—1.65; United States (1966)—3–9; Federal Republic of Germany (1969)—2.4; Japan (1971)—2.1–5.1; and Austria (1980)—0.80.

8. Sources of Arsenic in the Terrestrial Environment

The As flow (in tonnes) in the United States was summarized by the National Academy of Sciences (NAS, 1977) as follows: end products (steel, cast iron, and others)—24,055; dissipation to land (steel slag, pesticides, copper leach liquor, and others)—57,350; airborne emission (copper smelting, pesticides, coal combustion, and others)—8,880; and landfill wastes (copper flue dust, copper-smelting slag, coal fly ash, and others)—17,930. Of these sources, those disposed of on land and emitted to the atmosphere pose the greatest environmental threat. Those in the landfills are localized and therefore may be also of environmental concern when As in the leachates reaching the ground water approach levels above those of the background.

In Japan, the main thrust of the As pollution problem has been its toxicity to the rice plant (Takamatsu et al, 1982). Irrigation of paddy fields with water contaminated by mining wastes or waste water from geothermal electric power stations has frequently produced growth depression of rice. Toxicity can also occur on rice being grown in paddy fields converted from old apple orchards, where inorganic arsenicals were used as pesticides.

In general, a 10- to 20-fold increase or more of As levels above background in surface soils may result from agricultural and industrial input. About half of the mobile As inventory in the United States comes from pesticides. Arsenic has accumulated in the soil due to the past utilization of As-containing pesticides, desiccants, and defoliants. Arsenic residues on surface soils from 12 states averaged 165 ppm As as a result of As application versus only 13 ppm As in nontreated soils (Woolson et al, 1971). Similar surveys in Ontario, Canada (Frank et al, 1976) yielded comparable results, with soils from orchards having As levels about ten times the level from nontreated soils. Elevated levels of soil As were also detected in lowbush blueberry fields in Maine and Atlantic Canada from the use of calcium arsenate (Anastasia and Kender, 1973).

The cycling of As in an agro-ecosystem is depicted in Figure 2.3. The model shows the possible transfers to and from a field for the organo-arsenical herbicides. Sandberg and Allen (1975) concluded that transfers involving transformation to methylarsines, soil erosion, and crop uptake are the primary redistribution pathways in an agro-ecosystem, precluding hazardous accumulation in the soil. They indicate that As is mobile and nonaccumulative in the air, water, and plant phases of the ecosystem.

In Canada, smelting of arsenopyrite ores during the gold extraction process releases substantial amounts of As_2O_3 into the environment. White spruce trees grown on simulated contaminated soils, with As contents ranging from 44 to 1,780 ppm, were markedly retarded (Rosehart and Lee, 1973). This indicates that As from smelting operations can be phytotoxic

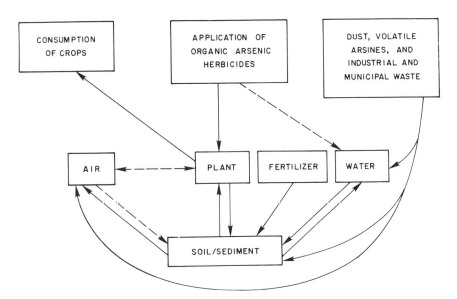

FIGURE 2.3. A proposed model for the arsenic cycle in an agronomic ecosystem. *Source:* Sandberg and Allen, 1975, with permission of the authors and the Am Chem Society.

TABLE 2.9. Estimates of global arsenic emissions from natural and anthropogenic sources. [a]

Source	Global production or consumption, 10^{12} g/yr	Arsenic emission factor, g/g source	Total arsenic emission (d≤5 μm), 10^9 g/yr
Natural			
Ocean			
Bubble bursting	1,000	5.7×10^{-7}	0.028
Gas exchange			0.11
Earth's crust			
Particle weathering	800	2.0×10^{-6}	0.24
Direct volatilization			0.0007
Volcanoes	25	2.8×10^{-4}	7.0
Forest wild fires	320	5.0×10^{-7}	0.16
Terrestrial biosphere			0.26
Total (natural)			7.8
Anthropogenic			
Coal	3,245	1.7×10^{-7}	0.55
Light fuels	585	1.2×10^{-10}	0.00007
Residual fuels	956	4.3×10^{-9}	0.0041
Wood fuel	1,200	5.0×10^{-7}	0.60
Agricultural burning	1,120	5.0×10^{-7}	0.56
Waste incineration	540	1.6×10^{-6}	0.43
Iron/steel production	1,220	7.0×10^{-6}	4.2
Copper production	8.7	3.0×10^{-3}	13.
Lead/zinc production	9	5.0×10^{-4}	2.2
Mining mineral ore	2,500	1.0×10^{-8}	0.013
Arsenic/chemicals	0.040	1.0×10^{-2}	0.20
Arsenic/agriculture	0.037	5.0×10^{-2}	1.9
Cotton ginning	14	3.3×10^{-6}	0.023
Total (anthropogenic)			23.6
GRAND TOTAL (natural and anthropogenic sources)			31.4

[a] *Source:* Walsh et al, 1979, with permission of the authors, copyright by the Am Geophysical Union.

to plants inhabiting areas near a smelter. Similar impacts of smelting operations on As levels in soils were reported in the United Kingdom (Thornton and Webb, 1977) and the United States (Ragaini et al, 1977).

An industrial complex near Lisbon, Portugal has been estimated to emit As in the range of 1,000 to 2,000 tonnes per year, most of which is coming from a pyrite roasting plant (Andreae et al, 1983). This release has contributed about 1 tonne of As per day into a nearby estuary of the Tejo River.

Because arsenicals are used in poultry ration, usage of poultry litter at excessive rates in land for fertilization purposes may elevate As levels in soil and plants (Isaac et al, 1978).

Estimates of global As emissions from both natural and anthropogenic sources indicate that metal production accounts for most of the latter sources (Table 2.9). Of the total anthropogenic releases, the use of arsenicals in agriculture represents about 8%.

References

Abernathy, J. R. 1983. In W. H. Lederer and R. J. Fensterheim, eds. *Arsenic— industrial, biomedical, environmental perspectives*, 57–62. Reinhold Publ Corp, New York.

Adriano, D. C., A. L. Page, A. A. Elseewi, A. C. Chang, and I. Straughan. 1980. *J. Environ. Qual.* 9:333–344.

Akins, M. B., and R. J. Lewis. 1976. *Soil Sci Soc Am J* 40:655–658.

Albert, W. B., and C. H. Arndt. 1931. S Carolina Agric Exp Sta 44th Annual Rep. Clemson, SC.

Alden, J. C. 1983. In W. H. Lederer and R. J. Fensterheim, eds. *Arsenic—industrial, biomedical, environmental perspectives*, 63–71. Reinhold Publ Corp, New York.

Anastasia, F. B., and W. J. Kender. 1973. *J Environ Qual* 2:335–337.

Andreae, M. O., J. T. Byrd, and P. N. Froelich, Jr. 1983. *Environ Sci Technol* 17:731–737.

Baker, R. S., W. L. Barrentine, D. H. Bowman, W. L. Hawthorne, and J. V. Pettiet. 1976. *Weed Sci* 24:322–326.

Batjer, L. P., and N. R. Benson. 1958. *Proc Am Soc Hort Sci* 72:74–78.

Benson, N. R. 1953. *Soil Sci* 76:215–224.

Benson, N. R., and H. M. Reisenauer. 1951. Washington Agric Exp Sta Circ175:1–3.

Berrow, M. L., and G. A. Reaves. 1984. In *Proc Intl Conf Environ Contamination*, 333–340. CEP Consultants Ltd, Edinburgh, UK.

Bishop, R. F., and D. Chisholm. 1966. Can J Plant Sci 46:225–231.

Bowen, H. J. M. 1979. *Environmental chemistry of the elements*. Academic Press, New York, New York. 333 pp.

Braman, R. S., and C. C. Foreback. 1973. *Science* 182:1247–1249.

Brenchley, W. E. 1914. *Ann Botany* 28:283–301.

Calvert, C. C. 1975. In E. A. Woolson, ed. *Arsenical pesticides*, 70–80. Am Chem Soc, Washington, DC.

Capar, S. G., J. T. Tanner, M. H. Friedman, and K. W. Boyer. 1978. *Environ Sci Technol* 7:785–790.

Carrow, R. N., P. E. Reike, and B. G. Ellis. 1975. *Soil Sci Soc Am Proc* 39:1121–1124.

Chang, S. C., and M. L. Jackson. 1957. *Soil Sci* 84:133–144.

Cheng, C. N., and D. D. Focht. 1979. *Appl Environ Microbiol* 38:494–498.

Chino, M. 1981. In K. Kitagishi and I. Yamane, eds. *Heavy metal pollution in soils of Japan*. Japan Sci Soc Press, Tokyo.

Clements, H. F., and J. Munson. 1947. *Pac Sci* 1:115–171.

Crafts, A. S., and R. S. Rosenfels. 1939. *Hilgardia* 12:177–200.

Cunningham, C. E., T. G. Eastman, and M. Goven. 1952. *Am Pot J* 29:8–16.

Davis, R. D., P. H. T. Beckett, and E. Wollan. 1978. *Plant Soil* 49:395–408.

Deuel, L. E., and A. R. Swoboda. 1972a. *J Environ Qual* 1:317–320.

—— 1972b. *Soil Sci Soc Am Proc* 36:276–278.

Dickens, R., and A. E. Hiltbold. 1967. *Weeds* 15:299–304.

Dickerson, O. B. 1980. In H. A. Waldron, ed. *Metals in the environment*, 1–24. Academic Press, New York.

Elkhatib, E. A., O. L. Bennett, and R. J. Wright. 1984a. *Soil Sci Soc Am J* 48:758–762.

—— 1984b. *Soil Sci Soc Am J* 48:1025–1030.

Epps, E. A., and M. B. Sturgis. 1939. *Soil Sci Soc Am Proc* 4:215–218.

Fassbender, H. W. 1974. *Zeitschrift für Pflanzenernährung und Bodenkunde* 137:188–203.

Ferguson, J. F., and J. Gavis. 1972. *Water Res* 6:1259–1274.

Fitzgerald, L. D. 1983. In W. H. Lederer and R. J. Fensterheim, eds. *Arsenic—industrial, biomedical, environmental perspectives*, 3–9. Reinhold Publ Corp, New York.

Frank, R., K. Ishida, and P. Suda. 1976. *Can J Soil Sci* 56:181–196.

Frans, R. E., C. R. Skogley, and G. H. Ahlgren. 1956. *Weeds* 4:11–14.

Frost, R. R., and R. A. Griffin. 1977. *Soil Sci Soc Am J* 41:53–57.

Furr, K. A., A. W. Lawrence, S. S. C. Tong, M. G. Grandolfo, R. A. Hofstader,C. A. Bache, W. H. Gutenmann, and D. J. Lisk. 1976. *Environ Sci Technol* 7:683–687.

Greaves, J. E. 1913. *Biochem Bull* 2:519–523.

—— 1934. *Soil Sci* 38:355–362.

Hess, R. E., and R. W. Blanchar. 1976. *Soil Sci Soc Am J* 40:847–852.

Hiltbold, A. E. 1975. In E. A. Woolson, ed. *Arsenical pesticides*. Am Chem Soc, Washington, DC.

Holm, T. R., M. A. Anderson, R. A. Stanforth, and D. G. Iverson. 1980. *Limnol Oceanogr* 25:23–30.

Huang, P. M. 1975. *Soil Sci Soc Am Proc* 39:271–274.

Hurd-Karrer, A. M. 1939. *Plant Physiol* 14:9–29.

Isaac, R. A., S. R. Wilkinson, and J. A. Stuedemann. 1978. In D. C. Adriano and I. L. Brisbin, eds. *Environmental chemistry and cycling processes*. CONF-740513, NTIS, Springfield, VA. 898 pp.

Jacobs, L. W., D. R. Keeney, and L. M. Walsh. 1970a. *Agron J* 62:588–591.

Jacobs, L. W., J. K. Syers, and D. R. Keeney. 1970b. *Soil Sci Soc Am Proc* 34:750–754.

Jacobs, L. W., and D. R. Keeney. 1970. *Commun Soil Sci Plant Anal* 1:85–93.

Johnson, L. R., and A. E. Hiltbold. 1969. *Soil Sci Soc Am Proc* 33:279–282.

Johnston, S. E., and W. M. Barnard. 1979. *Soil Sci Soc Am J* 43:304–308.

Juska, F. V., and A. A. Hanson. 1967. *Calif Turfgrass Cult* 17(4):27–29.

Kardos, L. T., S. C. Vandecaveye, and N. Benson. 1941. Washington Agric Exp Sta Bull 410:25.

Keaton, C. M., and L. T. Kardos. 1940. *Soil Sci* 50:189–207.

Krauskopf, K. B. 1979. *Introduction to geochemistry*. 2nd ed. McGraw-Hill, New York. 617 pp.

Liebig, G. F. 1965. In H. D. Chapman ed. *Diagnostic criteria for soils and plants*, 13–23. Quality Printing Co Inc, Abilene, TX.

Liebig, G. F., G. R. Bradford, and A. P. Vanselow. 1959. *Soil Sci* 88:342–348.

Livesey, N. T., and P. M. Huang. 1981. *Soil Sci* 131:88–94.

MacPhee, A. W., D. Chisolm, and C. R. MacEachern. 1960. *Can J Soil Sci* 40:59–62.

Mahaffey, K. R., P. E. Corneliussen, C. F. Jelinek, and T. A. Fiorino. 1975. *Environ Health Perspec* 12:63–69.

Marcus-Wyner, L., and D. W. Rains. 1982. *J Environ Qual* 11:715–719.

McGeorge, W. T. 1915. *J Agric Res* 5:459–463.

Mellanby, K. 1967. *Pesticides and pollution*. Collins Press, London. 221 pp.

National Academy of Sciences (NAS). 1977. In *Arsenic*. NAS, Washington, D. 276 pp.

National Research Council Canada (NRCC). 1978. In *Effects of arsenic in the Canadian environment*. NRCC 15391. Ottawa, Canada. 349 pp.

Olson, O. E., L. L. Sisson, and A. L. Moxon. 1940. *Soil Sci* 50:115–118.

Paden, W. R., and W. B. Albert. 1930. S C Agric Exp Sta Ann Rep 43:129. Clemson, S C.

Pfannhauser, W., and H. Woidich. 1980. *Toxicol Environ Chem Rev* 3:131–144.

Ragaini, R. C., H. R. Ralston, and N. Roberts. 1977. *Environ Sci Technol* 11:773–781.

Reed, J. F., and M. B. Sturgis. 1936. *J Am Soc Agron* 28:432–436.

Robinson, E. L. 1975. *Weed Sci* 23:341–343.

Rosehart, R. G., and J. Y. Lee. 1973. *Water, Air, Soil Pollut* 2:439–443.

Rumburg, C. B., R. E. Engel, and W. F. Meggitt. 1960. *Agron J* 52:452–453.

Russell, Jr., L. H. 1979. *In* F. W. Oehme, ed. *Toxicity of heavy metals in the environment,* 3–23. Dekker Inc, New York.

Sandberg, G. R., and I. K. Allen. 1975. In E. A. Woolson, ed. *Arsenical pesticides,* 124–127. ACS Sympos Ser 7. Amer Chem Soc, Washington, DC.

Schroeder, H. A., and J. J. Balassa. 1966. *J Chron Dis* 19:85–106.

Shacklette, H. T., J. A. Erdman, T. F. Harms, and C. S. E. Papp. 1979. In F.W. Oehme, ed. *Toxicity of heavy hetals in the environment,* 25–68. Dekker Inc, New York.

Shacklette, H. T., and J. G. Boerngen. 1984. *Element concentrations in soils and other surficial materials of the conterminous United States*. USGS Prof Paper 1270. US Govt Printing Office, Washington, DC.

Schweizer, E. E. 1967. *Weeds* 15:72–76.

Selby, L. A., A. A. Case, C. R. Dorn, and D. J. Wagstaff. 1974. *J Am Vet Med Assoc* 165:1010–1014.

Steevens, D. R., L. M. Walsh, and D. R. Keeney. 1972. *J Environ Qual* 1:301–303.

Stewart, J., and E. S. Smith. 1922. *Soil Sci* 14:119–126.

Stewart, J. 1922. *Soil Sci* 14:111–118.

Takamatsu, T., H. Aoki, and T. Yoshida. 1982. *Soil Sci* 133:239–246.

Tammes, P. M., and M. M. deLint. 1969. *Netherlands J Agric Sci* 17:128–132.

Tanner, J. T., and M. H. Friedman. 1977. *J Radioanal Chem* 37:529–538.

Thornton, I., and J. S. Webb. 1977. *Trace Subs Environ Health* 11:81–88.

Tsutsumi, M. 1980. *Soil Sci Plant Nutr* 26:561–569.

———— 1983. *Soil Sci Plant Nutr* 29:63–69.

US Bureau of Mines. 1973, 1978–79. *Minerals yearbook. Metals and minerals*. US Dept of Interior, Washington, DC.

US Environmental Protection Agency (US-EPA). 1976. *Quality Criteria for Water*. US-EPA, Washington, DC.

US Public Health Service (US-PHS). 1962. *Drinking water standards.* US Dept Health, Education, and Welfare, Washington, DC.

Vandecaveye, S. C., G. M. Horner, and C. M. Keaton. 1936. *Soil Sci* 42:203–215.

Veneman, P. L. M., J. R. Murray, and J. H. Baker. 1983. *J Environ Qual* 12:101–104.

Vincent, C. L. 1944. In Washington Agric Exp Sta Bull 437.

Vinogradov, A. P. 1959. *The geochemistry of rare and dispersed chemical elements in soils.* 2nd ed. Consultants Bureau, New York. 209 pp.

Von Endt, D. W., P. C. Kearney, and D. D. Kaufman. 1968. *J Agric Food Chem* 16:17–20.

Walsh, L. M., and D. R. Keeney. 1975. In E. A. Woolson, ed. *Arsenical pesticides,* 35–52. Am Chem Soc, Washington, DC.

Walsh, P. R., R. A. Duce, and J. L. Fasching. 1979. *J Geophys Res* 84:1719–1726.

Wauchope, R. D. 1975. *J Environ Qual* 4:355–358.

——— 1983. In W. H. Lederer and R. J. Fensterheim, eds. *Arsenic—industrial, biomedical, environmental perspectives,* 348–377. Reinhold Publ Corp, New York.

——— 1978. *J Agric Food Chem* 26:226–228.

Wauchope, R. D., and L. L. McDowell. 1984. *J Environ Qual* 13:499–504.

Wood, J. M. 1974. *Science* 183:1049–1052.

Woolson, E. A. 1973. *Weed Sci* 21:524–527.

Woolson, E. A., J. H. Axley, and P. C. Kearney. 1971. *Soil Sci Soc Am Proc* 35:101–105.

——— 1973. *Soil Sci Soc Am Proc* 37:254–259.

Woolson, E. A., and P. C. Kearney. 1973. *Environ Sci Technol* 7:47–50.

Woolson, E. A., and A. R. Isensee. 1981. *Weed Sci* 29:17–21.

World Health Organization (WHO). 1958. *International standards for drinking water.* Geneva, Switzerland.

——— 1984. *Guidelines for drinking-water quality. Vol. 1—Recommendations.* WHO, Geneva, Switzerland. 130 pp.

3
Boron

1. General Properties of Boron

Boron belongs to Group III-A of the periodic table and is the only nonmetal among the plant micronutrients. It has an atomic weight of 10.811, a melting point of 2,300°C, with specific gravity of crystals, 2.34. Boron has two stable isotopes in nature, ^{10}B (18.98%) and ^{11}B (81.02%). At room temperature, B is inert except to strong oxidizing agents, such as HNO_3. When fused with oxidizing alkaline mixtures, such as NaOH and $NaNO_3$, it forms borates. The only important oxide is boric oxide (B_2O_3), which is acidic, soluble in water, and forms boric acid $B(OH)_3$, a very weak acid. In nature, B is fairly rare and occurs primarily as the borates of Ca and Na. Borax ($Na_2B_4O_7 \cdot 10H_2O$) is the most common compound, along with boric or boracic acid. By far the most important source of B is the mineral kernite ($Na_2B_4O_7 \cdot 4H_2O$), an evaporite deposit found in the Mojave desert of California. It almost always occurs in chemical combination with O_2, as a borate, usually coordinated with three O atoms and occasionally with four. Boron, which has a constant oxidation number of III, never behaves as a cation.

2. Production and Uses of Boron

World production of B minerals in 1979 was about 2.6×10^6 tonnes, with about 55% (1.43×10^6 tonnes) of this total produced by the United States (US Bureau of Mines, 1980). Other major B-producing countries include Turkey, the USSR, Argentina, and Chile.

Domestically, California is the chief source of B minerals, which are mostly in the form of sodium borate, but also sometimes as calcium borate (colemanite) and sodium-calcium borates (ulexite). Recently, insulation products, fire retardants, and glass fiber-reinforced plastics are the most important consuming sectors of B (mostly borax pentahydrate and ulexite-probertite). Borates (colemanite, orthoboric acid, and anhydrous boric acid) are also extensively used in the manufacture of textile-grade glass fibers for use in aircraft, autos, and sports equipment.

Borates are also used in the manufacture of special borosilicate glasses (glasses, pottery, and enamels), and in cleaning and bleaching compounds, especially in the production of sodium perborate detergents. Boron compounds are used in the manufacture of biological growth control agents for use in water treatment, algicides, herbicides, insecticides, and also in fertilizers since B is one of the plant micronutrients. Other uses of B compounds include metallurgical work and electroplating, and nuclear industries. Boric acid is useful medicinally as a mild antiseptic, especially as an eyewash. In the past, it was used as a food preservative.

Technologically, B is very important because of the applications of B compounds to synthetic organic chemistry. Organoborane compounds are used as chemical intermediates in the manufacture of drugs, as well as in new classes of pesticides.

3. Natural Occurrence of Boron

Readily available sodium and calcium borates occur naturally in most soils. These come from the very slow dissolution of tourmaline, a mineral that contains about 3% to 4% B. Tourmaline, $[H_2MgNa_9Al_3(BO)_2Si_4O_{20}]$ is present in soils formed from acid rocks and metamorphosed sediments. Because of its slow dissolution, it is not a very good source of B for plants (Graham, 1957). Boron can substitute for tetrahedrally coordinated Si in some minerals, and it is likely that much of the B in rocks and soils is dispersed in the silicate minerals in this manner (Norrish, 1975).

The average concentration of total B in the earth's crust is about 10 ppm; it ranks 37th in abundance among the elements (Krauskopf, 1979). Concentrations in igneous rocks are in the same range but increase several times for sedimentary rocks (Table 3.1). Soils derived from marine clays and shales generally have sufficient B for plant production, since B and its compounds are soluble and accumulate in sea water. Some sedimentary rocks of marine origin can contain 500 ppm B or more (Aubert and Pinta, 1977). In the Byelorussia and Amur regions of the USSR, B levels in the range of 35 to 70 ppm were found in glacial clays, lacustrine alluvium, and stratified plain deposits. However, low levels of around 2 ppm B were reported for ancient alluvial sands in these regions. As a general rule, marine clays contain more B than clays accumulating in lakes or flood-plains.

In general, the B contents of soils are higher than the B contents of the rocks (Vinogradov, 1959). Weathering of B-containing rocks gives borate in solution, chiefly as the non-ionized boric acid, $B(OH)_3$. Boric acid is a weak acid (pK = 9.14) noted for its volatility even at low temperatures (Krauskopf, 1972). In natural environments, B solutions contain primarily

TABLE 3.1. Commonly observed boron concentrations
(ppm) in various environmental matrices.

Material	Mean (or Median)	Range
Basalt[a]	5	—
Granite[a]	15	—
Limestone[a]	20	—
Sandstone[a]	35	—
Shale[a]	100	—
Phosphorites[c]	<50	—
Soils[b]	—	2–100
Coal[b]	—	4–200
Petroleum	0.002	—
Lignite[c]	70	1–400
Fly ash	—	36–500
Animal manures	24	—
Sludge	—	20–2,000
Fertilizers	—	<1–395
Common crops	—	20–100
Herbaceous vegetables[c]	—	8–200
Dolomitic lime	4	<1–21
Lichens[c]	4.1	—
Fungi[c]	16	—
Ferns[c]	77	—
Woody gymnosperms[c]	—	11–52
Woody angiosperms[c]	—	12–140
Irrigation waters	—	<0.1–0.3
Freshwater[c]	15*	7–500
Sea water[c]	4,440	—

* µg/liter
Sources:
[a] Krauskopf (1972).
[b] Trudinger et al (1979).
[c] Bowen (1979).

$B(OH)_3$ and $B(OH)_4^-$ and the polymeric forms such as tetraborate, $B_4O_7^{2-}$, are likely to occur only in concentrated solutions.

In nature, B is found as a constituent of borax ($Na_2B_4O_7 \cdot 10H_2O$), kernite ($Na_2B_4O_7 \cdot 4H_2O$), colemanite ($Ca_2B_6O_{11} \cdot 5H_2O$), ulexite ($NaCaB_5O_9 \cdot 8H_2O$), tourmaline, and axinite.

4. Boron in Soils

4.1. Total Soil Boron

The total B content of normal soils ranges from 2 to 100 ppm (Swaine, 1955), with an average of about 30 ppm. The total B content of soils depends largely on the soil's parent material. The relationship between soil content and rock source was reported by Whetstone et al (1942), who did an extensive survey of US soils. They found an average content of 30 ppm for 200 soils and a range from 4 to 98 ppm. The average values of 14 and 40 ppm were found for soils derived from igneous and sedimentary rocks, respectively. The highest B values occurred in arid saline soils. Fine-textured humid soils had 30 to 60 ppm B, whereas sandy soils often had as low as 2 to 6 ppm. Soils of marine shale origin were particularly high in B. Similar relationships were noted in soils of California. Soils weathered from alluvium eroded from marine shales in Kern County had B contents ranging from 25 to 68 ppm, whereas soils from granitic material contained 10 ppm B or less (Bingham et al, 1970a). More recently, a national survey indicated that surface soils of the conterminous United States have an average B content of 33 ppm (range of < 20–300 ppm) (Appendix Table 1.17).

Low B content can be expected of soils derived from acid igneous rocks, fresh water sedimentary deposits, and in coarse-textured soils low in organic matter. Data for total B content of soils are available from several countries. The average B content of Russian soils, including all horizons but excluding salinized southern soils, is 6 ppm (Vinogradov, 1959). However, upon inclusion of the salinized southern soils, the average becomes 50 ppm. In the USSR, total B contents are highest in soils of the dry-steppe, semi-desert, and desert zones compared to soils in the central region, where levels of 200 to 400 ppm are not uncommon (Il'in and Anikina, 1974). Soils of China have higher B contents ($\bar{x}$ = 64 ppm; range of trace amounts to 500 ppm) than that of other countries (Liu et al, 1983). The highest content of total B was found in the Himalayan region, where soils were of marine sediment origin. However, tropical Asian paddy soils have a mean B content of 96 ppm, with the high of 197 ppm for Burmese soils and a low of 52 ppm for Indonesian soils (Appendix Table 1.18). Soils of central and northern Europe reportedly have lower B contents than soils of the United States, especially the western and central states. Soils of England and Wales have B contents in the range of 7 to 71 ppm, with a median of 33 ppm (Archer, 1980). A number of soils from eastern Canada have total B in the range of 45 to 124 ppm (Gupta, 1968).

4.2 Extractable (Available) Boron in Soils

Total soil B is an unreliable index of the phytoavailability of B in soil because of the extreme insolubility of indigenous soil B. Consequently, the extractable B is the universal form of this element used for diagnostic purposes. Hot water-soluble B is generally regarded as the best index of availability of B to plants (Wear, 1965). Only a small fraction of total B, usually less than 5%, is found in extractable form (Berger and Truog, 1940). The basic procedure for determining available soil B has remained the same since Berger and Truog (1939) adopted it. Boron is extracted by boiling a 1:2 soil/water suspension for five minutes and then filtering.

More recently, Cartwright et al (1983) found that extraction of some Australian soils by shaking for 1 hour at 20°C with 0.01 M CaCl$_2$ + 0.05 M mannitol was a more convenient soil test for plant-available B than the standard hot water-soluble method and to be as reliable in predicting the response of B uptake by plants. However, this method can be expected to extract less B from soils having low B contents but more B from soils having potentially toxic concentrations than the hot water-soluble method. Piland et al (1944) found that the average water-soluble B content of Coastal Plain and Piedmont soils were 0.24 and 0.27 ppm, respectively. Water-extractable B in several cropped Michigan soils ranged from 0.16 to 0.95 ppm, with the highest levels due to recent B fertilization (Robertson et al, 1975). Soils in Georgia had 0.01 to 0.65 ppm water-soluble B and Illinois soils, 0.20 to 1.22 ppm (DeTurk and Olson, 1941).

Gestring and Soltanpour (1984) found that Colorado soils having hot water-soluble B levels between 0.10 and 6.47 ppm did not produce any deficient or toxic symptoms in alfalfa grown in the greenhouse. Water-soluble B in soils of six geographically diverse areas of Egypt had 0.30 to 3.4 ppm B, with an average of 1.3 ppm (Elseewi and Elmalky, 1979).

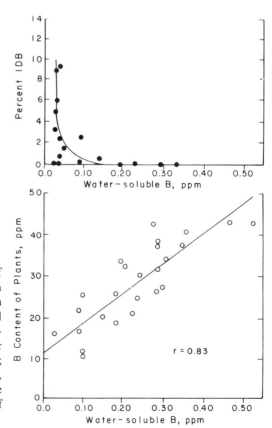

FIGURE 3.1. Relationships of hot water-soluble boron in soils to plant uptake of boron and to incidence of "internal damage" from boron deficiency (IDB) in peanuts. *Sources:* Upper—Hill and Morrill, 1974; Lower—Baird and Dawson, 1955, with permission of the authors and the Soil Sci Soc of America.

In eastern Canada, hot water-soluble B ranged from 0.38 to 4.67 ppm (Gupta, 1968). In England and Wales, hot water-soluble B ranged from 0.1 to 4.7 ppm, with a median of 1.0 ppm (Archer, 1980).

Another water-soluble form of B is determined from soil saturation extract. The values obtained by this procedure are usually lower than the hot water-soluble B, are variable, and generally give narrower ranges, thus making the tests more difficult to interpret. Robertson et al (1975) obtained an average of 0.50 ppm with saturation extract versus 0.74 ppm B with hot water extract. Similar results were obtained for Egyptian soils, where B from the former method averaged 0.6 ppm and from the latter method, 1.3 ppm (Elseewi and Elmalky, 1979).

The importance of levels of hot water-soluble B in soils in plant nutrition and disorders are demonstrated in Figure 3.1. Plant uptake of B can be predicted from water-soluble B in soil (lower, Fig. 3.1) as can the incidence of "internal damage" due to B deficiency (upper, Fig. 3.1). Highly significant positive correlations are usually obtained between plant B concentrations and hot water-soluble B concentrations in soils.

4.3 Boron in the Soil Profile

In general, concentrations of total B are usually higher in the surface horizons of the soil profile than in the underlying layers, partly because occurrence of B is often associated with organic matter. Organic matter is usually in the surface layer. Examination of 26 soil profiles in the Egyptian Nile Delta (Table 3.2), divided into 30-cm increments up to a depth of 90 cm, revealed that total B (ranging from 25 to 336 ppm) usually was high in the top 30 cm, and occasionally high in the next 30- to 60-cm layer (Elseewi and Elmalky, 1979). Total B, on the average, was twice as high in clayey samples as in sandy samples, and decreased by about 32% beyond the 60-cm depth of the soil profile. However, hot water-soluble and B in the saturation extract, were more or less evenly distributed in the soil profile. Additional studies of Egyptian soils (Awad and Mikhael, 1980) near the Nile Delta but more toward the western desert showed that water-soluble B, even at much higher concentrations (1.8 to 5.7 ppm), was practically evenly distributed in the soil profile (0–90 cm).

Others found different profile distribution of B. Soils supporting sugar beets in Michigan had water-soluble B ranging from nondetectable in the subsoil (120 cm and shallower) to 0.54 ppm in the surface soil (0–15 cm) (Robertson et al, 1975). Similarly, in the agricultural Polders region of the USSR, water-soluble B concentrations are usually higher in the surface soils, decreasing to trace levels in the subsoil (Shirokov and Panasin, 1972). The water-soluble B ranged from 0.13 to 6.80 ppm in the surface soils and from trace to 9.84 ppm in subsoils. This region is below or close to sea level with mostly peat bog and alluvial bog soils influenced by marine

TABLE 3.2. Forms and distribution of boron in the soil profiles of the Egyptian Nile Delta.[a]

Depth, cm	pH	Texture[b]	EC[d] mmho/cm	Sat. extract μg/ml	Boron[c] μg/g Hot water	Acid-soluble	Total
				Site No. 1			
0–30	8.0	SCL	2.5	0.6	1.3	10.8	113
30–60	8.1	SCL	2.4	0.5	1.1	10.1	139
60–90	8.3	CL	2.6	0.8	1.4	11.1	108
				Site No. 2			
0–30	7.5	SL	2.0	0.6	2.3	14.8	138
30–60	7.7	C	2.2	0.5	2.3	12.1	33
60–90	7.6	C	6.5	0.6	1.6	10.8	25
				Site No. 3			
0–30	7.9	C	5.5	1.9	1.3	2.7	323
30–60	7.9	C	7.3	1.0	1.4	5.2	214
60–90	8.0	C	8.9	0.7	1.7	6.4	50
				Site No. 4			
0–30	7.5	C	1.3	0.5	0.6	3.6	163
30–60	7.6	SC	4.3	0.3	0.3	1.6	336
60–90	7.5	SCL	3.8	0.6	1.2	5.6	43
				Site No. 5			
0–30	7.7	C	8.2	0.2	1.0	7.2	200
30–60	7.6	C	3.9	0.2	1.0	2.4	69
60–90	7.5	C	5.3	0.2	0.6	10.0	138
				Site No. 6			
0–30	7.8	LS	1.8	1.1	2.2	15.4	43
30–60	7.9	LS	1.6	0.7	1.4	11.6	38
60–90	8.1	LS	1.7	1.7	3.4	21.2	31

[a] *Source:* Elseewi and Elmalky, 1979, with permission of the authors and the Soil Sci Soc of America.
[b] S, L, and C refer to sand, loam, and clay.
[c] The extractant used for the acid-soluble fraction was 85% H_3PO_4 and the Na_2CO_3 fusion method was used for determining total B.
[d] Electrical conductivity of the saturation extract.

processes. In the vineyard region of Georgia, USSR, water-soluble B concentrations were higher in surface soils (0–20 cm) than in the subsoils, ranging from 0.34 to 2.4 ppm in the former layer (Bagdasarashvili et al, 1974).

In saline-alkaline soils of arid and semi-arid regions, B distribution in the soil profile is variable. In the Coachella Valley of southern California, B concentration in the saturation extract decreased from 50 ppm in the surface (0–15 cm) to 5 ppm B at the 120- to 150-cm depth (Reeve et al, 1955). In contrast, in the arid San Joaquin Valley of California, total and B in the saturation extract accumulate in subsoils in much higher levels than in the surface soils (Bingham et al, 1970a). In the arid regions of India, the pattern is also inconsistent, as in California. In the Haryana and Punjab regions, total B increased with depth, while water-soluble B decreased (Singh, 1970). However, in the Rajasthan area, the inverse pattern seems to occur in the profile distribution of B (Talati and Agarwal, 1974).

Most of the saline-alkaline soils are unfit for agricultural production because of the presence of excessive salts, undesirable soil structural properties, and occasionally excessive B levels. Reclaiming these soils, with leaching, can render them productive, since most of these salts move with the leaching water. In salt- and B-affected soils, about 30 cm of water or more are required per 30 cm of the soil profile for lowering salt to tolerable levels; B removal may require three or more times water (Reeve et al, 1955). Boron does not leach out of the profile as readily as chloride, nitrate, and sulfate salts (Bingham et al, 1972).

Natural leaching from precipitation can also mobilize B down the soil profile. Leaching of applied B on the soil surface was found by several investigators (Kubota et al, 1948; Ouellette, 1958; Rajaratnam, 1973; Winsor, 1952) to be fairly rapid, especially on well-drained sandy soils. Winsor (1952) reported that more than 85% of the added B, as borax, was leached below the 105-cm depth in Florida's Lakeland sand after 40 cm of rainfall in four months. Thus, it can be expected that B is higher in surface soils during dry periods and low during rainfall season (Bandyopadhya, 1974).

4.4 Forms and Speciation of Boron in Soils

There are four main forms of B in soils: water-soluble, adsorbed, organically bound, and fixed in the clay and mineral lattices. The water-soluble fraction is regarded as a plant-response indicator and is determined either in a soil saturation or hot water extract. The latter is usually designated as "available" B. Adsorbed B represents the fraction that is precipitated or adsorbed on surfaces of soil particles and is in equilibrium with soluble B. Under normal conditions, soils in the field contain small amounts of soluble and adsorbed B. Soils in arid and semi-arid regions could be expected to contain from 5% to 16% of total B in the water-soluble form (Aubert and Pinta, 1977). Soils from various countries were reported to contain about 10% of the total B in the water-soluble form (Vinogradov, 1959). In regions of high salinity, this form could rise to as much as 80% of the total B. Soils in the Nile Delta of Egypt (Table 3.2) had water-soluble B ranging up to only 6.4% of total (Elseewi and Elmalky, 1979). Similarly, in the Russian Central region, water-soluble B comprises only about 2% to 5% of the total B (Il'in and Anikina, 1974) rising to 5% to 10% of the total in saline soils.

In soils having unusually high B, such as the alluvial soils in Kern County, California, as much as 30% of the total B is in the adsorbed form (Bingham et al, 1970a). This fraction is considered to be easily leachable from the soil profile.

Boron is also contained in the organic fraction of the soil, although most of the B in soils is associated with minerals, such as tourmaline, that are resistant to weathering. In general, soils high in organic matter are usually high in B (Fleming, 1980). Little is known about the complexation and availability of B in organic matter. However, B is known to exhibit an affinity for alpha-hydroxy aliphatic acids and ortho-dihydroxy derivatives of aromatic compounds. Boron is also known to react with sugars, such as the type produced from microbial breakdown of soil polysaccharides. Another fraction, the "acid-soluble" fraction, represents precipitated B and that incorporated in organic matter (Jackson, 1965). It is otherwise known as "maximum" available B since it represents a capacity factor potentially capable of supplying soluble B to the plants through equilibrium between the various forms of B in soils (Eaton and Wilcox, 1939). In Egyptian soils (Table 3.2) this fraction ranged from traces to 68% of total B (Elseewi and Elmalky, 1979).

Boron may be entrapped in the clay lattice by substituting for Al^{3+} and/ or Si^{4+} ions. There are numerous borosilicate minerals in soils, but only tourmaline and axinite may be of significance (Krauskopf, 1972).

Only two soluble B species in soils can be expected (Lindsay, 1972). The non-ionized species, $B(OH)_3$, is the predominant species expected in soil solution. At pH greater than 9.2, $B(OH)_4^-$ becomes predominant. Polymeric forms of B are unstable in soils unless the concentrations exceed 10^{-4} M, which are seldom present in soils. Most of the B fertilizers are of the form $B_4O_7^{2-}$ and are expected to hydrolyze to $B(OH)_3$.

4.5 Adsorption and Fixation of Boron in Soils

There are several possible mechanisms for the chemical combination of B with soils: anion exchange, precipitation of insoluble borates with sesquioxides, sorption of borate ions or molecular boric acid, formation of organic complexes, and fixation of B in the clay lattice (Eaton and Wilcox, 1939; Hingston, 1964; Sims and Bingham, 1967, 1968a). The major inorganic adsorption sites for B in soils are: (1) Fe and Al-hydroxy compounds present as coatings on, or associated with, clay minerals; (2) Fe or Al-oxides in soils; (3) clay minerals, especially of the micaceous type; and (4) Mg-hydroxy coatings on the surfaces of ferromagnesian minerals (Ellis and Knezek, 1972). The adsorption sites are associated with broken Si-O and Al-O bonds exposed at edges of aluminosilicate minerals, and also with surfaces of amorphous hydroxide materials present in weathered soils, such as allophane (Bingham, 1973). In layer silicate clays, most of the adsorption can be ascribed to the sesquioxides coatings on the surface of the clay rather than to the exposed Si-O and Al-O bonds.

Sims and Bingham (1968a) showed that freshly precipitated oxides of Fe and Al adsorbed rather large amounts of B. Adsorption was pH-dependent, with maximum adsorption occurring at pH 7 and 8.5 for precipitated oxides of Al and Fe, respectively. Retention of B by hydroxy Al material was an order of magnitude greater than retention of B by hydroxy Fe material. The pH of maximum adsorption suggests that the adsorption is of the borate ion. The decrease in B retention at pH values above 8 or 9 would be expected partly from OH^- ion competition for sites but also from the fact that the hydrous oxides would begin to take on a negative charge resulting in repulsion of the borate ions and the dissolution of the adsorbent, such as by the formation of aluminate ions. Several investigators believe that the oxides of Fe and Al are the major soil constituents that sorb B (Hatcher et al, 1967; Sims and Bingham, 1968a). The mechanism of B retention by hydroxy Fe and Al compounds is still unclear, although Sims and Bingham (1967, 1968a,b) advanced two possible mechanisms:

1. Anion exchange of borate ions for hydroxyl ions:

or

2. Formation of borate-diol complex:

The pH dependence of B adsorption can be explained by the B hydrolysis reaction:

$$B(OH)_3 + 2\,H_2O \rightleftharpoons B(OH)_4^- + H_3O^+$$

and the assumption that $B(OH)_3$, $B(OH)_4^-$, and OH^- are all competing for the same adsorption sites (Keren and Gast, 1981). The pH at which maximum adsorption occurs is then a function of the relative affinities of $B(OH)_3$, $B(OH)_4^-$, and OH^- species for the clay surface. It has been shown that $B(OH)_4^-$ and OH^- have a much greater affinity for the surface than $B(OH)_3$ (Keren et al, 1981). The latter species predominates at lower pH however, and consequently there is relatively less total B adsorbed. As pH increases, the $B(OH)_4^-$ concentration increases rapidly, and since the OH^- concentration is still relatively low, the amount of B adsorbed dramatically increases. A further increase in pH results in increased OH^-

concentration relative to $B(OH)_4^-$, and B adsorption decreases rapidly due to the OH^- competition for the adsorption sites.

Adsorption of B has been described by the Langmuir equation for a wide range of soils, including amorphous soils, resulting in usual adsorption maxima in the range of 10 to 100 μg B/g of soil (Biggar and Fireman, 1960; Hatcher and Bower, 1958; Hingston, 1964; Okazaki and Chao, 1968; Bingham et al, 1971; McPhail et al, 1972; El-Damaty et al, 1970). Elrashidi and O'Connor (1982), however, noted that the Langmuir isotherm was applicable over only a limited B concentration range for New Mexico soils, whereas the Freundlich isotherm was applicable over a much wider B concentration range (0–100 μg/ml) for all the ten soils used. Adsorption maxima by amorphous soils were found to be considerably greater than those observed for other soils (Bingham et al, 1971; Schalscha et al, 1973). These soils, associated with volcanic ash deposits, contain amorphous materials such as allophane and hydrous oxides of Fe and Al, and can be found in the western portion of North and South America and on many islands in the Pacific Ocean. The fact that such soils from Mexico and Hawaii have adsorption maxima at pH 8 or greater (Fig. 3.2) suggests

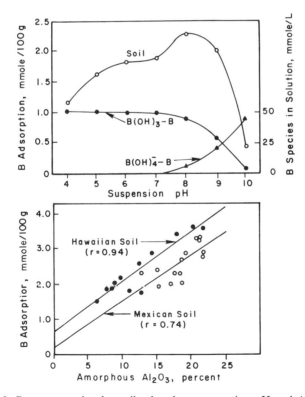

FIGURE 3.2. Boron retention by soil related to suspension pH and Al_2O_3 content of soil. *Source:* Bingham et al, 1971, with permission of the authors, and the Soil Sci Soc of America.

that these soils have a greater affinity for $B(OH)_4^-$ than for $B(OH)_3$. These soils increased in their B retention capacity as their content of amorphous Al_2O_3 increased (Fig. 3.2).

Although allophanic, amorphous soils are characterized for their strong affinity for other anions such as Cl^-, NO_3^-, SO_4^{2-}, and HPO_4^{2-}, B adsorption by these soils is more or less specific, being essentially independent of the presence of other anions (Bingham and Page, 1971).

In their review, Ellis and Knezek (1972) stated that clay minerals will adsorb B. In general, the micaceous-type clays—including vermiculite—will adsorb the most B, followed in decreasing order by kaolinite and montmorillonite. In this case, B is fixed in the clay lattice, possibly through the substitution of B for Al.

5. Boron in Plants

5.1 Essentiality of Boron in Plants

In spite of the essentiality of B for higher plants, its precise biochemical role is not well understood. Unlike the other micronutrients, it has not been shown to participate in any enzyme system (Jackson and Chapman, 1975). However, its roles can be deduced indirectly from physiological experiments, especially with B-deficient plants. Many bean plants suffering from B deficiency show decrease in root elongation, followed by degeneration of meristematic tissues, possibly due to limited cell division.

Since B is relatively immobile in plants, once B is utilized in the actively growing tissues of the plant, it cannot be retranslocated to other tissues. Therefore, it is necessary to have a continuous source of B available to the plant throughout its growth cycle. For example, sunflower plants grown in nutrient solution containing a total of 50 μg B per plant developed deficiency symptoms at about the 18th or 19th day (Skok, 1957). Boron is required for the formation of tissues but apparently is not required for the maintenance of older tissues. Thus, it is well known that actively growing plants require larger amounts of B than slowly growing or mature plants. Similarly, deficiency symptoms appear more readily in younger than in older tissues.

Boron is also believed to facilitate the transport of sugars through membranes, to be involved in auxin metabolism and in the synthesis of nucleic acids, and in protein and possibly phosphate utilization. It is also implicated in carbohydrate (sugar) metabolism, and in the synthesis of cell-wall components such as polyphenolic compounds (Jackson and Chapman, 1975; Price et al, 1972). It is also believed to play a part in photosynthetic and transport processes in plants (Gauch, 1972; Bonilla et al, 1980). Others include root growth (Wildes and Neales, 1969), regulation of seed dormancy (Cresswell and Nelson, 1972), and pectin synthesis, water balance relationships, and resynthesis of ATP (Sauchelli, 1969).

5.2 Uptake and Translocation of Boron in Plants

Boron uptake by plants exhibits the same two-phase pattern typical for other micronutrients. Early absorption is rapid, followed by a steady linear phase. Boron may be absorbed in one or more of its ionic forms, but since $B(OH)_3$ is more likely to exist at pH below 9.2 (at pH 9.2 there are equal concentrations of undissociated acid and dissociated anion), molecular uptake is also most likely. The $B(OH)_3$ species, being neutral, seems to encounter the least resistance upon passage through a charged root membrane. Thus, B is more favorably absorbed as $B(OH)_3$. In alkaline conditions, the reduced B absorption is possibly due in part to excessive OH^- concentration through a competitive action on the root surface, indicating absorption in an anionic form (Bingham et al, 1970b). Bingham et al (1970b) also found that B absorption by excised barley roots was noncumulative, rapid, physical, and a nonmetabolic process acting in response to B concentration gradient and unaffected by pH variations of the substrate in the acid range. However, increases in pH above 7.0 resulted in sharp reduction. At pH 6.0, changes in substrate temperature, salt composition, and level or addition of respiratory inhibitors (KCN and 2,4-DNP) failed to exert any influence on B uptake. Because of the ability of $B(OH)_3$ to form complexes with sugars and phenols containing cis-diol groups in *in vitro* studies, B has been implicated in facilitating sugar transport across cell membranes (Follett et al, 1981). Phenylboric acid forms complexes with sugars on a 1:1 mole basis. Because of their higher polarity, these complexes may be better able to move through the cell membrane.

Sesame plants grown in solution culture accumulated B in decreasing order: leaf blades > petioles > stems (Yousif et al, 1972). In sugar beet leaves, B accumulated in decreasing order: leaf margin > central part > petiole (Oertli and Roth, 1969). This pattern of B accumulation supports the view that B uptake is passive and that the B is transported in the transpiration stream to accumulate in the leaves. As water is lost to the atmosphere, residual B accumulates, and eventually a level is reached that is toxic to the plant. The mechanism of B injury is not known, although an osmotic mechanism has been hypothesized (Oertli and Roth, 1969).

5.3 Deficiency of Boron in Plants

Of the seven commonly recognized micronutrients, B deficiency is considered among the most widespread. Boron deficiency has been reported for one or more crops in 43 states (Sparr, 1970). However, there is no apparent geographic pattern to the occurrence of B deficiency in crops in the United States (Kubota and Allaway, 1972) and in the United Kingdom (Thornton and Webb, 1980). Field and vegetable crops are affected, as are fruit and nut trees. Boron deficiency in alfalfa has been reported in 38 of the United States, where it is of widespread occurrence in tree and foodcrops. Deficiency is also reported in certain soils of Nigeria

(Chude and Obigbesan, 1983), and B is one of the most common micro-nutrient deficiencies in *Pinus radiata* in New Zealand and eastern parts of Australia (Hopmans and Flinn, 1984).

The amount of B required by plants for normal nutrition varies (Tables 3.3 and 3.4). In general, monocotyledons require only about a fourth as much B for normal growth as do dicotyledons (Berger, 1949). Grasses do not seem to require as much B as do broadleaf plants, but there are variations within these categories (Bowen, 1977). Legumes seem to have an exceptionally large requirement for this nutrient. In contrast, most of the fruit and nut trees do not (Table 3.3). Amounts needed for responsive crops such as alfalfa, beets, and celery may cause damage to small grains, beans, peas, and cucumbers. The range between beneficial and toxic B concentrations in the substrate is narrow. For example, solution culture concentrations of 0.05 to 0.10 μg B/ml are ordinarily safe and adequate for many plants, whereas concentrations of 0.50 to 1.0 μg B/ml are often excessively high for B-sensitive plants (Bingham, 1973). Concentrations less than 0.05 μg B/ml may produce deficiency. Similarly, B level of 0.50 ppm in sand culture solution was adequate for sunflower growth, but 1.0 ppm was toxic (Eaton, 1940).

As a diagnostic tool, either soil test or plant test for B can be used. The hot water-soluble B in soils is the most commonly used. However, this level of B cannot be used universally because of the modifying effects of plant and soil factors and other nutrients on B availability. In general, there is a direct positive correlation between B content of plants and water-soluble B in soils. In their review, Cox and Kamprath (1972) reported a critical range of 0.10 to 0.70 ppm of hot water-soluble B in soils for most crops. At least two factors, soil texture and pH, must also be considered in the interpretation of soil tests based on this form of B. Consequently, critical levels of hot water-soluble B vary between locations and among crops. Stinson (1953) stated that heavier soils in Illinois containing less than 0.50 ppm water-soluble B seem to be deficient for alfalfa and possibly other legumes. For coarse sandy soils, the critical level was about 0.30 ppm. In contrast, soils in Alabama, which are of the sandy podzolic soils, containing at least 0.15 ppm hot water-soluble B were reported to support good growth of alfalfa and other legumes (Rogers, 1947). In Michigan, alfalfa needed B fertilization when the soil tests indicate less than 0.90 to 1.0 ppm hot water-soluble B (Baker and Cook, 1956).

In Oklahoma, no internal damage in peanuts due to B deficiency (Fig. 3.1) appeared when the hot water-soluble B was above 0.15 ppm (Hill and Morrill, 1974). With rutabaga, brown-heart symptoms due to B deficiency occurred when hot water-soluble B in soil ranged between 0.40 and 1.3 ppm (Gupta and Munro, 1969). For other crops, such as broccoli, Brussels sprouts, and cauliflower, levels of this form of B in the range of 0.34 to 0.49 ppm in soil were sufficient for optimum growth (Gupta and Cutcliffe, 1973). For apple, 0.50 ppm water-soluble B in soil would be sufficient for normal apple production (Woodbridge, 1937).

TABLE 3.3. Relative tolerance of plants to boron.[a]

Sensitive	Semi-tolerant	Tolerant
American elm	Alfalfa	Asparagus
Apple	Barley	Artichoke
Apricot	Birdsfoot trefoil	Athel
Avocado	Broccoli	Blueberry
Blackberry	Cabbage	Broadbean
Cherry	Calendula	Chard
Cowpea	California poppy	Cotton
Elm	Carrot	Cucumber
Fig	Cauliflower	Gladiolus
Grape	Celery	Mangel
Grapefruit	Clover	Muskmelon
Jerusalem artichoke	Corn	Oxalis
Kidney bean	Field pea	Palm
Kola	Hops	Pasture grass
Larkspur	Kentucky bluegrass	Peppermint
Lemon	Lettuce	Rye
Lupine	Lima bean	Sesame
Navy bean	Millet	Soybean
Orange	Milo	Spearmint
Pansy	Mustard	Sudangrass
Peach	Oats	Sugar beet
Pear	Olive	Sweet Clover
Pecan	Onion	Table beet
Persimmon, Japanese	Parsley	Turnip
Plum	Parsnip	
Strawberry	Peanut	
Violet	Pepper	
Walnut	Potato	
	Pumpkin	
	Radish	
	Rice	
	Rose	
	Rutabaga	
	Spinach	
	Sunflower	
	Sweet Corn	
	Sweet Pea	
	Sweet Potato	
	Timothy	
	Tobacco	
	Tomato	
	Vetch	
	Wheat	
	Zinnia	

[a] *Sources:* Bradford (1965); Il'in and Anikina (1974); Bingham (1973); Robertson et al (1976); USDA (1954).

In a review of plant diagnostic criteria, Jones (1972) reported that, for a wide spectrum of crops, B deficiency occurs when the B levels are less than 15 ppm in the dry matter. Adequate but not excessive B occurs between 20 to 100 ppm, and toxicity can be expected when the plant levels of B exceed 100 ppm, although much lower levels may cause toxicity to plants especially sensitive to B. In pine species (*P. radiata, P. elliottii,* and *P. pinaster*), levels below 10 ppm in the foliage were deficient (Will, 1971; Raupach, 1975). Specific examples of crops showing deficient, sufficient, and toxic levels of B in plant tissues are shown in Table 3.4. The data in this table indicate the importance of plant species and plant parts and age in interpreting plant tests.

TABLE 3.4. Deficient, sufficient, and toxic levels of boron in plants.[a]

Plant	Plant part	ppm B in dry matter		
		Deficient	Sufficient	Toxic
Rutabaga	Leaf tissue at harvest	20–38 <12 Severely deficient	38–140	>250
	Leaf tissue when roots begin to swell	32–40 Moderately deficient <12 Severely deficient	40	—
	Roots	<8 Severely deficient	13	—
Sugar beets	Blades of recently matured leaves	12–40	35–200	—
	Middle fully developed leaf without stem taken at end of June or early July	<20	31–200	>800
Cauliflower	Whole tops before the appearance of curd	3	12–23	—
	Leaves	23	36	—
	Leaf tissue when 5% heads formed	4–9	11–97	—
Broccoli	Leaves	—	70	—
	Leaf tissue when 5% heads formed	2–9	10–71	—
Brussels sprouts	Leaf tissue when sprouts begin to form	6–10	13–101	—
Carrots	Mature leaf lamina	<16	32–103	175–307
	Leaves	18	—	—
Tomatoes	Plants	14–32	34–96	91–415
	Mature young leaves from top of the plant	<10	30–75	>200
	63-day-old plants	—	—	>125
Celery	Petioles	16	28–75	—
	Leaflets	20	68–432	720
Potatoes	32-day old plants	—	12	>180
	Fully developed first leaf at 75 days after planting	<15	21–50	>50[c]

BLE 3.4. *Continued*

Plant	Plant part	ppm B in dry matter		
		Deficient	Sufficient	Toxic
ans	43-day-old plants	—	12	>160
arf kidney beans	Leaves and stems (plants cut 50 mm above the soil)	—	44	132
	Pods	—	28	43
ite pea beans	Aerial portion of plants 1 month after planting	—	36–94	144
cumber	Mature leaves from center of stem 2 weeks after first picking	<20	40–120	>300
nish peanuts	Young leaf tissue from 30-day-old plants	—	54–65 18–20[b]	>250
alfa	Whole tops at early bloom	<15	20–40 15–20[c]	200
	Top one-third of plant shortly before flowering	<20	31–80	>100
	Upper stem cuttings in early flower stage	—	30[b]	—
	Whole tops	<15	15–20	200
	Whole tops at 10% bloom	8–12	39–52	>99
	Whole tops	<20	—	—
clover	Whole tops at bud stage	12–20	21–45	>59
	Top one-third of plant at bloom	—	20–60	>60[c]
dsfoot trefoil	Whole tops at bud stage	14	30–45	>68
othy	Whole plants at heading stage	—	3–93	>102
ture grass	Aboveground part at first bloom at first cut	—	10–50	>800
n	Leaf at or opposite and below ear level at tassel stage	—	10[b]	—
	Total aboveground plant material at vegetative stage until ear formation	<9	15–90	>100
eat	Boot stage tissue	2.1–5.0	8	>16
	Straw	4.6–6.0	17	>34
ter wheat	Aboveground vegetative plant tissue when plants 40 cm high	<0.3	2.1–10.1	>10[c]
s	47-day old plants	—	—	>105
	Boot stage tissue	—	15–50	44–400
	Boot stage tissue	<1	8–30	>30[c]
	Boot stage tissue	1.1–3.5	6–15	>35
	Straw	3.5–5.6	14–24	>50
ey	Boot stage tissue	1.9–3.5	10	>20
	Straw	7.1-8.6	21	>46

urce: From various sources, as summarized by Gupta, 1979, with permission of the author and Academic
s.
nsidered critical.
nsidered high.

Boron deficiency may be expected to be more prevalent on leached acid soils. Soil types deficient in B are usually the sandy loams, fine-textured lake-bed soils, and acid organic soils (Vitosh et al, 1973). Sandy and silty loam soils in humid regions, such as the acid sandy soils of the Mississippi Valley and Atlantic Coastal Plain in the United States are often prone to B deficiency. Liming to above pH 6.5 increases susceptibility.

Since B is not translocated from old to young plant parts, the first symptoms of B deficiency will be in the growing points, such as the stem tips, the flower buds, and the axillary buds. Boron-deficient plants exhibit a breakdown of the growing tip tissue or a shortening of the terminal growth. This may show up as a rosetting of the plant, or twisting and distortion of the growing points and the youngest leaves of the plants. The growing point of stems generally show varying degrees of necrosis, sometimes accompanied by splitting along the length of the stem. More specifically, B deficiency in apical meristematic tissue of tomato can cause enlarged cell size, nuclear enlargement, and tissue disintegration (Brown and Ambler, 1973). In leaf tissue, B deficiency caused necrosis and collapse of the upper epidermal and palisade cells, and accumulation of phenolic materials. Internal tissues of beets, turnips, and rutabagas show breakdown and corky, dark coloration.

In pines, needles are fused and necrotic shoots suffer from dieback in late summer or autumn, with foliage immediately behind dieback abnormally short and yellowish (Will, 1971). Trees that are B-deficient can have branches that may crook or resin flowing from the stems (Raupach, 1975).

Boron-deficiency symptoms for specific crops are described by Gupta (1979) and Eaton (1944).

5.4 Toxicity of Boron in Plants

In a review on B deficiency and toxicity in plants, Bradford (1965) reported that the following soils are likely to cause excess B uptake by plants: soils of marine sediment and, in general, those derived from parent materials rich in B; arid soils; and soils derived from geologically young deposits. However, B usually does not exist in toxic quantities on most arable soils unless it has been added in excessive amounts.

As discussed in section 5.3, plants have varying degree of tolerance to B excess in the substrate (Tables 3.3 and 3.4). Members of the grass family are usually tolerant to high B levels in soils. In general, when the levels of B in plant dry matter exceed 100 ppm, B toxicity can be expected to occur. Gupta (1984) found that tissue B concentrations of 108, 92, and 95 ppm in field-grown alfalfa, red clover, and timothy, respectively, were related to B toxicity symptoms on the foliage but were not related to any substantial yield declines.

There are quite a few exceptions, however, even with some plant species classified as semi-tolerant (Table 3.4) when toxicities occur even at much lower B concentrations in plant tissues. Some tree species, such as red pine, white pine, and Scotch pine, are highly sensitive to excess B. Levels of about 55 to 100 ppm B in the needles of red pines produced toxic symptoms (Neary et al, 1975; Stone and Baird, 1956).

Boron is toxic to many plants when present in soil solutions in levels not much greater than the trace amounts needed for normal growth. Boron concentrations less than 0.50 μg/ml in soil solution are probably safe for most plants (Fig. 3.3), but many plants are adversely affected when B levels are in the range of 0.50 to 5.0 μg/ml (Eaton and Wilcox, 1939; Wilcox, 1960). Such toxic levels of B have been found in soils and irrigation waters of many arid regions of the world, including the irrigated regions in the western United States (Rhoades et al, 1970). For examples, toxic levels of B occur in several alkaline areas of eastern Oregon (Powers and Jordan, 1950) as well as many parts of California (Bingham, 1973). In general, irrigation waters with B levels above 4 ppm are unfit for plants.

According to Wilcox (1960), irrigation waters can be classified as to their effects on crop nutrition: sensitive crops, 0.30–1.0 μg/ml; semi-tolerant crops, 1.0–2.0 μg/ml; and tolerant crops, 2.0–4.0 μg/ml of B. Toxicity symptoms are similar in most plants, first showing up as a browning of leaf tips and margins. The yellowing (chlorosis) rapidly increases in severity, quickly followed by necrosis and tissue death. In severely affected

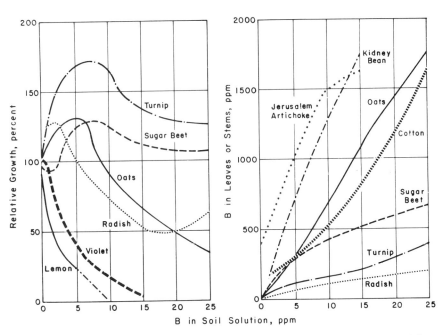

FIGURE 3.3. Plant growth and boron content of plant tissues as influenced by boron levels in soil solution. *Source:* Sprague, 1972, with permission of the author and the US Borax and Chemical Corp.

pine trees, necrosis is most pronounced on needles near the ends of shoots and in the upper half of the tree. The browning can be noticed also on the distal 1 to 2 cm of 1-year-old needles (Stone and Baird, 1956).

6. Factors Affecting the Mobility and Availability of Boron

6.1 Parent Material

As discussed in section 3, natural levels of B in soils largely depend upon the parent materials that are the precursors of soil. In general, soils derived from igneous rocks and those of tropical and temperate regions of the world are considerably lower in their B contents compared to soils derived from sedimentary rocks and those of arid and semi-arid regions. The content of total B in the latter may range up to 200 ppm, particularly in alkaline, calcareous soils, while that of the former is usually less than 10 ppm.

6.2 Clay Minerals and Soil Texture

In the pH range most commonly found in soils, illite adsorbed more B than kaolinite or montmorillonite, with kaolinite adsorbing the least (Fleet, 1965; Hingston, 1964). Frederickson and Reynolds (1959) proposed that most of the B in clay mineral fraction of sedimentary rocks is contained in the illite fraction. The adsorption by these soil-borne clays was quite pH-dependent, with maxima in the alkaline range (Hingston, 1964). Hingston proposed that his results support both a physical (molecular adsorption) and an anion adsorption mechanism.

In general, B is more readily leached to lower depths in sandy than in clayey soils (Kubota et al, 1948) and may be readily leached from the top 20 cm in sandy soils (Wilson et al, 1951). Thus, the possiblity of toxic B levels accumulating in soils is far less in coarse-textured than in fine-textured soils. In addition, the former soils—especially those of the humid regions—often are low in organic matter, and therefore are inherently low in B. Thus, B deficiency can be expected to be more prevalent in sandy than in clayey soils. For example, of the 266 alfalfa fields surveyed in Quebec, Canada, B deficiency was found in 119. Of that number, 61% were located on light-textured and only 14% on heavy-textured soils (Ouellette and Lachance, 1954). Similarly in Michigan, the occurrence of B deficiency in alfalfa was more prevalent in coarse-textured soils where the water-soluble B was lower than in fine-textured soils (Baker and Cook, 1956). In eastern Canada, Gupta (1968) found that greater amounts of hot water-soluble B were present in fine-than in coarse-textured soils. This can be attributed to the fact that more of the B is adsorbed on the clay and therefore, less subjected to leaching losses than B found in sandy soils. The relation between B content of plants and the hot water-soluble B in soil, showing the importance of soil texture, is illustrated in Figure 3.4.

6.3 pH

Soil reaction can influence the mobility and phytoavailability of B. In general, B becomes more available to plants with decreasing pH. Thus, liming acid soils reduces the B concentration in the plant tissues of several crops (Bartlett and Picarelli, 1973; Gupta, 1972a, b; Gupta and Cutcliffe, 1972; Gupta and MacLeod, 1973, 1981; Midgley and Dunklee, 1939; Olson and Berger, 1946) and in some cases, may induce B deficiency or diminish B toxicity in crops. Consequently, most of the reported studies relating to soil pH effects involve the interaction of B and a changing soil pH due to liming. However, there are a few exceptions. Peterson and Newman (1976) found that uptake of either indigenous or added B by tall fescue was not affected by the pH in the range of 4.7 to 6.3, but was substantially decreased at about pH 7.4. Similarly, Gupta and MacLeod (1977) did not get a definite trend on the relationship between soil pH and plant B below pH 6.5. Recently, there was no effect of pH over a range of 5.4 to 7.5 on B mobility in soils, but rather organic matter, clay, and free Fe and Al oxide contents can be more influential (Parker and Gardner, 1982).

Gupta and Cutcliffe (1972) obtained an interaction between soil pH and hot water-soluble B on the severity of brown-heart disorder in rutabaga. The degree of this disorder was more severe at high soil pH than at low pH. In contrast, no significant relationships were found between water-soluble B and pH in eastern Canada soils (pH 4.5 to 6.5) (Gupta, 1968) and Michigan soils (Robertson et al, 1975).

Boron absorption by excised barley roots was unaffected by pH variations of the substrate in the acid range. However, increases in pH above

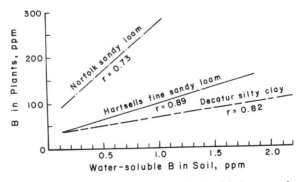

FIGURE 3.4. Importance of soil texture on the relationship between boron in plants and hot water-soluble boron in soil. *Source:* Wear and Patterson, 1962, with permission of the authors and the Soil Sci Soc of America.

7.0 resulted in sharp reductions in absorption (Bingham et al, 1970b; Oertli and Grgurevic, 1975). The mechanisms of B uptake by plants as related to pH were discussed in section 5.2.

Boron adsorption by soils and soil constituents was also shown to be largely pH-dependent, with the adsorption maxima in the alkaline range (see section 4.5).

6.4 Oxides of Iron and Aluminum

Sims and Bingham (1967, 1968b) found that B adsorption was greater for Fe- and Al-coated kaolinite or montmorillonite than for uncoated clays. They concluded that hydroxy Fe and Al compounds present in the layer silicates or as impurities dominate over clay mineral species *per se* in determining B adsorption characteristics. Bingham et al (1971) and Schalscha et al (1973) further concluded that B adsorption by certain soils was primarily due to their free Al oxide contents. The role of these compounds in B retention by soils is discussed in section 4.5.

6.5 Organic Matter

In his review, Gupta (1979) reported that organic matter is one of the main sources of B in acid soils. Several workers (Gupta, 1968; Berger and Truog, 1946; El-Damaty et al, 1970) found positive relationships between soil organic matter content and available B. Others (Olson and Berger, 1946; Martens, 1968; Berger and Pratt, 1963) have indicated the significance of organic matter as reservoirs for B that becomes available to plants upon decompositon of organic matter by soil microorganisms.

6.6 Interactions with Other Elements

The fact that B deficiency more frequently occurs on overlimed or alkaline soils than on acid soils suggests the possible effects of Ca^{2+}, and possibly Mg^{2+}, ions on B nutrition of plants. These cations are primary constituents of agricultural limestone. Thus, numerous investigators reported an apparent relationship between the level of Ca in the soil and other substrates and the phytoavailability of B (Cook and Millar, 1939; Midgley and Dunklee, 1939; Hill and Morrill, 1975; Powers, 1941; Fox, 1968; Jones and Scarseth, 1944). This often results in antagonistic interaction between Ca and B in plant nutrition, producing high Ca/B ratios in plant tissues. Consequently, Drake et al (1941) and Gupta (1972a) tried the possibility of using the Ca/B ratio in plants as a diagnostic criterion in B fertilization. Therefore, it can be expected that liming would increase the Ca/B ratios in plant tissues. Conversely, high hot water-soluble B in soils would decrease Ca/B ratios. Thus, the ideal Ca/B ratios were determined as 1,200 for tobacco; 500 for soybeans, and 100 for sugar beets (Jones and Scarseth,

1944). In rutabagas, Ca/B ratios over 3,300 were associated with very severe brown-heart conditions (Gupta and Cutcliffe, 1972). Gupta (1972a) reported that boot stage tissue Ca/B ratios of 1,370 were related to severe B deficiency in barley. A Ca/B ratio of about 180 appeared optimum, while ratios of 10 to 45 were in the severe B toxicity range. Oats showed B toxicity when the ratio was 200 (Jones and Scarseth, 1944).

Recently, Gupta and MacLeod (1981) suggested that the effect of lime in reducing B uptake by plants was due to soil pH effect rather than the Ca^{2+} level in the soil. They demonstrated this by growing barley and peas on three podzol soils limed with three different sources of Ca^{2+} ($CaSO_4$, $CaCO_3$, and $CaCl_2$). They noted that, at equivalent amounts of Ca^{2+}, tissue B concentrations were much higher using $CaSO_4$ rather than $CaCO_3$. The lower uptake with $CaCO_3$ source corresponded to much higher increases in soil pH. Similarly, no consistent effects of variable levels of Ca and K were noted on B absorption by excised barley roots (Bingham et al, 1970b). In contrast, Fox (1968) found that at high pH, high Ca solutions restricted B uptake by cotton and alfalfa by about 50%.

Interactions of B have also been observed with Mg (Wolf, 1940; Gupta and MacLeod, 1977), K (Hill and Morrill, 1975; Reeve and Shive, 1944; Heathcote and Smithson, 1974), N (Gupta et al, 1976; Heathcote and Smithson, 1974), and P (Murphy et al, 1981). The interactions of these elements with B are inconsistent—i.e., sometimes synergistic or antagonistic—as indicated by Gupta's (1979) review.

6.7 Plant Factors

As discussed in section 5.3, plant species vary in their B requirement for normal nutrition (Tables 3.3 and 3.4). Grasses require less B than do broadleaf plants; legumes seem to need large amounts of B. In general, tolerant species accumulate B at a much slower rate than the sensitive species (Oertli and Kohl, 1961).

As with other micronutrients, B nutrition varies between cultivars. An adequate supply of B for one cultivar may be toxic to another. Brown and Ambler (1973) noted differential uptake of B by two tomato genotypes in both soil and solution culture.

Boron also accumulates in varying degrees between plant parts, as noted in section 5.2 and in Table 3.4—i.e., B accumulates higher in leaf blades than in other plant parts and higher in the leaf margin than in other leaf parts. In general, B tends to accumulate in the margin of leaves (Touchton and Boswell, 1975; Jones, 1970).

6.8 Other Factors

Some environmental factors appear to influence the B nutrition of plants. Boron deficiency seems to be more prevalent under droughty conditions (Scott et al, 1975; Stewart and Axley, 1956). However, drought can also alleviate toxicity symptoms in plants (Gupta et al, 1976). This can be explained by the reduction of B uptake by plants, since mass flow in the rhizosphere is restricted. The loss of water from the plants via transpiration causes the mass flow of ions to the root surface.

Seasonal variations in B levels in plants have also been observed. In citrus, B levels tend to increase with the age of the leaves (Bradford and Harding, 1957; Labanauskas et al, 1959). Increases of B with tissue age were also observed in corn leaves (Clark, 1975) and cucumber leaves (Alt and Schwarz, 1973).

7. Boron in Drinking Water and Food

Boron concentrations in natural waters in many parts of the world have been reported at about 100 ppb or less. However, elevated levels have been reported for certain areas. For example, B content of the Los Angeles water supply varied from 0.50 to 1.5 ppm (Fairhall, 1941). Frequently, concentrations above 1 ppm, the desirable upper limit for drinking water, have been found in US community water supplies (US-PHS, 1970). An upper mandatory limit of 5.0 ppm has been set for drinking water; at levels of about 30 ppm and above, interferences with digestive processes may occur. The US-EPA (1976) has set a limit of 0.75 ppm B in water intended for long-term irrigation of sensitive crops.

The average daily human intake for B has been estimated at 3.0 mg (WHO, 1973). In the United Kingdom the daily intake from diet was estimated at 2.83 ± 1.55 mg compared with ICRP (International Commission on Radiological Protection) II's 6.0 mg intake value (Hamilton and Minski, 1972). Intake by sixth grade children from five geographic regions in the United States from Type A school lunches average 0.50 mg/lunch (Murphy et al, 1970).

Toxicologically, B content in foodstuff is of not great concern compared to other trace elements. In fact, in the past $B(OH)_3$ had been used as a food preservative but is now declared unacceptable by the FAO/WHO Expert Committee on Food Additives (WHO, 1973). Only at excessively high intake of 4,000 mg/day and higher can B prove toxic (Bowen, 1979).

8. Sources of Boron in the Terrestrial Environment

The global tropospheric B distribution and fluxes have been estimated (Table 3.5) by Fogg and Duce (1985), who concluded that the most important anthropogenic sources of B are coal combustion and agricultural burning. They pointed out that, on a regional scale, the anthropogenic inputs may be quite significant. Removal of B from the atmosphere by precipitation, particulate dry deposition, and gaseous absorption by the land and sea are estimated in Table 3.6. The estimated total removal of

TABLE 3.5. Estimates of global sources of atmospheric boron.[a,b]

Gaseous		Particulate	
Natural			
From sea salt	3.8–38 (EF = 1)[c]	Bubble bursting	
degassing	24–240 (EF = 3)	followed by B(OH)₃	
		volatilization	6–60
		Crustal weathering	0.52
Plant emissions	unknown	Plant emissions	0.4
Volcanism	210	Volcanism	0.03
Forest wild fires	0.75	Forest wild fires	0.75
Total	215–451		8–62
Anthropogenic			
Coal	12	Coal	1.2
Oil	0.00002		
Wood	5	Wood	5
Agricultural burning	18	Agricultural burning	18
	35		24
Total	250–284 (EF = 1)		32–86
	270–486 (EF = 3)		
	Grand Total ≈ 300–600		

[a] *Source:* Fogg and Duce, 1985, with permission of the authors, copyright by the Am Geophysical Union.
[b] Units of 10^{10} g/yr
[c] EF = enrichment factor

TABLE 3.6. Estimates of global sinks of atmospheric boron.[a,b]

Gas exchange	400	Wet deposition (ocean)	400
with ocean		(land)	100
Gaseous dry	20–200	Particulate dry (ocean)	200
deposition		deposition (land)	100
to land			
	420–600		800
	Grand Total ≈ 1,300		

[a] *Source:* Fogg and Duce, 1985, with permission of the authors.
[b] Units of 10^{10} g/yr.

B, which is $1,300 \times 10^{10}$ g/year, agrees within a factor of 2 to 4, with the total sources of B estimated as 300 to 600×10^{10} g/year. The data of Fogg (1983) support the belief that the sea is the net sink for atmospheric gaseous B.

Some of the sources mentioned by Fogg (1983), along with additional sources, are discussed below.

8.1 Soil Minerals and Parent Material

The most common natural source of B in soil is tourmaline. However, because of its slow dissolution, it is not a very good source of B for commercial plants. Soil B contents also vary according to its parent materials, with those originating from sedimentary rocks, marine clay, shale, and arid regions usually having high levels (see section 3).

8.2 Fertilizers

Borax (11% B) and sodium tetraborate ($Na_2B_4O_7 \cdot 5H_2O$) (14% B) are the most commonly used B fertilizers (Murphy and Walsh, 1972). Other sodium borates, such as sodium pentaborate ($Na_2B_{10}O_{16} \cdot 10H_2O$) (18% B) and Solubor ($Na_2B_4O_7 \cdot 5H_2O$ + $Na_2B_{10}O_{16} \cdot 10H_2O$) (20% B), are also used. Boric acid, $B(OH)_3$ (17% B), is used only sparingly.

Boron deficiency is usually corrected by application of B fertilizers to the soil, although some carriers like $B(OH)_3$ and Solubor are used as foliar spray. The optimum rate of soil-applied B depends on plant species, soil type, soil pH, as well as other factors. In general, numerous results indicate that a rate of 1.2 to 3.2 kg B/ha is needed for soil treatments on legumes and certain root crops, while lower rates (0.6 to 1.2 kg/ha) are needed for other crops (Murphy and Walsh, 1972). Applying 2 kg B/ha could maintain sufficient B levels in soils under eastern Canada conditions for alfalfa and red clover for two years (Gupta, 1984).

In Michigan, rates up to 3.4 kg B/ha are recommended for certain soil types and crops (Robertson et al, 1975). For soil applications, rates of 1.7 to 3.4 kg/ha are recommended for highly responsive crops while 0.6 to 1.1 kg/ha are recommended for low to medium responsive crops (Lucas, 1967). However, band applications of 5.0 kg B/ha on pea beans caused decreased yields and toxicity symptoms (Rieke and Davis, 1964). Because of these small amounts, B fertilizers are usually mixed in the desired proportion with other fertilizers (Mortvedt and Osborn, 1965; Mortvedt, 1968).

Foliar application of B has gained popularity recently for fruit crops, such as apples, because it can be mixed with pesticide sprays. Rates as low as 0.2 kg B/ha in 360 liters of water was as effective in providing enough B to apples as was application of 0.02 to 0.04 kg B/tree every three years (Burrell, 1958).

Boron is also present in various fertilizers from trace to several hundred ppm levels (Arora et al, 1975; Williams, 1977).

8.3 Irrigation Waters

Irrigation waters with B levels above 0.3 μg B/ml can damage certain crops, particularly sensitive species (see section 5.4). Although B levels in some waters may exceed this level, the majority of surface water supplies in the western United States have B concentrations in the range of

0.1 to 0.3 μg/ml (USDA, 1954). In arid and semi-arid regions, the B contents of irrigation waters, especially underground waters, are often elevated and in some cases may be as high as 5 μg/ml (Paliwal and Mehta, 1973). In other bodies of natural waters, B levels are in the neighborhood of 100 μg/liter, as shown for natural waters in Florida (Carriker and Brezonik, 1978), several streams in the southeastern United States (Boyd and Walley, 1972), and rivers of Norway and Sweden (Ahl and Jönsson, 1972).

8.4 Sewage Sludges and Effluents

Because of the use of borates and perborates in detergents as buffering, softening, and bleaching agents, B is usually present in sewage effluents and sludges. Irrigating forested areas with sewage effluents over an extended period was found injurious to pine species (Neary et al, 1975; Sopper and Kardos, 1973). Bradford et al (1975) showed that B levels in sewage effluents from several sewage treatment plants in southern California ranged from 0.3 to 2.5 μg/ml, with a mean of 1.0 μg B/ml.

Boron is usually more concentrated in sewage sludge, with levels ranging from a few to several hundreds ppm (Bradford et al, 1975; Berrow and Webber, 1972). However, when applied to soil, B did not significantly accumulate in plant tissues, even at rates above 112 tonnes/ha of sludge containing 13 ppm B (Dowdy and Larson, 1975).

8.5 Coal Combustion

Coal contains B that can range up to 400 ppm and higher, with an average of about 50 to 60 ppm (Swanson et al, 1976; Clark and Swaine, 1962). Upon combustion, B in coal is discharged into the various combustion residues produced, including fly ash, in which levels up to 600 ppm have been reported (Page et al, 1979; Furr et al, 1977). Losses in the neighborhood of 30% of total B in coal through emissions have been quoted (Gladney et al, 1978). Boron is considered one of the most highly enriched elements in power plant emissions, partly because of the volatility of $B(OH)_3$ and B-halides species. These levels are of environmental significance considering the fact that electric power utilities produce about 60 x 10^6 tonnes of fly ash annually. Because of the B-enriched status of fly ash, one of the main constraints of land utilization and/or disposal of fly ash is the B content, along with Mo and Se (Adriano et al, 1980).

James et al (1982), using precipitator fly ashes collected throughout the midwestern states, found that B in these ashes is leachable by water in the range of 17% to 64% of the total content. Since many plant species can only tolerate a narrow range of B concentrations in the soil, the toxic effects of B to plants from fly ash may be expected to be severe during the initial two or three years after land application, after which the B levels would have diminished substantially to tolerable levels.

References

Adriano, D. C., A. L. Page, A. A. Elseewi, A. C. Chang, and I. Straughan. 1980. *J Environ Qual* 9:333–344.
Ahl, T., and E. Jönsson. 1972. *Ambio* 1:66–70.
Alt, D., and W. Schwarz. 1973. *Plant Soil* 39:277–283.
Archer, F. C. 1980. *Minist Agric Fish Food* (Great Britain) 326:184–190.
Arora, C. L., V. K. Nayyar, and N. S. Randhawa. 1975. *Indian J Agric Sci* 45:80–85.
Aubert, H., and M. Pinta. 1977. *Trace elements in soils.* Elsevier, New York.
Awad, F., and M. I. Mikhael. 1980. *Egypt J Soil Sci* 20:89–98.
Bagdasarashvili, Z. G., K. A. Abashidze, N. V. Koberidze, D. V. Abashidze, and N. M. Korkotadze. 1974. *Sov Soil Sci* 7:576–580.
Baird, G. B., and J. E. Dawson. 1955. *Soil Sci Soc Am Proc* 19:219–222.
Baker, A. S., and R. L. Cook. 1956. *Agron J* 48:564–568.
Bandyopadhya, A. K. 1974. *Ann Arid Zone* 13:125–128.
Bartlett, R. J., and C. J. Picarelli. 1973. *Soil Sci 116:77–88.*
Berger, K. C. 1949. *Adv Agron* 1:321–351.
Berger, K. C., and E. Truog. 1939. *Ind Eng Chem Anal Ed* 11:540–545.
———— 1940. *J Am Soc Agron* 32:297–301.
———— 1946. *Soil Sci Soc Am Proc* 10:113–116.
Berger, K. C., and P. F. Pratt. 1963. In M. H. McVickar, G. L. Bridger, and L. B. Nelson eds. *Fertilizer technology and usage,* 287–340. Soil Sci Soc Amer Inc, Madison, WI.
Berrow, M. L., and J. Webber. 1972. *J Sci Food Agric* 23:93–100.
Biggar, J. W., and M. Fireman. 1960. *Soil Sci Soc Am Proc* 24:115–20.
Bingham, F. T. 1973. *Adv Chem Ser* 123:130–138.
Bingham, F. T., R. J. Arkley, N. T. Coleman, and G. R. Bradford. 1970a. *Hilgardia* 40:193–204.
Bingham, F. T., A. A. Elseewi, and J. J. Oertli. 1970b. *Soil Sci Soc Am Proc* 34:613–617.
Bingham, F. T., and A. L. Page. 1971. *Soil Sci Soc Am Proc* 35:892–893.
Bingham, F. T., A. L. Page, N. T. Coleman, and K. Flach. 1971. *Soil Sci Soc Am Proc* 35:546–550.
Bingham, F. T., A. W. Marsh, R. Branson, R. Mahler, and R. Ferry. 1972. *Hilgardia* 41:195–211.
Bonilla, I., C. Cadahia, and O. Carpena. 1980. *Plant Soil* 57:3–9.
Bowen, H. J. M. 1979. *Environmental chemistry of the elements.* Academic Press, New York.
Bowen, J. E. 1977. *Crops Soils* (Aug–Sept):12–14.
Boyd, C. E., and W. W. Walley. 1972. *Am Midl Nat* 88:1–14.
Bradford, G. R. 1965. In H. D. Chapman ed. *Diagnostic criteria for plants and soils,* 33–61. Quality Printing Co, Abilene, TX.
Bradford, G. R., and R. B. Harding. 1957. *Proc Am Soc Hort Sci* 70:252–256.
Bradford, G. R., A. L. Page, L. J. Lund, and W. Olmstead. 1975. *J Environ Qual* 4:123–127.
Brown, J. C., and J. E. Ambler. 1973. Soil Sci Soc Am Proc 37:63–66.
Burrell, A. B. 1958. *Proc Am Soc Hort Sci* 71:20–25.
Carriker, N. E., and P. L. Brezonik. 1978. *J Environ Qual* 7:516–522.

Cartwright, B., K. G. Tiller, B. A. Zarcinas, and R. L. Spouncer. 1983. *Aust J Soil Res* 21:321–332.

Chude, V., and G. O. Obigbesan. 1983. *Plant Soil* 74:145–147.

Clark, R. B. 1975. *Commun Soil Sci Plant Anal* 6:451–464.

Clark, M. C., and D. J. Swaine. 1962. CSIRO Tech Commun 45. Chatswood, Australia.

Cook, R. L., and C. E. Millar. 1939. *Soil Sci Soc Am Proc* 4:297–301.

Cox, F. R., and E. J. Kamprath. 1972. In J. J. Mortvedt, P. M. Giordano, and W. L. Lindsay, eds. *Micronutrients in agriculture.* 289–313. Soil Sci Soc Am, Inc, Madison, WI.

Cresswell, C. F., and H. Nelson. 1972. *Proc Grass Soc So Afr* 7:133–137.

DeTurk, E. E., and L. C. Olson. 1941. *Soil Sci* 52:351–357.

Dowdy, R. H., and W. E. Larson. 1975. *J Environ Qual* 4:278–282.

Drake, M., D. H. Sieling, and G. D. Scarseth. 1941. *J Am Soc Agron* 32:454–462.

Eaton, F. M. 1944. *J Agric Res* 69:237–277.

Eaton, S. V. 1940. *Plant Physiol* 15:95–107.

Eaton, F. M., and L. V. Wilcox. 1939. In *The behavior of boron in soils.* USDA Tech Bull 696. 58 pp.

El-Damaty, A. H., H. Hamdi, A. F. El-Kholi, and A. A. Hamdi. 1970. *UAR J Soil Sci* 10:39–58.

Ellis, B. G., and B. D. Knezek. 1972. In J. J. Mortvedt, P. M. Giordano, and W. L. Lindsay, eds. *Micronutrients in agriculture,* 59–78. Soil Sci Soc Am Inc, Madison, WI.

Elrashidi, M. A., and G. A. O'Connor. 1982. *Soil Sci Soc Am J* 46:27–31.

Elseewi, A. A., and A. E. Elmalky. 1979. *Soil Sci Soc Am J* 43:297–300.

Fairhall, L. T. 1941. *New England Water Works Assoc J* 55:400–410.

Fleet, M. E. L. 1965. *Clay minerals Bull.* 6:3–16.

Fleming, G. A. 1980. In B. E. Davis, ed. *Applied soil trace elements,* 155–197. Wiley, New York.

Fogg, T. R. 1983. *Sources and sinks of atmospheric boron.* PhD dissertation, Univ Rhode Island, Kingston. 319 pp.

Fogg, T. R., and R. A. Duce, 1985. American Geophysical Union Paper no. 4D1385. AGU, Washington, DC.

Follett, R. H., L. S. Murphy, and R. L. Donahue. 1981. *Fertilizers and soil amendments.* Prentice-Hall, Englewood Cliffs, NJ. 557pp.

Fox, R. H. 1968. *Soil Sci* 106:435–439.

Frederickson, A. F., and R. C. Reynolds, Jr. 1959. Clays Clay Miner. 8:203–213.

Furr, A. K., T. F. Parkinson, R. A. Hinrichs, D. R. van Campen, C. A. Bache, W. H. Gutenmann, L. E. St. John, I. S. Pakkala, and D. J. Lisk. 1977. *Environ Sci Technol* 11:1104–1112.

Gauch, H. G. 1972. In *Organic plant nutrition,* 243–259. Dowden, Hutchinson and Ross, Strousburg, PA.

Gestring, W. D., and P. N. Soltanpour. 1984. *Soil Sci Soc Am J* 48:96–100.

Gladney, E. S., L. E. Wangen, D. B. Curtis, and E. T. Jurney. 1978. *Environ Sci Technol* 12:1084–1085.

Graham, E. R. 1957. *Soil Sci Soc Am Proc* 21:505–508.

Gupta, U. C. 1968. *Soil Sci Soc Am Proc* 32:45–48.

——— 1972a. *Soil Sci Soc Am Proc* 36:332–334.

———— 1972b. *Commun Soil Sci Plant Anal* 3:355–365.

———— 1979. *Adv Agron* 31:273–307.

———— 1984. *Soil Sci* 137:16–22.

Gupta, U. C., and J. A. Cutcliffe. 1972. *Soil Sci Soc Am Proc* 36:936–939.

———— 1973. *Can J Soil Sci* 53:275–279.

Gupta, U. C., and J. A. MacLeod. 1973. *Commun Soil Sci Plant Anal* 4:389–395.

———— 1977. *Soil Sci 124:279–284.*

———— *1981. Soil Sci* 131:20–25.

Gupta, U. C., J. A. MacLeod, and J. D. E. Sterling. 1976. *Soil Sci Soc Am J* 40:723–726.

Gupta, U. C., and D. C. Munro. 1969. *Soil Sci Soc Am Proc* 33:424–426.

Hamilton, E. I., and M. J. Minski. 1972. *Sci Total Environ* 1:375–394.

Hatcher, J. T., and C. A. Bower. 1958. *Soil Sci* 85:319–323.

Hatcher, J. T., C. A. Bower, and M. Clark. 1967. *Soil Sci* 104:422–426.

Heathcote, R. G., and J. B. Smithson. 1974. *Exp Agric* 10:199–208.

Hill, W. E., and L. G. Morrill. 1974. *Soil Sci Soc Am Proc* 38:791–794.

———— 1975. *Soil Sci Soc Am Proc* 39:80–83.

Hingston, F. J. 1964. *Aust J Soil Res* 2:83–95.

Hopmans, P., and D. W. Flinn. 1984. *Plant Soil* 79:295–298.

Il'in, V. B., and A. P. Anikina. 1974. *Sov Soil Sci* 6:68–75.

Jackson, M. L. 1965. *Soil chemical analysis*. Prentice-Hall, Englewood Cliffs, NJ. 498 pp.

Jackson, J. F., and K. S. R. Chapman. 1975. In D. J. D. Nicholas and A. R. Egan, eds. *Trace elements in soil-plant-animal systems,* 213–225. Academic Press, New York.

James, W. D., C. C. Graham, M. D. Glascock, and A. G. Hanna. 1982. *Environ Sci Technol* 16:195–197.

Jones, J. B., Jr. 1970. *Commun Soil Sci Plant Anal* 1:27–34.

———— 1972. In J. J. Mortvedt, P. M. Giordano, and W. L. Lindsay, eds. *Micronutrients in agriculture,* 319–346. Soil Sci Soc Am Inc, Madison, WI.

Jones, H. E., and G. D. Scarseth. 1944. *Soil Sci* 57:15–24.

Keren, R., and R. G. Gast. 1981. *Soil Sci Soc Am J* 45:478–482.

Keren, R., R. G. Gast, and B. Bar-Yosef. 1981. *Soil Sci Soc Am J* 45:45–48.

Krauskopf, K. B. 1972. In J. J. Mortvedt, P. M. Giordano, and W. L. Lindsay, eds. *Micronutrients in agriculture,* 4–40. Soil Sci Soc Am Inc., Madison, WI.

Krauskopf, K. B. 1979. *Introduction to geochemistry.* 2nd ed. McGraw-Hill, New York. 617 pp.

Kubota, J., K. C. Berger, and E. Truog. 1948. *Soil Sci Soc Am Proc* 13:130–134.

Kubota, J., and W. H. Allaway. 1972. In J. J. Mortvedt, P. M. Giordano, and W. L. Lindsay, eds. *Micronutrients in agriculture,* 525–554. Soil Sci Soc Am Inc, Madison, WI.

Labanauskas, C. K., W. W. Jones, and T. W. Embleton. 1959. *Proc Am Soc Hort Sci* 74:300–307.

Lindsay, W. L. 1972. In J. J. Mortvedt, P. M. Giordano, and W. L. Lindsay, eds. *Micronutrients in agriculture,* 41–57. Soil Sci Soc Am Inc, Madison, WI.

Liu, Z., Q. Q. Zhu, and L. H. Tang. 1983. *Soil Sci* 135:40–46.

Lucas, R. E. 1967. Ext Bull E-486. Michigan State Univ, East Lansing, MI. 13 pp.

Martens, D. C. 1968. *Soil Sci* 106:23–28.

McPhail, M., A. L. Page, and F. T. Bingham. 1972. *Soil Sci Soc Am Proc* 36:510–514.

Midgley, A. R., and D. E. Dunklee. 1939. *Soil Sci Soc Am Proc* 4:302–307.

Mortvedt, J. J. 1968. *Soil Sci Soc Am Proc* 32:433–437.

Mortvedt, J. J., and G. Osborn. 1965. *Soil Sci Soc Am Proc* 29:187-191.

Murphy, L. S., and L. M. Walsh. 1972. In J. J. Mortvedt, P. M. Giordano, and W. L. Lindsay, eds. *Micronutrients in agriculture*, 347–387. Soil Sci Soc Am Inc, Madison, WI.

Murphy, E. W., B. K. Watt, and L. Page. 1970. *Trace Subs. Environ. Health* 4:194–205.

Murphy, L. S., R. Ellis, Jr., and D. C. Adriano. 1981. *J Plant Nutri* 3:593–613.

Neary, D. G., G. Schneider, and D. P. White. 1975. *Soil Sci Soc Am Proc* 39:981–982.

Norrish, K. 1975. In D. J. D. Nicholas and A. R. Egan, eds. *Trace elements in soil-plant-animal systems*, 55–81. Academic Press, New York.

Oertli, J. J., and H. C. Kohl. 1961. *Soil Sci* 92:243–247.

Oertli, J. J., and J. A. Roth. 1969. *Agron J* 61:191–195.

Oertli, J. J., and E. Grgurevic. 1975. *Agron J* 67:278–280.

Okazaki, E., and T. T. Chao. 1968. *Soil Sci* 105:255–259

Olson, R. V., and K. C. Berger. 1946. *Soil Sci Soc Am Proc* 11:216–220.

Ouellette, G. J. 1958. *Can J Soil Sci* 38:77–84.

Ouellette, G. J., and R. O. Lachance. 1954. *Can J Agric Sci* 34:494–503.

Page, A. L., A. A. Elseewi, and I. R. Straughan. 1979. *Residue Rev* 71:83–120.

Page, N. R., and W. R. Paden. 1954. *Soil Sci* 77:427–434.

Paliwal, K. V., and K. K. Mehta. 1973. *Indian J Agric Sci* 43:766–772.

Parker, D. R., and E. H. Gardner. 1982. *Soil Sci Soc Am J* 46:573–578.

Peterson, L. A., and R. C. Newman. 1976. *Soil Sci Soc Am J* 40:280–282.

Piland, J. R., C. F. Ireland, and H. M. Reisenauer. 1944. *Soil Sci* 57:75–84.

Powers, W. L. 1941. *Better Crops Plant Food* 15(6):17–19;36–37.

Powers, W. L., and J. V. Jordan. 1950. *Soil Sci* 70:99–107.

Price, C. A., H. E. Clark, and E. A. Funkhouser. 1972. In J. J. Mortvedt, P. M. Giordano, and W. L. Lindsay, eds. *Micronutrients in agriculture*, 231–242. Soil Sci Soc Am Inc, Madison, WI.

Rajaratnam, J. A. 1973. *Exp Agric* 9:233–240.

Raupach, M. 1975. In D. J. D. Nicholas and A. R. Egan, eds. *Trace elements in soil-plant-animal systems*, 353–369. Academic Press, New York.

Reeve, E., and J. W. Shive. 1944. *Soil Sci* 57:1–14.

Reeve, R. C., A. F. Pillsbury, and L. V. Wilcox. 1955. *Hilgardia* 24:69–91.

Rhoades, J. D., R. D. Ingvalson, and J. T. Hatcher. 1970. *Soil Sci Soc Am Proc* 34:871–875.

Rieke, P. E., and J. F. Davis. 1964. Mich Agric Exp Sta Quart Bull 46:401–406.

Robertson, L. S., B. D. Knezek, and J. O. Belo. 1975. *Commun Soil Sci Plant Anal* 6:359–373.

Robertson, L. S., R. E. Lucas, and D. R. Christenson. 1976. In Mich Coop Ext Bull E-1037.

Rogers, H. T. 1947. *J Am Soc Agron* 39:914–928.

Sauchelli, V. 1969. In *Trace elements in agriculture*, 81–106. Van Nostrand Reinhold, New York.

Schalscha, E. B., F. T. Bingham, G. G. Galindo, and H. P. Galvan. 1973. *Soil Sci* 116:70–76.

Scott, H. D., S. D. Beasley, and L. F. Thompson. 1975. *Soil Sci Soc Am Proc* 39:1116–1121.

Shirokov, V. V., and V. I. Panasin. 1972. *Sov Soil Sci* 4:341–344.

Sims, J. R., and F. T. Bingham. 1967. *Soil Sci Soc Am Proc* 31:728–732.

————— 1968a. *Soil Sci Soc Am Proc* 32:364–369.

————— 1968b. *Soil Sci Soc Am Proc* 32:369–373.

Singh, M. 1970. *J Indian Soc Soil Sci* 18:141–146.

Skok, J. 1957. *Plant Physiol* 32:648–658.

Sopper, W. E., and L. T. Kardos. 1973. In *Recycling treated wastewater and sludge through forest and cropland,* 271–294. Penn State Univ, Univ Park, PA.

Sparr, M. C. 1970. *Commun Soil Sci Plant Anal* 1:241–262.

Sprague, R. W. 1972. *The ecological significance of boron.* US Borax Corp, Anaheim, CA. 58 pp.

Stewart, F. B., and J. H. Axley. 1956. *Agron J* 48:259–262.

Stinson, C. H. 1953. *Soil Sci* 75:31–36.

Stone, E. L., and G. Baird. 1956. *J Forest* 54:11–12.

Swaine, D. J. 1955. *The trace elements content of soils.* Commonwealth Bureau of Soil Science (GB), Tech Commun 48.

Swanson, V. E., J. M. Medlin, J. R. Hatch, S. L. Coleman, G. H. Wood, Jr., S. D. Woodruff, and R. T. Hildebrand. 1976. USGS Report 76-468. 503 pp.

Talati, N. R., and S. K. Agarwal. 1974. *J Indian Soc Soil Sci* 22:262–268.

Thornton, I., and J. S. Webb. 1980. In B. Davies, ed. *Applied soil trace elements,* 381–434. Wiley, New York.

Touchton, J. T., and F. C. Boswell. 1975. *Agron J* 67:197–200.

Trudinger, P. A., D. J. Swaine, and G. W. Skyring. 1979. In P. A. Trudinger and D. J. Swaine, eds. *Biogeochemical cycling of mineral-forming elements,* 1–27. Elsevier, Amsterdam.

US Bureau of Mines. 1980. *Minerals yearbook. Metals and Minerals.* US Dept of Interior, Washington, DC.
of Interior, Washington, DC.

US Environmental Protection Agency (US-EPA). 1976. *Quality criteria for water.* US-EPA, Washington, DC.

US Public Health Service (US-PHS). 1970. *Community water supply study, Analysis of National Survey Findings.* US Dept Health, Education, and Welfare, Washington, DC.

Vinogradov, A. P. 1959. *The geochemistry of rare and dispersed chemical elements in soils.* Consultants Bureau, Inc, New York. 209 pp.

Vitosh, M. L., D. D. Warncke, and R. E. Lucas. 1973. Ext Bull E-486. Mich State Univ, East Lansing, MI. 19 pp.

Wear, J. I. 1965. In C. A. Black, ed. *Methods of soil analysis. Agron* 9:1059–1063.

Wear, J. I., and R. M. Patterson. 1962. *Soil Sci Soc Am Proc* 26:344–346.

Whetstone, R. R., W. O. Robinson, and H. G. Byers. 1942. USDA Tech Bull no 797:1–32.

Wilcox, L. V. 1960. In USDA Info Bull 211. 7 pp.

Wildes, R. A., and T. E. Neales. 1969. *J Exp Bot* 20:591–603.

Will, G. M. 1971. N Z Forest Serv Leaflet 32:1–4.

Williams, C. H. 1977. *J Aust Inst Agric Sci* (Sept-Dec):99–109.

Wilson, C. M., R. L. Lovvoron, and W. W. Woodhouse. 1951. *Agron J* 43:363–367.

Winsor, H. W. 1952. *Soil Sci* 74:459–466.

Wolf, B. 1940. *Soil Sci* 50:209–217.

Woodbridge, C. G. 1937. *Sci Agric* 18:41–48.

World Health Organization (WHO) 1973. In WHO Tech Rep Ser no 532, Geneva, Switzerland.

Yousif, Y. H., F. T. Bingham, and D. M. Yermanos. 1972. *Soil Sci Soc Am Proc* 36:923–926.

4
Cadmium

1. General Properties of Cadmium

Cadmium is a soft, ductile, silver-white, electropositive metal, with an atomic weight of 112.40, specific gravity of 8.642, and melting point of 320.9°C. It has eight stable isotopes in nature: ^{106}Cd, 1.22%; ^{108}Cd, 0.88%; ^{110}Cd, 12.39%; ^{111}Cd, 12.75%; ^{112}Cd, 24.07%; ^{113}Cd, 12.26%; ^{114}Cd, 28.86%; and ^{116}Cd, 7.58%. Like Zn and Hg, Cd is a transition metal in Group IIb of the periodic table. Cadmium and Zn, however, differ from Hg in that the latter forms particularly strong Hg-C bonds. Like Zn, Cd is almost always divalent in all stable compounds, and its ion is colorless. Its most common compound is CdS. It forms hydroxides and complex ions with ammonia and cyanide—e.g., $Cd(NH_3)_6^{4-}$ and $Cd(CN)_4^{2-}$. It also forms a variety of complex organic amines, sulfur complexes, and chelates. Cadmium ions form insoluble white compounds, usually hydrated, with carbonates, arsenates, phosphates, oxalates, and ferrocyanides.

Its low melting point is a valuable property to form important low melting alloys. Although the metal surface oxidizes readily, it is very resistant to rusting.

2. Production and Uses of Cadmium

Cadmium is produced commercially as a by-product of the Zn industry. The most important uses of Cd are as alloys, in electroplating (auto industry), in pigments, as stabliizers for polyvinyl plastics, and in batteries (Ni-Cd batteries). As an impurity in Zn, significant amounts of Cd are also present in galvanized metals. Consequently, Cd can be found in a wide variety of consumer goods, and virtually all households and industries have products that contain some Cd. Cadmium is also used in photography, lithography, process engraving, rubber curing, and as fungicides, primarily for golf course greens.

Cadmium is basically recovered as a by-product from the smelting and refining of Zn concentrates at a rate of about 3.5 to 3.9 kg/tonne of primary Zn (Förstner, 1980). No ores are mined and processed exclusively to provide Cd. Therefore, Cd production can be expected to more or less parallel Zn production. World production in 1979 was 18,280 tonnes, with the United States accounting for about 9% of the production (1,715 tonnes). This is in contrast to the 1940s, when the United States produced approximately 70% of the world's supply. Starting in the 1950s, United States dominance of the world production of Cd gradually diminished, partly due to large increases in production in Japan and the USSR. In 1969, the United States produced about 34% of the world's supply, while Japan and the USSR produced approximately 16% and 14%, respectively (Page and Bingham, 1973). While the rest of the world has practically maintained production levels in the 1970s, US production declined (Fig. 4.1). Domestic production of Cd metal in 1978 declined 17% from the production level of 1977, but output in 1979 was about 4% higher than in the preceding year (Lucas, 1980). The majority of Cd-producing countries, particularly Japan and western Europe, recorded significant increases in production in 1977, caused by expansion of Zn output (Förstner, 1980). As of 1977, Belgium, the Federal Republic of Germany, USSR, Canada, United States, and Japan were the major Cd producers; the major consumers were: Belgium, France, the Federal Republic of Germany, the United Kingdom, USSR, United States, and Japan. In 1974, the following comprise the major end uses (as percent of total use) in western world countries (Nriagu, 1980): United States—pigments, 18; stabilizers, 16; and plating, 48; Federal

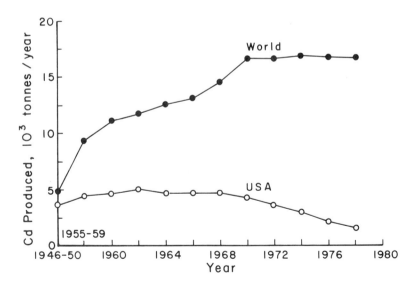

FIGURE 4.1. Trends in cadmium production in the United States and worldwide 1946–78. *Source: US Bureau of Mines, Minerals yearbook*, several issues.

Republic of Germany—pigments, 33; France—pigments, 45; plating, 24; United Kingdom—pigments, 29; plating, 32; and Japan—batteries, 35; pigments, 27.

Until recently, price index has been a major factor in Cd consumption, but environmental considerations have now assumed greater importance. For example: in Japan, Cd use dramatically declined following the incidence of Itai-Itai disease in the Jintsu River Basin in the late 1960s; in Sweden, the use of Cd was restricted, based on rising levels of Cd in human tissues, increase of its use in electroplating, as a stabilizer, and as a coloring agent (Environ Sci Technol, 1979). The Swedish government also ordered reduction in Cd levels in phosphate fertilizers. In addition, importation of Cd-containing goods, on commercial basis, was at least temporarily curtailed by the Swedish government (Nilsson, 1979).

TABLE 4.1. Commonly observed cadmium concentrations (ppm) in various environmental matrices.

Material	Average concentration	Range
Igneous rocks[a]	0.15	<0.001–1.6
Limestone[a]	0.10	—
Sandstone[a]	<0.03	—
Shales[a]	1.4	<0.3–11
Crude oil[b]	0.008	0.0003–0.027
Oil shale[b]	—	0.6–1.2
Coal[b]	0.10	0.07–0.18
Fly ash[b]	0.5	0.1–3.9
Phosphate rocks[b]	—	5–110
Superphosphate[b]	—	5–170
Phosphated fertilizers (mixed)[c]	4.3	1.5–9.7
Sewage sludges[d]	74	2–1,100
Soils[e]	0.35	0.01–2.0
Vegetables[f]	—	0.05–1.2
Grasses[f]	—	0.03–0.3
Crop grain[f]	—	0.1–0.5
Ferns[e]	0.13	—
Mosses[f]	—	0.7–1.2
Lichens[f]	—	0.1–0.4
Trees, deciduous[f]		
Leaves	—	0.1–2.4
Branches	—	0.1–1.3
Trees, coniferous[f]		
Leaves	—	0.1–0.9
Freshwater[e]	0.10*	0.01–3*
Sea water[e]	0.11*	<0.01–9.4*

*μg/liter
Sources:
[a]Fleischer et al (1974).
[b]Godbeer and Swaine (1979).
[c]Lee and Keeney (1975).
[d]For 57 sludges from locations in Michigan, Page (1974).
[e]Bowen (1979).
[f]Shacklette (1972).

3. Natural Occurrence of Cadmium

Cadmium found in soils, waters, plants, and other environmental matrices
(Table 4.1) not impacted by pollution can be regarded as natural, normal,
or background content. There is little difference among the igneous rocks
in Cd content (NRCC, 1979). Among the sedimentary rocks, the carbonaceous shales, formed under reducing conditions, contain the most Cd.
The Zn/Cd ratio in terrestrial rocks is about 250. The most often-quoted
average concentration of Cd in the earth's crust is 0.15–0.20 ppm (Fleischer
et al, 1974). Recently, this was revised to 0.098 ppm Cd (Heinrichs et al,
1980). Krauskopf (1979) ranks Cd 64th in crustal abundance among the
elements. Since Cd is closely related to Zn, it is found mainly in Zn, Pb-
Zn, and Pb-Cu-Zn ores. The amount in the principal Zn ore, sphalerite
(ZnS) varies markedly from a low of about 0.1% to a high of 5% and
sometimes even higher (Chizhikov, 1966). The Cd content of the majority
of Cu-Zn deposits is 0.3 part of Cd/100 parts of Zn, and in Pb-Zn deposits
it is 0.4 part of Cd/100 parts of Zn. Cadmium is also found in wurtzite,
another ZnS, and in trace amounts in galena, tetrahedrite, and a variety
of other sulfides and sulfosalts.

An atypical occurrence of Cd in soils was reported by Lund et al (1981)
in the Malibu Canyon area of California (Table 4.2). In this area, Cd in

TABLE 4.2. Cadmium concentration (ppm, dry wt
basis) in native vegetation from soils with different
natural levels of cadmium.[a]

Soil series	Soil pH	Total Cd in soil, ppm	ppm Cd Wild oats	Mustard
Millsholm	6.4	22	—[c]	2.0
Malibu	5.7	15	2.0	—
Castaic	6.0	12	7.6	—
Yolo	6.0	6.0	—	4.0
Los Osos	7.1	5.6	—	0.48
Linne	7.4	5.3	0.13	—
Salinas	6.7	4.2	1.0	1.0
Callegaus	7.4	3.7	0.51	1.6
Arnold	6.7	3.5	—	3.6
Cropley	6.2	2.9	1.3	—
Gazos	6.2	2.1	0.67	—
Lockwood	6.4	0.26	0.38	0.14
Elder	6.3	0.15	—	0.39
Corralitos	4.9	0.13	0.22	0.27
Huerhuero	5.4	0.10	0.11	—
Gilroy	6.0	0.01	0.09	—

[a]Soils and plants collected from Malibu Canyon, California.
[b]Source: Lund et al, 1981, with permission of the authors and the Am
Soc of Agronomy, Crop Sci Soc of America, Soil Sci Soc of America.
[c]—Indicates that the species was not found at that site.

soils ranged up to 22 ppm. The soil levels of Cd are also reflected by the Cd concentrations in native vegetation—wild oats and mustard. This initial survey was followed by an extensive one in the Santa Monica Mountains of Los Angeles and Ventura counties, California. Lund et al (1981) reported that residual soils developed from shale parent materials had the greatest Cd concentrations, with a mean of 7.5 ppm, whereas soils originating from sandstone and basalt had the lowest concentrations, with a mean of 0.84 ppm. Alluvial soils with parent materials from mixed sources had an intermediate Cd concentration of 1.5 ppm. They concluded that naturally high-Cd soils are widespread in the Santa Monica Mountains. However, because of unfavorable terrain for farming, Cd entry into the food chain is unlikely.

4. Cadmium in Soils

4.1 Total Soil Cadmium

In noncontaminated, noncultivated soils, Cd concentration is largely governed by the amount of Cd in the parent material. Based on the Cd concentrations reported for common rocks, one can expect that, on the average, soils derived from igneous rocks would contain the lowest Cd (<0.1–0.3 ppm), soils derived from metamorphic rocks would be intermediate (0.1–1.0 ppm), and soils derived from sedimentary rocks would contain the largest amount of Cd (0.3–11 ppm) (Page and Bingham, 1973). The study by Lund et al (1981) in California demonstrates the dramatic effect of parent material on the Cd content of soils (Table 4.3, see also section 3).

The terrestrial abundance of Cd is on the order of 0.30 ppm, as indicated by numerous data (Table 4.1). Normal Canadian soils were reported to contain from 0.01 to 0.10 ppm total Cd (mean = 0.07 ppm); normal glacial tills and other glacial materials had 0.01 to 0.70 ppm total Cd (mean = 0.07); whereas, soils and glacial tills near Cd-bearing deposits contained up to 40 ppm Cd (NRCC, 1979). For Arctic soils, Hutchinson (1979) reported an average of 0.67 ppm Cd, with much higher values (5.4 ppm) for areas containing Cd-mineral deposits. He also reported an average of 1.8 ppm Cd for northern Canadian surface soils. However, areas affected by smelting operations showed Cd concentrations ranging from 0.20 to 350 ppm in the surface soil. In agricultural soils, the background Cd level is about 1 ppm (or less) (John et al, 1972a; Mills and Zwarich, 1975; Whitby et al, 1978). Surface agricultural soils in Ontario, Canada had 0.56 ppm Cd (0.10–8.10 ppm range), with only 5% of all samples exceeding the 1.25 ppm level (Frank et al, 1976). Similar background levels were reported for rural nonmineralized areas of Wales (Bradley, 1980). A more extensive survey of the soils of England and Wales revealed a median value of <1.0 ppm Cd (0.08–10 ppm range, $n = 689$) (Archer, 1980).

TABLE 4.3. Cadmium concentrations of soils developed from shale, basalt, sandstone, or alluvial parent materials.[a]

			Parent Material					
Shale			Basalt and sandstone			Alluvial		
Soil series	Depth, cm	ppm Cd	Soil series (Parent Material)	Depth, cm	ppm Cd	Soil series	Depth, cm	ppm Cd
Saugus	0–38	1.4	Cibo	0–7	0.01	Lockwood	0–10	0.26
	38–93	0.59	(basalt)	7–84	0.01		10–61	0.27
	93+	0.74		84–116	0.01		61–105	0.16
	Rock	0.85	Gilroy	0–30	0.01		105–125	0.13
Millsholm	0–23	22	(basalt)				125+	0.15
	23–41	21	Arnold	0–13	3.5		Rock	0.29
	Rock	33	(sandstone)	13–25	1.0	Corralitos	0–18	0.13
Malibu	0–13	15		25–72	1.2		18–92	0.08
	13–25	13		72–87	4.5		92+	0.09
	25–41	13	Gaviota	0–5	1.1	Cropley	0–36	2.9
	41–66	7.0	(sandstone)	5+	1.1		36–92	3.1
	Rock	7.7		Rock	0.65		92–125	3.9
Callegaus	0–8	3.7	Hambright	0–7	0.05		125+	4.3
	8–16	3.8	(basalt)	7–18	0.02		Rock	4.5
	Rock	2.3		18–30	0.01	Yolo	0–30	6.0
Diablo	0–77	4.8		30+	0.07		30–45	6.2
	77–92	5.8		Rock	0.01		45+	6.5
	92–138	5.8	Sedimentary			Vina	0–33	0.15
	Rock	7.8	rockland	0–38	0.05		33+	0.37
Linne	0–54	5.3	(sandstone)				Rock	0.23
	54–82	4.9	Avg. of surface soil			Huerhuero	0–10	0.10
	82–120	4.7	samples (6)		0.79		10–50	0.05
	Rock	8.5	Avg. of rock samples (2)		0.33		50–92	0.01
Gazos	8–38	2.1					92–115	0.01
	38–80	1.6					115+	0.01
	Rock	1.0				Elder	0–20	0.15
Castaic	0–25	12					20–54	0.07
	25–72	12					54+	0.01
	72–92	11				Salinas	0–64	4.2
	Rock	5.6					64–108	3.2
Los Osos	0–25	5.6					Rock	2.9
	25–56	5.7				Sorrento	0–40	0.82
	56–69	5.8					40–90	0.91
	Rock	5.0					90–139	1.4
Avg. of surface soil samples (9)		8.0					139+	1.2
Avg. of rock samples (9)		8.0				Avg. of surface soil samples (9)		1.6
						Avg. of rock samples (4)		2.0

Source: Lund et al, 1981, with permission of the authors and the Am Soc of Agronomy, Crop Sci Soc of America, Soil Sci Soc of America.

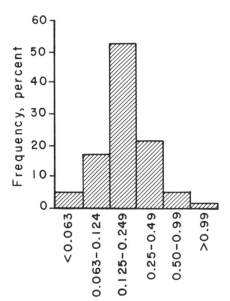

FIGURE 4.2. Frequency distribution of cadmium (in ppm) in Swedish soils ($n = 361$). *Source:* Andersson, 1977, with permission of the author.

In Denmark, Tjell and Hovmand (1978) found an average Cd concentration of 0.22 ppm (range of 0.03–0.90 ppm) for agricultural soils, and this was projected to increase at the rate of 0.6% per year due to atmospheric deposition and the use of phosphatic fertilizers (Tjell et al, 1980).

In an extensive survey of Swedish soils, Andersson (1977), who collected samples from both cultivated and noncultivated soils all over the country, found an average of 0.22 ppm Cd (0.03–2.3 ppm range) for all soils (Fig. 4.2). In addition, he reported the same value of 0.22 ppm for both cultivated and noncultivated soils.

In Japan, a nationwide survey on Cd content of nonpolluted paddy field soils revealed an average Cd concentration of 0.45 ± 0.23 ppm (Iimura, 1981). For world soils, Berrow and Reaves (1984) reported a mean of 0.40 ppm Cd as the background level.

4.2 Extractable (Available) Cadmium in Soils

Several chemical extractants have been tested to provide an index of Cd phytoavailability or Cd recovery from soils. They include weak acids, neutral salts, and chelating agents. Because the efficiency and predictability of a given extractant can be strongly influenced by soil and plant factors, there is no universal extractant for Cd. In using $CdCl_2$-treated soils, John et al (1972b) reported significant relationship between plant Cd (radish and lettuce) and Cd extracted from soil by neutral $1\ N$ NH_4OAc but not by $1\ N$ HCl and $1\ N$ HNO_3 solutions. The Cd extracted by the NH_4OAc was regarded as in the exchangeable form, whereas the HCl and HNO_3 extractions estimated closely the total amounts. Similar results were obtained by Haghiri (1974) with NH_4OAc solution using oats as the test plant. Andersson and Nilsson (1974) also favored neutral $1\ N\ NH_4OAc$ as the extractant most likely to predict the amount most available to plants. Although significant relationships were obtained for various crops (potatoes, cabbage, barley, parsley, and lettuce) using this extractant, Lag and El-sokkary (1978) had more reservations about this extractant since it extracted less than 20% of total Cd, potentially causing detectability problems.

In testing nine different soil extractants, Haq et al (1980) found that 0.5 N HOAc was the best extractant for predicting plant-available Cd when soil pH was included in the regression equation. Extractable Cd and soil pH accounted for 81% of the variability of plant Cd concentration. Using seven various extractants, Symeonides and McRae (1977) found the highest significant relationships ($r = 0.97$, statistically significant at the 0.1% level) between Cd concentration in radish and extractable Cd using unbuffered $1\ N\ NH_4NO_3$ solution. Significant relationships were also found using neutral $1\ N\ NH_4OAc$ ($r = 0.50$, statistically significant at the 0.1% level) and 5% HOAc (pH 2.5) (r = 0.36, significant at the 1% level). They ascribed the greater efficiency of the $1\ N\ NH_4NO_3$ solution to its ability in maintaining the natural pH of the soil, which has a marked influence on Cd phytoavailability. In testing industrially polluted soils in Norway, Lag and Elsokkary (1978) found the following extractants to give significant relationships in several vegetable crops: $1\ N$ HCl, $0.1\ N$ HCl, $1\ N\ HNO_3$, $0.1\ N\ HNO_3$, 2.5% HOAc, neutral $1\ N\ NH_4OAc$, $1\ N\ NH_4OAc$ (pH 4.8), ammonium lactate (pH 3.75), ammonium oxalate (pH 3.25), and 0.05 M EDTA (pH 7.0). They recommend $1.0\ N\ HNO_3$ solution as the extractant of choice. In metal-polluted soils from southern Ontario, Canada, Soon and Bates (1982) found that the exchangeable, complexed, and HNO_3-soluble fractions of Cd and the total of the three fractions may be used to predict Cd uptake by plants. The three fractions were extracted with $1\ M\ NH_4OAc$ (pH 7), $0.125\ M\ Cu(OAc)_2$, and $1\ M\ HNO_3$, respectively. Others (Jones et al, 1973; Takijima et al, 1973a) found 0.1 N HCl as a favorable extractant for predicting plant Cd. Recently, several workers (Bingham et al, 1976a; Keeney and Walsh, 1975; Street et al, 1978) favor the use of DTPA extracting solution in predicting Cd uptake and yield by crops. This extractant might gain wider acceptability over the others because of its popularity as a general soil test for several micronutrients.

Upon examination of the literature, Browne et al (1984) demonstrated that Cd accumulations in plants grown under greenhouse and aerobic conditions could be generally described by the equation:

$$\log Cd_{plant} = \alpha + \beta \log Cd_{DTPA}$$

where

Cd_{plant} is Cd concentration in plant tissues, in $\mu g/g$ dry weight,

Cd_{DTPA} is a measure of available soil Cd (μg/g soil), as determined by DTPA extraction, and

α and β are linear regression coefficients corresponding to the y intercept and slope, respectively.

The coefficient α was found to be primarily a characteristic of plant species, whereas β was found to be largely influenced by soil pH and CEC.

4.3 Cadmium in the Soil Profile

Cadmium is fairly immobile in the soil profile. In examining various soil profiles, Berrow and Mitchell (1980) observed that Cd, like Zn, was somewhat higher in soils on basic igneous rocks than soils formed on other rock types. The Cd level ($\leqslant 1$ ppm) is fairly uniform throughout the profile, and apparent mobilization also occurred in very poorly drained profiles. As indicated by EDTA-extractable Cd, most of the Cd is organically complexed in the upper horizons rich in organic matter. Similarly, Andersson (1977) observed higher accumulation of Cd (<1 ppm) in the upper part of the profiles, paralleling that of humus distribution. The levels of Cd in profiles of virgin muck soils in Canada (> 2 ppm) are much higher than the levels reported by Berrow and Mitchell (1980) in Scotland and by Andersson (1977) in Sweden, but nevertheless they showed increasing trends in the top layers (Hutchinson et al, 1974). Long-term cultivation tended to decrease Cd levels in the muck soil profile.

Contaminated soils also show the general immobility of Cd in the soil profile (Table 4.4). Soils contaminated by smelting operations showed Cd concentrations close to background level at a depth of about 30 to 40 cm. Andersson and Nilsson (1972) indicated that practically all of Cd remained in the surface 20 cm of soil following application of 84 tonnes/ha of sewage sludge over a 12-year period. Likewise, Boswell (1975) observed little Cd movement to 15 cm in an acid soil receiving surface application of 17

TABLE 4.4. Soil profile distribution of cadmium.[a]

Annaka City, Japan[b]		Fraser Valley, B. C., Canada[c]	
Soil depth, cm	ppm Cd	Soil depth, cm	ppm Cd
0–2	31	0–5	44.25
5	44	10	2.45
10	32	15	0.67
20	6.9	20	0.62
30	1.4	25	0.39
40	0.4	30	0.32
60	0.3		

[a]Sources: John et al (1972a); Kobayashi (1979).
[b]Near a Zn smelter.
[c]Near a battery smelter; Cd extracted with 1 N HNO_3.

FIGURE 4.3. Distribution of cadmium in the soil profile of a sandy loam soil, as influenced by the type and amount of sewage sludge. The sludges were obtained from Los Angeles county, CA where the liquid sludges were anaerobically digested. *Source:* Chang et al, 1984a, with permission of the authors and the Am Soc of Agronomy, Crop Sci Soc of America, Soil Sci Soc of America.

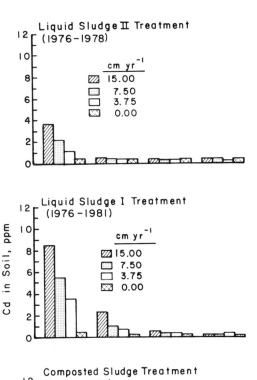
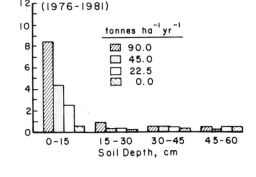

tonnes/ha of sewage sludge. Hinesly et al (1972), however, showed evidence of Cd movement below 15 cm in soil after application of 88 to 166 tonnes/ha of sludge over a three-year period. Lund et al (1976) observed Cd enrichment of profiles of coarse-textured soils below sludge and effluent disposal ponds. Although high concentrations were observed in the top soil layers, results indicate movement of Cd to a depth of at least 30 cm, and in one case, to at least a depth of 3.5 m.

Chang et al (1984a) have demonstrated that approximately 90% of the total Cd applied on a soil after six years of continued annual sewage sludge applications was found in the 0 to 15 cm soil depth (Fig. 4.3). This zone of accumulation corresponded to the depth where the sludges were incorporated. In addition, Williams et al (1984) found that Cd movement, within the profile, was limited to a depth of 5 cm below the zone of sludge incorporation.

Cadmium arising from phosphate fertilization of citrus groves over a 36-year period was proved to be less mobile than the P carrier itself in the soil profile (Mulla et al, 1980). About 71% of the accumulated Cd resided in the surface soil (0–15 cm) compared to only 45% for P.

4.4 Forms and Speciation of Cadmium in Soils

The mobility and phytoavailability of Cd largely depends on its chemical form and speciation in soils. Cadmium in soils may exist in one or more of the following sinks: *exchangeable* phase—adsorption of Cd by electrostatic attraction to negatively charge exchange sites on clays, organic particulates, and hydrous oxides; *reducible* (hydrous-oxide) phase—adsorption or coprecipitation with oxides, hydroxides, and hydrous oxides of Fe, Mn, and possibly Al present as coating on clay minerals or as discrete particles; *carbonate* phase—carbonate precipitation in soils high in free $CaCO_3$, bicarbonate, and alkaline in reaction (precipitation is also likely to occur with phosphate); *organic* phase—complexation with the organic fraction, chelated and/or organic bound. The complexes may vary in stability from immediately mobile, easily decomposable, and moderately resistant to resistant to decomposition; *lattice* phase—fixation within the crystalline lattices of mineral particles, sometimes known as *residual* fraction; *sulfide* phase—very insoluble and stable compounds of Cd sulfides occurring in poorly aerated soils, such as rice paddy soils; and *solution* phase—exists in soils solution either in ionic or complexed form. The distribution of Cd forms in soil is extremely variable since it is influenced by a suite of factors. Alloway et al (1979) found the following distribution (as percent of total concentration) of Cd in soils collected from sewage farms: water soluble—<1; exchangeable—21.3; organic bound—45.4; hydrous-oxide bound—20.1; and the rest were presumably in the residual form. The soils used had pH of 5.9, 24.0% organic matter, and 24.7 ppm total Cd.

In the analyses of three heavily polluted soils and one sediment, Hickey and Kittrick (1984) found that the greatest amount of Cd (i.e., 37%) in these materials was in the exchangeable fraction (Table 4.5). The Fe-Mn oxide and the residual fractions accounted for 23% and 15% of total Cd, respectively. The rest of the Cd was detected in the carbonate fraction, with the least and insignificant amounts in the organic fraction. Hickey and Kittrick (1984) indicated that the amounts of Cd in the residual fraction of these materials were about 50 times higher than typical for unpolluted soils. Using a different sequential extraction scheme, Chang et al (1984b) found that in soils treated with sewage sludge, every extracted fraction exhibited increases in Cd (as compared to nontreated soils), with the most

TABLE 4.5. Forms of cadmium (in ppm) in three soils and a sediment that had undergone massive additions of heavy metals.[a]

Form[b]	Samples[c]			
	Soil A	Soil B	Soil C	Sediment
Exchangeable	6.56 ± 0.42[d]	14.0 ± 0.2	6.18 ± 0.21	16.8 ± 0.4
Carbonate	2.92 ± 0.20	14.7 ± 0.1	1.19 ± 0.10	17.9 ± 0.7
Fe-Mn oxide	5.70 ± 0.37	9.16 ± 0.23	2.80 ± 0.08	10.8 ± 0.8
Organic	0.37 ± NS[e]	0.73 ± 0.07	0.39 ± 0.06	0.25 ± 0.01
Residual	3.85 ± 0.20	4.50 ± 0.02	2.93 ± 0.10	3.64 ± 0.04

[a]*Source:* Hickey and Kittrick, 1984, with permission of the authors and the Am Soc of Agronomy, Crop Sci Soc of America, Soil Sci Soc of America.
[b]The sequential extractants used were 1 M $MgCl_2$ (pH 7), 1 M NaOAc (pH 5), 0.04 M NH_2OH · HCl in 25% HOAc, (0.02 M HNO_3 + 30% H_2O_2 + 3.2 M NH_4OAc in 20% HNO_3), and finally (conc HF + conc $HClO_4$) for the following fractions: exchangeable, carbonate, Fe-Mn oxide, organic, and residual, respectively.
[c]Soil A was a surface soil taken within 1 km of a Cu smelter; soils B and C were surface soils from sewage sludge disposal sites; and the sediment was a contaminated sample.
[d]Values are mean ± SD (N = 4).
[e]Not significant, but not zero.

significant increases occurring in the carbonate and organic fractions. The two southern California soils were treated annually with a composted sludge and two liquid sludges for seven consecutive years and were annually cropped to barley.

In river sediments, (Khalid et al, 1978) found the following distribution of added Cd:

at pH 5.0—water-soluble > exchangeable >> DTPA-extractable ≈ reducible ≈ residual organic-bound, and

at pH 6.5—exchangeable > DTPA-extractable > residual organic-bound ≈ water-soluble > residual

In another river sediment experiment, Khalid et al (1981) found that in an alkaline (pH 8) sediment, essentially all the Cd extracted was associated with the DTPA-extractable, residual organic-bound, and reducible (oxides of Fe and Mn) fractions. Under acidic (pH 5) and oxidized conditions, Cd was primarily in the water-soluble and exchangeable forms.

Complexation of Cd with soluble inorganic and organic ligands was not a significant factor in keeping Cd^{2+} ions in solution. Essentially all the water-soluble Cd was present in a free ionic form. The lack of complexation with soluble organic material was attributed to low concentration of soluble humic material in relation to concentration of available Cd (Khalid et al, 1978).

With the advent of advanced computer technology, metal species in soil solutions can be calculated via computer modeling. Dr. Sposito and his group at the University of California, Riverside have developed the now widely used GEOCHEM computer program that calculates the chemical equilibrium in a soil solution (Sposito and Mattigod, 1979). In a

TABLE 4.6. Distribution of cadmium species in saturation extracts of two soils treated with sewage sludge enriched with variable amounts of cadmium.[a]

Soil	Cd application rate, μg Cd/g	pH_e[b]	I,[c] M	Cd_T[d] μM	Distribution, % Cd_T							
					Free	CO_3	SO_4	Cl	PO_4	NO_3	OH	Fulvate
Redding fine sandy loam; pH 5.1	0.1	5.1	0.0186	0.0361	71.6	0.1	16.0	1.6	0.2	0.3	0.0	10.2
	10	4.9	0.0186	1.43	71.0	0.0	18.4	1.4	0.2	0.1	0.0	8.7
	20	5.1	0.0210	3.30	72.8	0.1	14.9	1.4	0.2	0.4	0.0	10.0
	40	5.3	0.0241	8.04	67.9	0.1	13.9	2.9	0.2	0.5	0.0	14.4
	80	5.1	0.0352	20.1	70.7	0.0	13.9	6.4	0.2	1.0	0.0	7.7
	160	4.8	0.0700	87.5	67.2	0.0	11.0	15.6	0.1	2.0	0.0	4.0
Holtville clay; pH 7.6	0.1	7.6	0.0265	0.071	67.7	7.1	21.2	0.8	0.0	0.0	0.2	2.9
	10	7.6	0.0279	39.3	63.8	6.0	21.2	1.1	0.0	0.0	0.1	7.4
	20	7.5	0.0313	60.3	65.8	5.2	21.4	0.7	0.0	0.1	0.1	6.5
	40	7.6	0.0325	69.6	65.8	6.6	19.4	1.7	0.0	0.3	0.1	6.0
	80	7.4	0.0483	75.9	67.4	3.5	15.1	8.6	0.0	1.2	0.0	3.9
	160	7.2	0.0737	80.4	65.0	1.9	12.4	16.2	0.0	2.0	0.0	2.5

[a]*Source:* Mahler et al, 1980, with permission of the authors and the Am Soc of Agronomy, Crop Sci Soc of America, Soil Sci Soc of America.
[b]pH_e = pH of the extract.
[c]I = ionic strength.
[d]Cd_T = total concentration in the saturation extract.

typical soil solution, there may be 10 to 20 different metal cations (including trace elements) that can react with as many different inorganic and organic ligands to form 300 to 400 soluble complexes and up to 80 solid phases. This modern tool can have wide application in soil and plant nutrition, geochemistry, environmental chemistry, and other related fields (Sposito and Bingham, 1981). An example of the utility of the GEOCHEM program is presented in Table 4.6 in characterizing and quantifying the various Cd species in extracts of soils treated with Cd-enriched sewage sludge. It was found in this experiment that Cd uptake by the test plants (sweet corn, tomato, and Swiss chard) correlated equally well with the total Cd concentration in the saturation extract, free Cd^{2+} ions, or Cd^{2+} activity (Mahler et al, 1980).

In soils where very high Cl^- concentrations occur in solution, Garcia-Miragaya and Page (1976) predicted that Cd would be primarily complexed with chloride (e.g., $CdCl_2$, $CdCl_3^-$ and $CdCl_4^{2-}$) rather than as a free cation (Cd^{2+}).

4.5 Sorption, Fixation, and Complexation of Cadmium in Soils

Adsorption was found to be the operating mechanism of the reaction of low Cd concentrations with soils (Garcia-Miragaya and Page, 1977; Santillan-Medrano and Jurinak, 1975; Street et al, 1977). In most sorption studies, the results fitted either the Langmuir (Cavallaro and McBride, 1978; John, 1972; Levi-Minzi et al, 1976) or the Freundlich isotherm (Street et al, 1977; Garcia-Miragaya and Page, 1978; Jarvis and Jones, 1980). Kuo and McNeal (1984) found that sorption of Cd by hydrous iron oxides conformed to the Langmuir isotherm.

In addition to adsorption, precipitation can play a key role in controlling Cd levels in soil, as found by Santillan-Medrano and Jurinak (1975) and Street et al (1977). The former used equilibrium batch studies to obtain solubility data for Cd in three soils having different chemical properties. Cadmium solubility in soils decreased as pH increased. The lowest values were obtained in the calcareous soils (Nibley clay loam, pH 8.4). At low Cd concentrations, the equilibrium solution was undersaturated with regard to both $Cd_3(PO_4)_2$ and $CdCO_3$. However, at high Cd concentrations, the precipitation of $Cd_3(PO_4)_2$ and/or $CdCO_3$ controlled Cd solubility. Street et al (1977) found evidence of Cd adsorption onto soil surfaces and possible precipitation of Cd minerals. Based on current thermodynamic data, the solid phases $CdCO_3$ (Ksp = 12.07) and $Cd_3(PO_4)_2$ (Ksp = 32.61) most likely limited Cd^{2+} activities in soils. They observed $CdCO_3$ precipitation in sandy soils having low cation exchange capacities (<6 meq/100 g), low organic matter (< 1%), and alkaline pH (pH > 7.0).

Using two Danish soils (loamy sand and sandy loam), Christensen (1984a) demonstrated that soil sorption of Cd is a fast process. He found that >95% of the sorption took place within the first 10 minutes and that equilibrium was reached in an hour. He further found that the sorption capacity of the soil increased approximately three times per unit increase in pH within the pH range of 4.0 to 7.7 (Fig. 4.4).

In testing Cd adsorption characteristics in various horizons of two New York soils, Cavallaro and McBride (1978) obtained additional evidence that precipitation occurs at higher Cd^{2+} activities and that ion exchange predominates at the lower Cd^{2+} activities. They inferred $CdCO_3$ precipitation in a calcareous subsoil when they obtained a close fit of their pH-pCd data on the $CdCO_3$ solubility line. Based on the knowledge that carbonate surfaces chemisorb micronutrients, such as Zn^{2+} and Mn^{2+}, McBride (1980) elaborated on chemisorption of Cd^{2+} on calcite surfaces. He found that initial chemisorption of Cd^{2+} on $CaCO_3$ was very rapid, while $CdCO_3$ precipitation at higher Cd concentrations was slow. He concluded that chemisorption may regulate Cd^{2+} activity in some calcareous soils, producing solubilities much lower than predicted by the solubility product of $CdCO_3$.

Several factors influence the degree by which Cd is adsorbed on soil surfaces. In addition to pH, ionic strength and exchangeable cations also influence Cd adsorption. In studying Cd sorption by montmorillonite from solutions in the 15 to 120 ppb range, Garcia-Miragaya and Page (1976) found that the ionic strength of various salt solutions ($NaClO_4$, $NaCl$, and Na_2SO_4) affected the amount of Cd sorbed on the clay surfaces. The adsorption of Cd(II) from Cl^- solutions diminished with increasing salt con-

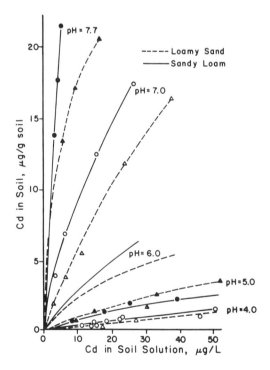

FIGURE 4.4. Cadmium sorption isotherms for two soils as influenced by soil texture and pH. *Source:* Christensen, 1984a, with permission of the author and D. Reidel Publ Co, Dordrecht, the Netherlands.

centration concomitant with the formation of uncharged ($CdCl_2^0$) and negatively charged complexes ($CdCl_3^-$, $CdCl_4^{2-}$, etc.) of Cd with Cl^- ligands. They showed that the Cl species of Cd are less strongly adsorbed than Cd^{2+}. Organic ligands can also influence the sorption of Cd by soils. Elliott and Denneny (1982) noted that, among several organic compounds tested, the ability to reduce Cd retention by soil followed the trend: EDTA > NTA (nitrilotriacetate) > oxalate ≈ acetate.

Chubin and Street (1981) demonstrated that addition of EDTA to dilute suspensions of $Al(OH)_3$, $Fe(OH)_3$, kaolinite, and montmorillonite inhibited Cd sorption in the pH range of 4 to 10. This inhibition was in the order kaolinite > montmorillonite > $Fe(OH)_3$ > $Al(OH)_3$. With citrate, there was little influence on Cd sorption except for the $Al(OH)_3$ system, in which sorption was reduced by about 25% in acidic solutions.

The presence of competing exchangeable cations in the soil-water system can also greatly affect Cd sorption by soils (Christensen, 1984a, b; Cavallaro and McBride, 1978; Garcia-Miragaya and Page, 1977; Bittell and Miller, 1974; Lagerwerff and Brower, 1972). Divalent ions, such as Ca^{2+}, greatly reduce the efficiency of Cd sorption by permanent charge clays (Bittell and Miller, 1974). The selectivity coefficient for Cd^{2+}-Ca^{2+} exchange on mineralogically pure clays is near unity—i.e., Cd^{2+} would compete on equal basis with Ca^{2+} for clay adsorption sites (Bittell and Miller, 1974). However, with soils Milberg et al (1978) found that Cd^{2+} was preferentially adsorbed over Ca^{2+} in Cecil, Winsum, and Yolo soils, with selectivity greatest in Yolo silt loam (pH 6.8, $CaCO_3$, 47% montmorillonite) and least in Cecil sandy loam (pH 4.3, 45% kaolinite). Apparently, the difference between the findings of Bittell and Miller (1974) and Milberg et al (1978) lies in the fact that soil colloids carry a number of specific exchange sites with higher bonding energy for Cd than are present on pure clays. The Al^{3+} ion demonstrates an even greater ability to exclude Cd^{2+} from clay surfaces (Garcia-Miragaya and Page, 1977). These workers found that, with montmorillonite, the following exchangeable cations affected Cd sorption, decreasing in the order Na > K > Ca > Al. Similarly, Lagerwerff and Brower (1972) observed that the adsorption of Cd was greater in the presence of Ca^{2+} than of Al^{3+} and decreased with increasing concentration of $AlCl_3$ or $CaCl_2$.

Christensen (1984a, b) found that Ca^{2+}, Zn^{2+}, or H^+ present in the soil solution can compete effectively with Cd for sorption sites in soil or can significantly desorb Cd from the soil.

5. Cadmium in Plants

5.1 Cadmium Uptake and Accumulation by Plants

Cadmium is a nonessential element in plant nutrition. Under normal conditions, plants take up small quantities of Cd from soil. In an extensive survey, Huffman and Hodgson (1973) collected 153 samples of wheat and perennial grasses in 19 states east of the Rocky Mountains. The sampling sites represented major agricultural soils with no known input and sources of Cd pollution except from fertilizers. They found that the levels of Cd were generally below 0.30 ppm (wheat = 0.20 ppm, n = 33; grasses = 0.17 ppm, n = 120) and that there was no distinct regional pattern of Cd concentration.

The levels of Cd in the tops of ten plant species grown without added Cd in a neutral soil ranged from 0.21 ppm in timothy to 0.71 ppm in soybean leaves (MacLean, 1976). However, in areas of suspected Cd contamination, as in rice fields in Japan where soil concentrations of 1 to 50 ppm have been reported, much higher Cd contents in plant tissues have been reported (Yamagata and Shigematsu, 1970).

If present in substrates in higher than background concentrations, Cd is readily taken up by roots and distributed throughout the plant. The amount of uptake is tempered by soil factors such as pH, CEC, redox potential, fertilization, organic matter, other metals, and other factors. Plant factors, such as species and cultivars, influence the total uptake. In addition to these factors, the level of Cd in the growth medium, as shown in Figure 4.5, influences uptake. Scientists are in agreement that in general, over a certain range, there is a positive, almost linear correlation between the levels of added Cd concentration in the substrate and the resulting Cd concentration in the plant tissues.

In addition, the Cd uptake-substrate relationship suggests that yields may be predicted on a basis of plant tissue analysis, as demonstrated by the greenhouse work of Dr. Bingham and his colleagues at the University

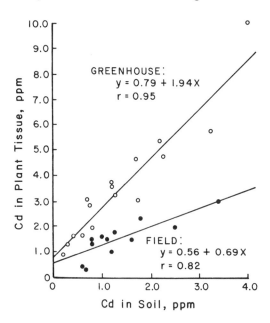

FIGURE 4.5. Relationships of cadmium concentration in soils from applications of sewage sludge to cadmium concentration in Swiss chard. *Source:* Page and Chang, 1978, with permission of the authors.

of California, Riverside (Fig. 4.6). Using this technique, one can predict
the yield of a given greenhouse-grown crop based on either the plant tissue
content of Cd or soil Cd content. This greenhouse technique should be
adaptable to field conditions, although the severity in yield decrements
may not be comparable.

Numerous studies add credentials to plants' tendency to accumulate
Cd. Radish tops can accumulate 5 ppm of Cd when grown on soil containing 0.6 ppm Cd (Lagerwerff, 1971). Leafy plants such as lettuce, spinach, and turnip greens accumulated 175 to 354 ppm Cd when grown on

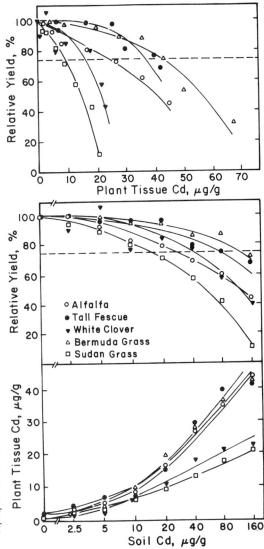

FIGURE 4.6. Predictability of plant yield based on soil cad-mium concentration (below) or plant tissue cadmium concen-tration (above). Broken hori-zontal lines indicate 25% yield decrement. *Source:* Bingham et al, 1976a, with permission of the authors and the Am Soc of Agronomy, Crop Sci Soc of America, Soil Sci Soc of America.

soils pretreated with sewage sludge enriched with Cd up to 640 ppm (Bingham et al, 1975). Fruits and seeds of other plants tested usually accumulated no more than 10 to 15 ppm. Uptake by several crop species (lettuce, spinach, broccoli, cauliflower, oats) grown on soil pretreated with Cd in variable amounts (from 40 to 200 ppm) ranged from 100 to 600 ppm in leaf tissues. In solution culture, Cd accumulation by plants can even be greater. Page et al (1972) found that the leaves of tomato, cabbage, pepper, barley, corn, lettuce, red beet, and turnips cultured on complete nutrient solutions with Cd concentrations of 0.1 to 0.5 ppm, contained up to 200 to 300 ppm. Turner (1973) noted excessive uptake, up to 158 ppm in tomato leaves, of Cd in assorted vegetable crops grown in solution culture containing as low as 0.1 ppm Cd.

Cadmium is readily translocated throughout the plant following its uptake by roots. Distribution between roots and tops differs with plant species, rooting media, and time of treatment. In rice, about 99% of the total Cd taken up by the plants was found in the shoots in a wide range of redox potential and pH (Reddy and Patrick, 1977). Some environmental factors, such as Cd concentration in the medium, ambient temperature, and light intensity can affect the distribution of the metal between the tops and roots of the rice plant (Chino and Baba, 1981). In general, roots contain at least twice the Cd concentration found in tops (Koeppe, 1977). Among aerial tissues of corn plants grown on the field, stems and leaves generally concentrated more Cd than did the husks, cobs, kernels, silks, or tassels (Peel et al, 1978). Koeppe believes that there are no cellular or subcellular Cd deposits in higher plants. Girling and Peterson (1981), using electrophoretic technique, demonstrated that Cd in the water-soluble extracts of the shoots and roots of various crop species, because of its similarity to Zn, occurred as the free divalent ion, Cd^{2+}. This indicates that Cd does not occur in anionic or neutral forms in the extracts.

In spite of the availability of Cd in soil to plants, only a small fraction of the Cd pool is recovered by plants. One reason is that Cd is phytotoxic up to certain levels, which can drastically reduce plant yield. Under field conditions, the recovery of Cd by crops from soils treated with sewage sludge was usually <1% of the total added (Kelling et al, 1977; Keeney and Walsh, 1975; Stenström and Lönsjö, 1974). However, under controlled experiments, total uptake from soil can be >1% of the total added (Williams and David, 1973; Andersson and Nilsson, 1974), with up to 6% recovered by tomatoes from added superphosphate. Williams (1977) reported the recovery of Cd added, at 0.2 ppm rate, to a podzolic soil using pot culture, to range from 0.9% for beans to 13.6% for tobacco. This is expected since the substrate in small containers can be fully exploited by the roots.

In a *Filipendula ulmaria* meadow ecosystem, Balsberg (1982) found that, of the added Cd to the soil, <10% of the total Cd in the ecosystem was found in the plant biomass. Root Cd concentration exceeded that in the soil and was several times the concentration in aboveground organs. The Cd concentrations in various plant parts decreased in the order: new roots > old roots > rhizomes > stem leaf-stalks > stems ≈ stem leaf-blades > reproductive organs.

5.2 Cadmium-Zinc Interactions in Plants

The association of Cd and Zn in geological deposits and the chemical similarity of the two elements carry over into biological systems. Cadmium has no known biological functions, but Zn is an essential element. Cadmium and Zn appear to compete for certain organic ligands *in vivo*, accounting in part for the toxic effects of Cd and the ameliorative effects of Zn on Cd toxicity (NAS, 1974). Cadmium supposedly competes with Zn in forming protein complexes, and perhaps a negative association should be expected (Vallee and Ulmer, 1972). For this reason, the ratio between these two elements within plant tissue is thought to be biologically important.

According to Chaney (1974), one way to ensure that the Cd contents of food crops grown in sludge-treated soil would not be hazardous in the food chain is to reduce the Cd content of sludge to 1.0% or less of the Zn content. The rationale behind this recommendation is that Zn toxicity to the crop would occur first, before the Cd content of the crop would reach levels that could constitute a health hazard. With lower Cd/Zn ratios in sludge, Chaney (1973) obtained marked reductions in this ratio in grains, fruit, and root crops relative to the ratio in the sludge. In most soils treated with sludge, the Cd/Zn ratio of tissues of plants grown thereon usually is 1% or less (Cunningham et al, 1975a, b; Kelling et al, 1977; Zwarich and Mills, 1979; Dowdy and Larson, 1975) although in some instances, higher percentage of about 3% for edible tissues was obtained (Schauer et al, 1980). Under natural agricultural conditions, crops would be expected to have much lower Cd/Zn ratios, based on the extensive survey of Huffman and Hodgson (1973) involving 153 immature wheat and grass samples expanding over 19 states. They obtained an average Cd/Zn ratio of 0.8%, with the ratio for wheat slightly higher (1.1%) due to both higher Cd and lower Zn levels in this crop. It appears that the ratio tends to be lower in the edible portions of crops, as indicated by the data of Dowdy and Larson (1975) for radish, potato, peas, and corn grown on sludge-treated soils, in which the ratio did not exceed that of the applied sludge (0.7%). Several investigators reported the possibility of the exclusion of Cd relative to Zn during the filling of the grain (Hinesly et al, 1972; Chaney, 1973; Zwarich and Mills, 1979), causing a drop in the Cd/Zn ratio. The net result is that the Cd/Zn ratio in grain is a tenth to a half the value found in leaves and stems.

The interaction between Cd and Zn is either antagonistic or synergistic. Bingham et al (1975) showed that additions of Cd-treated sewage sludge to a calcareous soil (pH 7.4) decreased the concentrations of Zn in tops of many but not all crops studied. Similarly, a depression of Cd concentrations of soybean shoots, relative to control, began to occur at the 100-ppm Zn level in soil and were further reduced by up to 400-ppm Zn level, especially at low temperatures (Haghiri, 1974). Lagerwerff and Biersdorf (1971) also observed that increasing concentrations of Zn suppresed Cd uptake at low Cd concentrations (2 and 20 ppb Cd) in the culture solution. However, at the 100 ppb level of Cd, increasing amounts of Zn increased Cd uptake. Japanese scientists reported that addition of Zn to the soil increases Cd uptake by the rice plants (Chino, 1981b). One possible explanation for this synergistic effect is that Zn dissociates Cd fixed to the binding sites in the soil as a result of competition for the sites and increases Cd in the soil solution. Likewise, Turner (1973) found that increased Cd concentrations produced increased concentration and total uptake of Zn in plant tops. He inferred that: (1) Cd caused root damage that subsequently enhanced Zn uptake, (2) Cd stimulated the uptake of Zn; and (3) redistribution of Zn occurred between roots and tops. Cunningham (1977) found that low levels of Cd in culture solution increased Zn concentration in foliage of soybeans but decreased it at high Cd treatments. In Dutch soils, Gerritse et al (1983) found that increasing Zn levels appeared to increase Cd uptake by crops at high soil solution concentrations of Cd and to decrease uptake at low solution concentrations.

5.3. Interactions of Cadmium with Other Ions

In addition to Cd-Zn interaction, Cd is known to interact in plants with other ions. Cadmium depressed the Mn uptake by plants (Wallace et al, 1977 a, b; Iwai et al, 1975; Patel et al, 1976; Root et al, 1975). Cadmium causes a type of Fe deficiency in plants (Root et al, 1975) and could also depress other ions (Ca, Mg, and N) uptake (Cunningham, 1977; Iwai et al, 1975). On the other hand, Se (Francis and Rush, 1983) and Ca (Tyler and McBride, 1982) have been shown to depress Cd uptake.

Often, Cd and Pb are closely scrutinized together because they arise from certain pollution sources, such as mining and smelting operations, sewage sludge, combustion of energy materials, and others. The two elements have been shown to act synergistically on plant growth. Cadmium content of rye grass and red fescue treated with Pb + Cd was greater than that of plants treated with Cd alone, with the effect being greater for fescue than for rye (Carlson and Rolfe, 1979). A synergistic response to Pb and Cd was found by Carlson and Bazzaz (1977) for root and shoot growth of American sycamore, and by Hassett et al (1976) for root growth of maize. Carlson and Bazzaz (1977) further observed that, while treatment with Cd or Pb alone caused a reduction in photosynthesis and transpiration of sycamore seedlings, the addition of Cd to Pb-stressed plants did not reduce rates of photosynthesis and transpiration below those observed for plants treated with Pb alone. Miller et al (1977) noted a tendency for soil Pb to increase both the plant Cd concentration and the total Cd uptake of the corn shoots. Both Cd (at 2.5 and 5 ppm levels in soil) and Pb (at 125 and 250 ppm levels in soil) reduced corn shoot yield, with a concomitant positive interaction of the two metals on growth. The enhancement of Cd uptake by soil Pb was attributed to Pb's ability to more effectively compete for exchange sites on colloidal surfaces than Cd, releasing Cd into the soil solution.

5.4 Toxicity of Cadmium in Plants

Cadmium is not only available to plants from soil and other substrates but is also known to be toxic to them at much lower concentrations than other metals like Zn, Pb, Cu, etc. Phytotoxicity has been observed to be dependent upon plant species as well as concentration of Cd in the substrate. For example, using nutrient solution culture, Bingham and Page (1975) found the following Cd concentrations (ppm) in solution associated with 50% yield decrements: beet, bean, and turnip, 0.2; corn and lettuce, 1.0; tomato and barley, 5; and cabbage, 9. In contrast, there is a wider range of Cd phytotoxicity in plants grown on soils (Table 4.7). For 25% yield decrement, the total soil Cd level ranged from 4 ppm for spinach to >640 ppm for paddy rice. However, these values correspond to 75 and 3 ppm in the diagnostic leaf tissues of spinach and rice, respectively. Haghiri (1973) noted that toxicity in soybean and wheat began to occur at soil Cd levels as low as 2.5 ppm (soil pH = 6.7). Roots of subterranean clover were depressed by the addition of 1 ppm Cd to soil, and at 5 ppm or higher, toxicity symptoms began to appear (Williams and David, 1977). Bingham et al (1976a) also found that there was close agreement between DTPA-extractable Cd, as well as Cd concentration in the soil saturation extract, and Cd uptake and crop yield data. Because of the highly toxic nature of Cd to plants and animals, it is necessary to obtain information on Cd contents of edible tissues of field and vegetable crops. The following Cd concentrations in edible tissues of greenhouse soil-grown plants were associated with 50% yield decrement (Bingham and Page, 1975): field bean, paddy rice, upland rice, sweet corn—≤3 ppm; zucchini squash, wheat, tomato, cabbage, beet, soybean—10–20 ppm; turnip, radish, carrot—20–30 ppm; romaine lettuce, Swiss chard, curlycress, spinach—>80 ppm. These levels, although much greater than what can be expected of field-grown crops, are well above the normal levels of Cd found in foodstuff, which is around 0.05 ppm (Friberg et al, 1971). In young spring barley, the critical level of Cd in plant tissues that affected their growth was about 15 ppm (Davis et al, 1978).

In solution culture, Cd concentration as low as 0.1 ppm in solution produced Cd accumulations varying between 9 ppm in bean leaves and 90 ppm in corn leaves (Page et al, 1972). At solution concentrations of 1 ppm

TABLE 4.7. Concentration of cadmium (ppm) in greenhouse soil-grown plants associated with a 25% yield depression.[a]

| Crop species | Yield component | Soil Cd producing a 25% yield decrease | | Tissue Cd level at 25% yield decrease | | Edible part Cd at 5 ppm soil Cd |
		Cd added, ppm	Cd DTPA extracted, ppm	Diagnostic leaf, ppm	Edible, ppm	ppm
Spinach	Shoot	4	2.4	75	75	91
Soybean	Dry bean	5	3.0	7	7	6.8
Curlycress	Shoot	8	4.8	70	80	55
Lettuce	Head	13	7.8	48	70	27
Corn	Kernel	18	10.8	35	2	0.5
Carrot	Tuber	20	12.0	32	19	8.2
Turnip	Tuber	28	16.8	121	15	4.9
Field bean	Dry bean	40	24.0	15	1.7	0.4
Wheat	Grain	50	30.0	33	11.5	2.8
Radish	Tuber	96	57.6	75	21	2.5
Tomato	Ripe fruit	160	96.0	125	7	1.6
Zucchini squash	Fruit	160	96.0	68	10	0.4
Cabbage	Head	170	102.0	160	11	1.3
Rice (paddy)	Grain	>640	>384.0	3	2	0.2
Sudangrass	Shoot	15	11	9	—	—
Alfalfa	Shoot	30	22	24	—	—
White clover	Shoot	40	29	17	—	—
Tall fescue	Shoot	95	71	37	—	—
Bermudagrass	Shoot	145	107	43	—	—

[a]Sources: Bingham et al, 1975, 1976a; Bingham, 1979, with permission of the authors and the Am Soc of Agronomy, Crop Sci Soc of America, Soil Sci Soc of America.

Cd, the range was 35 ppm in bean leaves to 469 ppm in turnip leaves. Reduced yield and chlorosis occurred in *Brassica chinensis* also at 1 ppm Cd level in the nutrient solution (Wong et al, 1984). In ryegrass cultured in nutrient solution spiked with Cd up to 82 ppm, the tops accumulated 100 to 500 ppm Cd, resulting in chlorotic plants and subsequent death (Dijkshoorn et al, 1974). In corn grown in nutrient solution, Iwai et al (1975) set 20 ppm as the critical level in plant tissues above which the plants would suffer from Cd toxicity.

In rice, the critical Cd content of plant tissues above which growth is retarded has been reported at about 10 ppm (Iimura and Ito, 1971). This level was later modified by Chino (1981a) to be 5 to 10 ppm for rice tops and 100 to 600 ppm for rice roots. Takijima et al (1973b) associate a concentration of 15.5 ppm Cd extracted with 0.1 N HCl from soil with excessive levels of Cd in the rice grain.

Cadmium toxicity symptoms in crops, in general, resemble Fe chlorosis (Haghiri, 1973; Mahler et al, 1978). In addition to chlorosis, plants may exhibit necrosis, wilting, red-orange coloration of leaves, and general reduction in growth (Bingham and Page, 1975). In rice, the number of tillers is usually reduced, accompanied by severely depressed root growth (Chino, 1981a). In radish, Cd toxicity was associated with the reduction of Zn levels in plant tissues to near the deficiency threshold value (Khan and Frankland, 1983).

Tree species are also sensitive to soil Cd, but apparently at much higher concentrations than levels phytotoxic to agronomic crops. In testing several tree species, Kelly et al (1979) obtained increased Cd content in roots and shoots in response to soil (pH 4.8) Cd levels, but the growth was drastically retarded only by the 100 ppm treatment (Fig. 4.7). Using sand

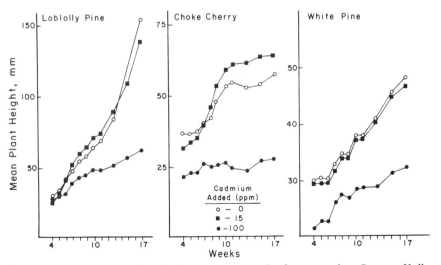

FIGURE 4.7. Effect of soil cadmium on the growth of tree species. *Source:* Kelly et al, 1979, with permission of the authors and the Am Soc of Agronomy, Crop Sci Soc of America, Soil Sci Soc of America.

culture, Mitchell and Fretz (1977) induced excessive Cd accumulation in white pine, red maple, and Norway spruce, resulting in reduced root initiation, poor development of root laterals, chlorosis, dwarfism, early leaf drop, wilting, and necrosis of current season's growth. Lamoreaux and Chaney (1977) partially attributed the severe reduction in the growth of silver maple with $CdCl_2$ application to sand culture to significantly reduced relative conductivity of stem, which in turn was caused by deterioration of xylem tissues in number and size.

In summary, reduced plant growth due to Cd addition to growth medium can be attributed to the following: (1) reduced photosynthetic rate, (2) internal water deficit in the vascular system caused by reduced conductivity of stems and poor root system development, (3) ion interactions in plants, and (4) possible inhibition of nutrient (N and P) mineralization in soil (Tyler, 1975).

6. Cadmium in Natural Ecosystems

Although the fate and behavior of Cd in agricultural ecosystems—primarily Cd from sewage sludge and phosphatic fertilizers—have been studied, very limited information is available concerning the fate of Cd in forest ecosystems. In his review, Shacklette (1972) indicated that Cd concentrations in deciduous foliage collected from polluted areas were in the range of 4 to 17 ppm, compared with Cd levels of 0.1 to 2.4 ppm in similar vegetation from normal environments. Cadmium concentrations in coniferous foliage were 0.05 to 1.0 ppm in polluted areas and 0.1 to 0.9 ppm in normal areas. Foliage of sugar maple growing in remote sections of northern New Hampshire and Vermont had Cd concentrations in the range of 0 to 5 ppm (Smith, 1973). In spruce forest sites in central Sweden that were subjected to industrial pollution, Cd levels were 0.4 to 1.0 ppm in spruce needles, compared to 0.2 to 0.4 ppm in spruce foliage from non-polluted areas (Table 4.8). Essentially all the components of this forest in Sweden showed high enrichment ratios, particularly in the litter component. Cadmium levels for pine trees in a mixed deciduous forest at the Walker Branch Watershed in eastern Tennessee (Van Hook et al, 1977) are generally lower than those in central Sweden. This eastern Tennessee forest has been considered as a relatively unpolluted ecosystem in spite of possible Cd input from three coal-fired electric generating stations with combined annual Cd aerial discharges of about 0.09 tonnes. The distribution of Cd in the vegetative and soil components of this forest is shown in Figure 4.8. In normal forest ecosystems, the concentration of Cd in vegetation followed the pattern: roots > foliage > branch > bole. However, this may not be the case in polluted ecosystems, as indicated by the data of Tyler (1972) in Table 4.8, which show branches having relatively high Cd levels. The major sites of Cd accumulation in a forest ecosystem are in the litter component, partly because of interception of the element

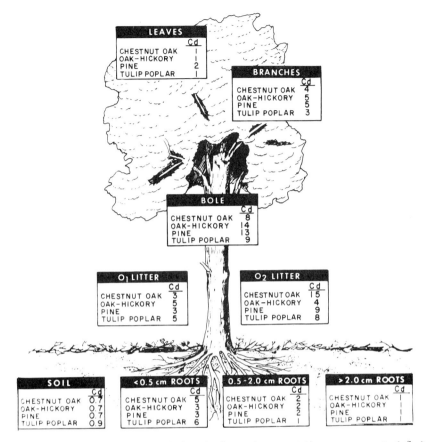

FIGURE 4.8. Distribution of cadmium in the major vegetative components (g/ha) and soils (kg/ha) in an eastern Tennessee mixed deciduous forest. *Source:* Van Hook et al, 1977, with permission of the authors.

by the litter from throughfall and by the immobilization of the element by the organic matter of the litter. However, the soil should still be expected as the major sink of metals (Van Hook et al, 1977; Heinrichs and Mayer, 1977).

In another relatively unpolluted forest ecosystem in the Bärhalde Watershed, southern Black Forest, Federal Republic of Germany, the retention of Cd is highest in the top soil, where it increases with increasing pH and humus content (Stahr et al, 1980). The watershed has Cd input of 400 μg/m²/year. Recent increases in atmospheric input of Cd resulted in an enrichment and rise of turnover of Cd, Cu, and Pb throughout this watershed.

Other studies also have indicated substantial enrichment of ecosystem components by atmospheric deposition from industrial activities. Forest components had elevated Cd levels due to smelting operations (Martin

TABLE 4.8. Concentration of cadmium in components of
forests affected by anthropogenic input of cadmium.

	Spruce forest Central Sweden[a]		Mixed deciduous forest Eastern Tennessee, USA[b]	
	ppm Cd	ER[c]		ppm Cd
Roots			Roots	
<5 mm diam	2.7	6.8	<5 mm diam	0.4
≥5 mm diam	1.5	7.5	5–20 mm diam	0.3
Bole	<0.1		>20 mm diam	0.2
Bark	2.5	13	Bole	0.21
Branch			Branch	0.30
1st year	5.4	14	Needles	0.30
2nd year	4.6	12	Litter	
3rd year	4.2	10	O1	0.3
4th year	3.3	7.5	O2	0.6
5th–7th year	2.7	6.0		
Needles				
1st year	0.6	3.0		
2nd year	0.4	1.5		
3rd year	0.5	1.4		
4th year	0.5	1.4		
5th–7th year	1.0	2.4		
Litter				
O1	24	63		
O2	44	40		
Moss	30	33		
Epiphytic lichens	12	30		

[a]Tyler (1972).
[b]Located at Walker Branch Watershed (Van Hook et al, 1977). Only pine
data presented.
[c]Enrichment ratio = ratio between the metal concentration in this site and in
a similar site with no local deposition.

and Coughtrey, 1975; Wixson et al, 1977). In a mixed oak forest ecosystem
close to a primary Pb-Zn smelter in Avonmouth, United Kingdom, the
vegetation had Cd concentrations from 2 to 18 ppm in polluted areas versus
0.8 to 2.2 in normal areas. Similar trends were observed for litter, soil,
and soil animals.

Comparing undisturbed forested ecosystems in urban and rural northwestern Indiana, Parker et al (1978) found much higher levels of Cd in
soils and vegetation in the urban site compared to a similar system in a
rural setting 67 km away. The urban area, which has been exposed for
about 100 years to contamination from industrial and other sources, has
about 10 ppm Cd in the top 2.5 cm of the soil versus 0.2 ppm in the rural
site. These higher soil levels were reflected in higher concentrations in
most plant species. Contrary to these findings, Smith (1973) found that
foliage of woody plants in New Haven, Connecticut (population, approximately 150,000) had Cd contents within the "normal" range of values
(0–5 ppm) found for foliage of red maples collected from remote areas in
Maine and New Hampshire.

Forest ecosystems are also potential recipients of municipal sewage sludges and waste waters because of their fertilizer values. Furthermore, contamination of the food chain is not of concern since the vegetation is not harvested for human consumption, except for the potential contamination of wildlife feeding on vegetation and contamination of ground water recharge. Data by Sidle and Sopper (1976) and Sidle and Kardos (1977) indicate that Cd levels were not significantly elevated in vegetation and soils of sites receiving either sludge or waste water effluent at loadings of <1 kg Cd/ha. However, at higher loading rates of Cd, the amount of Cd moving past the root zone could be of some concern when the Cd in the percolates exceeds the 10 ppb level set by the US Public Health Service.

7. Factors Affecting the Mobility and Availability of Cadmium

7.1 pH

Soil pH appears to be the most important single soil property that determines Cd availability to plants (CAST, 1980; Page et al, 1979). Consequently, it is recommended that soil pH be maintained at pH 6.5 or greater in land receiving solid wastes containing Cd. In general, Cd uptake by plants almost always increases with decreasing pH. Cadmium uptake by plants is usually higher in acidic than in alkaline or calcareous soils. Mahler et al (1978) obtained higher Cd concentrations in lettuce and Swiss chard grown on acid soils (pH 4.8–5.7) than on calcareous soils (pH 7.4–7.8). In wheat, liming reduced the grain Cd concentration by approximately 50% (Bingham et al, 1979). The liming potential of fly ash also significantly reduced Cd absorption by sudangrass grown on soils treated with composted sewage sludge having 31 ppm Cd (Adriano et al, 1982). The effect of liming in reducing Cd uptake by plants is demonstrated in Figure 4.9.

Andersson and Nilsson (1974) studied the effect of pH on Cd uptake by fodder rape and found that increase in soil pH caused by the addition of CaO decreased the Cd content of the rape. They attributed the pH effect on Cd uptake due partially to competition between Ca^{2+} and Cd^{2+} ions at the root surfaces. Lagerwerff (1971) studied Cd uptake by radish grown on Cd-contaminated soils and noted that increasing pH decreased the metal content of the plants. He suggested that this might be due to a decrease in solubility and/or mobility of Cd in the soil. McBride et al (1981) found that the "retention capacity" of northeastern US soils was strongly dependent on the exchangeable Ca^{2+} concentrations of the soils, i.e., the "retention capacity" increases with increasing levels of Ca^{2+} in the soil. In addition, they found that the best indicators of Cd availability to corn plants were the "retention capacity" and exchangeable base (mainly Ca^{2+}) content of the soil.

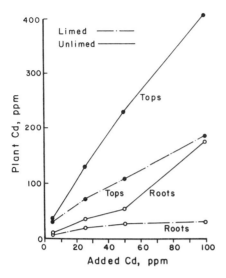

FIGURE 4.9. Cadmium concentration in radish tops and roots as influenced by liming the soil. *Source:* John, 1972, with permission from the Agricultural Institute of Canada, Ottawa.

Street et al (1978) found that increasing the pH of soil treated with $CdSO_4$ decreased the Cd concentrations of corn seedlings by approximately 67%. The reduction, however, was lowered to about 47% for a similar increase in pH treated with Cd-spiked sludge. Similarly, others (John, 1972; John et al, 1972b; Chaney, 1973; Miller et al, 1976, MacLean, 1976; Williams and David, 1976) observed negative effects of soil pH on Cd uptake by plants. However, there are exceptions. Liming soils that were treated with anaerobically digested sewage sludge to raise soil pH to 6.5 did not reduce Cd uptake by silage corn (Pepper et al, 1983). Thus, the current US-EPA requirement to lime soils to pH 6.5 (US-EPA, 1979) seems unjustified for sludge applied to these western Washington soils, if silage corn is to be grown.

Soil pH was also negatively correlated with Cd contents of rice (Bingham et al, 1980; Chino, 1981b; Takijima et al, 1973a; Reddy and Patrick, 1977). Total Cd uptake and shoot uptake increased with an increase in redox potential and a decrease in pH (Reddy and Patrick, 1977). They suggested that increased uptake by rice under low soil pH conditions may also be due to the increased solubility of solid phases of Cd such as carbonates, hydroxides, and phosphates, and increased concentration of Fe^{2+}, Mn^{2+}, Zn^{2+}, Cu^{2+}, and H^+, which may compete for exchange sites.

In addition to decreased solubility of Cd in soils associated with formation of $CdCO_3$ and $Cd_3(PO_4)_2$ with increasing pH (Santillan-Medrano and Jurinak, 1975; Street et al, 1977), raising the pH of soil solution can lead to the formation of hydrolysis products that have a different affinity for permanent charge and other exchange sites (Cavallaro and McBride, 1980). Hahne and Kroontje (1973) reported that the hydrolysis of Cd^{2+} becomes important above pH 8. Elevated pH can also change the nature of the exchange sites by hydrolyzing or precipitating Al^{3+} ions that occupy the exchange sites, thus creating more exchange sites. Street et al (1978) observed that water-soluble Cd decreased as much as 50% as pH was increased.

Thus, it is apparent that variations in plant uptake of Cd from soils of varying pH was due to changes in the solubility of Cd, physiological changes in ion uptake, and transport due to variation in soil pH. In the case of soils containing greater amounts of Ca^{2+}, the difference is considered due to ion competition between Ca^{2+} and Cd^{2+} ions at the root surface or interaction within the plant (Adriano et al, 1982; Iwai et al, 1975). In solution culture, however, the changes in the pH of the solution affects the membrane permeability of the metals and/or the competition between H^+ and Cd^{2+} ions.

7.2 Cation Exchange Capacity

Recommended limits on total cumulative added amounts of sludge-borne Cd are varied according to the soil CEC (US-EPA, 1979). The limits for the respective CEC ranges are as follows: 5 kg Cd/ha—for soils with CEC of <5 meq/100 g; 10 kg Cd/ha—for soils with CEC of 5 to 15 meq/100 g; and 20 kg Cd/ha—for soils with CEC of >15 meq/100 g. All limits pertain to soils maintained at a pH $\geq$ 6.5.

Results assessing the effect of soil CEC on the phytoavailability of Cd are variable. Haghiri (1974) found that Cd concentrations in the oat shoots decreased with increasing CEC of the soil. He increased the soil CEC by adding muck up to 7% by weight, producing a CEC of about 30.5 meq/100 g at this rate. The increase in CEC by the addition of organic matter also resulted in the increased growth of the oat shoots. This probably was due to suppression of Cd availability, which in turn diminished the deleterious effects of Cd on the growth. Others (John et al, 1972b; Miller et al, 1976) also observed a negative correlation between leaf Cd values and CEC. In contrast, Mahler et al (1978), by employing step-wise multiple regression analysis, found that with lettuce, soil CEC, in addition to Cd in the saturation extract, had a positive effect on leaf Cd. They used eight soil types with CEC ranging from 6.5 to 37.9 meq/100 g.

In a greenhouse pot culture study, Hinesly et al (1982) found that Cd uptake and yields of corn were inversely related to the soil CEC (range of 5.3 to 15.9 meq/100 g) on $CdCl_2$-treated soils but were not affected by CEC on soils treated with sewage sludge. They concluded that Cd source (soluble salt vs. sludge) exerted a far greater effect on Cd uptake than did the soil CEC.

In sorption studies, several investigators (Kuo and Baker, 1980; Levi-Minzi et al, 1976; Singh, 1979) found that CEC played an important role in the sorption of Cd by soils. In fact, Kuo and Baker (1980) observed that CEC was more important than organic matter in the sorption of this metal. It should be noted that the pH buffering capacity of soils increases with the CEC. Thus, the potential for increased Cd uptake by plants in acidic soils is less with high CEC than in those with low CEC.

7.3 Organic Matter

Soil organic matter is of special interest, since it is known to adsorb considerable amounts of inorganic cations, including toxic metal ions, by an ion-exchange mechanism. Haghiri (1974) stated that the retaining ability of organic matter for Cd is predominantly through its CEC property rather than chelating ability. In addition to CEC, soil humus has chelating ability, and certain metals have a tendency to combine with certain chelating groups and become fixed. Organic complexes of Cu^{2+} and Pb^{2+} were considerably more stable than those for Cd^{2+} (Stevenson, 1976). Bondietti and Sweeton (1973) found that the stability constants with humic acids were in the order $Cu > Pb > Cd$.

Petruzzelli et al (1977) observed that Cd added to soil was retained by alkali-soluble humic substances so strong that it was not taken up by wheat seedlings. The soil used was a histosol with 24.1% organic matter content, pH of 5.4, and CEC of 318 meq/100 g. White and Chaney (1980) showed that soil organic matter appears to be more important than hydrous oxides of Fe and Mn in limiting the uptake of Cd by soybean plants. Comparing two soils, they obtained higher binding with an organic matter-rich soil (3.8% organic matter; CEC of 16 meq/100 g) than with a soil with lower organic matter content (1.2% organic matter; CEC of 5.4 meq/100 g). Additional evidence of retention by organic matter was provided by Street et al (1977), when Cd added with sewage sludge was taken up less by corn seedlings than when added as inorganic Cd alone. Somers (1978) found that both fresh and partially decomposed residues of plant material from an old field, a deciduous hardwood forest, and from under pine trees exhibited high affinity for Cd^{2+}. This explains the usual tendency for metals to accumulate in the humus layer in forest ecosystems. Organic matter was also found important in sorption studies of Cd by soils (Levi-Minzi et al, 1976; Singh, 1979).

7.4 Redox Potential

In the Jintsu region of Japan, the cradle of Itai-Itai disease, Takijima et al (1973a, b) observed that the greater the number of days that the rice paddies were drained prior to harvest, the greater the uptake and accumulation of Cd in the rice grain. Bingham's group at the University of California, Riverside also found marked reduction in the availability of soil Cd to the rice plants upon flooding the soil (Bingham et al, 1976b). Other results (Chino and Baba, 1981; Chino, 1981b; Reddy and Patrick, 1977) also indicate greatly reduced solubility of Cd under flooded management causing anaerobic conditions (Fig. 4.10). This behavior of Cd

FIGURE 4.10. Effect of pH (above) and the redox potential (below) on the total cadmium uptake by the rice plant and on the concentration of water-soluble cadmium. *Source:* Reddy and Patrick, 1977, with permission of the authors, and the Am Soc of Agronomy, Crop Sci Soc of America, Soil Sci Soc of America.

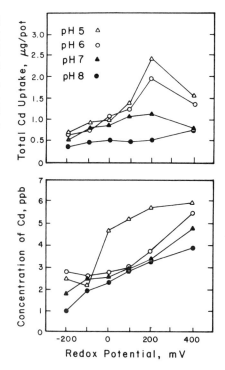

may be due to the precipitation of CdS at low redox potential (Bingham et al, 1976b; Reddy and Patrick, 1977; Krauskopf, 1956; Takijima et al, 1973b). Under low Eh-high pH conditions, the adsorption of Cd by amorphous reduced oxyhydroxides of Fe and Mn may also be important.

Bingham et al (1976b) observed that flood-managed rice has extreme tolerance to soil Cd compared to nonflooded rice grown in the greenhouse. For example, increasing soil Cd to a maximum rate of 640 ppm Cd resulted in a yield decrement of only 30% for flooded rice; the nonflooded rice did not survive treatments with $\geqslant 160$ ppm Cd. Chino (1981b) indicated that the activity of the rice roots may be involved in Cd absorption and translocation from flooded soil. Cadmium placed in the subsurface layer was absorbed by the roots in that layer but was not transported to the tops, whereas Cd placed in the surface layer was transported efficiently. This implies that roots in the reductive subsurface layer may be too inactive to transport Cd.

7.5 Plant Species, Cultivars, and Parts

Plant species exert marked differences with regard to Cd uptake, accumulation, and tolerance by plants. Sensitive crops grown in the greenhouse, such as spinach, curlycress, romaine lettuce, soybean, and field bean, had their yields reduced 25% by soil Cd additions of as low as 5 to 15 ppm (Bingham et al, 1975). More tolerant crops, such as tomato, zucchini squash, and cabbage required tenfold additions of Cd to produce similar yield reduction (Table 4.7). John (1973) found that among edible parts, highest Cd levels were found in lettuce and spinach leaves, followed by levels in *Brassica* tops, radish and carrot tubers, pea seeds, and oat grains. Similarly, Haghiri (1973) found Cd concentration for various crops in decreasing order: lettuce > radish top > celery stalk ≥ celery leaves > green pepper > radish roots. Using solution culture, Pettersson (1977) found Cd concentrations in the shoot decreased in the order: lettuce > sorrel, carrot, tomato > rape, kale, radish > mustard, corn > cucumber, sunflower, pea, bean > wheat and oats. Among forage species grown in potted soil under greenhouse conditions, the sensitivity increased in the order: bermudagrass < fescue < clover < alfalfa < sudangrass (Bingham et al, 1976a).

Because of genetic variation among cultivars within a plant species, differences in Cd uptake and accumulation in various plant parts have been observed for soybean (Boggess et al, 1978), wheat and barley (Pettersson, 1977; Chang et al, 1982), lettuce (CAST, 1980; John and Van-Laerhoven, 1976), rice (Chino, 1981a), and corn (CAST, 1980). In addition to the influence of cultivar on Cd uptake, translocation and accumulation among different plant parts vary. In general, Cd concentrations are lower in seed, tuber, and fruit tissues, relative to the concentrations in other parts such as roots and other leafy tissues as those in lettuce, spinach, and tobacco. A good example of differential uptake among crop species

TABLE 4.9. Cadmium concentration (ppm, dry wt basis) in various crops grown on a Grenville loam in the greenhouse.[a]

Crop and part	Cd, as CdCl$_2$, added, ppm			Crop and part	Cd, as CdCl$_2$, added, ppm		
	0	2.5	5.0		0	2.5	5.0
Oats, grain	0.21	1.50	2.07	Lettuce, tops	0.66	7.72	10.36
Oats, straw	0.29	2.30	3.70	Lettuce, roots	0.40	2.96	5.60
Oats, roots	0.81	5.06	8.72	Carrots, tops	0.46	5.66	7.70
Soybeans, grain	0.29	1.88	2.51	Carrots, roots	0.24	2.53	2.65
Soybeans, veg. part	0.71	3.95	4.88	Tobacco, leaves	0.49	5.41	11.57
Soybeans, roots	0.99	6.09	11.77	Tobacco, stems	0.28	2.42	4.80
Timothy, tops	0.21	1.04	1.41	Tobacco, roots	0.42	2.78	5.54
Timothy, roots	0.61	8.97	15.28	Potatoes, tops	0.58	3.46	7.35
Alfalfa, tops	0.28	1.34	1.72	Potatoes, tubers	0.18	0.89	1.09
Alfalfa, roots	0.52	7.99	10.51	Tomatoes, fruit	0.23	0.99	1.03
Corn, tops	0.22	1.84	2.68	Tomatoes, veg. parts	0.51	5.26	6.46
Corn, roots	0.73	10.47	17.02	Tomatoes, roots	0.59	5.08	10.35

[a]*Source:* MacLean, 1976, with permission of the Agricultural Institute of Canada, Ottawa.

and redistribution among plant parts within a species is demonstrated in Table 4.9.

7.6 Other Factors

Increases in soil temperature are also known to enhance Cd uptake by plants (Haghiri, 1973, 1974; Giordano et al, 1979). Soil type and the mineralogy of the clay fraction may also influence the sorption of Cd by soil. Higher sorption capacity or bonding energy is favored by fine-textured soils or soils high in organic matter or clay content (John, 1972; Andersson, 1977). When the clay fraction is dominated by 2:1 layer silicates, such as montmorillonite, the adsorptive capacity may be expected to be higher (Farrah and Pickering, 1977) than when the dominant clay mineral is a 1:1 type, such as kaolinite. The exception, however, is when soils are high in organic matter and oxides of Fe, Al, and Mn; then, higher affinity for Cd can be expected even in the presence of 1:1 layer silicates (Garcia-Miragaya and Page, 1978).

8. Cadmium in Drinking Water and Food

The US-EPA has set a standard of 10 ppb for Cd in drinking water for humans (US-EPA, 1976). In addition, the US-EPA water quality criteria, currently under revision are: 10 ppb Cd for irrigation water intended for continuous use on all soils, and 50 ppb Cd on neutral and alkaline soils for a 20-year period. For drinking water, the WHO (1984) has a limit of 5 ppb Cd. A survey of 969 community water supply systems in the United States showed the average Cd concentration to be 1.3 ppb (Craun and McCabe, 1975). Thus, drinking water contributes very little to the daily intake.

In food, there are basically four sources of Cd and other metal contamination: agricultural technology (e.g., pesticides, phosphate fertilizer, sewage sludge, etc.); industrial pollution; geological sources; and food processing (e.g., food additives, physical and chemical contact with equipment and vessels). Dietary intake of Cd varies from country to country, due to differences in eating habits, amount and type of food consumed, and levels of Cd residues. Estimation of the daily intake of Cd for several countries (United States, Federal Republic of Germany, Japan, Rumania, and Czechoslovakia) ranged from 25 to 60 µg/day per person (Friberg et al, 1971). This range may be compared with a WHO/FAO provisional tolerable weekly intake of 400 to 500 µg Cd, or about 57.1 to 71.4 µg/day per person (Mahaffey et al, 1975). The most recent estimate for the United States is about 39 µg/day (Table 4.10). There are no official limitations for Cd in food in the United States for adults. The intake for Canadians,

TABLE 4.10. Estimated dietary intakes and concentrations of cadmium in various food classes.

Class	Food intake[a]		Cd intake		Cd concentration[b]
	g food/day	% of total diet	μg Cd/day	% of total diet	ppb
Dairy products	769	22.6	4.4	11.3	5.7
Meat, fish & poultry	273	8.0	4.2	10.8	15.3
Grain and cereal	417	12.3	9.7	24.8	23.3
Potatoes	200	5.9	9.6	24.6	48.0
Leafy vegetables	63	1.9	2.6	6.7	40.5
Legume vegetables	72	2.1	0.4	1.0	6.3
Root vegetables	34	1.0	1.1	2.8	32.3
Garden fruits	89	2.6	1.3	3.3	14.7
Fruits	222	6.5	0.7	1.7	3.0
Oil and fats	51	1.5	0.8	2.0	15.3
Sugars and adjuncts	82	2.4	0.8	2.0	10.0
Beverages	1,130	33.2	3.4	8.7	3.0
TOTAL	3,402	100.0	39.0	100	

[a]Based on 1972–73 total diet survey by the US Food and Drug Administration for an adult person (Tanner and Friedman, 1977, with permission of Elsevier Sequoia S.A., Lausanne, Switzerland).
[b]Based on Pahren et al (1979).

based on 1970–71 survey, was about 67 μg/day (Somers, 1974). The "Total Diet Survey" of the US Food and Drug Administration between 1968 and 1974 gave a range of 26 μg/day in 1968 to 61 μg/day in 1969. In this survey, which involved "market baskets" of typical foods and beverages, of six metals (Pb, Cd, Hg, Zn, As, and Se), Cd has the most widespread distribution and Hg the most limited. Their results further show that the levels of these elements in foods do not vary significantly from one year to the next. Cadmium was most frequently detected in potatoes, leafy vegetables, grain and cereal products, and oils and fats; and least detected in dairy products, fruits, and beverages.

Traditionally, the mean concentrations for dairy products, red meats, and poultry are generally low (Mitchell and Aldous, 1974). However, the kidneys of animals are generally much higher than other parts of a carcass, averaging about 50 ppb, on wet weight basis (Neathery and Miller, 1975). Aquatic foods were reported to possess high mean concentration of about 20 ppb (wet weight basis), but some oysters were known to have concentrations as high as 2.0 ppm (Childs and Gaffke, 1974).

Because of the incidence of Itai-itai disease in some areas of Japan, partly caused by consuming unpolished rice with Cd levels as high as 3.4 ppm (Yamagata and Shigematsu, 1970), the Japanese government has set 1 ppm of Cd in rice grain as the maximum allowable limit (Chino, 1981a). In a survey of Cd content of rice involving 22 countries, polished and unpolished rice samples had similar contents of 29 ppb Cd (Masironi et al, 1977). In this survey, Japanese rice were found to contain high Cd levels, averaging about 65 ppb.

Cigarette tobacco contains about 1 ppm Cd. Data indicate that 0.1 to 0.2 μg of Cd are inhaled for each cigarette smoked (Friberg et al, 1971), giving a total intake of about 3 μg per pack of cigarettes smoked.

9. Sources of Cadmium in the Terrestrial Environment

Cadmium has been detected in air, water, and soil. Its natural level in soil is usually <1 ppm, including most agricultural soils. It can occur naturally in higher concentrations when associated with Zn ores, or in areas near Cd-bearing deposits. Because of its many uses in industries, significant amounts of Cd may be released to the environment. For example, in New York City, heavy metals, including Cd, have been found in residential waste water containing no electroplating or other industrial wastes (Varma and Katz, 1978). Of the 143 kg of total Cd discharged daily to the harbor, 50 kg were contributed by surface runoff, 27 kg by untreated waste water, 43 kg by plant effluents, and 23 kg by sludge. In addition, about 39% of total Cd entering the treatment plants in New York City came from industrial effluents and 49% from residential waste water, with the balance from runoff. Cadmium in runoff was attributed to fallout from fossil fuel combustion and corrosion of galvanized iron used by the construction industry.

As part of UNESCO's Man and the Biosphere Programme in Japan, Chino and Mori (1982) determined the Cd concentrations of domestic waste waters of some typical families. They found the Cd concentrations of waste waters from lavatory, kitchen, laundry, and bath to be 38, 0.4, 0.6, and 0.5 ppb Cd, respectively. The total amount of Cd discharged with such domestic waste waters was calculated to be 128 μg/day/person: 18 μg from the kitchen waste water, 42 μg from the laundry waste water, 18 μg from the bath waste water, 46 μg from the lavatory, and 4 μg from other sources.

The total release of Cd worldwide was estimated by Varma and Katz (1978) to be approximately 2,100 tonnes/year in 1980, a slight increase from 1,800 tonnes calculated for 1974–75. Of the total release of about 2,100 tonnes, 300 tonnes were through aerial emission, 15 tonnes via waterborne effluents, and 1,800 tonnes via land-destined wastes. In the latter category, it was estimated that about 250 tonnes arose from extraction, refining, and production of the metal, 75 tonnes came from industrial conversion, 460 tonnes from the consumption and disposal of Cd-containing products, and 1,060 tonnes via inadvertent sources. Regarding dispersal of Cd to the terrestrial environment, the inadvertent sources assume paramount significance. For example, phosphate fertilizers contributed about 130 tonnes worldwide in 1980; sewage sludge, 250 tonnes; and coal combustion, 680 tonnes. Cadmium arising from phosphate fertilizers and sewage sludge may reach man through the food chain. The sources described in the following subsections are identified as important contributors of Cd in agricultural and natural ecosystems.

9.1 Phosphatic Fertilizers

Cadmium occurs in ores used in the production of P fertilizers. The concentration of Cd in the ores can be as great as 980 ppm (USDA-USDI, 1977). Ores from the western United States contain considerably higher concentrations of Cd than ores from the southeastern states. Cadmium concentrations in concentrated superphosphate fertilizers (0-45-0) originating from western sources range from 50 to 200 ppm, whereas concentrations from southeastern sources range from 10 to 20 ppm (Mulla et al, 1980). In addition, diammonium phosphate from Idaho phosphate rock averaged 50 ppm, whereas the same fertilizer from North Carolina phosphate rock averaged 30 ppm (Mortvedt and Giordano, 1977). Only about 20% of all ores, approximately 10^7 tonnes, used for production of P fertilizers are derived from western sources (Harre et al, 1976). In 11 samples of diammonium phosphate (18-45-0), mostly of western United States origin, the Cd content ranged from 7.4 to 156 ppm, whereas 7 samples of concentrated superphosphate contained from 86 to 114 ppm Cd (US-EPA, 1974). In another survey of various phosphatic fertilizers of variable P_2O_5 contents, Lee and Keeney (1975) found that Cd concentration ranged from 1.5 to 9.7 ppm, with a median of 4.3 ppm.

The range of Cd concentration of several Australian commercial fertilizers was 18 to 91 ppm, with the superphosphates containing 38 to 48 ppm (Williams and David, 1973), whereas Swedish fertilizers were shown to contain <0.1 to 30 ppm Cd (Stenström and Vahter, 1974). Almost pure phosphate ores exist in the Kola Peninsula of the USSR (0.1 to 0.4 ppm), whereas ores in Senegal are known to contain 70 to 90 ppm Cd, and those in Togo, about 50 ppm (Nilsson, 1979).

With the widespread use of P fertilizers in agriculture, this source can potentially contribute to environmental pollution of Cd. Recent estimates indicate that this source accounts for a third to half of the total annual Cd introduced into Swedish agricultural soils (Nilsson, 1979). More specifically in Wisconsin (Lee and Keeney, 1975), commercial fertilizers added to soil may have enriched the farmlands by as much as 2,150 kg annually, compared to a potential of 1,700 kg, if waste water sludges from all municipal sewage treatment plants in the state were disposed of on land. These estimates were based on application rates of 50 kg/ha per year of phosphate fertilizer (median Cd content ≈ 17.5 ppm) and 9 tonnes/ha of sludge (Cd content ≈ 18 ppm), and incorporated into the surface at 0 to 15 cm. However, because of the higher application rates, sludge Cd on a soil concentration basis (0.20-ppm increase in the top 15 cm, based on 9 tonnes/ha per year application rate) is a much more concentrated source of Cd than that from phosphate fertilizers (0.001-ppm increase in the top 15 cm, based on 50 kg/ha phosphate fertilizer application).

Because of the large amounts of Cd that originate from these fertilizers, concern arises about food chain contamination from long-term accumulation of this element in soil. There is a direct correlation between P and Cd accumulation in soil and fertilizer (Fig. 4.11). Schroeder and Balassa (1963) reported that the concentration of Cd in several vegetable crops was increased by heavy applications of superphosphate. Similarly, Williams and David (1973) showed that application of superphosphate increased the Cd content of soils and of cereal and fodder crops. In addition, Reuss et al (1978) found that the uptake by horticultural crops under greenhouse conditions was a linear function of the Cd content of the fertilizer. The use of concentrated superphosphate containing 174 ppm Cd resulted in plant Cd levels of 2.4, 3.4, 6.3, 0.9, and 0.5 ppm for radish roots, radish tops, lettuce, pea seeds, and pea foliage, respectively. Mortvedt and Giordano (1977) obtained slight increases in uptake of Cd by forage from applications of commercial fertilizers, with the highest uptake resulting from western United States diammonium phosphate (18 μg/pot) and 10-15-0 fluid fertilizer (16 μg/pot). The nontreated soil had only 8 μg Cd/pot uptake. In a field experiment, Mortvedt et al (1981) observed that Cd concentrations in wheat grain and straw were significantly increased only with application of high-Cd diammonium phosphate (DAP) to the low-lime (pH 5.1) soil. Cadmium concentrations in the grain increased from 0.028 to 0.086 ppm, and those in the straw increased from 0.067 to

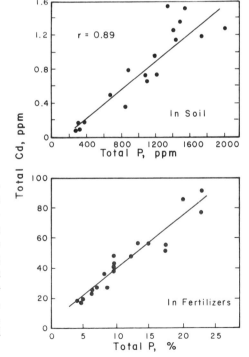

FIGURE 4.11. Relationships of cadmium and phosphorus in surface soil fertilized with phosphorus fertilizer for 36 years (above) and in various phosphatic fertilizers (below). *Sources:* Mulla et al, 1980; Williams and David, 1973, with permission of the authors and CSIRO, East Melbourne, Australia, and the Am Soc of Agronomy, Crop Sci Soc of America, Soil Sci Soc of America.

0.118 ppm (dry weight basis) with application of low-Cd DAP and high-Cd DAP, respectively, and were lower on the high-lime (pH 5.9) soil.

Mulla et al (1980) found that Swiss chard grown in the greenhouse on surface soils collected from a citrus grove in California that had been fertilized with about 175 kg P/ha per year, as concentrated superphosphate over a 36-year period, had significantly higher Cd levels (1.6 ppm) over those on the control soil (0.26 ppm). However, no significant rise in Cd concentration was observed in barley actually grown on the field where the soil used in the greenhouse study came from. If an annual application rate of 120 kg P/ha is assumed, a rate representative of maximum amounts applied to many horticultural crops, the annual input of Cd from concentrated superphosphate (0-45-0), containing 175 μg Cd/g of fertilizer, would be 0.10 kg Cd/ha. Annual suggested limits for the application of Cd to crops are currently 2 kg Cd/ha, and cumulative loading limits for Cd in soils, such as the one used by Mulla et al (1980), are 10 kg Cd/ha (Knezek and Miller, 1976). Thus, to exceed the suggested limits on cumulative total Cd for a soil with a CEC of 5 to 15 meq/100 g would require at least 200 years of consecutive annual application at high rates of P fertilizer containing the maximum concentration of Cd reported for western United States ores.

9.2 Sewage Sludges

Land application of municipal sewage sludges is becoming more popular because of constraints placed on alternative disposal methods, such as the proposed ban on ocean disposal in the United States, and air pollution problems and energy requirements involved with sludge burning. In the United States, of the approximately 6.0×10^6 tonnes (NAS, 1977) of municipal sewage sludge produced annually, 25% goes to landfills, 25% is disposed of on land, 15% is dumped in the ocean, and 35% is incinerated. Thus, if the constraints on the latter two methods become stringent, a greater proportion of sludge would have to be applied on lands used for farming, silviculture, and ornamental horticulture. Repeated applications of sewage sludge on agricultural lands increase the probability of Cd in foodstuff to exceed a given allowable limit, as reflected by the increases in Cd levels in soil concomitant with increases in Cd input (Fig. 4.12).

In the United Kingdom, about 45% of the sludge produced annually is utilized in agriculture (Davis and Coker, 1979). This is equivalent to 510 $\times 10^3$ dry tonnes per year, with a median Cd concentration of 22.6 ppm. Only about 1% of the sludge is used in the production of vegetable crops, which are generally considered as efficient accumulators of Cd from the soil. The UK guidelines allow a maximum increase in soil concentration of Cd resulting from sludge applications of 2.3 ppm to be reached over a period of at least 30 years.

In Denmark, Hansen and Tjell (1982) reported that, on a national basis,

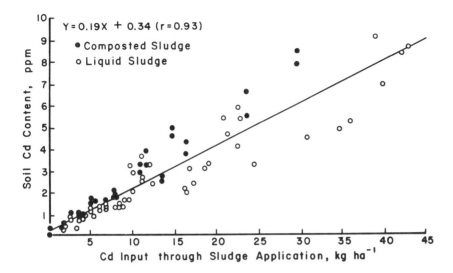

FIGURE 4.12. Relationship between cadmium concentration in the surface layer (0–15 cm) of a sandy loam soil and the amount of cadmium the soil received through repeated sludge application over a 6-year period. *Source:* Chang et al, 1983, with permission of the authors and the Am Soc of Agronomy, Crop Sci Soc of America, Soil Sci Soc of America.

sewage sludge contributed about 5% of the Cd burden in agriculture; the remainder resulted from aerial deposition (70%) and inorganic fertilizers (25%). However, on a local basis, the contribution from sludge accounted for about 90% (assuming that sludge with 7 ppm Cd was applied at a rate of 5 tonnes dry solid/ha per year), compared with 8% from aerial deposition and only 2% from fertilizers. Hansen and Tjell (1982) have quantitatively estimated Cd cycling in Danish agriculture (Fig. 4.13). Currently, the largest inputs of Cd to soil are from phosphate fertilizers and atmospheric deposition. They considered Cd loss through drainage and leaching as the only significant output. Losses via removal of crops, either for human consumption or animal feedstuff, are insignificant. The balance indicates that soil concentration of Cd in Denmark would rise at an average rate of about 0.6% per year.

In Sweden, the National Swedish Board of Health and Welfare has limited the use of sewage sludge on agricultural land to 1 tonne of dry solid/ha per year, with Cd concentration limited to 15 ppm dry weight basis (Stenström and Lönsjö, 1974). In the Netherlands, about 200,000 tonnes of sludge, on dry solid basis, are produced annually (Hemkes et al, 1980) with about half of this amount used in agriculture, horticulture, ornamental gardens, and landscapes. The remaining half is dumped.

Utilization of sewage sludge on agricultural lands is often associated with a potential risk of contaminating the food chain, with Cd posing the

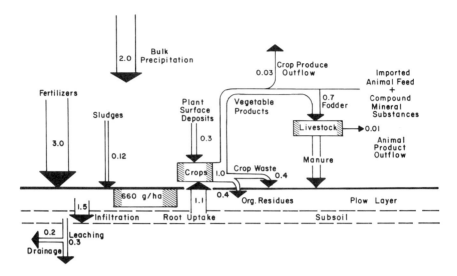

FIGURE 4.13. Current estimated flows of cadmium in Danish agro-ecosystems, involving the 0–25 cm soil depth. Flow units are in g Cd/ha per year. *Source:* Hansen and Tjell, 1982, with permission of the authors.

greatest concern. Because of the variable Cd contents of sewage sludges, the Cd concern appears to be more local than national, with big cities generating the greatest amounts and also, most likely, the sludges with the highest Cd concentrations (Appendix Tables 1.6, 1.7). Contamination of crops by sludge-borne Cd is well documented in the literature (Page, 1974; CAST, 1980).

Conceptually, the current land application of sludges somewhat varies from yesteryear. Instead of exploiting their plant nutrient contents, the practice today is often geared to maximize waste disposal. In the former approach, the sludge is applied to optimize or increase crop production. In the latter case, the practice integrates two otherwise separate systems, a crop production system and a waste disposal system. Sound practices of land application of sludges must satisfy the requirement of both systems.

In essence, this is the nucleus of most modern sludge land application studies. When sludge is used as a source of plant nutrients, usually N or P, the amount applied per year can be based on (CAST, 1980): (1) an annual limit on Cd, e.g. 2 kg Cd/ha per year; (2) the amount of N or P required by the crop grown; or (3) a combination of both criteria. The third option is typically used for privately owned farm land on which food or fuel crops are grown and the farmer uses a conventional soil testing program to monitor the soil after sludge application. The rationale for limiting sludge application on the basis of crop needs for N is that nitrate (NO_3^-) leaching and subsequent contamination of the ground water would not be greater than that caused by the use of commercial fertilizers.

The US-EPA (1979), in response to concern over Cd entering the food chain, established limits on both the annual and cumulative amounts of Cd that may be added to agricultural soils in the form of sewage sludge. However, the amount of N applied was not directly limited. The criteria mandate the following for all soils receiving solid wastes now being used or which in the future may be used to grow crops for the food chain: (1) the resulting pH of the treated soil is to be 6.5 or greater (no pH limitation if the waste contains ≤2 ppm Cd on a dry weight basis); (2) maximum annual application of 0.5 kg of Cd/ha for soils growing leafy vegetable, root crops, or tobacco; and (3) a maximum annual Cd applicaton for other crops of 2 kg/ha until June 1984, thereafter decreasing to 1.25 kg/ha until December 1986, and further decreasing to 0.5 kg/ha after 1986.

On a cumulative basis, the US-EPA (1979) established limits of 5 kg of Cd/ha for soils having an initial pH of <6.5. For soils with initial pH > 6.5, or for soils that will be maintained at pH 6.5 or above whenever food chain crops are grown, the cumulative amounts of Cd permitted increase with increasing soil CEC, as outlined in section 7.2.

The effect of sludge on some vegetables grown over a three-year period is shown in Table 4.11. The results indicate that increasing amounts of sludge applied to soil increases the Cd uptake by plants. In addition, cumulation of Cd in soil over time tends to elevate Cd uptake.

TABLE 4.11. Effect of composted sewage sludge application to a calcareous soil on the cadmium concentration (ppm) of field-grown vegetables[a,b]

Cumulative sludge applied,[c] tonnes/ha	Cd added, kg/ha	Lettuce	Swiss chard	Radish		Carrot		Turnip	
				tops	tubers	tops	tubers	tops	tubers
				1st year					
0	0	0.71	0.24	0.20	0.08	0.94	0.69	—	0.11
22.5	0.8	0.92	0.43	0.63	0.16	1.59	0.91	1.01	0.17
45	1.5	1.44	0.51	0.77	0.32	1.45	1.51	0.61	0.12
90	2.9	2.40	0.41	1.07	0.42	2.10	0.83	0.65	0.18
180	3.8	2.97	1.95	1.53	0.73	1.70	1.17	—	—
				2nd year					
0	0	0.50	0.20	0.48	0.19	0.54	0.44	0.98	0.16
45	1.6	2.20	0.50	0.70	0.36	1.64	0.91	1.24	0.27
90	2.9	3.10	0.70	1.86	0.63	2.52	1.64	2.66	0.27
180	5.7	4.40	2.10	2.81	0.88	3.89	2.44	2.59	0.36
360	11.4	6.40	2.60	5.01	1.87	3.91	2.61	2.25	0.65
				3rd Year					
0	0	0.60	0.27	0.25	0.12	0.60	0.30	0.30	0.10
67.5	2.9	1.50	0.60	0.95	0.50	1.20	0.60	0.60	0.20
135	5.6	1.70	1.60	1.37	0.72	1.60	0.80	0.60	0.03
270	11.1	3.80	2.60	2.03	1.16	2.20	0.90	0.90	0.40
540	22.2	7.90	2.90	3.16	1.76	3.50	1.40	1.80	0.70

[a]Source: Courtesy of Dr. A. C. Chang, University of California, Riverside.
[b]All samples were washed to remove soil and dust particles prior to preparation.
[c]Applied biannually over a 3-year period.

9.3 Atmospheric Fallout

Primary sources of air pollution probably rank as follows: smelter > incineration of plastics and Cd pigments > fossil fuel, including coking > steel mills > metallurgical (NAS, 1974). Numerous studies here and abroad (Blackmer et al, 1976; John et al, 1972a; Kobayashi, 1971; Munshower, 1977; Lagerwerff et al, 1972; Wixson et al, 1977; Smet, 1974; Buchauer, 1973; Jordan, 1975; Burkitt et al, 1972; Ragaini et al, 1977; Nwankwo and Elinder, 1979; Dorn et al, 1975; Little and Martin, 1972) have demonstrated significant Cd contaminations in soils, plants, and animals in the vicinity of lead-, zinc-, and battery-smelting complexes. For example, Wixson et al (1977) noticed increased concentrations of Cd and other metals in the forest near the AMAX lead smelter in southeast Missouri. Aerial deposition of Cd from smelting caused decomposing leaf litter to contain Cd from 10 ppm at the 4.4 km location to 200 ppm at the 0.4 km location along forest ridges. In Annaka City, Japan, where the country's largest Zn refinery is located, Kobayashi (1971) found that the content of Cd and other metals in mulberry leaves, wheat, barley, and vegetables (especially in Chinese cabbage—40 ppm) were enriched in areas near the refinery. He attributed the accumulation of metals in these plants to root absorption from the polluted soils rather than to direct deposition of the metals on the plant surfaces, since the samples were washed with redistilled water prior to analysis. The surface soils of farms between 150 and 250 m south of the refinery had 23 to 88 ppm of Cd, whereas in a highland area 900 m east of the chimney, the surface soil (0–10 cm) contained 31 to 44 ppm Cd. Near the Avonmouth industrial complex in the United Kingdom, which includes one of the largest Pb and Zn smelting plants in the world, elm leaves collected close to the complex had 50 ppm Cd; this decreased to background level (0.25 ppm) 10 to 15 km away (Little and Martin, 1972). In Deer Lodge Valley, about 24 km downwind from a Cu-Zn smelter complex in southwestern Montana, Munshower (1977) reported the following average Cd concentrations in contaminated vs. noncontaminated areas: grasses *(Agropyron cristatum and Stipa comata),* 1.72 and 0.07; alfalfa, 0.83 and 0.06; and barley grain, 0.65 and 0.08 ppm, respectively. Average levels in animal tissues were: cattle liver, 0.34 and 0.06; cattle kidney, 1.67 and 0.22; swine liver, 0.24 and 0.14; and swine kidney, 0.99 and 0.39 ppm, from contaminated and noncontaminated areas, respectively. Much higher enrichments of surface soils and grasses in the vicinity of the Pb smelting complex in Kellogg, Idaho were also observed (Ragaini et al, 1977).

Other sources of atmospheric Cd originate from the combustion of coal, oil, paper, and urban organic trash. Lagerwerff and Specht (1970) reported that the amount of burnable trash produced in the United States is staggering—at least 300,000 tonnes per day. Near roads, motor oils and tread wear from vehicular tires are sources of Cd and other metals. By growing leafy vegetables (0.8–9.1 ppm Cd on dry weight basis) in 14 gardens in the Boston metropolitan area, Preer and Rosen (1977) demonstrated the influence of automobile emissions and possibly other industrial input on the soil contamination in urban areas.

9.4 Mining Operations

Although usually localized, pollution of river and stream waters by Cd in waste waters from mining operations can affect a broad area and can afflict hundreds of its inhabitants. Such is the case in the Jintsu River Basin in Toyama Prefecture, Japan, where the Itai-itai disease was discovered by Dr. J. Kobayashi of Okayama University (1971, 1979). The residents had been ingesting Cd and other metals over a 30-year period, in both their drinking water and in their rice, in which Cd had accumulated through the river water used for irrigation. Similarly, in areas in Fukui Prefecture in Japan, Takijima and Katsumi (1973) found that unpolished rice harvested along the Kuzuryu River had Cd concentrations ranging from 0.02 to 1.82 ppm, with rice containing >1 ppm found near a Zn mining station. Cadmium in soils ranged from 0.20 to 10.4 ppm.

References

Adriano, D. C., A. L. Page, A. A. Elseewi, and A. C. Chang. 1982. *J Environ Qual* 11:197–203.

Alloway, B. J., M. Gregson, S. K. Gregson, R. Tanner, and A. Tills. 1979. *In Management and control of heavy metals in the environment*, 545–548. CEP Consultants Ltd, Edinburgh, UK.

Andersson, A. 1977. *Swedish J Agric Res* 7:7–20.

Andersson, A., and K. O. Nilsson. 1972. *Ambio* 1:176–179.

———— 1974. *Ambio* 3:198–200.

Archer, F. C. 1980. *Minist Agric Fish Food* (Great Britain) 326:184–190.

Balsberg, A. M. 1982. *Oikos* 38:91–98.

Berrow, M. L., and R. L. Mitchell. 1980. Trans Royal Soc Edinburgh: *Earth Sci.* 71:103–121.

Berrow, M. L., and G. A. Reaves. 1984. In *Proc Intl Conf Environ Contamination*, 333–340. CEP Consultants Ltd, Edingurgh, UK.

Bingham, F. T. 1979. *Environ Health Perspec* 28:39–43.

Bingham, F. T., and A. L. Page. 1975. In *Proc Intl Conf on Heavy Metals in the Environment*, 433–441. Toronto, Ontario, Canada.

Bingham, F. T., A. L. Page, R. J. Mahler, and T. J. Ganje. 1975. *J Environ Qual* 4:207–211.

———— 1976a. *J Environ Qual* 5:57–60.

———— 1976b. *Soil Sci Soc Am J* 40:715–719.

Bingham, F. T., A. L. Page, G. A. Mitchell, and J. E. Strong. 1979. *J Environ Qual* 8:202–207.

Bingham, F. T., A. L. Page, and J. E. Strong. 1980. *Soil Sci* 130:32–38.

Bittell, J. E., and R. J. Miller. 1974. *J Environ Qual* 3:250–253.

Blackmer, G. L., G. L. Geary, T. M. Vickrey, and R. H. Reynolds. 1976. *J Environ Sci Health* All(6):357–366.

Boggess, S. F., S. Willavize, and D. E. Koeppe. 1978. *Agron J* 70:756–760.

Bondietti, E. A., and F. H. Sweeton. 1973. *Agron Abst.* 89–90.

Boswell, F. C. 1975. *J Environ Qual* 4:267–272.

Bowen, H. J. M. 1979. *Environmental chemistry of the elements.* Academic Press, New York. 333 pp.

Bradley, R. I. 1980. *Geoderma* 24:17–23.

Browne, C. L., Y. M. Wong, and D. R. Buhler. 1984. *J Environ Qual* 13:184–188. ·

Buchauer, M. J. 1973. *Environ Sci Technol* 7:131–135.

Burkitt, A., P. Lester, and G. Nickless. 1972. *Nature* 238:327–328.

Carlson, R. W., and F. A. Bazzaz. 1977. *Environ Pollut* 12:243–253.

Carlson, R. W., and G. L. Rolfe. 1979. *J Environ Qual* 8:348–352.

Cavallaro, N., and M. B. McBride. 1978. *Soil Sci Soc Am J* 42:550–556.

——— 1980. *Soil Sci Soc Am J* 44:729–732.

Chaney, R. L. 1973. In *Recycling municipal sludges and effluents on land,* 129–141. Natl Assoc State Univ and Land-Grant Colleges, Washington, DC.

——— 1974. In *Factors involved in land application of agricultural and municipal wastes,* 67–120. USDA-ARS, Beltsville, Maryland.

Chang, A. C., A. L. Page, K. W. Foster, and T. E. Jones. 1982. *J Environ Qual* 11:409–412.

Chang, A. C., A. L. Page, J. E. Warneke, M. R. Resketo, and T. E. Jones. 1983. *J Environ Qual* 12:391–397.

Chang, A. C., J. E. Warneke, A. L. Page, and L. J. Lund. 1984a. *J Environ Qual* 13:87–91.

Chang, A. C., A. L. Page, J. E. Warneke, and E. Grgurevic. 1984b. *J Environ Qual* 13:33–38.

Childs, E. A., and J. N. Gaffke. 1974. *J Food Sci* 39:453–454.

Chino, M. 1981a. In K. Kitagishi and I. Yamane, eds. *Heavy metal pollution in soils of Japan,* 65–80. Japan Sci Soc Press, Tokyo.

——— 1981b. In K. Kitagishi and I. Yamane, eds. *Heavy metal pollution in soils of Japan,* 81–94. Japan Sci Soc Press, Tokyo.

Chino, M., and A. Baba. 1981. *J Plant Nutr* 3:203–214.

Chino, M., and T. Mori. 1982. In *Researches related to the UNESCO's Man and Biosphere Programme in Japan,* 106–109. Tokyo.

Chizhikov, D. M. 1966. *Cadmium.* Pergammon, New York.

Christensen, T. H. 1984a. *Water, Air, Soil Pollut* 21:105–114.

——— 1984b. *Water, Air, Soil Pollut* 21:115–125.

Chubin, R. G., and J. J. Street. 1981. *J Environ Qual* 10:225–228.

Council for Agricultural Science and Technology (CAST). 1980. In *Effects of sewage sludge on the cadmium and zinc content of crops.* CAST Rep. no 83, Ames, Iowa. 77 pp.

Craun, G. F., and L. J. McCabe. 1975. *J Am Water Works Assoc* 67:593–599.

Cunningham, L. M. 1977. *Trace Subs Environ Health* 11:135–145.

Cunningham, J. D., D. R. Keeney, and J. A. Ryan. 1975a. *J Environ Qual* 4:448–454.

Cunningham, J. D., J. A. Ryan, and D. R. Keeney. 1975b. *J Environ Qual* 4:455–460.

Davis, R. D., P. H. T. Beckett, and E. Wollan. 1978. *Plant Soil* 49:395–408.

Davis, R. D., and E. G. Coker. 1979. In *Proc Intl Conf Management and Control of Heavy Metals in the Environment,* 553–556. CEP Consultants Ltd, Edinburgh, UK.

Dijkshoorn, W., J. E. M. Lampe, and A. R. Kowsoleea. 1974. *Netherlands J Agric Sci* 22:66–71.

Dorn, C. R., J. O. Pierce, G. R. Chase, and P. E. Phillips. 1975. *Environ Res* 9:159–172.

Dowdy, R. H., and W. E. Larson. 1975. *J Environ Qual* 4:278–282.

Elliott, H.A., and C. M. Denneny. 1982. *J Environ Qual* 11:658–662.

ES & T. 1979. *Currents. Environ Sci Technol* 13:1447.

Farrah, H., and W. F. Pickering. 1977. *Aust J Chem* 30:1417–1422.

Fleischer, M., A. F. Sarofim, D. W. Fassett, P. Hammond, H. T. Shacklette,I. C. Nisbet, and S. Epstein. 1974. *Environ Health Perspec* 7:253–323.

Förstner, U. 1980. In O. Hutzinger, ed. *Handbook of environmental chemistry.* Vol. 3, part A. *Anthropogenic compounds,* 59–107. Springer-Verlag, New York.

Francis, C. W., and S. G. Rush. 1973. *Trace Subs Environ Health* 7:75–81.

Frank, R., K. Ishida, and P. Suda. 1976. *Can J Soil Sci* 56:181–196.

Friberg, L., M. Piscator, and G. Nordberg. 1971. *Cadmium in the environment.* CRC Press, Cleveland, OH. 166 pp.

Garcia-Miragaya, J., and A. L. Page. 1976. *Soil Sci Soc Am J* 40:658–663.

———— 1977. *Soil Sci Soc Am J* 41:718–721.

———— 1978. *Water, Air, Soil Pollut* 9:289–299.

Gerritse, R. G., W. van Driel, K. W. Smilde, and B. van Luit. 1983. *Plant Soil* 75:393–404.

Giordano, P. M., D. A. Mays, and A. D. Behel, Jr. 1979. *J Environ Qual* 8:233–236.

Girling, C. A., and P. J. Peterson. 1981. *J Plant Nutr* 3:707–720.

Godbeer, W. C., and D. J. Swaine. 1979. *Trace Subs Environ Health* 13:254–260.

Haghiri, F. 1973. *J Environ Qual* 2:93–96.

———— 1974. *J Environ Qual* 3:180–183.

Hahne, H. C. H., and W. Kroontje. 1973. *J Environ Qual* 2:444–450.

Hansen, J. A., and J. C. Tjell. 1982. In R. D. Davis, G. Hucker, and P. L'Hermite, eds. *Environmental effects of organic and inorganic contaminants in sewage sludge,* 91–112. Proc of workshop sponsored by Comm European Communities, Reidel Publ Co, Dordrecht, the Netherlands.

Haq, A. U., T. E. Bates, and Y. K. Soon. 1980. *Soil Sci Soc Am J* 44:772–777.

Harre, E., M. N. Goodson, and J. D. Bridges. 1976. *Fertilizer trends.* Natl Fert Dev Center, Muscle Shoals, AL.

Hassett, J. J., J. E. Miller, and D. E. Koeppe. 1976. *Environ Pollut* 11:297–302.

Heinrichs, H., and R. Mayer. 1977. *J Environ Qual* 6:402–407.

Heinrichs, H., B. Schulz-Dobrick, and K. H. Wedepohl. 1980. *Geochim Cosmochim Acta* 44:1519–1532.

Hemkes, O. J., A. Kemp, and L. W. van Broekhoven. 1980. *Neth J Agric Sci* 28:228–237.

Hickey, M. G., and J. A. Kittrick. 1984. *J Environ Qual* 13:372–376.

Hinesly, T. D., R. L. Jones, and E. L. Ziegler. 1972. *Compost Sci* 13:26–30.

Hinesly, T. D., K. E. Redborg, E. L. Ziegler, and J. D. Alexander. 1982. *Soil Sci Soc Am J* 46:490–497.

Huffman, E. W. D., and J. F. Hodgson. 1973. *J Environ Qual* 2:289–291.

Hutchinson, T. C., M. Czuba, and L. Cunningham. 1974. *Trace Subs Environ Health* 8:81–93.

Hutchinson, T. C. 1979. In *Effects of cadmium in the Canadian environment.* NRCC 16743, Ottawa, Canada.

Iimura, K. 1981. In K. Kitagish and I. Yamane, eds. *Heavy metal pollution in soils of Japan,* 19–26. Japan Sci Soc Press, Tokyo.

Iimura, K., and H. Ito. 1971. *Res Bull Science Soil Manure of Chuba Branch, Japan* 34:1–15.

Iwai, I., T. Hara, and Y. Sonoda. 1975. *Soil Sci Plant Nutr* 21:37–46.

Jarvis, S. C., and L. H. P. Jones. 1980. *J Soil Sci* 31:469–479.

John, M. K. 1972. *Can J Soil Sci* 52:343–350.

—— 1973. *Environ Pollut* 4:7–15.

John, M. K., H. H. Chuah, and C. J. VanLaerhoven. 1972a. *Environ Sci Technol* 6:555–557.

John, M. K., C. J. VanLaerhoven, and H. H. Chuah. 1972b. *Environ Sci Technol* 6:1005–1009.

John, M. K., and C. J. VanLaerhoven. 1976. *Environ Pollut* 10:163–173.

Jones, R. L., T. D. Hinesly, and E. L. Ziegler. 1973. *J Environ Qual* 2:351–353.

Jordan, M. J. 1975. *Ecology* 56:78–91.

Keeney, D. R., and L. M. Walsh. 1975. In *Proc Intl Conf on Heavy Metals in the Environment,* 379–401. Toronto, Ontario, Canada.

Kelling, K. A., D. R. Keeney, L. M. Walsh, and J. A. Ryan. 1977. *J Environ Qual* 6:352–358.

Kelly, J. M., G. R. Parker, and W. W. McFee. 1979. *J Environ Qual* 8:361–364.

Khalid, R. A., R. P. Gambrell, and W. H. Patrick, Jr. 1978. In D. C. Adriano and I. L. Brisbin, Jr., eds. *Environmental chemistry and cycling processes,* 417–433. CONF-760429, NTIS, Springfield, VA.

—— 1981. *J Environ Qual* 10:523.528.

Khan, D. H., and B. Frankland. 1983. *Plant Soil* 70:335–345.

Knezek, B. D., and R. H. Miller. 1976. In *Application of sludges and waste waters on agricultrual land: A planning and educational guide.* NCRR Public 235, Wooster, OH. 88 pp.

Kobayashi, J. 1971. *Trace Subs Environ Health* 5:117–128.

—— In F. W. Oehme, ed. *Toxicity of heavy metals in the environment.* Part I, 199–260. Dekker Inc, New York.

Koeppe, D. E. 1977. *Sci Total Environ* 7:197–206.

Krauskopf, K. B. 1956. *Geochim Cosmochim Acta* 9:1–32.

Krauskopf, K. B. 1979. *Introduction to geochemistry.* 2nd ed. McGraw-Hill, New York. 617 pp.

Kuo, S., and A. S. Baker. 1980. *Soil Sci Soc Am J* 44:969–974.

Kuo, S., and B. L. McNeal. 1984. *Soil Sci Soc Am J* 48:1040–1044.

Lag, J., and I. H. Elsokkary. 1978. *Acta Agric Scan* 28:76–80.

Lagerwerff, J. V. 1971. *Soil Sci* 111:129–133.

Lagerwerff, J. V., and A. W. Specht. 1970. *Trace Subs Environ Health* 4:85–92.

Lagerwerff, J. V., and G. T. Biersdorf. 1971. *Trace Subs Environ Health* 5:515–522.

Lagerwerff, J. V., and D. L. Brower. 1972. *Soil Sci Soc Am Proc* 36:734–737.

Lagerwerff, J. V., D. L. Brower, and G. T. Biersdorf. 1972. *Trace Subs Environ Health* 6:71–78.

Lamoreaux, R. J., and W. R. Chaney. 1977. *J Environ Qual* 6:201–205.

Lee, K. W., and D. R. Keeney. 1975. *Water, Air, Soil Pollut* 5:109–112.

Levi-Minzi, R., G. F. Soldatini, and R. Riffaldi. 1976. *J Soil Sci* 27:10–15.

Little, P., and M. H. Martin. 1972. *Environ Pollut* 3:241–254.

Lucas, J. M. 1980. In US Bureau of Mines, *Minerals yearbook,* 1978–79, 139–145. US Dept of Interior, Washington, DC.

Lund, L. J., A. L. Page, and C. O. Nelson. 1976. *J Environ Qual* 5:330–334.

Lund, L. J., E. E. Betty, A. L. Page, and R. A. Elliott. 1981. *J Environ Qual* 10:551–556.

MacLean, A. J. 1976. *Can J Soil Sci* 56:129–138.

Mahaffey, K. R., P. E. Corneliussen, C. F. Jelinek, and J. A. Fiorino. 1975. *Environ Health Perspect* 12:63–69.

Mahler, R. J., F. T. Bingham, and A. L. Page. 1978. *J Environ Qual* 7:274–281.

Mahler, R. J., F. T. Bingham, G. Sposito, and A. L. Page. 1980. *J Environ Qual* 9:359–363.

Martin, M. H., and P. J. Coughtrey. 1975. *Chemosphere* 3:155–160.

Masironi, R., S. R. Koirtyohann, and J. O. Pierce. 1977. *Sci Total Environ* 7:27–43.

McBride, M. B. 1980. *Soil Sci Soc Am J* 44:26–28.

McBride, M.B., L. D. Tyler, and D. A. Hovde. 1981. *Soil Sci Soc Am J* 45;739–744.

Milberg, R. P., D. L. Brower, and J. V. Lagerwerff. 1978. *Soil Sci Soc Am J* 42:892–894.

Miller, J. E., J. J. Hassett, and D. E Koeppe. 1976. *J Environ Qual* 5:157–160.

———— 1977. *J Environ Qual* 6:18–20.

Mills, J. G., and M. A. Zwarich. 1975. *Can J Soil Sci* 55:295–300.

Mitchell, D. G., and K. M. Aldous. 1974. *Environ Health Perspec* 7:59–64.

Mitchell, C. D., and T. A. Fretz. 1977. *J Am Soc Hort Sci* 102:81–84.

Mortvedt, J. J., and P. M. Giordano. 1977. In H. Drucker and R. A. Wildung, eds. *Biological implications of heavy metals in the environment.* CONF-750929. NTIS, Springfield, VA.

Mortvedt, J. J., D. A. Mays, and G. Osborn. 1981. *J Environ Qual* 10:193–197.

Mulla, D. J., A. L. Page, and T. J. Ganje. 1980. *J Environ Qual* 9:408–412.

Munshower, F. F. 1977. *J Environ Qual* 6:411–413.

National Academy of Sciences (NAS). 1974. In W. L. Petrie, ed. *Geochemistry and the environment.* Vol. 1, 43–56. Washington, DC.

———— 1977. *Multimedia management of municipal sludge.* NAS-NRC, Vol. 9, Washington, DC.

National Research Council Canada (NRCC). 1979. *Effects of cadmium in the Canadian environment.* NRCC no. 16743. Ottawa, Canada. 148 pp.

Neathery, M. W., and W. J. Miller. 1975. *J Dairy Sci* 58:1767–1781.

Nilsson, R. 1979. *Ambio* 8:275–277.

Nriagu, J. O. 1980. In J. O. Nriagu, ed. *Cadmium in the environment. Part 1 - Ecological cycling,* 35–70. Wiley, New York.

Nwankwo, J. N., and C. G. Elinder. 1979. *Bull Environ Contamin Toxicol* 22:625–631.

Page, A. L., F. T. Bingham, and C. Nelson. 1972. *J Environ Qual* 1:288–291.

Page, A. L., and F. T. Bingham. 1973. *Residue Rev* 48:1–44.

Page, A. L. 1974. *Fate and effects of trace elements in sewage sludge when applied to agricultural lands.* EPA- 670/2-74-005. US-EPA, Cincinnati, OH. 98 pp.

Page, A. L., and A. C. Chang. 1978. In *Acceptable sludge disposal techniques*. Info Transfer, Inc, Rockville, MD. 239 pp.

Page, A. L., A. C. Chang, and F. T. Bingham. 1979. In *Management and controls of heavy metals in the environment*, 525–528. CEP Consultants Ltd, Edinburgh, UK.

Pahren, H. R., J. B. Lucas, J. A. Ryan, and G. K. Dotson. 1979. *J Water Pollut Cont Fed* 51:2588–2601.

Parker, G. R., W. W. McFee, and J. M. Kelly. 1978. *J Environ Qual* 7:337–342.

Patel, P. M., A. Wallace, and R. T. Mueller. 1976. *J Am Soc Hort Sci* 101:553–556.

Peel, J. W., R. J. Vetter, J. E. Christian, W. V. Kessler, and W. W. McFee. 1978. In D. C. Adriano and I. L. Brisbin, eds. *Environmental chemistry and cycling processes*, 628–636. CONF-760429. NTIS, Springfield, VA.

Pepper, I. L., D. F. Bezdicek, A. S. Baker, and J. M. Sims. 1983. *J Environ Qual* 12:270–275.

Petruzzelli, G., G. Guidi, and L. Lubrano. 1977. *Water, Air, Soil Pollut* 8:393–399.

Pettersson, O. 1977. *Swedish J Agric Res* 7:21–24.

Preer, J. R., and W. G. Rosen. 1977. *Trace Subs Environ Health* 11:399–404.

Ragaini, R. C., H. R. Ralston, and N. Roberts. 1977. *Environ Sci Technol* 11:773–781.

Reddy, C. N., and W. H. Patrick, Jr. 1977. *J Environ Qual* 6:259–262.

Reuss, J. O., H. L. Dooley, and W. Griffis. 1978. *J Environ Qual* 7:128–133.

Root, R. A., R. J. Miller, and D. E. Koeppe. 1975. *J Environ Qual* 4:473–476.

Santillan-Medrano, J., and J. J. Jurinak. 1975. *Soil Sci Soc Am Proc* 39:851–856.

Schauer, P. S., W. R. Wright, and J. Pelchat. 1980. *J Environ Qual* 9:69–73.

Schroeder, H. A., and J. J. Balassa. 1963. *Science* 140:810–820.

Shacklette, H. T. 1972. *Cadmium in plants*. Geol Surv Bull 1314-G. US Govt Printing Office, Washington, DC. 28 pp.

Sidle, R. C., and W. E. Sopper. 1976. *J Environ Qual* 5:419–422.

Sidle, R. C., and L. T. Kardos. 1977. *J Environ Qual* 6:431–437.

Singh, S. S. 1979. *Can J Soil Sci* 59:119–130.

Smet, S. Denaeyer-De. 1974. *Oecologia Plant*. 9:169–182.

Smith, W. H. 1973. *Environ Sci Technol* 7:631–636.

Somers, E. 1974. *J Food Sci* 39:215–217.

Somers, G. F. 1978. *Environ Pollut* 17:287–295.

Soon, Y. K., and T. E. Bates. 1982. *J Soil Sci* 33:477–488.

Sposito, G., and F. T. Bingham. 1981. *J Plant Nutr* 3:35–49.

Sposito, G., and S. V. Mattigod. 1979. *GEOCHEM: A computer program for calculating chemical equilibria in soil solutions and other natural water systems*. Kearny Foundation of Soil Science, Univ. of California, Riverside.

Stahr, K., H. W. Zöttl, and Fr. Hädrich. 1980. *Soil Sci* 130:217–224.

Stenström, T., and H. Lönsjö. 1974. *Ambio* 3:87–90.

Stenström, T., and M. Vahter. 1974. *Ambio* 3:91–92.

Stevenson, F. J. 1976. *Soil Sci Soc Am J* 40:665–672.

Street, J. J., W. L. Lindsay, and B. R. Sabey. 1977. *J Environ Qual* 6:72–77.

Street, J. J., B. R. Sabey, and W. L. Lindsay. 1978. *J Environ Qual* 7:286–290.

Symeonides, C., and S. G. McRae. 1977. *J Environ Qual* 6:120–123.

Takijima, Y., and F. Katsumi. 1973. *Soil Sci Plant Nutr* 19:29–38.

Takijima, Y., F. Katsumi, and K. Takezawa. 1973a. *Soil Sci Plant Nutr* 19:173–182.

Takijima, Y., F. Katsumi, and S. Koizumi. 1973b. *Soil Sci Plant Nutr* 19:183–193.

Tanner, J. T., and M. H. Friedman. 1977. *J Radioanal Chem* 37:529–538.

Tjell, J. C., and M. F. Hovmand. 1978. *Acta Agric Scand* 28:81–89.

Tjell, J. C., J. A. Hansen, T. H. Christensen, and M. F. Hovmand. 1980. In P. L'Hermite and H. Ott, eds. *Characterization, treatment, and use of sewage sludge.* 652–664. Proc. of 2nd Comm European Communities Workshop, Reidel Publ Co, Dordrecht, the Netherlands.

Turner, M. A. 1973. *J Environ Qual* 2:118–119.

Tyler, G. 1972. *Ambio* 1:52–59.

––––– 1975. *Nature* 255:701–702.

Tyler, L. D., and M. B. McBride. 1982. *Plant Soil* 64:259–262.

US Bureau of Mines. 1978–79. *Minerals yearbook. Metals and minerals.* US Dept of Interior, Washington, DC.

USDA-USDI. 1977. In *Development of phosphate resources in S E Idaho: Final environmental impact statement.* Vol. 1, 52–54.

US Environmental Protection Agency (US-EPA). 1974. EPA-440/1-74-011-A. Washington, DC.

––––– 1976. *Quality criteria for water.* US-EPA, Washington, DC.

––––– *Criteria for classification of solid waste disposal facilities and practices.* 1979. *Fed Regist.* 44:53438–53468.

Vallee, B. L., and D. D. Ulmer. 1972. *Ann Rev Biochem* 41:91–128.

Van Hook, R. I., W. F. Harris, and G. S. Henderson. 1977. *Ambio* 6:281–286.

Varma, M. M., and H. M. Katz. 1978. *J Environ Health* 40:324–329.

Wallace, A., E. M. Romney, G. V. Alexander, R. T. Mueller, S. M. Soufi, and P. M. Patel. 1977a. *Agron J* 69:18–20.

Wallace, A., E. M. Romney, G. V. Alexander, and S. M. Soufi. 1977b. *Commun Soil Sci Plant Anal* 8:765–772.

Whitby, L. M., J. Gaynor, and A. J. MacLean. 1978. *Can J Soil Sci* 58:325–330.

White, M. C., and R. L. Chaney. 1980. *Soil Sci Soc Am J* 44:308–313.

Williams, C. H. 1977. *J Aust Inst Agric Sci* (Sept–Dec):99–109.

Williams, C. H., and D. J. David. 1973. *Aust J Soil Res* 11:43–56.

––––– 1976. *Soil Sci* 121:86–93.

––––– 1977. *Aust J Soil Res* 15:59–68.

Williams, D. E., J. Vlamis, A. H. Pukite, and J. E. Corey. 1984. *Soil Sci* 137:351–359.

Wixson, B. G., N. L. Gale, and K. Downey. 1977. *Trace Subs Environ Health* 11:455–461.

Wong, M. K., G. K. Chuah, L. L. Koh, K. P. Ang, and C. S. Hew. 1984. *Environ Exp Bot* 24:189–195.

World Health Organization (WHO). 1984. *Guidelines for drinking-water quality. Vol. 1 - Recommendations.* WHO, Geneva, Switzerland, 130 pp.

Yamagata, N., and I. Shigematsu. 1970. *Inst Public Health* 19(1):1–27. Tokyo, Japan.

Zwarich, M. A., and J. G. Mills. 1979. *Can J Soil Sci* 59:231–239.

5
Chromium

1. General Properties of Chromium

Chromium, a member of Group VI-B of the periodic table, has an atomic no. of 24, an atomic weight of 51.996, a specific gravity of 7.18 at 20°C, a melting point of 1,903°C and four stable isotopes with the following percent abundance: ^{50}Cr (4.31%); ^{52}Cr (83.76%); ^{53}Cr (9.55%); and ^{54}Cr (2.38%). It has five radioactive isotopes, but only ^{51}Cr, with a halflife of 27.8 days is the most commonly used for tracer studies. Chromium is a steel-gray, lustrous, hard, brittle metal that takes a high polish. It dissolves readily in nonoxidizing mineral acids but not in cold aqua regia or HNO_3. In other words, it is resistant to attack by oxidizing acids and a range of other chemicals, hence its use in corrosion-resistant alloys.

Chromium can occur in any of the oxidation states from $-II$ to VI, but it is not commonly found in oxidation states other than 0, III, and VI, the III being the most stable (Mertz, 1969). Within the range of pH and redox potential found in soils, it has the capability of existing in four states—two trivalent III forms, the Cr^{3+} cation and the CrO_2^- anion, and two hexavalent VI anion forms, $Cr_2O_7^{2-}$ and CrO_4^{2-} (Bartlett and Kimble, 1976a). The trivalent form has a great tendency for coordination with oxygen- and nitrogen-containing ligands (Mertz, 1969). Chromium compounds with oxidation states below III are reducing, and those with greater than III are oxidizing. The chemistry of Cr of relevance to biology, ecology, and health is that of Cr(III) and Cr(VI) present in various environmental matrices. Of the two forms found in nature, the trivalent form is relatively benign, and the hexavalent form is relatively toxic.

2. Production and Uses of Chromium

World production of chromite has been increasing since the mid-1960s, peaking at about 9.55×10^6 tonnes in 1979 (Fig. 5.1). The principal producers of chromite, in decreasing order, are: South Africa, the USSR,

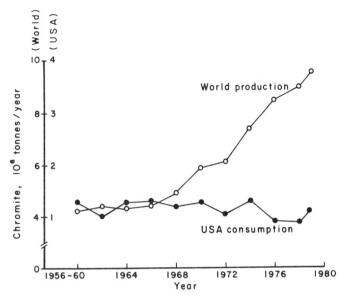

FIGURE 5.1. World production of chromite and the comparable US chromite consumption. *Source:* US Bureau of Mines, *Minerals yearbook,* several issues.

the Philippines, Southern Zimbabwe, and Turkey (Matthews and Morning, 1980). Chromite deposits are also found in several other countries, including the United States, Albania, Cuba, Brazil, Japan, India, New Caledonia, Pakistan, Iran, and the Malagasy Republic (NAS, 1974). Mine production of chromite ceased in the United States in 1961, since it was not economically feasible to recover the chromite from the mine. While the United States is not a producer of chromite, it consumes about 12% of the world's output (Fig. 5.1).

The only important Cr ore is chromite, with the formula [(Fe, Mg)O (Cr, Al, Fe)$_2$O$_3$], The end member, FeCr$_2$O$_4$ (also called chromite), would contain 68% chromic oxide, Cr$_2$O$_3$, and 32% ferrous oxide, FeO. The highest grades of ore contain about 52% to 56% Cr$_2$O$_3$ and 10% to 26% FeO (NAS, 1974).

The big users of Cr are the metallurgical, refractory, and chemical industries (Langård, 1980). For example, of the total amount of chromite consumed in the United States in 1979, the metallurgical industry used 63%; the refractory industry, 17%; and the chemical industry, 20% (Matthews and Morning, 1980). Metallurgic-grade chromite is usually converted into one of several types of ferrochromium or chromium metal that are alloyed with Fe, Ni, or Co (NAS, 1974). Over 60% of the Cr used in the metallurgic industry is used in making stainless steel. The chemical industry uses ore containing about 45% chromic oxide for preparation of sodium chromate and sodium dichromate, from which most other Cr chemicals are produced. Chromium chemicals are used as tanning agents, catalysts, pigments, and plating and wood preservatives.

3. Natural Occurrence of Chromium

Chromium is omnipresent in the environment, found in varying concentrations in air, soil, water, and all biological matter. Chromium is abundant in the earth's crust, more so than Co, Cu, Zn, Mo, Pb, Ni, and Cd (NAS, 1974). It ranks 21*st* among the elements in crustal abundance (Krauskopf, 1979).

The levels of Cr in soil have been reported to vary from trace to as high as 5.23% (NAS, 1974). Soils derived from serpentines usually contain high Cr. Serpentines, a type of ultramafic igneous rock, contains an average of 1,800 ppm Cr and 2,000 ppm Ni (Cannon, 1978). In addition, shales and phosphorites usually contain high concentrations of Cr (Table 5.1). In rocks, Cr is often present as chromite, $FeCr_2O_4$. In the United States, soils have a geometric mean of 37 ppm (n = 863) (Shacklette et al, 1971). Canadian soils have an arithmetic mean of 43 ppm (n = 173) and a range of 10 to 100 ppm (McKeague and Wolynetz, 1980). There are contrasting values reported for world soils: 200 ppm by Vinogradov (1959); 100 to 300 ppm by Aubert and Pinta (1977); and 40 ppm (10 to 150 ppm range) by NAS (1974).

TABLE 5.1. Commonly observed chromium concentrations (ppm) in various environmental matrices.

Material	Average concentration	Range
Continental crust	125	80–200
Basic igneous rocks[a]	680	2–3,400
Limestone[a]	10	<1–120
Sandstone[b]	35	—
Shale[a]	120	30–590
Petroleum[c]	0.3	—
Coal[c]	20	10–1,000
Fly ash:[d]		
Bituminous	172	—
Sub-bituminous	50	—
Lignite	43	—
Phosphate fertilizers[a]	—	30–3,000
Soils[c] (normal)	40	10–150
Herbaceous vegetables[b]	—	<0.05–14
Ferns[e]	1.9	—
Fungi[e]	2.6	—
Lichens[e]	—	0.6–7.3
Woody angiosperm[e]	—	0.03–10
Woody gymnosperm[e]	—	0.1–0.5
Freshwater[e]*	1	0.1–6
Sea water[e]*	0.3	0.2–50

* µg/liter
Sources: Extracted from
[a] NRCC (1976).
[b] Cannon (1978).
[c] NAS (1974).
[d] Adriano et al (1980).
[e] Bowen (1979).

4. Chromium in Soils

4.1 Total Soil Chromium

The Cr concentration of soil is largely determined by the parent material. Soils derived from igneous rocks can be expected to contain high levels of Cr (Table 5.1); so are soils derived from shales. In Minnesota, Pierce et al (1982) examined 16 soil series to establish baseline levels of several heavy metals. The soils were representative of those formed on seven major parent materials in the state and comprised a broad range of soil properties. They found that, in general, concentrations of Cr were highest in the parent materials of all soils (48 ppm in parent materials vs. 39 ppm in soil surface samples). The highest concentration of total Cr (and Cu and Ni) were found in soils developed from the Rainy Lobe till. In Byelorussia, USSR, where soil parent materials include various clay and sandy deposits mainly of glacial, meltwater, and fluvial origin, Cr was found least in eolian deposit (11 ppm) and highest in the alluvial deposit (54 ppm) (Lukashev and Petukhova, 1975). In the Nakhichevan region of the USSR, among the parent materials examined, Cr was lowest in marl, sandstone, and granite; Cr was highest in serpentinite, gabbro, diorite, basalt, and diabase; and in between in shale and clay (Shakuri, 1978).

In his summary of the literature, Mitchell (1964) stated that most soils contain between 5 and 1,000 ppm Cr, but soils containing less than 5 ppm and up to a few percent Cr are also known. Shacklette et al (1971), however, indicated that the majority of soils in the United States contain between only 25 and 85 ppm Cr. Lisk (1972) reported Cr levels of 1 to 5,000 ppm in United States soils; Morley (1975) reported a range of 20 to 125 ppm in Canadian soils. Furthermore, Frank et al (1976) reported that the average natural background level of Cr in agricultural soils of Ontario

TABLE 5.2. Mean total chromium and nickel concentrations of serpentine soils (as ppm air-dry soil).[a]

Location	Cr	Ni
Zimbabwe	125,000	5,500
New Caledonia	33,460	1,490
Zimbabwe	20,250	n.d.[b]
Portugal	4,930	3,220
Finland	4,000	1,200
Guyana	3,800	5,000
New Zealand	3,670	2,430
Shetland	1,800	3,330
Maryland (USA)	1,160	24
Australia	634	3,410

[a] *Source:* Peterson, 1975, with permission of the author.
[b] Not detectable.

(Canada) was 14.3 ppm. Soils derived from serpentine rocks commonly have the highest Cr contents, as indicated in Table 5.2. Values reported by Proctor (1971) for British and Swedish serpentine soils ranged from 2,500 to 4,000 ppm Cr. It should be noted that serpentine soils also have high Ni contents (see Table 5.2 and the chapter on nickel).

More recently, Berrow and Reaves (1984) reported a mean value of 50 ppm Cr for world soils, whereas Domingo and Kyuma (1983) reported a mean value of 136 ppm (range of 46–467 ppm) for tropical Asian paddy soils.

4.2 Extractable (Available) Chromium in Soils

Extractable Cr, just like any extractable trace element, is the form most commonly used to indicate Cr availability to plants. Often, the extractable form of any trace element is not directly related to the total content in the soil. In general, only a very small but variable fraction of the total Cr in soil is extractable with the ordinary extractants. In extracting several heavy metals from a polluted soil by a 1 N NH_4OAc containing 0.02 M EDTA, Nakos (1982) found that only 0.30% of the total Cr was extractable by this reagent in contrast to 13% for Zn and 17% for Pb. In Scottish peaty podzols, Mitchell (1971) found that from 0.34% to 0.80% of the total Cr fraction was extractable by HOAc reagent. Schueneman (1974) found <2% of the total (400 ppm) Cr added was extractable by 1 N NH_4OAc and <25% of the total amount extractable by 0.1 N HC1. MacLean and Langille (1980) found 0.1 N HC1 extractions from two agriculturally important Nova Scotia soils to contain 0.102 to 2.9 ppm Cr. Wild (1974) reported <2 ppm Cr extracted by 2.5% HOAc in Zimbabweian serpentine soils. Swaine and Mitchell (1960) found from <0.03 to 2.5 ppm Cr were extractable by 2.5% HOAc from four Scottish podzols. Mitchell et al (1957) found no apparent relationship between EDTA- or NH_4OAc-extractable Cr and plant uptake of Cr.

In his review, Mitchell (1964) reported that Cr extracted from surface layers of arable Scottish soils are generally in the following ranges: <0.01 to 1.0 ppm with 2.5% HOAc and 0.1 to 4.0 ppm with 0.05 M EDTA. Extraction of Scottish soils by 0.1 N HCl may yield up to tenfold increases over the 2.5% HOAc extractant. Schueneman (1974) found that among four extractants (distilled water, 1 N NH_4OAc, 0.005 M DTPA, and 0.1 N HC1), toxicity symptoms in several crop species became more severe as 1 N NH_4OAc extractable Cr increased.

4.3 Chromium in the Soil Profile

Reports on the pattern of profile distribution of Cr appear to be inconsistent. In some podzols in the United States, Connor et al (1957) found that Cr and most of the trace elements they examined tended to be higher in the B or C horizon than in the A horizon. Chromium was about 2.5 to 4 times higher in the C than in the A horizon. In addition, these investigators found that the majority of the trace elements in the soil were associated with the clay fraction. Similar trends in the soil profile accumulation of Cr were observed in some soils in the United Kingdom. Butler (1954) observed that in six Lancashire gleyed soils supporting permanent pasture, the maximum concentration of Cr occurred in the lowest horizon. He observed a somewhat regular decrease of this element in the top horizon, which seemed to be unaffected by the redox conditions. Similarly, Mitchell (1971) observed an increasing pattern of Cr accumulation with increasing soil depth in some Scottish peaty podzols.

Conversely, some soil profiles displayed accumulations of Cr in the surface horizons and depletion with depth. This trend is particularly pronounced in some soil profiles in Papua, New Guinea (Bleeker and Austin, 1970). Chromium in these soil profiles decreased from 946 ppm in the A1 horizon to 294 ppm in the C horizon, and to 283 ppm in the parent material.

Some soils tend to exhibit a more uniform distribution in the soil profile, as in some agricultural soils in Ontario (Ap = 53; B = 55; and C = 49 ppm) (Whitby et al, 1978) and in serpentine soils in western Newfoundland (Roberts, 1980).

The mobility of Cr in the soil profile was addressed by Lund et al (1976) when they found metal enrichment to depths as great as 3 m under some sewage sludge disposal ponds. They hypothesized that the metals moved as soluble metal-organic complexes. Thus, in instances where Cr could be complexed or in hexavalent (VI) form, it can move deeper in soil profiles.

4.4 Sorption and Complexation of Chromium in Soils

The immobilization of Cr in soils apparently depends on several factors: oxidation state, pH, clay minerals, competing ions, complexing agents, and others. Detection of Cr(VI) in drinking water and ground water from wastes disposed of through recharge basins, diffusion wells, or landfills indicates that perhaps this form of Cr is not effectively sorbed by the soil (Griffin et al, 1977).

Cary et al (1977b) found that the downward movement of Cr(VI) in soil columns was deeper in the alkaline than in acid soil, indicating lower sorption of this Cr species with higher pH. Griffin et al (1977) determined the effects of oxidation state, pH, and complexing agent, represented by landfill leachate, on the sorption of Cr by clay minerals. Their data plots fit the Langmuir adsorption isotherm, which appeared typical for anion adsorption, such as for phosphate and arsenate. They found that the adsorption of Cr from either the landfill leachate or Cr salt solutions by montmorillonite or kaolinite was highly dependent upon the pH of the clay suspension and other chemical properties of the clay minerals. Ad-

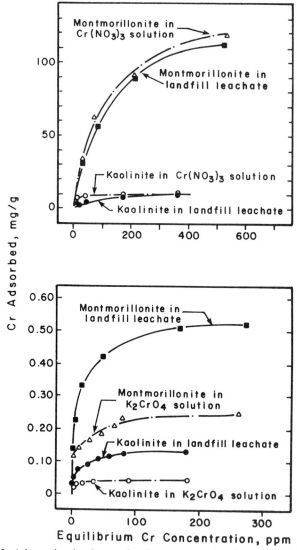

FIGURE 5.2. Adsorption isotherms for Cr(III) (above) and Cr(VI) (below) at pH 4.0 and 25° C. *Source:* Griffin et al, 1977, with permission of the authors.

sorption of Cr(VI) decreased as pH increased, with no further sorption occurring above pH 8.5. The $HCrO_4^-$ ion was the Cr(VI) species predominantly adsorbed. More Cr(VI) was adsorbed from the leachate solutions than from the K_2CrO_4 solutions. Griffin et al (1977) attributed this trend to the formation of polynuclear complexes in the leachate solutions that can be sorbed (Fig. 5.2) and possibly also to the higher ionic strength (7.2 mmho/cm) of the leachate (i.e., via depression of the diffuse double-layer surrounding the clay particles) compared to the pure K_2CrO_4 solution.

No precipitation of Cr(VI) occurred in either solutions over the pH range of 1.0 to 9.0. The amounts adsorbed by montmorillonite were generally two to three times higher than the amounts adsorbed by kaolinite.

In the case of trivalent Cr, Griffin et al (1977) found that the adsorption of Cr(III) increased as the pH of the suspensions increased. Chromium (III) is known to be extensively hydrolyzed in acid solutions to species such as $Cr(OH)^{2+}$, $Cr_2(OH)_4^{2+}$, or $Cr_6(OH_{12})^{6+}$. They explained the increased adsorption of Cr(III) as pH increased due to cation exchange-adsorption of these hydrolyzed species. About 30 to 300 times more Cr(III) than Cr(VI) was adsorbed by both clays. In contrast to Cr(VI) the adsorption of Cr(III) is 3% to 14% lower in the landfill leachate than in pure $Cr(NO_3)_3$ solutions (Fig. 5.2). This was attributed to the presence of cations in the leachate competing with the cationic Cr(III) species for exchange-adsorption sites, resulting in lower amounts sorbed compared to the pure $Cr(NO_3)_3$ solutions.

Bartlett and Kimble (1976b) found that an excess of orthophosphate in the equilibrating solution can, in some cases, totally prevent the adsorption of Cr(VI). They attributed this to phosphates competing with Cr(VI) for the same adsorption sites. Sorption of Cr(VI) by certain soils appears to be of specific type (Gebhardt and Coleman, 1974). This indicates that chromate is tightly bound compared with anions such as chloride, nitrate, or sulfate, but it can be desorbed by reaction of the soil with other specifically adsorbed anions, such as phosphates. Amacher (1981) found that $0.01\ M$ KH_2PO_4 solution was sufficient to extract Cr(VI) from soil, indicating that $H_2PO_4^-$ ions replaced $HCrO_4^-$ ions.

Amacher (1981) stated that sorption by organic matter and iron oxides is the most likely explanation for immobilization of Cr(III) in some soils, since humic acids have a high affinity for Cr(III). Grove and Ellis (1980a) have shown that when Cr(III) was added to soil, a large fraction of added Cr was extracted with the free iron oxides.

Bartlett and Kimble (1976a) have obtained evidence that Cr(III) can be complexed by organic matter components, especially fulvic acid. In their soil extraction by $Na_4P_2O_7$, they found that considerably more Cr(III) was removed from soils collected from horizons (primarily B horizon) where fulvic acid was more predominant than humic acid. The $Na_4P_2O_7$ extractant is supposedly capable of removing up to about 80% of the organic matter, consisting largely of fulvic acid. In one soil, they found that Na-citrate allowed about 90% of the Cr to stay in solution. James and Bartlett (1983a, b) reported that organic ligands can affect the solubility of Cr and noted that free Cr(III) ions would quickly become adsorbed and/or hydrolyzed and precipitated in the absence of soluble, complexing ligands. Citric acid, DTPA, fulvic acids, and water-soluble organic matter could keep most of Cr(III) in solution above pH 5.5. Gerritse et al (1982) found that, in general, sludge solutions appear to have increased the mobility of elements in soil due to a combination of complexation by dissolved organic compounds and high ionic strength of the soil solutions. Several animal manures had varying effects, but all less than the citrate, in solubilizing Cr (Bartlett and Kimble, 1976a). Apparently, the effectiveness of the manures depends on their degree of decomposition. However, organic matter may not necessarily have any effect on the solubility, and therefore, the phytotoxicity of Cr in soils, since Mortvedt and Giordano (1975) found that even up to 1,360 ppm Cr applied in sewage sludge did not affect the growth of corn.

4.5 Forms and Transformations of Chromium in Soils

Chromium, when added to soil, has several possible fates. It can be oxidized or reduced, remain in solution, be adsorbed on the mineral and organic exchange complex or on the Fe and Mn hydrous oxides coating of soil particles, become chelated by an organic ligand, or precipitated as sparingly soluble or highly insoluble compounds. Soluble Cr may be converted to insoluble forms when added to soil. One process is via reduction of soluble and relatively toxic Cr(VI) to Cr(III). Hexavalent Cr can be reduced to Cr(III) in environments with a ready source of electrons according to:

$$Cr_2O_7^{2-} + 14\ H^+ + 6\ e^- \rightleftharpoons 2\ Cr^{3+} + 7\ H_2O$$

Cary et al (1977b) found that Cr(VI) was reduced to Cr(III) in soil with the rate of conversion being slower in alkaline soils than in acid soils. These investigators found that the insoluble forms of Cr had the properties of mixed hydrous oxides of Cr(III) and Fe(III), and further found somewhat similar chemistry between Cr(III) and Fe(III) in soils. Similarly, Bartlett and Kimble (1976a, b) found analogous behavior between Cr(III) and Al(III) and also between CrO_4^{2-} and orthophosphate.

The equilibrium distribution of different forms of Cr in aqueous solution has been calculated as a function of pH and redox potential by Huffman (1973). He predicted that at neutral pH, Cr(VI) could be a predominant form of Cr in some natural systems. At pH 7 in well-aerated soils, the concentration of CrO_4^{2-} could be equal to the concentration of $Cr(OH)_2^+$.

Reduction of Cr(VI) in soil can be enhanced by the presence of organic matter (Cary et al, 1977b; Bartlett and Kimble 1976b; James and Bartlett, 1983c). Bloomfield and Pruden (1980) found that Cr(VI) was extensively reduced in soils of near neutral pH under anaerobic conditions, particularly when the soil contained organic matter. Amacher (1981) concluded that Cr from organic sources was not oxidized in soil and remained associated with the soil organic matter where it was resistant to oxidation. While reduction of hexavalent Cr is likely to occur in most soils, Cr oxidation in soils could also occur. Bartlett and his group have observed that oxidation of Cr(III) could occur under most field soil conditions. Bartlett and James (1979) found that for the oxidation of Cr to proceed, the presence in the soil of oxidized Mn is essential because it serves as the electron acceptor in the oxidation reaction (Fig. 5.3). They developed a test to

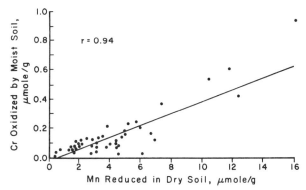

FIGURE 5.3. Amounts of oxidized chromium formed in 50 fresh moist soils related to the amount of manganese reduced in the dried samples of the same soils. *Source:* Bartlett and James, 1979, with permission of the authors and the Am Soc of Agronomy, Crop Sci Soc of America, Soil Sci Soc of America.

determine the potential of a soil to oxidize Cr by measurement of Mn reducible by hydroquinone, or the potential may be determined directly by means of a quick test in which Cr(III) is added to a fresh moist soil sample. James and Bartlett (1983b) found that oxidation of Cr(III) in tannery wastes can proceed despite the high levels of reducing compounds in the sludge, effluent, and the soil used. However, the rate of oxidation of Cr(III) from this source was slower than those of pure chemical form used, such as $Cr(OH)_3$. Sheppard et al (1984) found that Cr(III) applied to alkaline soils remained reduced for considerable lengths of time.

Grove and Ellis (1980b) proposed the following reaction for soluble Cr(III) compounds added to soil:

$$Cr^{3+} + 6H_2O \rightleftharpoons Cr(H_2O)_6^{3+} \rightleftharpoons [Cr(OH)_x(H_2O)_{6-x}]^{3-x} + xH^+ \rightleftharpoons 1/2Cr_2O_3$$
$$\cdot zH_2O$$

This reaction can account for the lower pH found on Cr(III)-treated soils and the poor extractability of the Cr compounds by distilled water, 1 M NH_4Cl solution (exchangeable fraction), and 0.1 M $CuSO_4$ solution (organic-bound fraction). However, the oxalate and dithionite-citrate extractants removed large quantities of Cr because of the known similarities between Fe(III) and Cr(III) chemistry. James and Bartlett (1983c) found that Cr(VI) was adsorbed on the amorphous fraction of $Fe(OH)_3$ and was solubilized by oxalate.

Cary et al (1977b) found that, in soils incubated for 41 days prior to $^{51}CrCl_3$ or $K_2^{51}CrO_4$ additions and then further incubated up to 90 days, in no case did the water-soluble Cr [determined by 0.01 M $Ca(NO_3)_2$ extraction] exceed more than 6% of the added Cr, and most values were 3% or less. The fact that only trace amounts of Cr in the water-soluble form were detected even after only one day of Cr addition, indicate that

Cr was quickly converted to forms that were not water-soluble. Using isotopic exchange techniques (0.01 M $CrCl_3$ and 0.01 M K_2CrO_4 extractants), Cary et al (1977b) found that most of the added Cr(VI) may have been converted to Cr(III) by day 7, and that less than 5% of the added [51]Cr(III) was oxidized to Cr(VI) within 90 days. Similarly, Ross et al (1981) observed a rapid decrease of extractable Cr(VI) during a three-week incubation study, which indicates significant occurrence of reduction of Cr(VI). Cary et al (1977b) obtained evidence that soluble organic complexes of Cr (extracted by 0.1 N NH_4OH) were present in soils they used. Thus, results indicate that although Cr(III) could be oxidized to Cr(VI) in soil, this is probably not all that common. On the other hand, when Cr(VI) is added to soil, it can be expected to rapidly reduce to Cr(III). Therefore, Cr in most soils probably occurs as Cr(III).

5. Chromium in Plants

5.1 Essentiality of Chromium in Plants

Interest in Cr contents of plants is spurred by the known essentiality of Cr for animal and human nutrition (Mertz, 1969). Pratt (1966) reported stimulatory effects of Cr in plant growth, although they were generally small, inconsistent, and mostly inconclusive. The stimulatory effect of Cr on plant growth has been interpreted as a limited substitution of chromate for molybdate (Warington, 1946). Huffman (1973) reported that, although there are numerous reports on the stimulation of growth or yield of plants grown either in the field, solution, or sand cultures, it is difficult to explain responses in plant growth to Cr additions, since there is insufficient evidence to justify Cr as an essential element or even as a generally beneficial element. This difficulty exists despite its experimental use to increase the yield of potatoes, enhance the growth of citrus and avocado trees, and increase the sugar content and ripening of grapes. Nevertheless, the non-essentiality of Cr for normal plant growth has been demonstrated conclusively (Huffman and Allaway, 1973a).

5.2 Uptake and Translocation of Chromium in Plants

Controlled studies have shown that Cr(III) and Cr(VI) are practically equally available to plants grown in culture solution (Huffman and Allaway, 1973b; Breeze, 1973). However, Hewitt (1953) observed that the effect of Cr in causing toxicity in sugar beets grown on sand culture depended on oxidation state, with greater effects observed for Cr(VI). Similarly, Cr(VI) was found to be slightly more toxic than Cr(III) to bush beans (Wallace et al, 1977) and to corn grown on pot culture (Mortvedt and Giordano, 1975).

Apparently, absorbed Cr is poorly translocated in plants. Huffman and Allaway (1973b) found that Cr absorbed by plants grown in culture solutions remained primarily in the roots and is poorly translocated to the leaves. At maturity, bean plants contained about 55% and wheat about 81% of the added Cr. However, bean roots contained about 92% and wheat roots 95% of the total plant Cr. Myttenaere and Mousny (1974) found that >99% of the Cr(III) and Cr(VI) absorbed remained bound in the roots of rice grown in culture solutions. Although greater removal of Cr(III) and Cr(VI) than of CrEDTA was observed, there was greater translocation to the tops of the latter than was the case for Cr(III) and Cr(VI). Similar accumulations of absorbed Cr in roots were found by Cary et al (1977a), Wallace et al (1976), and Ramachandran et al (1980). Cary et al (1977a) reported that the barrier, believed to be the cell wall (Myttenaere and Mousny, 1974), to translocation of Cr from roots to tops was not circumvented by supplying Cr in the form of organic acid complexes, Cr(III), or Cr(VI), or by increasing the Cr(III) concentration in the nutrient solution. They further found that plants or plant tissues that tend to accumulate Fe also accumulate Cr. Lyon et al (1969) found that, even with a Cr accumulator species, *Leptospermum scoparium,* most of the absorbed Cr was in the roots. Shewry and Peterson (1974) suggested that most of the Cr retained in the roots is present in the soluble form in vacuoles of root cells, and more specifically in the protoplasmic fractions of the roots (Myttenaere and Mousny, 1974).

James and Bartlett (1984) found that adding $Cr(OH)_3$ to a neutral soil increased the Cr contents of bean shoots and roots. However, bean shoots grown in soil amended with Cr-enriched tannery effluent and sewage sludge did not contain more Cr than the control, but the roots did. Soluble Cr(VI) was not detected in soils amended with Cr(III) in tannery effluent or sewage sludge, indicating the lack of oxidization, probably due to the presence of high organic matter in these wastes. In unplanted soil amended with $Cr(OH)_3$, soluble Cr(VI) levels were higher than in planted soil, but levels of soluble Cr(III) were higher in the rhizosphere soil. This indicates the influence of plant roots and its metabolites on the form of Cr and its subsequent availability to plants—i.e., plant root exudates could influence the reduction of Cr(VI) to Cr(III) and the formation of Cr(III)-organic complexes, facilitating its uptake by the roots.

5.3 Plant Accumulators of Chromium

In "normal" soils, Cr concentrations in plants are usually in the <1 ppm range and seldom exceed 5 ppm on dry weight basis (Pratt, 1966). However, some plant species, commonly termed "accumulators," can accumulate appreciable quantities of Cr. For definition of this term and since Cr correlates highly with Ni (Table 5.2) in these species, refer to the chapter on Ni. Peterson (1975) reported that many plant species have become adapted to serpentine soils, which have relatively high levels of total Cr but are characterized to have usually low available Cr. Some species, however, can contain appreciable amounts of Cr despite the low solubility of Cr. For example, Peterson (1975) reported, in ppm ash weight: 48,000 ppm for *Sutera fodina;* 30,000 ppm for *Dicoma niccolifera;* and 2,470 ppm for *Leptospermum scoparium.* Their plant to soil concentration ratios averaged about 0.30. Plant to soil concentration ratios can vary widely, with some values reported as low as 0.01. Peterson concluded that the concentration in plants is below the soil concentration.

5.4 Toxicity of Chromium in Plants

Most of the reported Cr toxicity has been observed under controlled experimental conditions, either in solution culture or in soils spiked with high levels of Cr. Under field conditions, Cr toxicity is essentially nonexistent, except possibly in soils derived from ultrabasic or serpentinic rocks. These soils are characteristically high in Ni and Cr, and it is most likely Ni, rather than Cr, that could cause toxicity. Numerous workers (DeKock, 1956; Haas and Brusca, 1961; Hewitt, 1953; Hunter and Vergnano, 1953; Vergnano, 1959; Walker and Grover, 1957) have demonstrated toxicity to plants associated with high levels of Cr. Information from recent reviews (Pratt, 1966; NRCC 1976) indicates that 1 to 5 ppm Cr present in the available form in the soil solution, either as Cr(III) or Cr(VI), is the critical level for a number of plant species. When added to soil, the toxic threshhold level for Cr(VI) varied from 5 ppm (on air-dried soil basis) for tobacco grown in sandy soil (Soane and Saunder, 1959) to 500 ppm for L. *perenne* grown in a potting compost (Breeze, 1973). Toxic limits for Cr(III) varied from 8 ppm for sugar beets grown in sand (Hewitt, 1953) to 5,000 ppm for L. *perenne* grown in potting compost (Breeze, 1973). Apparently the high organic matter in the potting soil bound the Cr or enhanced the reduction of Cr(VI) to Cr(III). It is apparent from Table 5.3 that Cr may prove to be toxic to plants at about 5 ppm or higher in nutrient solutions and at about 100 ppm when added to a mineral soil.

Pratt (1966) reported that the range of high Cr concentrations in plant tissues before toxicity symptoms could be observed was from about 5 ppm for barley, corn, oats, and citrus to 175 ppm for tobacco. Concentrations in plant tissues definitely associated with toxicity symptoms are usually in the several hundreds ppm range. For the rice plants, 35 to 177 ppm of Cr in stem and leaf can cause a 10% reduction in yield (Chino, 1981). Hunter and Vergnano (1953) pointed out that while the toxic effects of most heavy metals (e.g., Ni, Co, Zn, etc.) are associated with high concentrations of the element in the leaf tissue, this is not the case with Cr.

TABLE 5.3. Effects of oxidation state and level of chromium in growth medium on plant growth.

Plant species	Oxidation state of Cr	Cr concentration in growth media, ppm	Growth medium	Effects	Source
Corn	Cr(III)	0.5	Solution culture	Stimulation of growth	NRCC (1976)
Corn	Cr(III)	5	Solution culture	Moderate toxicity	"
Corn	Cr(III)	50	Solution culture	Severely stunted growth	"
Ryegrass (perennial)	Cr(III)	10	Solution culture	Increased plant mortality	"
Ryegrass (perennial)	Cr(VI)	10	Solution culture	Increased plant mortality	"
Barley	Cr(VI)	50	Soil	Severely stunted growth	"
Barley	Cr(VI)	500	Soil	Death	"
Oats	Cr(VI)	5	Sand	Iron chlorosis	"
Sugar beet	Cr(VI)	8	Sand	Iron chlorosis	"
Tobacco	Cr(VI)	5	Sand	Retarded stem development	"
Corn	Cr(VI)	10	Sand	Stunted growth	"
Sweet orange seedling	Cr(VI)	75	Soil	No toxicity	"
Sweet orange seedling	Cr(VI)	150	Soil	Observed toxicity	"
Soybean	Cr(VI)	5	Loam soil	Inhibition of uptake of Ca, K, Mg, P, B, Cu	"
Barley	Cr(III)	8	Solution culture	Chlorotic leaves	Davis et al (1978)
Bush beans	Cr(III)	0.5	Solution culture	Decreased yield	Wallace et al (1976)
Corn, beans, and tomatoes	Cr(III)	100	Soil	Stunted growth	Schueneman (1974)

Several explanations as to the cause of Cr toxicity have been proposed, but the most plausible mechanism is not clear. Excess Cr may interfere with Fe (Hewitt, 1953; Cannon, 1960; DeKock, 1956; Walker and Grover, 1957; Anderson et al, 1973), Mo (Hewitt, 1954), P (Soane and Saunder, 1959; Spence and Millar, 1963; Vergnano, 1959), and N (Hunter and Vergnano, 1953) metabolisms in plants. Soane and Saunder (1959) indicated that Cr(III) and Fe(III) acted analogously in causing the acute P deficiency symptoms they observed in oats. Grove and Ellis (1980b) explained Cr toxicity in terms of Cr(III) effects on Mn and Fe chemistry in soil; severe Fe chlorosis associated with Cr(III) toxicity can be a result of greater quantities of available Mn and lesser amounts of soluble Fe, indicating a Cr(III)-induced Fe-Mn interaction in plants. Hence, at lower rates of application, Cr(III) could stimulate growth and yield of plants where Mn was a limiting factor in the plant's mineral nutrition. Turner and Rust (1971) also found that the uptake of several nutrient elements (K, Mg, P, Mn, Ca, and Fe) by plants was affected by high Cr levels. However, Jaiswal and Misra (1984) found substantial increases in available Fe concentrations in soils treated with Cr(VI), which was converted to Cr(III). They attributed the increase in available Fe partly to the decreased pH of the treated soil. Chromium is known to inhibit various enzymatic reactions *in vitro* (Mertz, 1969), although direct cause of Cr toxicity to a specific enzyme system has not been conclusively shown.

The visual symptoms of Cr toxicity commonly observed are: stunted growth, poorly developed root system, and curled and discolored leaves (Pratt, 1966). Some plants may exhibit brownish-red leaves containing small necrotic areas or purpling of basal tissues.

6. Factors Affecting the Mobility and Availability of Chromium

6.1 pH

The pH level will affect the solubility of the Cr forms and therefore its sorption by soil and its availability to plants. Figure 5.4A indicates that the oxidation of Cr(III), as $CrCl_3$, equilibrated in a northern Vermont soil, can be affected by pH. At pH 3.2, all of the Cr(III) present was oxidized to Cr(VI) after only 18 hours. On the other hand, Cary et al (1977b) found that reduction is more rapid in acid than in alkaline soils. The same soil from Vermont and another soil from Rothamsted, the United Kingdom (Fig. 5.4A, B) both exhibit decreased sorption with increasing pH. Desorption of the Rothamsted soil at pH 10 to 11 gave a quantitative recovery of Cr(VI) sorbed between pH 6 and 9; the small losses of sorbed Cr(VI)

FIGURE 5.4. Influence of pH on oxidation of Cr(III) and soil sorption of Cr(VI). Part A depicts the percentage of chromium in a 2,000:1 (solution/soil ratio) equilibrium of 10^{-6} M CrCl$_3$ with soil that was oxidized or adsorbed as Cr(VI), as pH was varied by addition of HCl or KHCO$_3$. Part B depicts sorption of Cr(VI) on NaOH-extracted (to remove extractable organic matter) subsoil and desorption at pH 10–11. *Source:* Bartlett and James, 1979 (Part A), Bloomfield and Pruden, 1980 (Part B), with permission of the authors and the Am Soc of Agronomy, Crop Sci Soc of America, Soil Sci Soc of America and Applied Science Publishers.

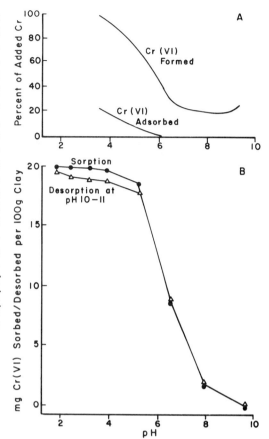

at pH <5 may be due to reduction by residual (i.e., not extracted by dilute NaOH) organic matter during the sorption process.

Bloomfield and Pruden (1980) showed that reduction of Cr(VI) by water-soluble soil organic matter in soil equilibration study occurred appreciably only below pH 4, and that within experimental error there was no reduction between pH 5 and 9 in 40 days. With topsoil, the extent of reduction after three weeks' aerobic incubation increased from about 5 mg Cr(VI) reduced per 100 g soil at pH 6.7 to 14 mg per 100 g soil at pH 4.1. They concluded that the organic constituents of the soil became more effective in reducing Cr(VI) with decreasing pH. They cautioned that in raising the soil pH, organic matter can be extracted and thus may increase the reduction of Cr(VI) during its determination with the acid S-diphenyl-carbazide reagent.

Griffin et al (1977) found that water-soluble Cr(VI) decreased as the pH of kaolinite and montmorillonite suspensions decreased. They attributed this phenomenon to HCrO$_4^-$ adsorption by pH-dependent anion exchange sites on clay colloids. However, it is more likely that the anaerobic system maintained by Griffin et al (1977) favored Cr(VI) reduction to Cr(III) and then the cation Cr(III) sorption by the clay exchange sites followed.

Bartlett and Kimble (1976a) and James and Bartlett (1983a) found that the solubility of Cr(III), in pure Cr solutions, decreased as the solution pH was raised above 4, with essentially complete precipitation occurring at pH 5.5. They presumed that the precipitate was composed of macro-molecules with Cr ions in six coordination complexes with water and hydroxy groups, and can be likened to the formation of Al polymers. When soil was added, the solubility of Cr(III) decreased above pH 2.5, with virtually complete precipitation occurring at about pH 4.5.

There are indications in plant uptake studies that Cr from both trivalent and hexavalent sources can be taken up more from alkaline than from acid soil (Cary et al, 1977b), although this was contradicted by Rama-chandran et al (1980), who observed that the uptake of ^{51}Cr(III) by bean shoots decreased as the pH of the nutrient solution increased.

6.2. Oxidation State

The oxidation state of Cr is very important relative to its mobility and role in plant and human nutrition. The hexavalent form is more toxic in plant and human nutrition and also known to be more mobile in the soil environment than the trivalent form. Griffin et al (1977) concluded that landfill disposal of Cr(III) would not present a pollution problem, but in case oxidation of Cr(III) to Cr(VI) occurs, it can present a serious problem because of the known mobility of Cr(VI), even at relatively lower pH.

In soil column studies simulating landfill leaching, Artiole and Fuller (1979) found that in anaerobic conditions, Cr(VI) was eluted prior to Cr(III). They explained this difference as due to the anionic nature of the Cr(VI) species, while Cr(III) exists as a cation and its hydroxy forms are prone to precipitate at pH starting at about 4.5.

Long (1983) indicated that the fate of Cr in the environment is more complex than that of the other metals (e.g., Fe, Mn, Cu, Zn, Ni, and Pb) they studied partly because of its more variable oxidation state. Chromium would tend to be fairly immobile in environments with a complex mixture of substrate and in the pH range of 6 to 7.5, allowing maximum adsorption of all Cr species.

6.3 Organic Matter and Electron Donors or Acceptors

Reduction of Cr(VI) to Cr(III) may occur substantially under certain soil conditions. While the presence of Mn appears to be the key to oxidation, serving as the electron acceptor in the reaction, electron donors are like-wise necessary for reduction to proceed. Organic matter and Fe(II) can serve as electron donors in soils and waste disposal systems. Thus, re-duction may not occur in soils low in organic matter and/or soluble Fe

(Bartlett and Kimble, 1976b; Korte et al, 1976). Long (1983) reported that if Fe(II) is present, Cr(VI) could be reduced, which then will tend to immobilize Cr as a hydroxide.

Other organic compounds, such as citric acid, gallic acid, acetic acid, etc., may serve a dual role as chelators for Cr(III) or electron donors for reducing Cr(VI) (James and Bartlett, 1983a, b, c). These compounds may exist in soils *per se* as degradation products of organic wastes, or as metabolites excreted by plant roots.

6.4. Oxides of Iron and Manganese

Bartlett and James (1979) proposed that adsorption of Cr(III) by Mn oxides is a first step in its oxidation by Mn. In aerobic soils, Mn oxides typically occur as surface coatings of clays and Fe oxides. Manganese oxides have high adsorptive capacities for heavy metals because of their large surface areas and high negative charges. Korte et al (1976) observed that soils having high "free" Fe and Mn oxides would significantly retard migration of Cr(VI). Long (1983) found that Mn oxide, and to a lesser extent Fe oxide, are very important in the adsorption of Cr by sediments.

6.5 Redox Potential

Artiole and Fuller (1979) felt that conversion of Cr(III) to Cr(VI) would be slow in the anaerobic soil environment of municipal landfills. It can be partly due to the absence of electron acceptors, such as oxidized Mn and Fe, in reducing conditions. It can be predicted from the Eh-pH diagram presented by Bartlett and Kimble (1976a) that at low redox potential, and particularly at relatively lower pH, Cr would exist in the Cr(III) state.

Perhaps a good way to summarize the fate and behavior of Cr in the environment is through Figure 5.5. In short, the model describes the following pathways for Cr(VI) that enters a stream environment as an industrial waste. Much of the Cr(VI) would be quickly reduced to Cr(III) by organic matter. This Cr(III) could exist as the kinetically stable species, $Cr(H_2O)_4(OH)_2^+$, or (1) be adsorbed by clays (minor importance); (2) precipitate as an insoluble hydroxide if concentrations are high enough (importance unknown); (3) adsorb onto or be complexed with organic matter (probably important); (4) adsorb onto Fe oxides (extent unknown); (5) be oxidized by O_2 (kinetically slow); or (6) adsorb, oxidize, and desorb at the Mn oxide surface (extent and rate not well defined). Chromium in the hexavalent state can (1) exist as the thermodynamically stable ion CrO_4^{2-} as a free species; (2) be adsorbed by Fe oxides; or (3) be reduced by organic matter. Any natural system could be expected to show wide variations within this model as a function of time or location.

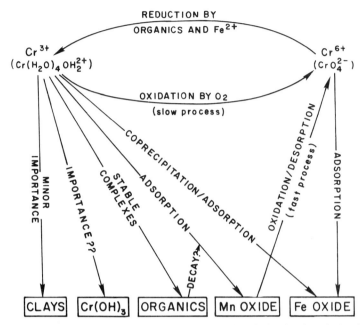

FIGURE 5.5. A proposed qualitative model of chromium behavior in oxic freshwater sediment system at pH 6–7.5. *Source:* Long, 1983, with permission of the author.

7. Chromium in Drinking Water and Food

Because microgram amounts of Cr are essential in human nutrition for maintenance of normal glucose metabolism, the amounts of Cr in drinking water and food are of interest. A study of more than 1,500 surface waters showed a maximum Cr content of 110 µg/liter, with a mean of 10 µg/liter (NAS, 1977). About 96% of Canadian stream and river samples ($n = 4,163$) had <10 µg/liter (NRCC, 1976). A 1962 survey of municipal drinking water in the United States indicates Cr levels ranging from below detection limits up to 35 µg/liter, with a median of 0.43 µg/liter (NAS, 1974). Waters from large rivers of North America have a median Cr content of 5.8 µg/liter and a range of 0.72 to 84 µg/liter (Durum and Haffty, 1963). Maximum Cr content in cold water taps in 50 homes in the eastern United States was only 11 µg/liter, well below the US-EPA limit of 50 µg/liter (Strain et al, 1975). The US-EPA has set a standard of 50 µg Cr/liter for drinking water and 100 µg/liter for freshwater aquatic life (US-EPA, 1976). Similarly, the WHO has set a 50 µg/liter standard for drinking water (WHO, 1984).

Chromium contents of various foodstuffs in the United States and Canada are presented in Table 5.4. Duke (1970) reported the following Cr values (ppm in dry matter): breadfruit, 2.0; coconut, 0.15; taro, 8.0; yam,

TABLE 5.4. Chromium contents of various foodstuffs (μg/g fresh weight).[a]

Foodstuffs	Ottawa/Hull, Canada	Vancouver, Canada	Halifax, Canada	USA
Milk and dairy products	0.11	0.06	0.05	0.23
Meat, fish, and poultry	0.18	0.07	0.06	0.23
Cereals	0.15	0.17	0.06	0.22
Potatoes	0.26	0.05	0.04	0.24
Leafy vegetables	0.09	0.11	0.09	0.10
Legumes	0.10	0.16	0.06	0.04
Root vegetables	0.15	0.06	0.09	0.09
Garden fruits	0.23	0.10	0.05	—
Fruits	0.07	0.06	0.05	0.09
Oils and fats	0.09	0.09	0.03	0.18
Sugars and candy	0.33	0.17	0.34	—
Drinks	0.07	0.03	0.03	—
Condiments and spices	—	—	—	0.33

[a] *Source:* National Research Council of Canada (1976).

0.2; cassava, 0.15; banana, 0.5; rice grain, 0.6; avocado, 0.03; beans, 0.05; and corn grain, 0.25. Masironi et al (1977) reported 0.011 ppm for polished and unpolished rice grain.

Meager information is available on the average daily intake of Cr, even for major countries of the world. The following intakes, in μg Cr/day, have been reported: United States, 60–280 (NAS, 1977); Japan, 130–254 (Murakami et al, 1965); Canada, 136–152 (Kirkpatrick and Coffin, 1974); United Kingdom, 320 ± 162 (Hamilton, 1979); Italy, 50 (Clemente, 1976); and India, 150 (Soman et al, 1969). The International Commission on Radiological Protection set 150 μg/day for an average diet. In general, variations in average daily intake are wide, from 5 μg well into the hundreds of micrograms (NAS, 1974). This is simply caused by dietary preference by peoples around the world. It has been suggested that diets containing mostly processed foods may be Cr deficient (NAS, 1977).

8. Sources of Chromium in the Terrestrial Environment

Chromium can originate from the following major industries: paper products (pulp, paper mills, paperboard, building paper, board mills); organic chemicals, petrochemicals; alkalis, chlorine, inorganic chemicals; fertilizers; petroleum refining; basic steel work, foundries; basic nonferrous metal works, foundries; motor vehicles, aircraft-plating, finishing; flat glass, cement, asbestos poducts, etc.; textile mill products; leather tanning, finishing; and power plants (NRCC 1976).

8.1 Sources via Air Emission

Major portions of overall Cr emissions to the air come from the following industries, with total values of emission after control in 10^3 kg/year: ferrochrome produced by electric furnace—12,360; refractory brick (noncast)—1,630; Cr steel production—520; and coal combustion—1,564 (NRCC, 1976). Deposition of atmospheric particulates from point sources diminish within a few hundred meters (Bourne and Rushin, 1950).

8.2 Sources via Aqueous Emission

Table 5.5 shows the various processes that release uncontrolled waste waters, which can contaminate surface and ground waters. However, methods are available to reduce the levels of Cr in waste waters to trace amounts. Advanced electroplating technology now allows the use of Cr(III) in the plating process instead of the more toxic Cr(VI).

Wastes from Cr smelting can also be a source of water pollution from soluble chromates and other soluble salts (Gemmell, 1973). Chromium could leach from disposal pits (Davids and Lieber, 1951) and landfills (Griffin et al, 1977; Artiole and Fuller, 1979) and contaminate the ground waters.

8.3 Sources via Land Disposal

Typical fertilizer materials contain from trace to sometimes several thousand ppm of Cr. The following values, in ppm, were reported (NRCC, 1976) for: N fertilizers, <5–3,000; P fertilizers, 30–3,000; superphosphates, 60–250; bone meal, <20–500; and limestone, <1–200. Williams (1977) reported a range of 25 to 100 ppm Cr in Australian superphosphates. Mortvedt and Giordano (1977) reported a range of 175 to 485 ppm Cr in P fertilizers. They concluded, however, that with the possible exception of

TABLE 5.5. Aqueous releases of chromium from various industrial processes.

Process	Uncontrolled emissions, ppm	Oxidation state of Cr
Electroplating	trace–600	Cr(VI)
Metal pickling	600	Cr(VI)
Metal bright dip	10,000–50,000	Cr(VI)
Leather tanning	40	Cr(III)
Cooling tower blowdown	10–60	Cr(VI)
Animal glue manufacture	475–600	Cr(VI)
Textile dyeing	1	Cr(III)
Fur dressing and dyeing	20	Cr(III)
Laundry (commercial)	1.2	Cr(VI)

[a] *Source:* National Research Council of Canada (1976).

Zn, plant uptake of heavy metals, including Cr, from fertilizers is not of concern at the usual rates of P applied under field conditions.

Because of the increasing trend in the use of municipal sewage sludge in agricultural lands, heavy metal transfer from this source to the food chain may become a concern (CAST, 1976). Sommers (1977) reported a mean value of 2,620 ppm Cr with a range of 10 to 99,000 ppm ($n = 180$) for the eight states in the north-central and eastern regions. Page (1974) reported that, for several countries, Cr in sewage sludge is extremely variable, ranging from trace amounts to several thousand ppm. Although repeated use or high application rates of sewage sludge can enrich the soil with heavy metals (Andersson and Nilsson, 1972; El-Bassam et al, 1979), Cr has been found to be relatively unavailable to plants from this source (Williams et al, 1980; Zwarich and Mills, 1979; Bradford et al, 1975; Cunningham et al, 1975).

The trend of land applications of fly ash from coal combustion may also increase due to more emphasis on coal as an energy source. However, the concern from this waste is from the phytotoxicity from B and transfer of Mo and Se to the food chain and not from Cr uptake (Adriano et al, 1980).

References

Adriano, D. C., A. L. Page, A. A. Elseewi, A. C. Chang, and I. Straughan. 1980. *J Environ Qual* 9:333–344.

Amacher, M. C. 1981. *Redox reactions involving chromium, plutonium and manganese in soils.* PhD Dissertation. The Pennsylvania State Univ, 166 pp.

Anderson, A. J., D. R. Meyer, and F. K. Mayer. 1973. *Aust J Agric Res* 24: 557–571.

Andersson, A., and K. O. Nilsson. 1972. *Ambio* 1:176–179.

Artiole, J., and W. H. Fuller. 1979. *J Environ Qual* 8:503–510.

Aubert, H., and M. Pinta. 1977. *Trace elements in soils.* Elsevier, New York.

Bartlett, R. J., and J. M. Kimble. 1976a. *J Environ Qual* 5:379–383.

―――― 1976b. *J Environ Qual* 5:383–386.

Bartlett, R., and B. James. 1979. *J Environ Qual* 8:31–35.

Berrow, M. L., and G. A. Reaves. 1984. In *Proc Intl Conf Environ Contamination,* 333–340. CEP Consultants Ltd, Edinburgh, UK.

Bleeker, P., and M. P. Austin. 1970. *Aust J Soil Res* 8:133–143.

Bloomfield, C., and G. Pruden. 1980. *Environ Pollut* A23:103–114.

Bourne, H. G., and W. R. Rushin. 1950. *Ind Med* 19:568–569.

Bowen, H. J. M. 1979. *Environmental chemistry of the elements.* Academic Press, New York. 333 pp.

Bradford, G. R., A. L. Page, L. J. Lund, and W. Olmstead. 1975. *J Environ Qual* 4:123–127.

Breeze, V. G. 1973. *J Appl Ecol* 10:513–525.

Butler, J. R. 1954. *J Soil Sci* 5:156–166.

Cannon, H. L. 1960. *Science* 132:591–598.

―――― 1978. *Geochem Environ* 3:17–31.

Cary, E. E., W. H. Allaway, and O. E. Olson. 1977a. *J Agric Food Chem* 25:300–304.

——— 1977b. *J Agric Food Chem* 25:305–309.

Chino, M. 1981. In K. Kitagishi and I. Yamane, eds. *Heavy metal pollution in soils of Japan,* 66–80. Japan Sci Soc Press, Tokyo.

Clemente, G. F. 1976. *J Radioanal Chem* 32:25–41.

Connor, J., N. F. Shimp, and J. C. F. Tedrow. 1957. *Soil Sci* 83:65–73.

Council for Agricultural Science and Technology (CAST). 1976. *Application of sewage sludge to cropland: appraisal of potential hazards of the heavy metals to plants and animals.* EPA-430/9-76-013. General Serv Admin, Denver. 63 pp.

Cunningham, J. D., D. R. Keeney, and J. A. Ryan. 1975. *J Environ Qual* 4:448–454.

Davids, H. W., and M. Lieber. 1951. *Water Sewage Works* 98:528–534.

Davis, R. D., P. H. T. Beckett, and E. Wollan. 1978. *Plant Soil* 49:395–408.

DeKock, P. C. 1956. *Ann Bot* (London) 20:133–141.

Domingo, L. E., and K. Kyuma. 1983. *Soil Sci Plant Nutr* 29:439–452.

Duke, J. A. 1970. *Econ Bot* 24:344–366.

Durum, W. H., and J. Haffty. 1963. *Geochim Cosmochim Acta* 27:1–11.

El-Bassam, N., C. Tietjen, and J. Esser. 1979. In *Management and control of heavy metals in the environment,* 521–524. CEP Consultants Ltd, Edinburgh, UK.

Frank, R., K. Ishida, and P. Suda. 1976. *Can J Soil Sci* 56:181–196.

Gebhardt, H., and N. T. Coleman. 1974. *Soil Sci Soc Am Proc* 38:263–266.

Gemmell, R. P. 1973. *Environ Pollut* 5:181–197.

Gerritse, R. G., R. Vriesema, J. W. Dalenberg, and H. P. de Roos. 1982. *J Environ Qual* 11:359–364.

Griffin, R. A., A. K. Au, and R. R. Frost. 1977. *J Environ Sci Health* A12(8):431–449.

Grove, J. H., and B. G. Ellis. 1980a. *Soil Sci Soc Am J* 44:238–242.

Grove, J. H., and B. G. Ellis. 1980b. *Soil Sci Soc Am J* 44:243–246.

Haas, A. R. C., and J. N. Brusca. 1961. *Calif Agric* 15(2):10–11.

Hamilton, E. I. 1979. *Trace Subs Environ Health* 13:3–15.

Hewitt, E. J. 1953. *J Exp Bot* 4:59–64.

Hewitt, E. J. 1954. *J Exp Bot* 5:110–118.

Huffman, E., Jr. 1973. *Chromium: essentiality to plants, forms and distribution in plants and availability of plant chromium to rats.* PhD Dissertation. Cornell University. 107 pp.

Huffman, E. W. D., Jr., and W. H. Allaway. 1973a. *Plant Physiol* 52:72–75.

——— 1973b. *J Agric Food Chem* 21:982–986.

Hunter, J. G., and O. Vergnano. 1953. *Ann Appl Biol* 40:761–767.

Jaiswal, P. C., and S. G. Misra. 1984. *J Plant Nutr* 7:541–546.

James, B. R., and R. J. Bartlett. 1983a. *J Environ Qual* 12:169–172.

——— 1983b. *J Environ Qual* 12:173–176.

——— 1983c. *J Environ Qual* 12:177–181.

——— 1984. *J Environ Qual* 13:67–70.

Kirkpatrick, D. C., and D. E. Coffin. 1974. *J Inst Can Sci Technol Aliment* 7:56–58.

Korte, N. E., J. Skopp, W. H. Fuller, E. E. Niebla, and B. A. Alesii. 1976. *Soil Sci* 122:350–359.

Krauskopf, K. B. 1979. *Introduction to geochemistry.* 2nd ed. McGraw-Hill, New York. 617 pp.

Langård, S. 1980. In H. A. Waldron, ed. *Metals in the environment,* 111–132. Academic Press, New York.

Lisk, D. J. 1972. *Adv Agron* 24:267–325.

Long, D. T. 1983. In *Geochemical behavior of chromium/water-sediment systems.* Preliminary Report, EPA Grant no. R808306010, Mich State Univ, East Lansing.

Lukashev, K. I., and N. N. Petukhova. 1975. *Sov Soil Sci* 7:429–439.

Lund, L. J., A. L. Page, and C. O. Nelson. 1976. *J Environ Qual* 5:330–334.

Lyon, G. L., P. J. Peterson and R. R. Brooks. 1969. *Planta* (Berl) 88:282–287.

MacLean, K. S., and W. M. Langille. 1980. *Commun Soil Sci Plant Anal* 11:1041–1049.

Masironi, R., S. R. Koirtyohann, and J. O. Pierce. 1977. *Sci Total Environ* 7:27–43.

Matthews, N. A., and J. L. Morning. 1980. In US Bureau of Mines, *Minerals yearbook 1978–79,* Vol. 1. *Metals and minerals,* 193–205. US Dept of Interior, Washington, DC.

McKeague, J. A., and M. S. Wolynetz. 1980. *Geoderma* 24:299–307.

Mertz, W. 1969. *Physiol Rev* 49:163–239.

Mitchell, R. L. 1964. In F. E. Bear, ed. *Chemistry of the soil,* 320–366. Reinhold, New York.

Mitchell, R. L. 1971. In Tech Bull no. 21. Her Majesty's Sta Office, London.

Mitchell, R. L., J. W. S. Reith, and I. M. Johnson. 1957. *J Sci Food Agric* 8:551–559.

Morley, H. V. 1975. As cited in *Effects of chromium in the Canadian environment.* NRCC no. 15017, Ottawa, Ontario.

Mortvedt, J. J., and P. M. Giordano. 1975. *J Environ Qual* 4:170–174.

―――― 1977. In H. Drucker and R. E. Wildung, eds. *Biological implications of metals in the environment,* 402–416. CONF-750929. NTIS, Springfield, VA.

Murakami, Y., Y. Suzuki, T. Yamagata, and N. Yamagata. 1965. *J Rad Res* 6:105–110.

Myttenaere, C., and J. M. Mousny. 1974. *Plant Soil* 41:65–72.

Nakos, G. 1982. *Plant Soil* 66:271–277.

National Academy of Sciences (NAS). 1974. In *Chromium.* NAS, Washington, DC. 155 pp.

―――― 1977. In *Drinking water and health,* 205–488. NAS, Washington, DC.

National Research Council of Canada (NRCC). 1976. In *Effects of chromium in the Canadian environment.* Publ. no. 15017. Ottawa, Canada. 168 pp.

Page, A. L. 1974. *Fate and effects of trace elements in sewage sludge when applied to agricultural lands.* EPA- 670/2-74-005. US-EPA, Cincinnati, OH. 98 pp.

Peterson, P. J. 1975. In *Proc Intl Conf on Heavy Metals in the Environment,* 39–54. Toronto, Ontario, Canada.

Pierce, F. J., R. H. Dowdy, and D. F. Grigal. 1982. *J Environ Qual* 11:416–422.

Pratt, P. F. 1966. In H.D. Chapman, ed. *Diagnostic criteria for plants and soils,* 136–141. Quality Printing Co, Inc, Abilene, TX.

Proctor, J. 1971. *J Ecology* 59:827–842.

Ramachandran, V., T. J. D'Souza, and K. B. Mistry. 1980. *J Nucl Agric Biol* 9:126–128.

Roberts, B. A. 1980. *Can J Soil Sci* 60:231–240.

Ross, D. S., R. E. Sjogren, and R. J. Bartlett. 1981. *J Environ Qual* 10:145–148.

Schueneman, T. J. 1974. *Plant response to and soil immobilization of increasing levels of Zn^{2+} and Cr^{3+} applied to a catena of sandy soils.* PhD Dissertation, Mich State Univ, East Lansing. 134 pp.

Shacklette, H. T., J. C. Hamilton, J. G. Boerngen, and J. M. Bowles. 1971. In US Geol Survey Paper 574-D. Washington, DC. 71 pp.

Shakuri, E. K. 1978. *Sov Soil Sci* 10:189–194.

Sheppard, M. I., S. C. Sheppard, and D. H. Thibault. 1984. *J Environ Qual* 13:357–361.

Shewry, P. R., and P. J. Peterson. 1974. *J Exp Bot* 25:785–797.

Soane, B. D., and D. H. Saunder. 1959. *Soil Sci* 88:322–330

Soman, S. D., V. K. Panday, K. T. Joseph, and S. J. Raut. 1969. *Health Phys* 17:35–40.

Sommers, L. E. 1977. *J Environ Qual* 6:225–232.

Spence, D. H. N., and E. A. Millar. 1963. *J Ecol* 51:333–343.

Strain, W. H., A. Flynn, E. G. Mansour, F. R. Plecha, W. J. Pories, and O. A. Hill. 1975. In *Proc of Intl Conf on Heavy Metals in the Environment*, 1003–1011. Toronto, Ontario, Canada.

Swaine, D. J., and R. L. Mitchell. 1960. *J Soil Sci* 11:327–368.

Turner, M. A., and R. H. Rust. 1971. *Soil Sci Soc Am Proc* 35:755–758.

US Environmental Protection Agency (US-EPA). 1976. *Quality criteria for water.* US-EPA, Washington, DC.

Vergnano, O. 1959. *Agrochimica* 3:262–269.

Vinogradov, A. P. 1959. *The geochemistry of rare and dispersed chemical elements in soils.* 2d ed. Consultants Bureau, New York. 209 pp.

Walker, R. B., and R. Grover. 1957. *Plant Physiol* 32:Supp 23.

Wallace, A., S. M. Soufi, J. W. Cha, and E. M. Romney. 1976. *Plant Soil* 44:471–473.

Wallace, A., G. V. Alexander, and F. M. Chaudry. 1977. *Commun Soil Sci Plant Anal* 8:751–756.

Warington, K. 1946. *Ann Appl Biol* 33:249–254.

Whitby, L. M., J. Gaynor, and A. J. MacLean. 1978. *Can J Soil Sci* 58:325–330.

Wild, H. 1974. *Kirkia* 9:233–242.

Williams, C. H. 1977. *J Aust Inst Agric Sci* (Sept–Dec):99–109.

Williams, D. E., J. Vlamis, A. H. Pukite, and J. E. Corey. 1980. *Soil Sci* 129:119–132.

World Health Organization (WHO). 1984. *Guidelines for drinking-water quality.* Vol. 1—*Recommendations*. WHO, Geneva, Switzerland, 130 pp.

Zwarich, M. A., and J. G. Mills. 1979. *Can J Soil Sci* 59:231–239.

6
Copper

1. General Properties of Copper

Copper (atomic no. 29), one of the most important metals to man, is reddish colored, takes on a bright metallic luster, and is malleable, ductile, and a good conductor of heat and electricity (second only to silver in electrical conductivity). It belongs to Group I-B of the periodic table, has an atomic weight of 63.546, a melting point of 1,083°C, and specific gravity of 8.96. It consists of two natural isotopes: ^{63}Cu and ^{65}Cu with relative abundances of 69.09% and 30.91%, respectively. The radioactive isotope, ^{64}Cu, with a halflife of 12.8 hours, is the most suitable for tracer work.

In nature, Cu occurs in the I and II oxidation states with ionic radii of 0.96Å and 0.72Å, respectively. In the II state, it is isomorphous with Zn^{2+}, Mg^{2+}, and Fe^{2+} ions. It is found in minerals such as cuprite, malachite, azurite, chalcopyrite, and bornite. The most important Cu ores are the sulfides, oxides, and carbonates. More frequently Cu occurs as a primary sulfate mineral, such as bornite (Cu_5FeS_4) or chalcopyrite ($CuFeS_2$).

2. Production and Uses of Copper

The principal uses of Cu are in the production of wire and its brass and bronze alloys. Copper is alloyed with Sn, Pb, Zn, Ni, Al, and Mn. The electrical industry is one of the major users of Cu in the production of electrical wires and other electrical apparatus. Because of its high thermal conductance and relative inertness, Cu is extensively used in containers such as boilers, steam pipes, automobile radiators, and cooking utensils. The metal is widely used in water delivery systems. It is also extensively used in agriculture in the form of fertilizers, bactericides, and fungicides, and as an algicide in water purification. It is used as a feed additive, such as in antibiotics, drugs, and selected chemical compounds, as a growth promoter, and as an agent for disease control in livestock and poultry production.

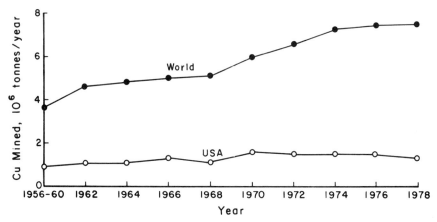

FIGURE 6.1. Trends in copper production in the United States and worldwide, 1956–1978. *Source:* US Bureau of Mines, *Minerals yearbook,* several issues.

The United States has been the world's leader in mine output of Cu (Fig. 6.1) with about 19% of the total in 1979, followed by Chile (14%), the USSR (12%), Canada (9%), Zambia (8%), Peru (5%), Zaire (5%), Poland (4%), the Philippines (4%), Australia (3%), and South Africa (2%). World consumption of refined Cu rose to 9.50×10^6 tonnes in 1978 and to 9.86×10^6 tonnes in 1979, the highest ever recorded (Jolly, 1980).

3. Natural Occurrence of Copper

In nature, Cu forms sulfides, sulfates, sulfosalts, carbonates, and other compounds and also occurs as the native metal. Chalcopyrite ($CuFeS_2$, 34% Cu) is by far the most abundant Cu mineral, being found widely dispersed in rocks and concentrated in the largest Cu ore deposits (Cox, 1979).

Estimates for the average crustal abundance of Cu range from 24 to 55 ppm (Cox, 1979); Bowen (1979) cited 50 ppm. It is ranked 26th among elements in crustal abundance, behind Zn and Ce (Krauskopf, 1979).

The cycling of Cu in the pedosphere is shown in Figure 6.2. The value for soil was based on the assumption that world soils had an average concentration of 20 ppm, lower than the more recent reported value of 30 ppm (Table 6.1). The Cu associated with the soil organic matter represents about 36% of the Cu burden in soil. Shown also in Figure 6.2 are the primary import pathways of Cu to soil: waste disposal, fertilizer application, and atmospheric deposition. The major export pathway is via runoff and erosion.

McKeague and Wolynetz (1980) list the following mean values, in ppm, for world soils: Canada—22 (range of 5–50); United States—25; world soils—20. These are considered background levels of Cu in soils, derived

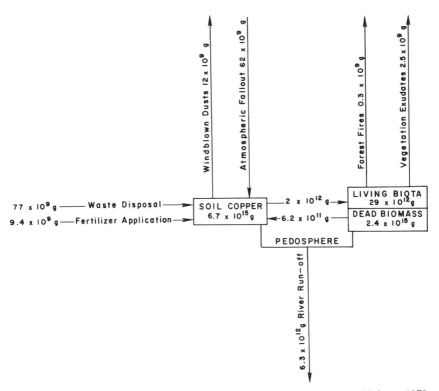

FIGURE 6.2. The annual cycling of copper in the pedosphere. *Source:* Nriagu, 1979, with permission of the author and John Wiley & Sons, New York.

from the soil parent material and by the redistribution of Cu in the profile due to pedogenesis. Levels of Cu in various geologic materials, plants, and other environmental matrices are given in Table 6.1.

4. Copper in Soils

4.1 Total Soil Copper

Recent reported average value for total Cu content of world soils is 30 ppm, with values ranging from 2 to 250 ppm (Table 6.1). For normal agricultural soils, total Cu is expected to range from 1 to about 50 ppm (Gilbert, 1952). In the United States, the severely weathered, leached, and more acid soils of the South Atlantic states averaged much lower in total Cu than soils from western and midwestern states. Total Cu averaged 25 ppm (range of <1–700 ppm) in surface soils of the conterminous United States (Appendix Table 1.17). A mean of 33 ppm total Cu was reported for tropical Asian paddy soils, ranging from 11 ppm for East Malaysian

TABLE 6.1. Commonly observed copper concentrations
(ppm) in various environmental matrices.

Material	Average concentration	Range
Igneous rocks[a]	—	10–100
Limestone & dolomite[b]	6	0.6–13
Sandstone[b]	30	6–46
Shale and clay[b]	35	23–67
Petroleum[c]	1.3	—
Coal[b]	17	1–49
Fly ash:[d]		
Bituminous	132	—
Sub-bituminous	45	—
Lignite	75	—
Sewage sludges[e]	690	100–1,000
Soils[c]	30	2–250
Common crops[f]	—	6–40
Common fruit trees[g]	—	4–20
Ferns[c]	—	15
Fungi[c]	—	7–160
Lichens[c]	—	9–24
Woody angiosperm[c]	—	6–14
Woody gymnosperm[c]	—	5–13
Freshwater[c]*	3	0.2–30
Sea water[c]*	0.25	0.05–12

* µg/liter
Sources: Extracted from
[a] Krauskopf (1972).
[b] Cox (1979).
[c] Bowen (1979).
[d] Adriano et al (1980).
[e] Page (1974).
[f] Follett et al (1981).
[g] Reuther and Labanauskas (1965).

soils to 50 ppm for Sri Lanka soils (Appendix Table 1.18). In the European part of the USSR, total Cu content of soils in the plow layer (0–20 cm) ranged from 4.9 ppm in sod-podzolic sandy soils to 55.2 ppm in meadow-chernozemic soils (Zborishchuk and Zyrin, 1978). Most of the values for the other soil groups fall within 10 to 30 ppm. In China, total Cu content of main soil types fluctuates widely, from 1 to 166 ppm, with an average of 22 ppm (Liu et al, 1983). The Cu content of most soils generally falls in the 20 to 40 ppm range, but values up to 100 ppm or more have occasionally been reported for lateritic soils. For Japanese soils, Shiha (1951) reported an average of 93.1 ppm total Cu, the range being 26.4 to 150.8 ppm. In Canada, the mean value for total Cu in soils is 22 ppm, ranging from 5 to 50 ppm (McKeague and Wolynetz, 1980). Similar mean values were reported for the soils in Manitoba (Mills and Zwarich, 1975) and in Ontario (Whitby et al, 1978). The average total Cu content of British surface soils has been reported as 20 ppm (Swaine and Mitchell, 1960). Similarly, an average value of 20 ppm total Cu has been determined for soils of France (Aubert and Pinta, 1977).

Copper contents of world soils have been characterized by Aubert and Pinta (1977) according to various climatic zones as follows: (1) temperate and boreal regions—total Cu range from trace to several hundred ppm, with the lowest contents (trace to 8 ppm) reported on the fluvio-glacial and alluvial sands of the lower Volga Valley in the USSR; some of the highest values (50 to 200 ppm) in the USSR are found in the Kola Peninsula; (2) arid and semi-arid regions—soils in these regions generally contain from average to high total Cu; and (3) tropical humid regions—soils have fluctuating total Cu contents, with values from trace to 200–250 ppm.

4.2 Extractable (Available) Copper in Soils

Extractable, sometimes called "available," Cu refers to the amount of this element in soil that is readily absorbed and assimilated by plants. It is widely accepted that trace element uptake can be correlated with some extractable fraction of the element in the soil. However, variability in trace element extractability with time, chemical extractant employed, and other extraction conditions must be taken into account. Cox and Kamprath (1972) summarized the various extraction techniques for Cu into three groups: (1) water soluble and exchangeable (e.g., hot water and salts of NH_4^+, Mg^{2+}, and Ca^{2+} ions such as NH_4NO_3, etc.); (2) dilute acid (e.g., dilute HCl and HNO_3); and (3) chelating agents (e.g., EDTA, DTPA, etc.). While total Cu in normal agricultural soils varies from 1 to 50 ppm, available Cu may range from only 0.1 to 10 ppm, depending on the kind of extractant used (Baker, 1974). For example, Follett and Lindsay (1970) found that Cu determined by the DTPA method ranged from 0.14 to 3.18 ppm, while total Cu ranged from 2 to 92 ppm. In this case, total Cu and DTPA-extractable Cu were correlated ($r = 0.67$ to 0.76). Mitchell (1964) found that Scottish soils contained between 0.3 and 10.0 ppm Cu extractable by 0.05 M EDTA.

Often, extractable Cu is used as a diagnostic criterion to identify Cu-deficient conditions for plants. The critical levels for some extractants have been given as: 0.5 M $CuCl_2$—0.18 ppm (Sabet et al, 1975); 0.005 M DTPA—0.20 ppm (Lindsay and Norvell, 1978); NH_4HCO_3—DTPA solution—0.5 ppm (Soltanpour and Schwab, 1977); 1 N HCl—1.9 ppm for citrus culture (Fiskell and Leonard, 1967); 0.05 M EDTA—0.6 ppm (Mitchell, 1964). For Atlantic Coastal Plain soils of North Carolina, Makarim and Cox (1983) found the following critical levels of soil Cu: double acid (Nelson et al, 1953)—0.26 ppm; Mehlich-Bowling reagent (Mehlich and Bowling, 1975)—0.62 ppm; NH_4HCO_3–DTPA (Soltanpour and Schwab, 1977)—0.53 ppm; and Mehlich-3 (Makarim and Cox, 1983)—0.37 ppm. Critical levels of soil-extractable Cu for other extractants are summarized by Robson and Reuter (1981).

Because of the overriding importance of organically-bound Cu in regulating Cu supply to plant roots, chelating agents are becoming more widely

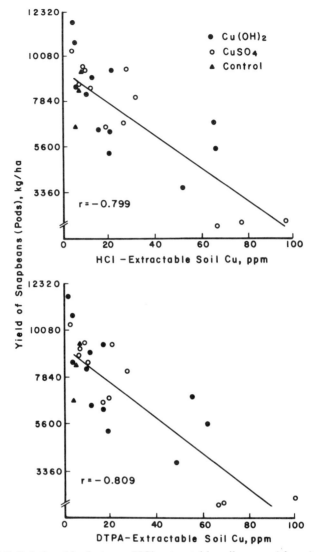

FIGURE 6.3. Relationships between HCl-extractable soil copper (above) and DTPA-extractable soil copper (below) and yield of snapbeans (pods). *Source:* Walsh et al, 1972, with permission of the authors and the Am Soc of Agronomy, Crop Sci Soc of America, Soil Sci Soc of America.

accepted as extractants to measure available Cu (Baker, 1974; Robson and Reuter, 1981). However, other extractants may also prove useful in predicting plant uptake. For example, Kruglova (1962) found that boiling water-soluble Cu correlated best with Cu concentrations in cotton leaves. It should be pointed out, however, that water does not extract sufficient Cu to represent adequately the labile fraction available to plants. Walsh et al (1972) found highly significant negative correlations between 0.1 N HCl, EDTA, and DTPA and the yield of snapbeans (Fig. 6.3). In other

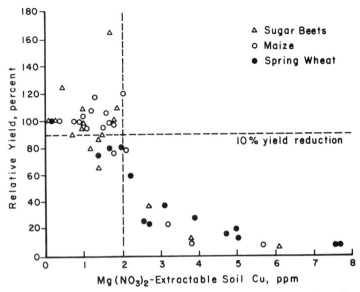

FIGURE 6.4. Relationships between Mg(NO$_3$)$_2$-extractable copper in soil and crop yield. *Source:* Lexmond and de Haan, 1977, with permission of the authors.

words, they found highly significant positive correlations between these extractants and Cu levels in plant tissues up to phytotoxic levels. Gupta and MacKay (1966) recommended the use of 0.2 *M* ammonium oxalate (pH 3) since they found that this extractant was superior to 0.1 *N* HCl. Fiskell (1965) used citrate-EDTA for a quick test to determine toxic Cu levels exceeding 50 ppm, even if the extractant NH$_4$OAc (pH 4.8) was the one generally used. For Wisconsin soils, the extractant 1 *N* NH$_4$OAc (pH 7)/0.01 *M* EDTA shows promise in soil tests for the simultaneous availability of Cu, Zn, and Mn (Dolar and Keeney, 1971; Dolar et al, 1971). It should be pointed out that acidic extractants, such as HCl, may extract Cu from pools that are not available to plants, are less reproducible, and are unsuitable for extracting calcareous soils (Robson and Reuter, 1981).

In the Netherlands, the extractant 0.05 *M* Mg(NO$_3$)$_2$ looks promising especially in soils having high Cu contents (Lexmond and de Haan, 1977). When the level of Cu extracted by this solution exceeds 2 ppm, yield depressions of some crops can be expected to occur (Fig. 6.4).

In paddy soils, the EDTA extractant was recommended to be the most suitable for Cu, followed by NH$_4$OAc and HCl (Selvarajah et al, 1982). There are few instances, however, when plant uptake of Cu could not be predicted by extractable Cu in soil, even if chelating agents were to be used (Jarvis, 1981b; Gough et al, 1980).

Caution was indicated by Beckett et al (1983) and Williams and McLaren (1982) in equating "extractable" to "available" since increased extractability may not necessarily lead to increased uptake and in some cases, uptake may increase even if extractability has not.

Robson and Reuter (1981) proposed that, in order to standardize soil testing procedures, three tests may prove useful for identifying Cu responsive conditions:

1. For acid and near-neutral soils (containing no free $CaCO_3$)—0.02 M EDTA in 1 M NH_4OAc (pH 7), as recommended by Borggaard (1976);
2. For neutral and calcareous soils—0.005 M DTPA containing 0.01 M $CaCl_2$ and 0.1 M triethanolamine (pH 7.3), as recommended by Lindsay and Norvell (1978); and
3. For polluted or potentially toxic soils—0.1 M EDTA (disodium salt) adjusted to pH 6 with NH_4OH, as recommended by Clayton and Tiller (1979).

4.3 Copper in the Soil Profile

Applied or deposited Cu will persist in soil because it is strongly fixed by organic matter, oxides of Fe, Al, and Mn, and clay minerals (Baker, 1974; Gilbert, 1952; Schnitzer, 1969). Thus, it is one of the least mobile of the trace elements, thereby rendering it uniformly distributed in many soil profiles. For instance, Jones and Belling (1967) observed virtually no downward movement of Cu on silty and clayey soils and only slight movement (1–3 cm) in sandy soils with a low cation exchange capacity. Copper tends to accumulate in the litter overlying the soil profile or within the top few centimeters of the soil profile, as observed in soils near a Cu smelter (Kuo et al, 1983) or in soils heavily impacted by industrial emissions (Miller and McFee, 1983). Similarly, in agricultural soils, Cu may be expected to accumulate in the surface from fungicide applications, soil amendments, and by accumulation from crop residues (Wright et al, 1955; Pack et al, 1953; Walsh et al, 1972; Thornton, 1979). However, considerable leaching of Cu in the soil profile can occur in exceptional cases. This can happen in humus-poor, acidic peat (histosol) upon application of high levels of Cu (1,500 ppm or more) (Mathur et al, 1984) or in very acidic mineral soils, such as those around the Ni and Cu smelters in Coniston, Ontario, Canada (Hazlett et al, 1984). Soils around these smelters are acidified to as low as pH 2.4, caused by the aerial emission of sulfur-containing compounds during smelting.

The level and distribution of total and extractable Cu in the soil profile can be expected to vary with soil type and parent materials from which the soils are derived (Vinogradov, 1959; Bleeker and Austin, 1970; Swaine and Mitchell, 1960). Distribution of total Cu can be expected to be fairly uniform in profiles developed *in situ* on igneous rock or on unsorted sediment such as sandstone or loess (Mitchell, 1964). Little variation can also be expected in freely drained soils (Thornton and Webb, 1980), but Cu may be mobile in profiles with restricted drainage, with extractable Cu possibly accumulating in gleyed horizons of these soils (Swaine and Mitchell, 1960). However, profile distribution of Cu can be altered by various pedological factors—physical, chemical, and biological in nature.

Thornton (1979) reported that there is a strong evidence that Cu is redistributed in the profile as a result of podzolization, with depletion in the A2 horizon and accumulation in the B. Examples have been cited from podzols in the United States, the United Kingdom, and Canada. Other factors, such as organic matter accumulation and pH can also influence the distribution of Cu in the soil profile (Holmes, 1943; Follett and Lindsay, 1970; Thornton, 1979).

4.4 Forms and Speciation of Copper in Soils

Copper in soil can occur in several forms (Fig. 6.5): (1) in the soil solution, ionic and complexed; (2) in normal exchange sites; (3) on specific sorption sites; (4) occluded in soil oxide material; (5) in organic residues and living organisms; and (6) in the lattice structure of primary and secondary minerals. From the standpoint of plant nutrition, the dissolved and exchangeable forms are of special importance. These various forms are estimated by sequential extraction, similar to the scheme shown in Figure 6.5. Copper in the soil solution and exchangeable, specifically adsorbed, and organically-bound Cu are considered in equilibrium and represent the available forms for plant uptake; Cu in the oxide-bound and residual forms is relatively unavailable to plants (Viets, 1962; McLaren and Crawford, 1973a).

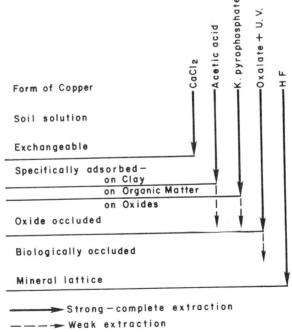

FIGURE 6.5. An example of a fractionation scheme for soil copper. *Source:* McLaren and Crawford, 1973a, with permission of the authors and Blackwell Scientific, Oxford, England.

The amount of Cu in a given form can vary according to extraction time and chemical extractant. McLaren and Crawford (1973a) found that the distribution of Cu between the main soil constituents can be influenced considerably by the presence of free manganese oxides. They found that the bulk of the "available" soil Cu reserves resides in the organically-bound fraction. In addition, they (McLaren and Crawford, 1974) found that not all organically-bound Cu is readily isotopically exchangeable, but it is the one most ready to diffuse from soil. Organic forms of Cu can be expected to be the dominant form in histosols (Mathur and Levesque, 1983) and in soils treated with sewage sludge (Sposito et al, 1982; Alloway et al, 1979). The organic-bound form of Cu also can be expected to dominate in ordinary arable soils (Shuman, 1979; Alloway et al, 1979) and in other soils (Harrison et al, 1981; Kishk et al, 1973). Apparently, enrichment of the soil with Cu, from either organic or inorganic sources, can be reflected by increases in most fractions, with the exception of the residual form, with most of the increases occurring in the organic fraction (Mullins et al, 1982a, b). In some cases, however, Cu associated with Fe and Mn oxides may constitute the major fractions of total Cu in soils (Kuo et al, 1983; Miller and McFee, 1983).

McBride (1981) indicated that, among the anions (nitrate, chloride, sulfate, and phosphate) in soil solutions, sulfate and chloride may form complexes with Cu^{2+} in saline soils. He further indicated that only the hydroxy and carbonate complexes are expected to exist as important species in soil solution. Mattigod and Sposito (1977) suggested that the major inorganic form of complexed Cu^{2+} in neutral and alkaline soil solutions would be $CuCO_3^0$. In highly alkaline soils, $Cu(OH)_4^{2-}$ and $Cu(CO_3)_2^{2-}$ become the predominant soluble species.

In soil solutions from sewage sludge-amended soils, the GEOCHEM computer program predicted that only small percentages (2%–19%) of the total Cu occurred as free-ionic form (Emmerich et al, 1982; Behel et al, 1983); even smaller amounts were complexed with sulfate and phosphate, while up to about 80% of the total amount of Cu occurred in the fulvate-metal complexes (Behel et al, 1983). McBride and Bouldin (1984) estimated that at least 99.5% of the Cu in soil solution of a surface loamy soil (5.3% organic matter content) was in an organically-complexed form. Similarly, Kerven et al (1984) found that most of the Cu in the soil solution of acid peaty soils was present as organic complexes. However, in solutions containing relatively high concentrations of chlorides, upwards to 80% of Cu^{2+} can be complexed with this ligand (Doner et al, 1982).

4.5 Adsorption and Fixation of Copper in Soils

In his comprehensive reviews of trace element sorption by sediments and soils, Jenne (1968, 1977) indicated that the most important sinks for trace elements are the oxides of Fe and Mn, organic matter, sulfides, and carbonates. Of lesser importance are the phosphates, iron salts, and the clay-size aluminosilicate minerals. In addition, he proposed that the most significant role of clay-size minerals in trace element sorption by soils and sediments is as a mechanical substrate for the precipitation and flocculation of organics and secondary minerals. Clay minerals may contain trace elements as structural components, as exchangeable cations, and by specific sorption. Since $CaCO_3$ is abundant in soils of arid and semi-arid regions, it may be an important sink, particularly for Zn and Cd, in these soils. Sulfides, which are prevalent in soils under reducing conditions, are an important sink because they effectively precipitate trace elements, including Cu.

Copper can be "specifically" adsorbed by layer silicate clays, oxides of Fe, Mn, and Al, and organic matter. In specific adsorption, ions are held much more strongly by the surface since these ions penetrate the coordination shell of the structural atom and are bonded by covalent bonds via O atoms (e.g., Cu-O-Al or Cu-O-Fe bond) or OH groups to the structural cations (Huang, 1980). More simply, specific adsorption refers to adsorption in the presence of excess amounts of Ca^{2+} (e.g., strong background concentration of $CaCl_2$) or some other electrostatically bonded metal ion that is capable of preventing significant Cu^{2+} adsorption by simple ion exchange (McBride, 1981). In this regard, Jenne (1977) and McLaren et al (1981) concluded that the oxides of Fe and Mn and organic matter are the most important soil constituents. They ranked the adsorption maxima among various soil constituents as follows: Mn oxide > organic matter > Fe oxide > clay minerals. However, in soils where clay minerals and free iron oxides are dominant, their gross contribution can override those of organic matter and free Mn oxides (McLaren and Crawford, 1973b). Stevenson and Fitch (1981) contended, however, that organic colloids and clay play a major role in Cu retention by soil. In most mineral soils, Cu may be bound as clay-metal-organic complex, since in these soils, organic matter is intimately bound to the clay. They implied through pesticide retention results that, in soils containing up to 8% organic matter, both organic and mineral surfaces are involved in adsorption; at higher organic matter contents, adsorption occurs mostly on organic surfaces. McLaren and Crawford (1973b) also contended that for the relatively small amounts of Cu normally present in soils, specific adsorption is a more important process than the nonspecific adsorption (through cation exchange process) in controlling the concentration of Cu in the soil solution, even if the latter (i.e., CEC) was greater than the specific adsorption maximum.

The relative adsorptive capacities of the various soil constituents are depicted in Figure 6.6, which demonstrates that Mn oxide and organic matter are the most likely materials to bind Cu^{2+} in a nonexchangeable form in soils. With the possible exception of Pb^{2+}, Cu^{2+} is the most strongly adsorbed of all the divalent transition and heavy metals on Fe and Al oxides and oxyhydroxides (McBride, 1981).

The Cu sorption capacity of soils may follow closely the extent of CEC

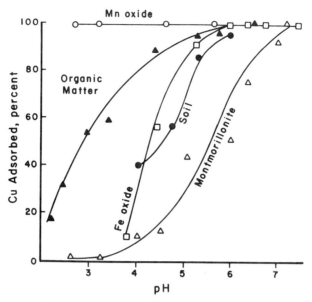

FIGURE 6.6. Specific adsorption of Cu^{2+} by soil constituents as a function of pH (One gram sample equilibrated with 200 ml of 0.05 M $CaCl_2$ containing 5 ppm copper). *Source:* McLaren and Crawford, 1973b, with permission of the authors and Blackwell Scientific, Oxford, England.

and the amounts of the oxides of Fe and Mn and organic matter in the soil. More importantly, organic matter alone may not be a good indicator of the sorption capacity of soil (Kuo and Baker, 1980; Harter, 1979; Dhillon et al, 1981).

The sorption of Cu by soils and/or soil constituents has been shown to follow both the Langmuir (McLaren and Crawford, 1973b; Cavallaro and McBride, 1978; Petruzzelli et al, 1978; Harter, 1979; Dhillon et al, 1981) and the Freundlich (Jarvis, 1981a; Sidle and Kardos, 1977; Sanders, 1980) isotherms for adsorption. From Langmuir isotherm plots, Takahashi and Imai (1983) deduced that the montmorillonite surface seems to provide three types of adsorption sites for Cu. Adsorption onto site 1 corresponds to ion-exchange reaction, which accounts for 70%–85% of the adsorption maximum. They explained that precipitation of the metal hydroxides in the bulk solution corresponds to site 3. Adsorption onto site 2 also seemed to be due to formation of metal-hydroxy species, although it occurred even at pH 4. The adsorption maximum for site 2 exceeded the CEC value of montmorillonite.

Cavallaro and McBride (1978) summarized their findings on the relative adsorption of Cu and Cd by mineral soils:

1. Cu^{2+} ion is much more strongly adsorbed than Cd^{2+} in soils, so that Cd^{2+} is likely to be more mobile than Cu^{2+} in the soil profile.

2. The soil solutions are generally undersaturated with respect to the least soluble mineral phases of these heavy metals, indicating that precipitation phenomena do not control metal solubility in the surface soils.
3. Competing ions such as Ca^{2+} shift the adsorption equilibria for Cd^{2+} drastically, with less effect upon Cu^{2+}, suggesting that ion exchange is responsible for Cd^{2+} adsorption, but Cu^{2+} may be bonded more specifically.
4. Low pH soils are much less effective in removing Cu^{2+} and Cd^{2+} from solution than are neutral soils or soils containing calcium carbonates.
5. Soluble metal-organic complexes of Cu^{2+} are more prevalent than those of Cd^{2+} in soil solution, but acid soils demonstrate much less evidence of these soluble complexes.

In acidic soils, a greater fraction of functional groups in the soil organic matter is associated with protons or Al^{3+}, thereby reducing the ability to adsorb Cu^{2+}.

Hodgson et al (1966) showed that more than 98% of the Cu in solution was in an organic complexed form, suggesting that only small quantities of free Cu^{2+} ions are available for adsorption in neutral soils. The hydrolysis constant for Cu^{2+} (for the reaction $Cu^{2+} + H_2O \rightleftharpoons CuOH^+ + H^+$) is $10^{-7.6}$, indicating that at pH levels above 7, the hydrolyzed species, $CuOH^+$, would be more important than Cu^{2+} for adsorption (Ellis and Knezek, 1972). Although $CuOH^+$ has been suggested as the adsorbed hydroxy ion (Menzel and Jackson, 1950), solution hydrolysis data suggest that $Cu_2(OH)_2^{2+}$ and $Cu(OH)_2^0$ are probably more important species (Baes and Mesmer, 1976).

4.6 Complexation of Copper in Soils

Coleman et al (1956) early predicted that, because of the stability of Cu-organic complex (peat was the source of the organic ligand), especially at low salt concentrations, only extremely small quantities of free Cu^{2+} ions would exist in soil solutions. A number of studies have shown that Cu^{2+} in soil solutions, especially at higher pH, exists primarily in a form complexed with soluble organics (Hodgson ct al, 1965, 1966). In soils amended with organic residues, such as sewage sludge, organically-complexed Cu may be expected to be the predominant form (Behel et al, 1983). Stevenson and Fitch (1981) reported that, based on a commonly used soil Cu fractionation scheme, organically-bound Cu accounts for about 20% to 50% of the total soil Cu. The amount of organically-complexed Cu in solution generally increases above pH 7 because of the greater solubility of soil organic matter at higher pH (McBride and Blasiak, 1979), while the concentration of free ionic Cu at higher pH is much lower, usually in the range of 10^{-9} to 10^{-8} M (Hodgson et al, 1965; McBride and Blasiak, 1979). Total native Cu in soil solution is often in the range 10^{-7} to 10^{-6} M (Bradford et al, 1971), so that the ratio of total dissolved Cu to free ionic

Cu can be greater than 100. However, this ratio may be only as low as 4 in soil solutions of organically-enriched soils (Behel et al, 1983).

In section 4.3, it was indicated that organic-enriched surface horizons usually contain higher concentrations of Cu than the lower horizons that usually contain less organic matter. Thus complexation by organic matter, in the form of humic and fulvic acid, has long been recognized as an effective mechanism of Cu retention in soils (McBride, 1981). Ample evidence exists for the complexation of Cu^{2+} by humic and fulvic acids (Stevenson and Fitch, 1981):

1. Inability of K^+ and other monovalent cations to replace adsorbed Cu^{2+} from mineral and organic soils;
2. Correlation between Cu^{2+} retention and humus content;
3. Ability of known chelating agents to extract Cu while solubilizing part of the soil humus; and
4. Selective retention of Cu^{2+} by humic and fulvic acids in the presence of a cation-exchange resin.

In addition, several investigators have shown that, among the metals, including the transition and heavy metals, Cu is the most extensively complexed by humic materials (Bloomfield and Sanders, 1977; Baker, 1974). Bloom (1981) proposed the following preference series for divalent ions for humic acids and peat: Cu > Pb >> Fe > Ni = Co = Zn > Mn = Ca. The complexing ability of humic and fulvic acids is due to their high oxygen-containing functional groups, such as carboxyl, phenolic hydroxyl, and carbonyls of various types (Stevenson, 1972; Schnitzer, 1969). Because of high acidities and relatively low molecular weights, metal complexes of fulvic acids are more mobile than those of humic acids. Fulvic acids are also more efficient in complexing metals than humic acids (Stevenson, 1972; Kawaguchi and Kyuma; 1959) and may also be more available to plants than humic acid-complexed metals (Stevenson and Fitch, 1981).

As discussed in section 4.5, colloidal organic matter in soil provides "specific" sorption sites for Cu, thereby causing Cu deficiency in organic soils. The fraction of soil Cu associated with organic matter would be expected to be high in soil rich in organic matter (e.g., histosols and mollisols).

Stevenson and Fitch (1981) summarized the following effects on soil of the formation of Cu-organic matter complexes:

1. The soil solution Cu concentration can be decreased through complexation to clay-humus or to the formation of insoluble complexes with humic acids. Soluble ligands may be of considerable importance in transforming solid-phase forms of Cu into dissolved forms;
2. In high pH soils (e.g., calcareous soils) complexation will promote maintaining Cu in dissolved forms;

3. In the presence of excess Cu, complexation may reduce the concentration of Cu^{2+} to nontoxic levels; and
4. Natural complexing agents may be involved in the transport and mobility of Cu in soils.

Like the natural complexing agents (humic and fulvic acids) that enhance the migration of metals, synthetic chelating agents (e.g., EDTA, DTPA, etc.) combine with trace metals to increase the total levels of these metals in soil solutions, thereby increasing the nutrient availability by increasing both diffusion and mass-flow of these metals to plant roots (Lindsay, 1974). However, the stability of metal-synthetic chelating agents depends largely on soil pH. Copper-DTPA, like ZnDTPA, is expected to be unstable in acid soils, moderately stable in slightly acid soils, and stable in alkaline and calcareous soils (Norvell and Lindsay, 1972; Lindsay and Norvell, 1969; Norvell, 1972). In contrast, CuEDTA is most stable in slightly acidic to near neutral soils (pH 6.1–7.3) (Norvell and Lindsay, 1969). In acid soils (pH 5.7 and below) CuEDTA becomes unstable, since Fe displaces Cu, while in alkaline soils (pH 7.85 and above), Ca displaces Cu.

5. Copper in Plants

5.1 Essentiality of Copper in Plants

Copper is one of the seven micronutrients (Zn, Cu, Mn, Fe, B, Mo, and Cl) essential for normal plant nutrition. In the 1930s, primarily through the work of Sommer (1931) and Lipman and MacKinney (1931) with nutrient cultures, the essentiality of Cu was firmly established. During the ensuing 20 years, a great mass of evidence was gathered confirming the essentiality of Cu for the growth of plants (Reuther and Labanauskas, 1965).

Copper is required in very small amounts: 5 to 20 ppm in plant tissue is adequate for normal growth (Jones, 1972), while less than 4 ppm is considered deficient and 20+ ppm is considered toxic. Copper has been established as a constituent of a number of plant enzymes. It occurs as part of the prosthetic groups of enzymes, as an activator of enzyme systems, and as a facultative activator in enzyme systems (Gupta, 1979). Bussler (1981) reported that a lack of Cu affects a multitude of physiological processes in plant: carbohydrate metabolism (photosynthesis, respiration, and carbohydrate distribution), N metabolism (N_2 fixation and protein synthesis and degradation), cell wall metabolism (especially lignin synthesis), water relations, seed production (especially pollen viability), and disease resistance. It may also affect ion uptake and plant differentiation in early developmental stages.

5.2 Deficiency of Copper in Plants

A micronutrient survey conducted between 1961 and 1967 revealed that 16 states recognized and made recommendations for Cu application for major field, forage, vegetable, fruit, and nut crops (Sparr, 1970). The world distribution of Cu deficiency follows closely that of Zn deficiency, occurring in the Americas, Africa, Australia, Europe, and other parts of the world (Bould et al, 1953). In other parts of the world, Cu deficiency has also been known as "reclamation disease," "wither-tip," "yellow-tip," or "blind ear" for grain crops. In woody species, particularly in citrus, the malady is known as "dieback" or "exanthema." Deficiencies have been reported in the Netherlands, Norway, the USSR, and Germany (Gartrell, 1981); also in Denmark (Steenbjerg and Boken, 1950), Australia (Piper, 1942; King, 1974), Sweden (Lundblad et al, 1949), India (Grewal et al, 1969), Egypt (Elrashidi et al, 1977), the United Kingdom (Purves and Ragg, 1962; Caldwell, 1971; Davies et al, 1971; Reith, 1968), Poland (Liwski, 1963), and New Zealand (Cunningham et al, 1956).

Reuther and Labanauskas (1965) and Reuter (1975) reported the prevalent occurrence of Cu deficiency in:

1. Peat and muck soils
2. Alkaline and calcareous soils, especially sandy types or those with high levels of free $CaCO_3$
3. Highly leached sandy soils, such as the acid sands of Florida
4. Highly leached acid soils, in general
5, Calcareous sands as in the south coast of Australia
6. Soils heavily fertilized with N, P, and Zn
7. Old corrals
8. Droughty soils

Copper deficiency in plants has frequently appeared in many parts of the world whenever acid histosols were brought into agricultural production (Kubota and Allaway, 1972). The term "reclamation disease" refers to severe Cu deficiency that appears in newly drained and developed acid histosols. In addition, Gartrell (1981) reported that peats and mucks have commonly produced Cu-deficient crops throughout northern and western Europe, some midwestern and eastern states in the United States, and in New Zealand and Australia. Furthermore, Barnes and Cox (1973) reported that, in general, soils most commonly found to be deficient in Cu are poorly drained mineral soils, mineral soils high in organic matter, and histosols. Thus, application of Cu fertilizers on organic soils is usually recommended for normal plant growth. Lucas and Knezek (1972) indicated that the total Cu in the soil should exceed 4 to 6 ppm on mineral soils and 20 to 30 ppm on organic soils to sustain maximum yields of responsive crops. The responsiveness of crops to Cu fertilization is indicated in Table 6.2, with some major cereal crops noticeably highly responsive. Crops that are highly sensitive to Cu deficiency can serve as "indicator" species

TABLE 6.2. Relative sensitivity of crops to copper
deficiency.[a]

Low	Moderate	High
Beans	Barley	Wheat
Peas	Broccoli	Lucerne
Potato	Cabbage	Carrots
Asparagus	Cauliflower	Lettuce
Rye	Celery	Spinach
Pasture grasses	Clover	Table beets
Lotus spp.	Parsnips	Sudangrass
Soybean	Radish	Citrus
Lupine	Sugar beets	Onions
Rape	Turnips	Alfalfa
Pines	Pome and stone fruits	Oats
Peppermint	Vines	Barley
Spearmint	Pineapple	Pangolagrass
Rice	Cucumber	Millet
Rutabagas	Sugar beets	Sunflower
	Corn	Dill
	Cotton	
	Sorghum	
	Sweet corn	
	Tomato	
	Apple	
	Peaches	
	Pears	
	Blueberries	
	Strawberries	
	Tung-oil	
	Mangels	
	Swiss chard	

[a] *Sources:* Gartrell (1981); Lucas and Knezek (1972); Follett et al (1981).

for detecting Cu-deficient soils. Some of the sensitive species, with their
Cu concentrations are shown in Table 6.3.

Copper deficiency in conifers could occur in trees growing in sandy
soils low in Cu contents, and in soils likely to immobilize Cu, such as
peat soils, sandy podzols, and calcareous soils. Turvey (1984) reported
that Cu deficiency in *Pseudotsuga menziesii* occurred on acid humus podzols in the Netherlands, in *Pinus radiata* on alkaline soils in Greece, and
on alkaline soils in south Australia. Turvey (1984) found that Cu deficiency
in *Pinus radiata* in Victoria, Australia was associated with acid, organic-
rich sandy podzols. Very acid, infertile, humus-podzols in forests in
northeastern Netherlands are also prone to cause Cu deficiency in conifers
(van den Burg, 1983). In general, conifers are more susceptible to Cu
deficiency than broadleaf species.

In most plant species, Cu deficiency is characterized by chlorosis, necrosis, leaf distortion, and terminal dieback, with symptoms occurring
first in young shoot tissues (Robson and Reuter, 1981). Once absorbed,

TABLE 6.3. Deficient, sufficient, and toxic levels of copper in various crops.[a]

Plant species	Part or plant tissue sampled	Deficient	Sufficient	Toxic
		ppm Cu in dry matter		
Spring wheat	Boot stage tissue	1.9	3.2	
	Kernels	0.8	2.3	
	Straw	2.4	3.9	
	Grain	2.5[b]	—	
Winter wheat	Aboveground vegetative plant tissue		5–10	>10[c]
Barley	Boot stage tissue	2.3	4.8	
	Kernels	0.5	2.0	
	Straw	1.5	3.0	
	Top 4 leaves at bloom		6–12	>12[c]
Oats	Boot stage tissue	2.3	3.3	
	Kernels	0.7	1.8	
	Straw	1.2	2.3	
	Leaf lamina	<4.0	6.8–16.5	
Soybeans	Upper fully mature trifoliate leaves before pod set	<4	10–30	>50
Corn	Leaf at or opposite and below ear level at tassel stage	5[b]	—	
	Ear leaf sampled when in initial silk	<2	6–20	>50
	Middle of first ear leaf at tasseling	<2	6–50	>70
Alfalfa	Upper stem cuttings in early flower stage	7[b]	—	
	Top 15 cm of plant sampled before bloom	<5	11–30	>50
	Top ⅓ of plant shortly before flowering	<2	8–30	>60
Red clover	Tops		7–16.4	
	Top of plants at bloom	<3	8–17	>17[c]
	Top of plants at bloom		9.8–11.5	
Timothy	Tops	—	5.7–11.7	
Pasture grasses	Aboveground part at first cut	<5	5–12	>12[c]
Potato	Aboveground part 75 days from planting	<8	11–20	>20[c]
Cucumber	Mature leaves from center of stem 2 weeks after picking	<2	7–10	>10[c]
Tomato	Mature leaves from top of plant	<5	8–15	>15[c]
	Mature leaves		3.1–12.3	

[a] *Source:* From several sources, summarized by Gupta, 1979, with permission of the author and the publisher.
[b] Considered critical level.
[c] Considered high but not toxic level.

Cu is poorly translocated. Hence, the terminal growth of most plants is the first to be affected. Specific symptoms often depend on plant genotypes and the stage of deficiency. In general, deficiencies in crops produce abnormal coloring and development, lowered quality in fruit and grain, and lowered grain yields (Murphy and Walsh, 1972).

5.3 Accumulation and Toxicity of Copper in Plants

Absorption rates of Cu by plant roots are among the lowest for the essential elements (Graham, 1981). The absorption process has been characterized as active. Graham further pointed out that there exists either a linear relationship between absorption rate and external Cu concentration, or a curvilinear relationship in the high concentration range. Apparently, once absorbed, Cu accumulates in roots, even in cases where roots have been damaged by toxicity (Jarvis, 1978; Brams and Fiskell, 1971; Andrew and Thorne, 1962; Osawa and Ikeda, 1974; Dykeman and de Sousa, 1966).

In a national survey (n = 2,399), Kubota (1983) found a median concentration of 8.4 ppm Cu in the dry matter for US legumes (range of 1–28 ppm); the median value for grasses was much lower—4 ppm (range of 1–16 ppm). Most of the forage samples were from soils with about 25 ppm total Cu (range of 1–200 ppm total Cu). Davis et al (1978) found that 20 ppm Cu in the shoots of spring barley grown in sand culture (4 ppm Cu in solution) was critical for its growth. Chino (1981) reported that for rice, toxic levels in tops were in the range of 20 to 30 ppm Cu; in roots, toxic range was from 100 to 300 ppm. Agarwala et al (1977) ranked the following heavy metals in inducing visual toxicity symptoms in barley as follows:

$$Ni^{2+} > Co^{2+} > Cu^{2+} > Mn^{2+} > Zn^{2+}$$

In rice, Chino and Kitagishi (1966) found that the toxicity of metals follow the order: Cu > Ni > Co > Zn > Mn, which correspondingly follows the order of the metal electronegativity. Hewitt (1953) observed that Cu consistently induced Fe chlorosis in crops that were susceptible to Cu toxicity.

Phytotoxicity of Cu could be predicted by Cu concentrations in soils, either on total or extractable basis. Copper can accumulate in soils from continued applications of Cu in excess of the need for normal plant growth. In Florida, problems of excessive Cu in citrus soils are a much greater problem than Cu deficiency (Leonard, 1967). Reuther and Smith (1952) reported that the top 15 cm of soil in some citrus groves in Florida contained Cu in excess of 300 ppm, which had accumulated from repeated use of Bordeaux sprays. Generally, Cu toxicities have been associated with soil Cu levels of 150 to 400 ppm (Baker, 1974).

In a field study, significant reductions in yield of snapbeans were noted when more than 20 ppm of Cu was extracted from soil with HCl or DTPA and when more than 15 ppm Cu was extracted with EDTA (Walsh et al, 1972). In Japan, the limit of Cu in the soil has been set at 125 ppm extractable by 0.1 N HCl for rice culture (Chino, 1981). Lexmond and de Haan (1980) suggested that, in soils with pH of 4.5 to 6.5, soil Cu may become phytotoxic when 20 to 30 ppm of HNO_3-extractable Cu is present for each 1% of soil organic C. Delas (1980) noted that the threshold of soil-Cu phytotoxicity is about 25 ppm of NH_4-exchangeable Cu for sandy soils and 100 ppm in clay soils. Similarly, Drouineau and Mazoyer (1962) reported the limit at 50 ppm NH_4-exchangeable Cu in soils with pH 5 and 100 ppm in soils with pH 6 to 7. Mathur and Levesque (1983) proposed that soil Cu could become phytotoxic when its total level in certain organic soils is equivalent to more than 5% of the soil's CEC—in their case, 16 ppm Cu for each meq of CEC per 100 g soil. The soil's CEC was determined by neutral NH_4OAc method.

Reuther and Smith (1954) recognized the importance of organic matter and clay in Cu retention and found that, in general, phytotoxicity in orchard trees began when total soil-Cu level reached about 1.6 mg for each meq of CEC per 100 g of soil having a pH of 5.0. Assuming their results can be extrapolated to other soils, the maximum loading capacity for Cu would be 36 kg/ha for each meq of CEC per 100 g soil. In soils with pH of about 5, the loading limit for total soil-Cu can be estimated to equal 5% of soil CEC (Leeper, 1978).

The most common Cu toxicity symptoms include reduced growth vigor, poorly developed and discolored root system, and leaf chlorosis (Robson and Reuter, 1981). In addition, Cu toxicity causes stunting, reduced branching, thickening, and unusually dark coloration in the rootlets of many plants (Reuther and Labanauskas, 1965). The chlorotic symptoms in shoots often resemble those of Fe deficiency.

6. Copper in Natural Ecosystems

It is well recognized that long-distance transport of air pollutants can affect remote areas including forest ecosystems. Nriagu (1979b) reported that there has been substantial enrichment of Cu in atmospheric particulates collected at several remote locations. This has been confirmed by Reiners et al (1975), who observed high depositions of heavy metals in the New England mountain region, and by Heinrichs and Mayer (1980), who concluded that the Solling mountains in central Germany are subject to the input of airborne pollutants transported over long distances. Similarly, Rühling and Tyler (1971) in Sweden demonstrated distinct regional gradients for some heavy metals, including Cu, which indicated that the airborne metals originated from outside Sweden.

The total aerial burden of metals is presumed to originate from both present-day urban-industrial sources (particulates, aerosols, smokes, automobile exhaust, etc.) and power generation (combustion of fossil fuel), and former industry (via resuspension of contaminated soil and industrial wastes). Lead, Cd, Zn, and Cu are usually the principal metals of concern (see chapters on Zn, Pb, and Cd) in forest ecosystems because they are often the main metal constituents of air pollutants and because of their known toxicities to biota. It should be pointed out, however, that the primary route of Cu into forest ecosystems is from the soil itself and its parent material (Lepp, 1979).

Aerial Cu apparently is originating from coal combustion in urban areas (Kneip et al, 1970). In the Solling mountains, Germany, the annual precipitation input of Cu is about two times greater than at an urban Indiana (United States) site (Parker et al, 1978).

Studies in forest ecosystems indicate that metals accumulate primarily in the forest floor humus and that atmospheric input exceeds stream output (Van Hook et al, 1977; Siccama and Smith, 1978). In the Solling forest ecosystem (Heinrichs and Mayer, 1980) Cu accumulated primarily in the organic matter, with only about 8% in the total ecosystem stored in the biomass, and 2% in the organic topsoil. Similarly, in the Bärhalde watershed in central Europe, increased atmospheric input of metals has resulted in an enrichment and rise of turnover for Cu, as well as Cd and Pb (Stahr et al, 1980). These three metals are adsorbed primarily by the humus and oxides of Fe in the soil.

The enrichment of Cu in various components of a spruce forest in central Sweden due to atmospheric deposition is indicated in Table 6.4. Copper enrichments of various forest ecosystem compartments due to atmospheric input have also been reported near urban northwestern Indiana (Parker

TABLE 6.4. Concentrations of copper (ppm) in components of a spruce forest being polluted in central Sweden.[a]

	Concentration, ppm	Enrichment ratio[b]
Spruce:		
Roots <5 mm diam	21	5.4
≥5 mm diam	8.0	2.6
Wood	2.0	
Bark	25	5.9
Twigs, 1st year	70	11
2nd year	76	11
3rd year	73	13
4th year	69	13
5th–7th year	42	8
Needles, 1st year	5.9	2.3
2nd year	4.4	1.7
3rd year	4.5	1.9
4th year	4.9	2.0
5th–7th year	8.1	3.5
Needle litter	260	34
Cowberry (aboveground biomass)	28	5.6
Bilberry (aboveground biomass)	48	5.8
Hairgrass:		
leaves	45	4.7
leaf litter	70	7.5
roots + rhizomes	70	7.1
Epiphytic lichens	145	29
Moss	580	48
Humus layer (raw humus)	660	63

[a] Source: Tyler, 1972, with permission of the author and the publisher.
[b] Ratio between Cu concentration at the site vs. those at unpolluted site.

et al, 1978) and in a forested watershed near a lead smelter in Missouri (Jackson and Watson, 1977). This is not surprising, since the ratio of Cu in urban air to that in rural air usually ranges from 2 to 10 (Nriagu, 1979). Areas with major metal industries can be expected to have higher atmospheric levels of Cu. In contrast, in a relatively rural forest ecosystem in central Massachusetts, there were no statistically significant changes in total Cu, as well as Zn content of the forest floor over a 16-year period (Siccama et al, 1980). Similarly, Smith (1973) found "normal" levels of Cu, Cd, and Mn in woody plants in New Haven (Connecticut) as well as in sugar maple sampled from remote areas in southern New Hampshire and Vermont.

Lepp (1979) concluded that, in spite of the role of Cu as an important micronutrient for trees, the cycling of this element in woodlands has received only little attention.

7. Factors Affecting the Mobility and Availability of Copper

7.1 pH

The solubility, mobility, and availability of Cu to plants are largely dependent on soil pH. Copper availability is drastically reduced at a soil pH above 7 and is most readily available below pH 6 (Locascio, 1978), and especially at pHs below 5 (Lucas and Knezek, 1972). Harmer (1945) stated that the more acidic an organic soil, the greater will be the relative response to Cu and the greater will be the number of crop species that are likely to respond to Cu application. Cavallaro and McBride (1980) and Jarvis (1981b) pointed out that increasing the pH of a soil solution, which can be effected by liming, can lead to the formation of hydrolysis products that have a different affinity for permanent charge and other exchange sites. Increasing the pH can also hydrolyze or precipitate Al^{3+} ions that occupy the exchange sites, thus decreasing the fraction of sites occupied by the strongly held Al^{3+} ions and creating more available exchange sites. Furthermore, soil constituents with a pH-dependent charge, such as Fe and Al oxides and organic matter, could cause very highly pH-dependent adsorption of Cu. In fact, Cavallaro and McBride (1980) found that Cu^{2+} solubility in soils showed a high dependence on pH due to the hydrolysis of Cu^{2+} at pH values above 6 and to the removal of Al^{3+} and H^+ ions from exchange sites as the pH was increased in more acidic soils.

Soil pH also influences the degree of complexation of Cu in soil solution. Several investigators (McBride and Bouldin, 1984; Kerven at al, 1984) found that increasing the pH of the soil solution or peat extract increased the degree of complexation of soluble Cu (Fig. 6.7). Increasing the pH of the peat extracts to about 6 resulted in almost complete complexation of

FIGURE 6.7. Influence of pH on the amount of total soluble copper present in soil solution (upper) and peat extracts (below) as uncomplexed Cu^{2+} ions. The peat extracts were adjusted to have a total solution copper concentration of 2 μM. *Sources:* Top figure from McBride and Bouldin, 1984, with permission of the authors and Soil Sci Soc of America; bottom figure from Kerven et al, 1984, with permission of the authors and Williams & Wilkens Co., Baltimore, MD.

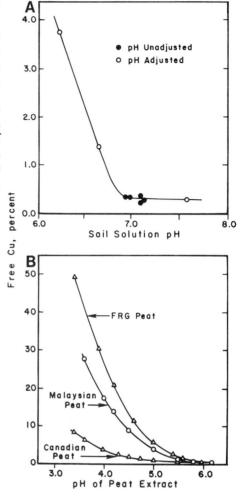

Cu (Fig. 6.7B). In contrast, only 60% to 90% of total Cu (0.14–0.2 μM) in solution of these acidic peats was complexed at its natural pH values of 3.4 to 3.5. However, the total soluble Cu concentration in the soil solution should remain relatively insensitive to pH change (Jeffery and Uren, 1983; McBride and Bouldin, 1984).

In general, liming lowers Cu concentrations in plants (Peech, 1941; Younts and Patterson, 1964). There is usually a negative correlation between Cu uptake and soil pH (Tiwari and Kumar, 1982; Lundblad et al, 1949; Piper, 1942) and a negative correlation usually exists between available Cu and soil pH (Dragun and Baker, 1982; Neelakantan and Mehta, 1961). The importance of pH on Cu availability in soils was stressed by Dolar and Keeney (1971a), who concluded that the predictability of Cu uptake can best be improved by inclusion of soil pH. There are exceptions, however, to these generalities. For example, Caldwell (1971) found that the response to foliarly applied Cu by winter wheat was enhanced by liming an acid peaty soil. He partially attributed the response to increased root growth due to liming. Another unusual exception occurs if there is a significant interaction of Al ions with Cu for plant uptake in highly acidic soils, where Cu would be more available at higher pH range (pH 3.9 to 5.1) (Lutz et al, 1972). Copper sorption increased with increasing equilibrium pH in soil sorption studies (Fig. 6.6). Positive correlations between soil sorption of Cu and equilibrium pH have been shown by McLaren and Crawford (1973a), Jarvis (1981a), Kishk and Hassan (1973), and Tirsch et al (1979).

7.2 Organic Matter

The roles of organic matter in the incidence of Cu deficiency, complexation, and sorption of Cu have already been discussed. Peats and mucks (histosols) commonly produce Cu-deficient crops.

The high sorption capacity of organic matter for Cu (Fig. 6.6) indicates that, among soil constituents, it is probably as important as the oxides of Mn and Fe in sorbing Cu. Of the divalent cations, Cu forms the most stable complexes with humic acid and peat, more so than Pb, Fe, and Zn.

Since organic matter represents a "specific" sorption site in soil, Lucas (1948) found that Cu in organic soils could be extracted in significant amounts only with 1 N HNO_3 or HCl solutions, but not with NH_4OAc extractant. Others (Kuo et al, 1983; Kishk et al, 1973) also have demonstrated appreciable association of Cu with the soil organic fraction. Fractionation studies have shown that most of the soil Cu is associated with organic matter (Jarvis, 1981b), and it has been calculated that, on a global basis, 36% of the total Cu in soils is in this form (Nriagu, 1979). King and Dunlop (1982) concluded that high organic matter content in soils can substitute for high pH in immobilizing metals and thus sewage sludge can be applied to organic soils that have pH values lower than the currently suggested value of pH 6.5 for metal application from municipal sewage sludges.

The ecological importance of organic matter on the cycling of Cu was demonstrated by Dykeman and de Sousa (1966) in the Tantramar copper swamp, located on the isthmus between New Brunswick and Nova Scotia, Canada. Soils in the swamp, which support a luxurious growth of larch, black spruce, and ground cover species, contain up to 7% Cu, which is believed to have come from the bedrock beneath a 60-m layer of glacial drift through deep circulating water and upward percolation to the ground surface. They attributed the lush growth of vegetation in the swamp, in spite of the very large amounts of total Cu, either to Cu fixation by the organic matter or Cu immobilization in plant roots and poor translocation of toxic amounts of Cu to the leaves.

7.3 Oxides of Iron, Manganese, and Aluminum

The role of the oxides of Fe, Mn, and Al in the environmental chemistry of trace metals is discussed in section 4.5. Briefly, it has been concluded that the hydrous oxides of Mn and Fe and organic matter are the principal soil constituents controlling the fixation of heavy metals (Fig. 6.6). These soil constituents are probably more important than clay minerals in the sorption of heavy metals, as indicated in section 4.5. More recently, Cavallaro and McBride (1984) and McLaren et al (1983) emphasized the important role of the oxides in metal retention in soil. The former investigators concluded that the microcrystalline and noncrystalline oxides in the clay fraction of an acid silty loam soil in New York, representing <20% of the clay by weight, provided chemisorption sites for Cu and Zn. Korte et al (1975) indicated that, based on their leaching studies of diverse soil types, the oxides of Fe and Mn are very important in estimating metal release from these soils. In their study of trace metal adsorption characteristic of estuarine particulate matter, Lion et al (1982) indicated that Mn-Fe oxide coatings may play a relatively greater role in the binding of Pb, while adsorption of Cu and Cd may be controlled to a greater extent by organic coatings. Since these hydrous oxides are nearly ubiquitous in soils, clays, and sediments (Jenne, 1968) their significance in metal fixation cannot be overlooked.

7.4 Soil Type and Clay Mineralogy

Copper deficiencies occur most commonly on histosols and soils high in organic matter, but also occur on coarse-textured mineral soils. In general, soils derived from coarse-grained materials (i.e., sands and sandstone) or from acid igneous rocks contain lower concentrations of Cu than those developed from fine-grained sedimentary rocks (i.e., shales and clays) or from basic igneous rocks (Jarvis, 1981b).

In England and Wales, Cu deficiencies are primarily restricted to crops grown on organic or peaty soils, on highly leached and podzolic sands, or on shallow rendzina soils developed on chalk (Caldwell, 1971). In Australia, Cu deficiencies occur on calcareous sands and on lateritic podzolic soils (Donald and Prescott, 1975).

Soil textures, particularly the clay or silt fractions, have been positively correlated with the sorption capacities of soils for Cu (Dhillon et al, 1981), with Cu availability (Dragun and Baker, 1982; Rai et al, 1972), and with total soil Cu content (Neelakantan and Mehta, 1961; Zborishchuk and Zyrin, 1978).

Among the clay minerals, their sorption capacities generally relate to their CEC values (i.e., montmorillonite > illite > kaolinite) (McLaren et al, 1981; Riemer and Toth, 1970; Kishk and Hassan, 1973).

7.5 Interactions with Other Elements

Detrimental effects of N or P interactions with micronutrients on plant growth and nutrition have been reported frequently during the past 20 years. In general, interactions of the metals with a macronutrient are expressed as an intensification of micronutrient deficiency when supplemental macronutrients are applied to soils where micronutrients are deficient or near-deficient. Good examples of these macro-micronutrient interactions are the N-Cu and P-Cu couples.

Nitrogen-Cu interactions have been shown for several crops (Cheshire et al, 1982; Harris, 1947; Fleming and Delaney, 1961; Reilly and Reilly, 1973; Davies et al, 1971; Gilbert, 1952; and Chaudry and Lonegran, 1970). Gilbert (1951) indicated that increased protein concentrations in roots, as a result of increased N application, could lead to increased retention of Cu by increased formation of protein-Cu complexes. Similarly, Smith (1953) mentioned the possible role of protein-Cu complexes when he found strong correlations between Cu and N concentrations in roots of citrus plants grown in nutrient solutions and between N concentrations in the roots and the amount of Cu they absorbed. Nambiar (1976b) obtained evidence showing that cereal genotypes with higher grain protein contents are potentially more susceptible to Cu deficiency than those with lower grain protein. Chaudry and Lonegran (1970) indicated that N-induced Cu deficiency may be explained by the two effects of N on growth—a large increase in total growth, and a marked increase in top relative to root growth, both contributing to the dilution of Cu contents of plant tissues.

Phosphorus-Cu interactions have also been shown in several crops (Murphy et al, 1981; Labanauskas et al, 1958; Bingham, 1963; Locascio et al, 1968; Wallace et al, 1978; Singh and Swarup, 1982; Spratt and Smid, 1978; Bingham and Martin, 1956) and even in tree species (Smilde, 1973). Smilde (1973) reported that in the Netherlands, P can accentuate Cu deficiency in trees, especially in the presence of N.

Copper has also been shown to interact antagonistically with Zn (Gilbert, 1952; Adriano et al, 1971; Kausar et al, 1976; Levesque and Mathur, 1983), with Mo (Gilbert 1952; MacKay et al, 1966), and with Mn (Gilbert, 1952; Gupta, 1972).

7.6 Other Factors

Genotypic differences in Cu nutrition among cereal crops have been reported by Brown (1965), Smilde and Henkens (1967), and Nambiar (1976a). Jarvis and Whitehead (1983) reported that the concentration of Cu is often greater in pasture legumes than in grasses growing in the field under similar conditions.

Copper nutrition by crops can be affected also by the method of placement of Cu fertilizers (band, broadcast, or foliar) and type of fertilizer (organic vs. inorganic), although the results are inconsistent because of confounding by soil type, crop species, application rates, and other factors (Varvel, 1983; Barnes and Cox, 1973; Murphy and Walsh, 1972).

8. Copper in Drinking Water and Food

The interim Cu limit for drinking water in the United States is 1 ppm, which is based on taste consideration, rather than toxicity (US-EPA, 1976). The European standards for drinking water are set at 0.05 ppm at the pumping station and 3 ppm after 16 hours of contact with plumbing; the USSR has a 0.05-ppm acceptable limit; the international acceptable limit is 0.05 ppm, with a maximum of 1.5 ppm (NAS, 1977). There are instances when drinking water, as is the case in well waters in Sri Lanka, can be contaminated by industrial wastes (Dissanayake and Jayatilaka, 1980). Although the average in the Colombo area was 0.038 ppm, this value was much higher than the lake (0.020 ppm), stream (0.024 ppm), and river water (0.026 ppm) concentrations and approached the level in the waste water (0.050 ppm). A survey of household tapwater in Dallas (Texas) showed an average concentration of 0.037 ppm Cu ($n = 43$), with a range of 0.004 to 0.164 ppm (NAS, 1977). It should be noted that household plumbing could elevate Cu and Zn concentrations in water. The median Cu concentration of finished water in public water supplies of the 100 largest cities in the United States was 0.0083 ppm (range of <0.0006 to 0.250 ppm).

The Cu content of leafy plants consumed as food rarely exceeds 25 ppm and usually ranges between 10 and 15 ppm of the dry matter (Underwood, 1973). Whole cereal grains normally contain 4 to 6 ppm Cu, and white flour and bread, 1 to 2 ppm. Root vegetables, nuts, and fruits have usually <10 ppm Cu (dry weight basis) (Quarterman, 1973). Dairy products usually contain <1 ppm Cu (dry weight basis). Crustacea and shellfish, especially oysters, and organ meats (liver and kidney) can contain as high as 200 to 400 ppm Cu on dry basis, but these food items comprise a very small proportion of the total diet (Underwood, 1973). Most all other foods of animal origin contain between 2 and 4 ppm (Russell, 1979).

The dietary intake of Cu in the western world is reported to range between 1 and 5 mg/day (Pier and Bang, 1980). The following intake values (in mg/day) have been reported: the United Kingdom—3.1 ± 0.76 (Hamilton and Minski, 1972); India—5.80 (Soman et al, 1969). The International Commission on Radiological Protection reported an intake value of 3 mg/day. The recommended daily dietary requirement for Cu has been placed at 2 mg, which can be met by most normal diets.

The joint FAO/WHO Expert Committees on Food Additives have established a maximum daily intake of 30 mg Cu for a 60-kg man (Somers, 1974).

Klevay (1975) reported that the ratio of Zn to Cu has some clinical significance in that it has been associated with a variety of epidemiological features of coronary heart disease and with the metabolism of cholesterol. Therefore, it is possible that diets deficient in Cu contribute to the increase in risk of coronary heart disease associated with a high ratio of Zn to Cu.

9. Sources of Copper in the Terrestrial Environment

9.1 Copper Fertilizers

The most widely used source of Cu in agriculture is $CuSO_4 \cdot 5H_2O$ (25.5% Cu), which is water soluble. Other sources, such as copper oxides (CuO, 75% Cu; Cu_2O, 89% Cu), basic copper sulfate [$CuSO_4 \cdot 3Cu(OH)_2$ (13%– 53% Cu)], and copper carbonate [$CuCO_3 \cdot Cu(OH)_2$ (55%–57%)] have been used widely also. A single application of 2.25 to 9.0 kg/ha of Cu on mineral soils or 22.5 to 45.0 kg/ha on a peat or muck soil is generally adequate for several years (Locascio, 1978).

Copper chelates are another source of Cu in agriculture that was proven more effective in alkaline than in acid soils. Chelates are fairly stable in soil, keeping Cu in soluble and available forms to the plants. Examples of synthetic Cu chelates are $Na_2CuEDTA$ (13% Cu) and NaCuHEDTA (9% Cu), and natural organic complexes (ligninsulfonates and polyflavinoid), which are highly efficient Cu sources.

Foliar sprays of chelated Cu or low rates of $CuSO_4$ (0.30 kg/ha) can also be used to prevent or alleviate deficiencies. However, multiple sprays may be required to obtain maximum production, since Cu is immobile in plants and does not move from older to younger tissues. Murphy and Walsh (1972) listed the sources of Cu used for soil and foliar applications and the various application rates used by different investigators. Like Zn, Cu is most commonly applied to the soil, either by broadcast or banded applications.

In a soil survey in Ontario, Frank et al (1976) found the highest total Cu levels in organic soils (65 ± 27 ppm) due to the production practices of adding $CuSO_4$ once every three to four years to prevent Cu deficiencies and using Cu fungicides on vegetables.

9.2 Fungicides and Bactericides

Copper compounds have been used for many years as fungicides and bactericides. In the past, $CuSO_4$ was mixed with lime (Bordeaux mixture) to control many plant pathogens, while recently $Cu(OH)_2$, has been used in fairly large amounts in sprays as a blanket preventive treatment (Walsh et al, 1972). These fungicides are used on pome, stone, and citrus fruit as well as on grapevines, hops, and vegetables to counter downy mildew (Tiller and Merry, 1981). For example, Cu sprays have been recommended for disease control on each of three vegetable crops (snapbeans, potatoes, and cucumbers) grown on irrigated sands in central Wisconsin. These sprays can enrich Cu in the soil by as much as 11.2 to 16.8 kg/ha per year. Copper toxicity could develop in these sandy soils because they have a low CEC (<5 meq/100 g) and often contain <1% organic matter.

Copper fungicides have been used extensively in Kenya since the 1930s to control leaf rust and since the 1960s to control the berry disease in coffee (Dickinson et al, 1984). In a large-scale coffee plantation near Nairobi, Cu concentrations as high as 236 ppm were found in the surface 10 cm of the soil. Established in 1915, the plantation is currently being sprayed 10 to 12 times per year at the rate of 5 kg/ha Cu, as copper oxychloride.

Continued overuse of Cu-containing fungicides and bactericides can result in accumulation of Cu in soil to phytotoxic levels (Delas et al, 1960; Hirst et al, 1961).

9.3 Livestock Manures

Copper, as $CuSO_4$, is a typical feed additive for swine and is used to increase feed efficiency, weight gains, and to control dysentery (Cromwell et al, 1978; Lucas et al, 1962; Stahly et al, 1980). Dietary Cu levels of 125 to 250 ppm are usually sufficient to stimulate growth of swine and poultry, but higher levels can be toxic (Dalgarno and Mills, 1975). Zinc and Fe are usually added simultaneously with Cu to prevent a possible Cu intoxication to the animals and to ensure prevention of Fe- and Zn-deficiency symptoms in animals (Lexmond and de Haan, 1977). Zinc, as $ZnSO_4$, is added at a level of 80 ppm. Most (80%–95%) of this dietary Cu is excreted in the manure. Purves (1977) reported that concentrations up to 800 ppm Cu (dry basis) are not unusual in swine wastes. Therefore, the potential accumulation and toxic effects of Cu from high rates of manure application to croplands are of environmental concern. Here and abroad, the continual use of swine manure on land has been shown to have the potential to elevate soil Cu levels and eventually cause phytotoxicity (Batey et al, 1972; Kornegay et al, 1976; Sutton et al, 1983). For example, repeated applications of swine manure on grassland in the Netherlands caused Cu accumulation in the sod layer (Lexmond and de Haan, 1977) There is evidence of Cu leaching in the profiles of these Dutch soils, especially in podzols due to their low organic matter contents. Surface application on pastures of manure slurry from pigs fed 250 ppm of $CuSO_4$ in the diet resulted in Cu toxicity to grazing sheep (Blaxter, 1973). Underwood (1973) reported that sheep are very susceptible to Cu poisoning (a level of 25 ppm in diet can be toxic to sheep), while monogastric animals, such as swine and poultry, can tolerate much higher levels of Cu.

9.4 Sewage Sludges

Metal concentrations in sewage sludges vary widely. Levels of metals in sludges have been reported by Page (1974) and Sommers (1977). Among the heavy metals, Cu, Ni, Zn, and Cd are the ones usually singled out as critical to agricultural use of sewage sludge. Although sludges often contain appreciable levels of Cu, application of sludge to soils generally results in only slight to moderate increases in the Cu contents of plants (CAST, 1976). In general, good management practices would render Cu in sludges harmless to plants and pose no hazard to the food chain.

TABLE 6.5. Sources of copper in the environment.[a]

A: Copper emissions from natural sources

Source	Global production[b] (10^9 kg/ year)	Emission factor[b] ($\mu g/g$)	Copper emitted (10^6 kg/year)
Windblown	500	24	12.0
Forest fires	36	8	0.3
Volcanogenic particles	10	360	3.6
Vegetation	75	33	2.5
Seasalt sprays	1000	0.082	0.08
Total			18.5

B: Worldwide anthropogenic emissions of copper

Source	Annual handling (10^9 kg)	Emission factor	Copper emitted (10^6 kg)
Copper production			
Mining	7.5	0.1 kg/10^3 kg Cu mined	0.75
Metallurgical processing	7.9	2.5 kg/10^3 kg Cu produced	19.8
Secondary production	2.2	0.15 kg/10^3 kg Cu produced	0.33
Iron and steel production	1304	4.5 g/10^3 kg produced	5.9
Foundries	163	2.3 g/10^3 kg fabricated	0.37
Lead production	4.0	73 g/10^3 kg Pb produced	0.29
Zinc production	5.6	0.14 kg/10^3 kg Zn produced	0.78
Industrial applications	—	—	4.9
Waste incineration	1500	3.5 $\mu g/g$ wastes	5.3
Coal combustion	3100	1.8 g/10^3 kg coal burned	5.6
Fuel oil combustion	1060	0.7 g/10^3 kg oil burned	0.74
Wood combustion	640	18 g/10^3 kg wood burned	11.5
Grain handling	330	0.08 g/10^3 kg grain handled	0.03
Total			56.3

[a] *Source:* Nriagu, 1979, with permission of the author and John Wiley & Sons, New York.
[b] For definitions of global production and emission factor, see Nriagu (1979).

9.5 Industrial and Other Sources

Worldwide anthropogenic emissions of Cu are listed in Table 6.5. Of the various sources, metallurgical processing (smelting) for Cu, iron and steel production, and coal combustion are among the major sources of Cu.

Duce and Windom (1976) estimated the annual natural fluxes of Cu to the atmosphere to be 14.3×10^6 kg. This is in reasonable agreement with Nriagu's (1979) estimate of 18.5×10^6 kg/year total Cu emitted from natural sources.

Environmental contamination due to heavy metals from industrial emission including smelting is already well documented (see the chapters on Cd, Zn, and Pb).

References

Adriano, D. C., G. M. Paulsen, and L. S. Murphy. 1971. *Agron J* 63:36–39.

Adriano, D. C., A. L. Page, A. A. Elseewi, A. C. Chang, and I. Straughan. 1980. *J Environ Qual* 9:333–344.

Agarwala, S. C., S. S. Bisht, and C. P. Sharma. 1977. *Can J Bot* 55:1299–1307.

Alloway, B. J., M. Gregson, S. K. Gregson, R. Tanner, and A. Tills. 1979. In *Management and control of heavy metals in the environment*, 545–548. CEP Consultants Ltd, Edinburgh, UK.

Andrew, C. S., and P. M. Thorne. 1962. *Aust J Agric Res* 13:821–835.

Aubert, H., and M. Pinta. 1977. *Trace elements in soils*. Elsevier, New York.

Baes, C. F., and R. E. Mesmer. 1976. *The hydrolysis of cations*. Wiley, New York.

Baker, D. E. 1974. *Proc Fed Am Soc Exp Biol* 33:1188–1193.

Barnes, J. S., and F. R. Cox. 1973. *Agron J* 65:705–708.

Batey, T., C. Berryman, and C. Line. 1972. *J Br Grassl Soc* 27:139–143.

Beckett, P. H. T., E. Warr, and R. D. Davis. 1983. *Plant Soil* 70:3–14.

Behel, D., Jr., D. W. Nelson, and L. E. Sommers. 1983. *J Environ Qual* 12:181–186.

Bingham, F. T. 1963. *Soil Sci Soc Am Proc* 27:389–391.

Bingham, F. T., and J. P. Martin. 1956. *Soil Sci Soc Am Proc* 20:382–385.

Blaxter, K. L. 1973. *Vet Rec* 92:383–386.

Bleeker, P., and M. P. Austin. 1970. *Aust J Soil Res* 8:133–143.

Bloom, P. R. 1981. In R. H. Dowdy, J. A. Ryan, V. V. Volk, and D. E. Baker, eds. *Chemistry in the soil environment*, 129–150. ASA Spec Publ no. 40, Madison, WI.

Bloomfield, C., and J. R. Sanders. 1977. *J Soil Sci* 28:435–444.

Borggaard, O. K. 1976. *Acta Agric Scand* 26:144–149.

Bould, C., D. J. D. Nicholas, J. A. H. Tolhurst, and J. M. S. Potter. 1953. *J Hort Sci* 28:268–277.

Bowen, H. J. M. 1979. *Environmental chemistry of the elements*. Academic Press, New York. 333 pp.

Bradford, G. R., F. L. Bair, and V. Hunsaker. 1971. *Soil Sci* 112:225–230.

Brams, E. A., and J. G. A. Fiskell. 1971. *Soil Sci Soc Am Proc* 35:772–775.

Brown, J. C. 1965. *Agron J* 57:617–621.

Bussler, W. 1981. In J. F. Lonegran, A. D. Robson, and R. D. Graham, eds. *Copper in soils and plants,* 213–234. Academic Press, New York.

Caldwell, T. H. 1971. In *Trace elements in soils and crops,* 73–87. Tech Bull no 21, Her Majesty's Sta Office, London.

Cavallaro, N., and M. B. McBride. 1978. *Soil Sci Soc Am J* 42:550–556.

—— 1980. *Soil Sci Soc Am J* 44:729–732.

—— 1984. *Soil Sci Soc Am J* 48:1050–1054.

Chaudry, F. M., and J. F. Lonegran. 1970. *Aust J Agric Res* 21:865–879.

Cheshire, M. V., W. Bick, P. C. DeKock, and R. H. E. Inkson. 1982. *Plant Soil* 66:139–147.

Chino, M., and K. Kitagishi. 1966. *J Sci Soil Manure* 37:342–347.

Chino, M., 1981. In K. Kitagishi and I. Yamane, eds. *Heavy metal pollution in soils of Japan,* 65–80. Japan Sci Soc Press, Tokyo.

Clayton, P. M., and K. G. Tiller. 1979. Div Soil Tech Paper no. 41, CSIRO, Australia (as cited by Robson and Reuter, 1981).

Coleman, N. T., A. C. McClung, and D. P. Moore. 1956. *Science* 123:330–331.

Council for Agricultural Science and Technology (CAST). 1976. *Application of sewage sludge to cropland: appraisal of potential hazards of the heavy metals to plants and animals.* EPA-430/9-76-013. General Serv Admin, Denver. 63 pp.

Cox, D. P. 1979. In J. O. Nriagu, ed. *Copper in the environment,* 19–42. Wiley, New York.

Cox, F. R., and E. J. Kamprath. 1972. In J. J. Mortvedt, P. M. Giordano, and W. L. Lindsay, eds. *Micronutrients in agriculture,* 289–313. Soil Sci Soc Am Inc, Madison, WI.

Cromwell, G. L., V. W. Hays, and T. L. Clark. 1978. *J Anim Sci* 46:692–698.

Cunningham, I. J., K. J. Hogan, and J. N. Green. 1956. N Z J Sci Tech A38:225–238.

Dalgarno, A. C., and C. F. Mills. 1975. *J Agric Sci* 85:11–18.

Davies, B. D., L. J. Hooper, R. R. Charlesworth, R. C. Little, C. Evans, and B. Wilkinson. 1971. In *Trace elements in soils and crops,* 88–101. Tech Bull no. 21, Her Majesty's Sta Office, London.

Davis, R. D., P. H. T. Beckett, and E. Wollan. 1978. *Plant Soil* 49:395–408.

Delas, J. 1980. In *Problems encountered with copper.* Europ Econ Commun Workshop, Bordeaux, France, Oct 8–10.

Delas, J., J. Delmas, and C. Dimias. 1960. *Ann Agron* 13:31–53.

Dhillon, S. K., P. S. Sidhu, and M. K. Sinha. 1981. *J Soil Sci* 32:571–578.

Dickinson, N. M., N. W. Lepp, and K. L. Ormand. 1984. *Environ Pollut* (B) 7:223–231.

Dissanayake, C. B., and G. M. Jayatilaka. 1980. *Water, Air, Soil Pollut* 13:275–286.

Dolar, S. G., and D. R. Keeney. 1971a. *J Sci Fd Agric* 22:273–278.

Dolar, S. G., D. R. Kenney, and L. M. Walsh. 1971b. *J Sci Fd Agric* 22:282–286.

Donald, C. M., and J. A. Prescott. 1975. In D. J. D. Nicholas and A. R. Egan, eds. *Trace elements in soil-plant-animal systems,* 7–37. Academic Press, New York.

Doner, H. E., A. Pukite, and E. Yang. 1982. *J Environ Qual* 11:389–394.

Dragun, J., and D. E. Baker. 1982. *Soil Sci Soc Am J* 46:921–925.

Drouineau, G., and R. Mazoycr. 1962. *Ann Agron Paris* 13:31–53.

Duce, R. A., and H. Windom, eds. 1976. In *Marine pollutant transfer,* 77–119. Lexington Books, Lexington, MA.

Dykeman, W. R., and A. S. de Sousa. 1966. *Can J Bot* 44:871–878.

Ellis, B. G., and B. D. Knezek. 1972. In J. J. Mortvedt, P. M. Giordano, and W. L. Lindsay, eds. *Micronutrients in agriculture,* 59–78. Soil Sci Soc Am, Madison, WI.

Elrashidi, M. A., A. Shehata, and M. Wahab. 1977. *Agrochimica* 21:226–234.

Emmerich, W. E., L. J. Lund, A. L. Page, and A. C. Chang. 1982. *J Environ Qual* 11:182–186.

Fiskell, J. G. A. 1965. In C. A. Black, ed. *Methods of soil analysis.* Part 2, 1078–1089. Amer Soc Agron, Madison, WI.

Fiskell, J. G. A., and C. D. Leonard. 1967. *J Agric Food Chem*15:350–353.

Fleming, G. A., and J. Delaney. 1961. *Irish J Agric Res* 1:81–82.

Follett, R. H., and W. L. Lindsay. 1970. In *Profile distribution of zinc, iron, manganese, and copper in Colorado soils.* Colorado Agric Exp Sta Tech Bull 110:1–78.

Follett, R. H., L. S. Murphy, and R. L. Donahue. 1981. *Fertilizers and soil amendments.* Prentice-Hall, Englewood Cliffs, NJ. 557 pp.

Frank, R., K. Ishida, and P. Suda. 1976. *Can J Soil Sci* 56:181–196.

Gartrell, J. W. 1981. In J. F. Lonegran, A. D. Robson, and R. D. Graham, eds. *Copper in soils and plants,* 313–349. Academic Press, New York.

Gilbert, G. S. 1951. *Plant Physiol* (Lancaster) 26:398–405.

Gilbert, F. A. 1952. *Adv Agron* 4:147–177.

Gough, L. P., J. M. McNeal, and R. C. Severson. 1980. *Soil Sci Soc Am J* 44:1030–1036.

Graham, R. D. 1981. In J. F. Lonegran, A. D. Robson, and R. D. Graham, eds. *Copper in soils and plants,* 141–163. Academic Press, New York.

Grewal, J. S., D. R. Bhumbla, and N. S. Randhawa. 1969. *J Indian Soc Soil Sci* 17:27–31.

Gupta, U. C. 1972. *Soil Sci* 114:131–136.

―――― 1979. In J. O. Nriagu, ed. *Copper in the environment,* 255–288. Wiley, New York.

Gupta, U. C., and D. C. MacKay. 1966. *Soil Sci* 101:93–97.

Hamilton, E. I., and M. J. Minski. 1972. *Sci Total Environ* 1:375–394.

Harmer, P. M. 1945. *Soil Sci Soc Am Proc* 10:284–294.

Harris, H. C. 1947. *Soil Sci Soc Am Proc* 12:278–281.

Harrison, R. M., D. P. H. Laxen, and S. J. Wilson. 1981. *Environ Sci Technol* 15:1378–1383.

Harter, R. D. 1979. *Soil Sci Soc Am J* 43:679 683.

Hazlett, P. W., G. K. Rutherford, and G. W. van Loon. 1984. *Geoderma* 32:273–285.

Heinrichs, H., and R. Mayer. 1980. *J Environ Qual* 9:111–118.

Hewitt, E. J. 1953. *J Exp Bot* 4:59–64.

Hirst, J. M., H. H. LeRiche, and C. L. Bascomb. 1961. *Plant Pathol* 10:105–108.

Hodgson, J. F., H. R. Geering, and W. A. Norvell. 1965. *Soil Sci Soc Am Proc* 29:665–669.

Hodgson, J. F., W. L. Lindsay, and J. F. Trierweiler. 1966. *Soil Sci Soc Am Proc* 30:723–726.

Holmes, R. S. 1943. *Soil Sci* 56:359–370.

Huang, P. M. 1980. In O. Hutzinger, ed. *Handbook of environmental chemistry.* Vol. 2., part A, 47–59. Springer-Verlag, New York.

Jackson, D. R., and A. P. Watson. 1977. *J Environ Qual* 6:331–338.

Jarvis, S. C. 1978. *J Sci Fed Agric* 29:12–18.

―――― 1981a. *J Soil Sci* 32:257–269.

―――― 1981b. In J. F. Lonegran, A. D. Robson, and R. D. Graham, eds. *Copper in soils and plants,* 265–285. Academic Press, New York.

Jarvis, S. C., and D. C. Whitehead. 1983. *Plant Soil* 75:427–434.

Jeffery, J. J., and N. C. Uren. 1983. *Aust J Soil Res* 21:479–488.

Jenne, E. A. 1968. *Adv Chem* 73:337–387.

―――― 1977. In W. R. Chappel and K. L. Petersen, eds. *Molybdenum in the environment.* Vol. 2, 425–553. Marcel Dekker, Inc., New York.

Jolly, J. H. 1980. In US Bureau of Mines. *Minerals yearbook, 1978–79,* 271–311. US Dept of Interior, Washington, DC.

Jones, J. B., Jr. 1972. In J. J. Mortvedt, P. M. Giordano, and W. L. Lindsay, eds. *Micronutrients in agriculture,* 319–346. Soil Sci Soc Am Inc, Madison, WI.

Jones, G. B., and G. B. Belling. 1967. *Aust J Agr Res* 18:733–740.

Kausar, M. A., F. M. Chaudry, A. Rashid, A. Latif, and S. M. Alam. 1976. *Plant Soil* 45:397–410.

Kawaguchi, K., and K. Kyuma. 1959. *Soil Plant Food* 5:54–63.

Kerven, G. L., D. G. Edwards, and C. J. Asher. 1984. *Soil Sci* 137:91–99.

King, P. M. 1974. *J Agric Sci Aust* 77:96–99.

King, L. D., and W. R. Dunlop. 1982. *J Environ Qual* 11:608–616.

Kishk, F. M., and M. N. Hassan. 1973. *Plant Soil* 39:497–505.

Kishk, F. M., M. N. Hassan, I. Ghanem, and L. El-Sissy. 1973. *Plant Soil* 39:487–496.

Klevay, L. M. 1975. *Nutr Reports Intl* 11:237–242.

Kneip, T. J., M. Eisenbud, C. D. Strehlow, and P. C. Freudenthal. 1970. *J Air Pollut Cont Assoc* 20:144–149.

Kornegay, E. T., J. D. Hedges, D. C. Martens, and C. Y. Kramer. 1976. *Plant Soil* 45:151–162.

Korte, N. E., J. Skopp, E. E. Niebla, and W. H. Fuller. 1975. *Water, Air, Soil Pollut* 5:149–156.

Krauskopf, K. B. 1972. In J. J. Mortvedt, P. M. Giordano, and W. L. Lindsay, eds. *Micronutrients in agriculture,* 7–40. Soil Sci Soc Am, Inc., Madison, WI.

―――― 1979. *Introduction to geochemistry.* 2nd edition. McGraw-Hill, New York. 617 pp.

Kruglova, Y. K. 1962. *Sov Soil Sci* 5:516–521.

Kubota, J., and W. H. Allaway. 1972. In J. J. Mortvedt, P. M. Giordano, and W. L. Lindsay, eds. *Micronutrients in agriculture,* 525–554. Soil Sci Soc Am, Inc, Madison, WI.

Kubota, J. 1983. *Agron J* 75:913–918.

Kuo, S., and A. S. Baker. 1980. *Soil Sci Soc Am J* 44:969–974.

Kuo, S., P. E. Heilman, and A. S. Baker. 1983. *Soil Sci* 135:101–109.

Labanauskas, C. K., T. W. Embleton, and W. W. Jones. 1958. *Proc Am Soc Hort Sci* 71:285–291.

Leeper, G. W. 1978. *Managing the heavy metals on the land.* Dekker, New York.

Leonard, C. D. 1967. *Farm Technol* 23, no. 6.

Lepp, N. W. 1979. In J. O. Nriagu, ed. *Copper in the environment,* 289–323. Wiley, New York.

Levesque, M. P., and S. P. Mathur. 1983. *Soil Sci* 135:88–100.

Lexmond, T. M., and F. A. M. de Haan. 1977. In Proc. Intl Seminar on Soil Environment and Fertility Management in Intensive Agriculture (Tokyo), 383–393. Soc Sci Soil Manure, Japan.

Lexmond, T. M., and F. A. M. de Haan. 1980. In J. K. R. Gasser, ed. *Effluents from livestock,* 410–419. Applied Sci Publishers, London.

Lindsay, W. L., and W. A. Norvell. 1969. *Soil Sci Soc Am Proc* 33:62–68.

Lindsay, W. L. 1974. In E. W. Carson, ed. *The plant root and its environment,* 507–524. Univ Virginia Press, Charlottesville.

Lindsay, W. L., and W. A. Norvell. 1978. *Soil Sci Soc Am J* 42:421–428.

Lion, L. W., R. S. Altmann, and J. O. Leckie. 1982. *Environ Sci Technol* 16:660–666.

Lipman, C. B., and G. MacKinney. 1931. *Plant Physiol* 6:593–599.

Liu, Z., Q. Q. Zhu, and L. H. Tang. 1983. *Soil Sci* 135:40–46.

Liwski, S. 1963. *Polish Agric Ann,* Ser A, no. 3.

Locascio, S. J. 1978. *Solutions,* 30–42.

Locascio, S. J., P. H. Everett, and J. G. A. Fiskell. 1968. *Proc Am Soc Hort Sci* 92:583–589.

Lucas, R. E. 1948. *Soil Sci* 66:119–129.

Lucas, I. H. M., R. M. Livingston, A. W. Boyne, and I. McDonald. 1962. *J Agric Sci* 58:201–208.

Lucas, R. E., and B. D. Knezek. 1972. In J. J. Mortvedt, P. M. Giordano, and W. L. Lindsay, eds. *Micronutrients in agriculture,* 265–288. Soil Sci Soc Am Inc, Madison, WI.

Lundblad, K., O. Svanberg, and P. Ekman. 1949. *Plant Soil* 1:277–302.

Lutz, J. A., Jr., C. F. Genter, and G. W. Hawkins. 1972. *Agron J* 64:583–585.

MacKay, D. C., E. W. Chipman, and U. C. Gupta. 1966. *Soil Sci Soc Am Proc* 30:755–759.

Makarim, A. K., and F. R. Cox. 1983. *Agron J* 75:493–496.

Mathur, S. P., and M. P. Levesque. 1983. *Soil Sci* 135:166–176.

Mathur, S. P., R. B. Sanderson, A. Belanger, M. Valk, E. N. Knibbe, and C. M. Preston. 1984. *Water, Air, Soil Pollut.* 22:277–288.

Mattigod, S. V., and G. Sposito. 1977. *Soil Sci Soc Am J* 41:1092–1097.

McBride, M. B. 1981. In J. F. Lonegran, A. D. Robson, and R. D. Graham, eds. *Copper in soils and plants,* 25–45. Academic Press, New York.

McBride, M. B., and J. J. Blasiak. 1979. *Soil Sci Soc Am J* 43:866–870.

McBride, M. B., and D. R. Bouldin. 1984. *Soil Sci Am J* 48:56–59.

McKeague, J. A., and M. S. Wolynetz. 1980. *Geoderma* 24:299–307.

McKenzie, R. M. 1980. *Aust J Soil Res* 18:61–73.

McLaren, R. G., and D. V. Crawford. 1973a. *J Soil Sci* 24:172–181.

——— 1973b. *J Soil Sci* 24:443–452.

——— 1974. *J Soil Sci* 25:111–119.

McLaren, R. G., R. S. Swift, and J. G. Williams. 1981. *J Soil Sci* 32:247–256.

McLaren, R. G., J. G. Williams, and R. S. Swift. 1983. *Geoderma* 31:97–106.

Mehlich, A., and S. S. Bowling. 1975. *Commun Soil Sci Plant Anal* 6:113–128.

Menzel, R. G., and M. L. Jackson. 1950. *Soil Sci Soc Am Proc* 15:122–124.

Miller, W. P., and W. W. McFee. 1983. *J Environ Qual* 12:29–33.

Mills, J. G., and M. A. Zwarich. 1975. *Can J Soil Sci* 55:295–300.

Mitchell, R. L. 1964. In F. E. Bear, ed. *Chemistry of the soil,* 320–368. Am Chem Soc no. 160. Reinhold, New York.

Mullins, G. L., D. C. Martens, W. P. Miller, E. T. Kornegay, and D. L. Hallock. 1982a. *J Environ Qual* 11:316–320.

Mullins, G. L., D. C. Martens, S. W. Gettier, and W. P. Miller. 1982b. *J Environ Qual* 11:573–577.

Murphy, L. S., and L. M. Walsh. 1972. In J. J. Mortvedt, P. M. Giordano, and W. L. Lindsay, eds. *Micronutrients in agriculture*. 347–387. Soil Sci Soc Am Inc, Madison, WI.

Murphy, L. S., R. Ellis, Jr., and D. C. Adriano. 1981. *J Plant Nutri* 3:593–613.

Nambiar, E. K. S. 1976a. *Aust J Agric Res* 27:453–463.

—— 1976b. *Aust J Agric Res* 27:465–477.

National Academy of Sciences (NAS). 1977. *Drinking water and health*. NAS, Washington, DC. 939 pp.

Neelakantan, V., and B. V. Mehta. 1961. *Soil Sci* 91:251–256.

Nelson, W. L., A. Mehlich, and E. Winters. 1953. In W. H. Pierre and A. G. Norman, eds. *Soil and fertilizer phosphorus in crop nutrition,* 153–188. Academic Press, New York.

Norvell, W. A. 1972. In J. J. Mortvedt, P.M. Giordano, and W. L. Lindsay, eds. *Micronutrients in agriculture,* 115–138. Soil Sci Soc Am, Inc., Madison, WI.

Norvell, W. A., and W. L. Lindsay. 1969. *Soil Sci Soc Am Proc* 33:86–91.

Norvell, W. A., and W. L. Lindsay. 1972. *Soil Sci Soc Am Proc* 36:778–783.

Nriagu, J. O., ed. 1979. *Copper in the environment*. Wiley, New York.

Osawa, T., and H. Ikeda. 1974. *J Japan Soc Hort Sci* 43:267–272.

Pack, M. R., S. J. Toth, and F. E. Bear. 1953. *Soil Sci* 75:433–441.

Page, A. L. 1974. *Fate and effects of trace elements in sewage sludge when applied to agricultural lands.* EPA-670/2-74-005. US-EPA Cincinnati, OH. 98 pp.

Parker, G. R., W. W. McFee, and J. M. Kelly. 1978. *J Environ Qual* 7:337–342.

Peech, M. 1941. *Soil Sci* 51:473–486.

Petruzzelli, G., G. Guidi, and L. Lubrano. 1978. *Water, Air, Soil Pollut* 9:263–269.

Pier, S. M., and K. M. Bang. 1980. In N. M. Trieff, ed. *Environment and health,* 367–408. Ann Arbor Sci Publ Inc, Ann Arbor, MI.

Piper, C. S. 1942. *J Agric Sci* 32:143–178.

Purves, D. 1977. *Trace element contamination of the environment*. Elsevier, Amsterdam.

Purves, D. and J. M. Ragg. 1962. *J Soil Sci* 13:241–246.

Quarterman, J. 1973. *Qual Plant—Pl Fds Hum Nutr* 23:171–190.

Rai, M. M., J. M. Dighe, and A. R. Pal. 1972. *J Indian Soc Soil Sci* 20:135–142.

Reilly, A., and C. Reilly. 1973. *Plant Soil* 38:671–674.

Reiners, W. A., Marks, R. H., and P. M. Vitousek. 1975. *Oikos* 26:264–275.

Reith, J. W. S. 1968. *J Agric Sci* (Camb) 70:39–45.

Reuter, D. J. 1975. In D. J. D. Nicholas and A. R. Egan, eds. *Trace elements in soil-plant-animal systems,* 291–324. Academic Press, New York.

Reuther, W., and P. F. Smith. 1952. *Proc Fla State Hort Soc* 65:62–69.

—— 1954. *Proc Soil Sci Soc Fla* 14:17–23.

Reuther, W., and C. K. Labanauskas. 1965. In H. D. Chapman, ed. *Diagnostic criteria for plants and soils,* 157–179. Quality Print Co, Abilene, TX.

Riemer, O. N., and S. J. Toth. 1970. *Am Water Works Assoc J* 62:195–197.

Robson, A. D., and D. J. Reuter. 1981. In J. F. Loneragan, A. D. Robson, and R. D. Graham, eds. *Copper in soils and plants,* 287–312. Academic Press, New York.

Rühling, A., and G. Tyler. 1971. *J App Ecol* 8:497–507.

Russell, L. H., Jr. 1979. In F. W. Oehme, ed. *Toxicity of heavy metals in the environment*, 3–23. Dekker Inc, New York.

Sabet, S. A., M. A. Omar, F. Makled, and M. M. Wassif. 1975. *Egypt J Soil Sci* 15:51–65.

Sanders, J. R. 1980. *J Soil Sci* 31:633–641.

Schnitzer, M. 1969. *Soil Sci Soc Am Proc* 33:75–81.

Selvarajah, N., V. Pavanasasivam, and K. A. Nandasena. 1982. *Plant Soil* 68:309–320.

Shiha, K. 1951. *J Sci Soil Manure* 22:26–28.

Shuman, L. M. 1979. *Soil Sci* 127:10–17.

Siccama, T. G., and W. H. Smith. 1978. *Environ Sci Technol* 12:593–594.

Siccama, T. G., W. H. Smith, and D. L. Mader. 1980. *Environ Sci Technol* 14:54–56.

Sidle, R. C., and L. T. Kardos. 1977. *J Environ Qual* 6:313–317.

Singh, D. V., and C. Swarup. 1982. *Plant Soil* 65:433–436.

Smilde, K. W. 1973. *Plant Soil* 39:131–148.

Smilde, K. W., and C. H. Henkens. 1967. *Netherlands J Agric Sci* 15:249–258.

Smith, P. F. 1953. *Bot Gaz* 114:426–436.

Smith, W. H. 1973. *Environ Sci Technol* 7:631–636.

Soltanpour, P. N., and A. P. Schwab. 1977. *Commun Soil Sci Plant Anal* 8:195–207.

Soman, S. D., V. K. Panday, K. T. Joseph, and S. J. Raut. 1969. *Health Phys* 17:35–40.

Somers, E. 1974. *J Food Sci* 39:215–217.

Sommer, A. L. 1931. *Plant Physiol* 6:339–345.

Sommers, L. E. 1977. *J Environ Qual* 6:225–232.

Sparr, M. C. 1970. *Commun Soil Sci Plant Anal* 1:241–262.

Sposito, G., L. J. Lund, and A. C. Chang. 1982. *Soil Sci Soc Am J* 46:260–264.

Spratt, E. D., and A. E. Smid. 1978. *Agron J* 70:633–638.

Stahly, T. S., G. L. Cromwell, and H. J. Monegue. 1980. *J Anim Sci* 51:1347–1351.

Stahr, K., H. W. Zöttl, and Fr. Hädrich. 1980. *Soil Sci* 130:217–224.

Steenbjerg, F., and E. Boken. 1950. *Plant Soil* 2:195–211.

Stevenson, F. J. 1972. *BioScience* 22:643–650.

Stevenson, F. J., and A. Fitch. 1981. In J. F. Lonegran, A. D. Robson, and R. D. Graham, eds. *Copper in soils and plants*, 69–95. Academic Press, New York.

Sutton, A. L., D. W. Nelson, V. B. Mayrose, and D. T. Kelly. 1983. *J Environ Qual* 12:198–203.

Swaine, D. J., and R. L. Mitchell. 1960. *J Soil Sci* 11:347–368.

Takahashi, Y., and H. Imai. 1983. *Soil Sci Plant Nutr* 29:111–122.

Thornton, I. 1979. In J. O. Niriagu, ed. *Copper in the environment*, 171–216. Wiley, New York.

Thornton, I., and J. S. Webb. 1980. In B. E. Davies, ed. *Applied soil trace elements*, 381–439. Wiley, New York.

Tiller, K. G., and R. H. Merry. 1981. In J. F. Lonegran, A. D. Robson, and R. D. Graham, eds. *Copper in soils and plants*, 119–137. Academic Press, New York.

Tirsch, F. S., J. H. Baker, and F. A. DiGiano. 1979. *J Water Pollut Cont Fed* 51:2649–2660.

Tiwari, R. C., and B. M. Kumar. 1982. *Plant Soil* 68:131–134.

Turvey, N. D. 1984. *Plant Soil* 77:73–86.

Tyler, G. 1972. *Ambio* 1:52–59.

Underwood, E. J. 1973. In *Toxicants occurring naturally in foods,* 43–87. NAS, Washington, DC.

US Environmental Protection Agency (US-EPA). 1976. *Quality criteria for water.* US-EPA, Washington, DC.

Van den Burg, J. 1983. *Plant Soil* 75:213–219.

Van Hook, R. I., W. F. Harris, and G. S. Henderson. 1977. *Ambio* 6:281–286.

Varvel, G. E. 1983. *Argon J* 75:99–101.

Viets, F. G. 1962. *J Agric Food Chem* 10:174–177.

Vinogradov, A. P. 1959. *The geochemistry of rare and dispersed chemical elements in soils.* Consultants Bureau Inc, New York. 209 pp.

Wallace, A., R. T. Mueller, and G. V. Alexander. 1978. *Soil Sci* 126:336–341.

Walsh, L. M., W. H. Erhardt, and H. D. Seibel. 1972. *J Environ Qual* 1:197–200.

Whitby, L. M., J. Gaynor, and A. J. MacLean. 1978. *Can J Soil Sci* 58:325–330.

Williams, J. G., and R. G. McLaren. 1982. *Plant Soil* 64:215–224.

Wright, J. R., R. Levick, and H. J. Atkinson. 1955. *Soil Sci Soc Am Proc* 19:340–344.

Younts, S. E., and R. P. Patterson. 1964. *Agron J* 56:229–232.

Zborishchuk, Y. N., and N. G. Zyrin. 1978. *Sov Soil Sci* 10:27–33.

7
Lead

1. General Properties of Lead

Lead (atomic no. 82) is a bluish-white metal of bright luster, is soft, highly malleable, ductile, and a poor conductor of electricity. It is very resistant to corrosion. It belongs to Group IV-A of the periodic table, has an atomic weight of 207.19, melting point of 327.5°C, boiling point of 1,744°C, and specific gravity of 11.35. It has oxidation states of II or IV. Lead has four stable isotopes, ^{204}Pb(1.48%), ^{206}Pb(23.6%), ^{207}Pb(22.6%), and ^{208}Pb(52.3%). Two radioactive isotopes ^{210}Pb (t½ = 22 years) and ^{212}Pb (t½ = 10 hours) are used in tracer experiments. The chloride and bromide salts are slightly soluble (about 1%) in cold water, whereas carbonate and hydroxide salts are almost insoluble.

2. Production and Uses of Lead

Current world mine production of Pb is about 3.5×10^6 tonnes/year, with about 15% (5.3×10^5 tonnes) coming from the United States (Rathjen and Rowland, 1980). While world production of Pb has been somewhat constant during the last several years, that of the United States has been declining (Fig. 7.1). The recent leading producers of Pb, in descending order are: United States, USSR, Australia, Canada, Peru, Mexico, Yugoslavia, and China.

There is a long history of man's production and uses of Pb (Fig. 7.1, insert). Ancient world technologies for smelting Pb-Ag alloys from sulfide (PbS) ores and cupeling Ag from the alloys were developed at least 5,000 years ago (Settle and Patterson, 1980). The need for Ag propelled Pb production in ancient times. In addition, the low melting point, the ease with which it can be worked and its durability partially contributed to its use as a construction material during this era. According to Settle and Patterson (1980), world Pb production averaged about 145 tonnes per year from 4,000 years ago until about 2,700 years ago, rose to about 9,100 tonnes

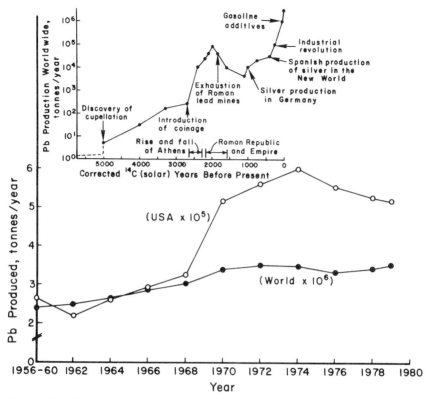

FIGURE 7.1. Trends in lead production in the USA and worldwide 1956–1979. In inset is the historical world production during the past 5,500 years. *Source:* (inset) Settle and Patterson (1980); US Bureau of Mines, *Minerals yearbook,* several issues.

with the introduction of silver coinage, and rose again to about 73,000 tonnes during the era of the Roman Empire, 2,000 years ago. Lead production declined during medieval times but dramatically rose again during the onset of the industrial revolution—from 91,000 tonnes annually 300 years ago to about 1×10^6 tonnes 50 years ago.

Lead is a very vital metal in any industrial economy. The only nonferrous metals that are used in greater amounts than Pb are Al, Cu, and Zn (Robinson, 1978). The major consuming areas in the western world are the United States, Europe, and Japan, which account for approximately 85% of refined Pb consumption. The Eastern block countries are currently net importers of Pb (approximately 46×10^3 tonnes/year).

The automobile and construction industries are the primary users of Pb. In western countries, an estimated 60% is consumed in the auto industry for solder in the building of autos, trucks and buses, in vehicular batteries, and as a gasoline additive (Robinson, 1978). The production of Pb-acid electrical storage batteries now accounts for over 40% of all the

Pb consumed by major industrial nations, with most of it destined for automobiles. The industry uses approximately equal amounts of metallic Pb (for battery grids and lugs) and Pb oxides (PbO, Pb_3O_4, and PbO_2), which are used in battery plates. Lead used in batteries can be easily recycled, and about 80% of all scrap Pb comes from this source (Waldron, 1980).

The other large use of Pb in the auto industry is in the production of alkyl lead compounds for use as anti-knock agents in petrol. The amount of Pb added to gasoline varies from 0.4 g/liter for "regular" to 0.6 g/liter for "premium" (Varma and Doty, 1980)). In 1975, about 17% of all the Pb used in the United States was used for alkyl Pb production (mostly tetraethyl and tetramethyl Pb). On a world basis, this use amounted to about 10% of the total Pb consumed. However, the use of Pb as a gasoline additive has been declining since the mid-1970s as a result of legislation in many countries responding to environmental concern about the release of Pb from auto exhausts.

Other important uses of Pb are in the manufacture of cable sheathings, pigments, pipe, foil and tubes, in various alloys with Sn, Cu, Sb, etc., ammunition and type metal, caulking, and varnishes (as red lead, white lead, litharge, chromates, sulfate, and titanate), and in the production of various inorganic compounds. Red Pb is used as a protective coating for steel structures; Pb chromate is used as a yellow pigment. Lead chemicals are used in glassware, ceramics, and in TV tubes. The metal is used to shield against radioactivity. Other miscellaneous uses include annealing, galvanizing, lead plating, weights, and ballasts. Uses of other Pb pigments, especially in interior paints, have been greatly reduced because of concern over their toxicity. Lead arsenate also has been banned as an ingredient in pesticides.

3. Natural Occurrence of Lead

In the earth's crust, Pb is the most abundant among the heavy metals with an atomic number >60. The average Pb content of the earth's crust has been set to range from 13 to 16 ppm (Swaine, 1978; NRCC, 1973). However, the most accurate estimate of the average abundance of Pb in the continental crust is probably those of Heinrichs et al (1980) at 14.8 ppm.

Although there are more than 200 minerals of Pb, few are common, with galena (PbS, 87% Pb by weight), cerussite ($PbCO_3$), and anglesite ($PbSO_4$) as the most economically important. Lead occurs in rocks as a discrete mineral, or, since it can replace K, Sr, Ba, and even Ca and Na, it can be fixed in the mineral lattice (Nriagu, 1978). Among the silicate minerals, potassium feldspars and pegmatites are notable accumulators of Pb.

TABLE 7.1. Commonly observed lead concentrations
(ppm) in various environmental matrices.

Material	Average concentration	Range
Igneous rocks[a]	15	2–30
Limestone[a]	9	—
Sandstone[a]	7	<1–31
Shale[a]	20	16–50
Carbonates[b]	9	—
Coal[b,c]	16	up to 60
Fly ash:[c]		
Bituminous	15	—
Sub-bituminous	3.1	—
Lignite	12	—
Phosphate fertilizers[d]	—	4.4–488
Animal waste[d]	—	2.1–3.4
Sewage sludges[d]	1,832	136–7,627
Soils[b]	20	2–200
Plants[b]	—	0.1–30
Vegetables and fruits[a]	1.5	<1.5–18
Forage legumes[a]	2.5	trace–3.6
Ferns[e]	2.3	—
Fungi[e]	—	0.2–40
Lichens[e]	—	1–78
Woody angiosperm[e]	—	1–8
Woody gymnosperm[e]	—	0.9–13
Municipal water*[a]	3.7	trace–6.2
Freshwater*[e]	3	0.06–120
Sea water*[e]	0.03	0.03–13

* μg/liter
Sources: Extracted from
[a] Cannon (1974).
[b] Swaine (1978).
[c] Adriano et al (1980).
[d] Appendix Tables.
[e] Bowen (1979).

The most commonly observed concentrations of Pb in various environmental matrices are given in Table 7.1. Lead content is higher in coal and shale, particularly organic shales, than in other types of rocks. It also accumulates with Zn and Cd in ore deposits. Lead contents of agricultural soils are reported to range from 2 to 200 ppm, with soils remote from human activity averaging about 5 to 25 ppm (Waldron, 1980). Much higher values can be expected in areas near densely populated centers, near industrial facilities, and in areas of geochemical deposits.

Background levels of Pb in 173 samples (53 soils) widely dispersed in Canada average 20 ppm (McKeague and Wolynetz, 1980), with values ranging from 5 to 50 ppm. This is comparable to the mean value of 19 ppm for United States soils (Appendix Table 1.17), and 10 ppm (Vinogradov, 1959), 15 to 25 ppm (Aubert and Pinta, 1977), and 15 ppm (Berrow and Reaves, 1984) values reported for world soils.

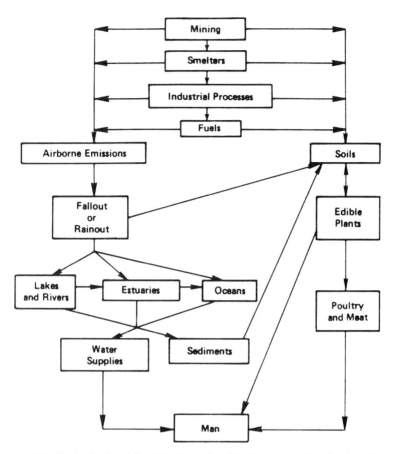

FIGURE 7.2. Ecological and food chain cycle of lead. *Source:* Newland and Daum, 1982, with permission of Springer-Verlag, New York.

The ecological and food chain cycle of Pb is depicted in Figure 7.2. It shows pools of Pb interconnected by pathways for the transfer of the metal. The majority of Pb in the lithosphere resides in soils, sediments, and rocks. Ecologically, the atmospheric Pb pool is the most important since it accounts for the majority of transfer of Pb between other pools.

4. Lead in Soils

4.1 Total Soil Lead

Lead is present in all soils at levels that range from <1.0 ppm in normal soils to well over 10% in ore materials. Nriagu (1978) reported the following Pb contents (in ppm) of noncontaminated soils for several regions: United States—18; USSR—12; Canada—12; Europe—from 10 for Spanish to 42

TABLE 7.2. Means and ranges of concentrations of lead for
Scottish soils and soil parent materials.

Sample	No. of data	Mean, ppm	Range, ppm
Soil			
All samples	3,944	14	2.6–83
All mineral samples	3,491	13	3.0–59
All organic samples	437	30	1.9–440
All L, F, and H horizon samples[b]	244	38	2.2–640
All A horizon samples	456	17	3.1–98
All S (surface) horizon samples	506	23	6.0–91
All B horizon samples	1,413	12	3.2–44
All C horizon samples	1,094	11	2.7–48
All samples from leached profiles	2,415	15	2.5–85
All samples from gley soils	1,114	15	2.9–75
Soil parent material			
Basic igneous rocks	153	9	1.4–57
Sandstones and mudstones	133	10	3.8–31
Slates, schists, and sandstones	78	11	2.9–44
Sandstones, conglomerates, and schists	101	11	3.2–41
Greywackes, shales, and sandstones	203	13	3.6–45
Sandstones, mudstones, limestones	97	13	2.2–78
Quartzites and quartzose schists	52	14	3.8–51
Mixed acid and basic igneous rock	92	14	4.5–44
Alluvium	73	15	2.7–83
Fluviatile gravels	117	15	3.8–62
Intermediate igneous	115	15	3.8–62
Schist	212	17	4.1–74
Slates and sandstones	152	20	6.0–68
Granite and granitic gneiss	67	22	6.7–69

[a] Source: Reaves and Berrow, 1984a, with permission of the authors.
[b] L = relatively undecomposed plant litter; F = partially decomposed litter; H = well decomposed organic matter.

for Scottish soils; Australia and New Zealand—15; Japan—11; Africa—from 12 for South African to 21 ppm for Egyptian soils; and Antarctica—8. For major soil groups, organic soils ($\bar{x}$ = 44 ppm) appear to have three times as much Pb as the other groups. Reaves and Berrow (1984a) found a mean concentration of 14 ppm Pb for 3,944 soil samples collected from 896 Scottish soil profiles, with a range of 2.5 to 85 ppm (Table 7.2). Their results indicate that the average Pb content of organic soil samples is much higher than that of the mineral soil samples (i.e., 30 ppm vs. 13 ppm), but it is also much more variable. Davies (1983) reported that concentrations greater than 110 ppm total Pb should not occur naturally in soils. Values greater than this should be considered anomalous, with a high possibility that they may have originated from contamination. Swaine's (1955) 2 to 200 ppm range for total Pb is widely used and is regarded as a reasonable assessment of typical soil values.

Agricultural soils can have a wide range of Pb content, depending on a number of factors, such as parent material and anthropogenic input. In England and Wales, the median Pb content was 42 ppm, with range of 5 to 1,200 ppm (Archer, 1980). These values are based on the analysis of 752 samples collected from 226 farms. In Ontario, Canada, Pb content of agricultural soils (n = 296) averaged 46 ppm, with range of 1.5 to 888 ppm (Frank et al, 1976). Soils from fruit orchards had the highest average contents of 123 ppm (range of 4.4 to 888 ppm) from the use of primarily lead arsenate pesticides. Other cropped soils had only 14 ppm Pb, with ranges of 1.5 to 50 ppm. In addition, Chisholm and Bishop (1967) noted that the Pb content of orchard soil samples from Nova Scotia usually had >50 ppm and non-orchard soils had concentrations below this level. The Pb values reported for normal agricultural soils in Canada are in the range of <1.0 to 12 ppm (Warren et al, 1969). Other major anthropogenic sources of Pb that can enrich agricultural soils are auto emissions, industrial deposition, and sewage sludge.

4.2 Extractable (Available) Lead in Soils

As with most trace elements, total Pb in soils is not a good indicator of Pb phytoavailability. Therefore, extractable Pb is normally used as the indicator of amounts available for plant uptake. A number of extractants (HCl, HNO_3, NH_4OAc, $CaCl_2$, organic acids, and others) have been used to extract soil Pb to predict its availability to plants. The success of predictability seems to depend on several factors, such as type of extractant, soil properties including pH, plant species, as well as other factors.

In testing four extractants, Misra and Pandey (1976) found that soil Pb extractability was in the order: Grigg's reagent* > 0.1 N HCl > 0.02 M EDTA > 1 N NH_4OAc. Both Grigg's reagent and NH_4OAc extractants gave significant correlations with Pb concentrations in wheat plants. John (1972) suggested that the degree of Pb contamination may affect the choice of a particular extractant since amounts extracted by the weaker extractants may be small and hence, difficult to analyze by atomic absorption. Consequently, he compared several strengths of HNO_3 (1 N to 0.01 N) with 1 N NH_4OAc as soil Pb extractants. As expected, the amounts of Pb extracted from the treated soils (500 ppm of Pb added to each of 29 soils) decreased with reduced normality of HNO_3. The 1 N HNO_3 extractant recovered most ($\bar{x}$ = 96% recovery) of the Pb added to soils, compared to 44% recovery for 0.1 N HNO_3, 4% for 0.01 N HNO_3, and 11% for 1 N NH_4OAc. The predictability of Pb uptake depended on plant species; only 1 N HNO_3 gave significant correlations with Pb in lettuce, but 1 N NH_4OAc, 0.10 N HNO_3 and 0.01 N HNO_3 all gave significant correlations with Pb in oat shoots.

---

* Grigg's reagent is an acid ammonium oxalate buffered solution.

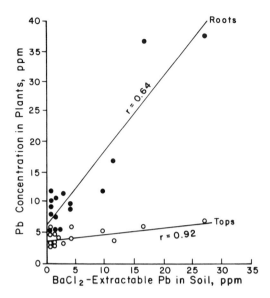

FIGURE 7.3. Relationships between extractable lead in soils and lead concentrations in ryegrass plants. *Source:* Jones et al, 1973b, with permission of the authors and Martinus Nijhoff Publishers, Dordrecht, Netherlands.

Jones et al (1973b) noted that soil Pb extracted by 0.5 M BaCl$_2$ or 0.05 M EDTA produced significant relationships (Fig. 7.3) with Pb contents of ryegrass than either total soil Pb or Pb extracted by 2.5% HOAc. On the average, the solutions of HOAc, BaCl$_2$, and EDTA extracted 1.0%, 16.3%, and 32.7%, respectively, of the total Pb in soils. They attributed the higher efficiency of extraction by BaCl$_2$ over HOAc to the greater degree of replacement of Pb^{2+} by Ba^{2+} than by H$^+$, due to the valence and similar radii of the ions (1.32 and 1.43Å for Pb and Ba, respectively).

The HOAc extractant has been successfully used to estimate the exchangeable and dilute-acid-soluble fractions of trace elements such as Pb, Co, Ni, and Zn (Berrow and Mitchell, 1980). Reaves and Berrow (1984b) reported that Scottish soils have a mean HOAc-extractable Pb content of 0.24 ppm, with a range of 0.016 to 3.4 ppm. Their results indicate that extractable Pb concentration was positively and linearly correlated with that of total Pb. Extractable Pb was found to be generally higher in surface horizons than in underlying horizons and that extractable levels were higher in poorly drained soils.

Other extractants used for soil Pb were: 0.05 M CaCl$_2$ (Karamanos et al, 1976), and Riehm and Egner reagent* (Kerin, 1975), and varying strengths of HCl and HNO$_3$ solutions.

The extractability of Pb in soils is influenced by several soil properties. Soils limed to high pH yielded less extractable Pb (Misra and Pandey, 1976; John, 1972; MacLean et al, 1969). Also, soils with high phosphate (MacLean et al, 1969), organic matter, or clay contents (Karamanos et al, 1976; Scialdone et al, 1980) tend to reduce the extractability of Pb from soils.

* A solution of 0.1 N ammonium lactate and 0.4 N HOAc, buffered at pH 3.7.

4.3 Lead in the Soil Profile

Several factors—for example, pedogenic processes, climatic and topographic effects, and microbial activities—influence the distribution of Pb in the soil profile. In general, Pb accumulates in the soil surface, usually within the top few centimeters, and diminishes with depth. In Scotland, Swaine and Mitchell (1960) observed that the surface horizons of most soils have higher, often considerably higher, Pb than the subsurface horizons. This pattern of Pb accumulation was attributed to organic matter accumulation on the surface due to plant dry matter recycling rather than to anthropogenic sources. Other investigators have shown this general pattern in other areas having diverse soil types (Mitchell, 1971; Butler, 1954; Bradley et al, 1978; Whitby et al, 1978). Butler (1954) suggested that formation of fairly insoluble PbS may partially explain the immobility of Pb in the soil profile, as he noted a preferential leaching of Ba^{2+} (Pb and Ba have the same valence and similar ionic radii) as the more relatively soluble $BaSO_4$. The declining trend of Pb with soil depth has been observed also in farmed organic soils (>85% organic matter) in the Holland Marsh Area, near Toronto, Canada (Czuba and Hutchinson, 1980). Lead levels decreased from surface values of 22 to 10 ppm at a depth of 48 cm. In comparison, Pb did not decrease with depth in mineral soils with very low Pb levels.

Lead of anthropogenic origin generally exhibits the same accumulation pattern on the surface layers, but also may migrate to deeper layers in some cases. In Cape Cod, Massachusetts, excess ^{210}Pb from atmospheric deposition has migrated to a depth of 40 cm in an undisturbed soil profile (Fisenne et al, 1978). This penetration depth of deposited material has been confirmed independently by ^{90}Sr measurements in the same soil profile (^{210}Pb and ^{90}Sr have some chemical similarity with comparable half-lives). Field plots at the University of Illinois (Table 7.3) amended with

TABLE 7.3. Lead distribution (in ppm) in the soil profile of cropped field plots, 6.5 years after soil treatment with lead, as lead acetate.[a,b]

Soil depth, cm	Applied Pb, kg/ha			
	0	800	1,600	3,200
0–15	37.1	275	325	650
15–30	16.8	32	46	320
30–45	17.3	22.8	20.2	28.0
45–60	18.5	21.2	19.2	22.8
60–75	17.5	19.8	18.5	21.7
75–90	17.3	18.0	17.5	20.5

[a] Source: Adapted from Stevenson and Welch, 1979, with permission of the authors and the Am Chem Society.
[b] The plots were planted to corn during the initial 2 years, then planted to soybeans during the rest of the study period.

variable amounts of Pb (0 to 3,200 kg Pb/ha, as lead acetate) are shown to have had Pb migrating down to at least the 90-cm depth, 6.5 years after the soil treatment (Stevenson and Welch, 1979). The soil in the plots, a Drummer silty clay loam with a pH of 5.9 and CEC of 30.3 meq/100 g, is known to effectively bind Pb in nonexchangeable forms with sorption capacity for Pb exceeding 20,000 μg/g. Downward movement of Pb was attributed to leaching as soluble chelate complexes with organic matter, transfer of soil particles by earthworms and other faunal organisms, translocation in plant roots, or a combination of these. Downward movement below the 30-cm depth could not have been caused by mechanical mixing of the soil during tillage operations. However, the retention of Pb in the surface layers (Table 7.3) was postulated due to reactions involving insoluble organic matter (Zimdahl and Skogerboe, 1977). In addition to vertical movement, significant horizontal movement to adjacent plots also occurred, primarily due to physical transfer of soil particles during tillage operations, or as windblown soil or plant particles. Rains (1975) also showed the close association of Pb with organic matter in soil profiles of grassland ecosystems impacted by smelting operations. Similarly, Andersson (1977) found close associations between Pb and organic matter in profiles of forest ecosystems.

Another evidence of downward movement of Pb in the soil profile was presented by Rolfe and Jennett (1975) when, in comparing the Pb contents of soil profiles from urban versus rural areas, higher Pb contents were detected in the 40- to 80-cm depths in the former areas. Other investigators (Page and Ganje, 1970; Milberg et al, 1980) found that essentially all of the Pb that has accumulated from automobile emissions remained in the surface few centimeters of the soil profile. Chang et al (1984b) found that, in cropped soils treated continuously for six years with municipal sewage sludge, more than 90% of the applied heavy metals, including Pb, was found in the surface 0 to 15 cm, the zone of application. No statistically significant increase in heavy metal contents was detected below the surface 30 cm of the soil profile. Similarly, Williams et al (1984) found that metal movement within the profile was limited to a depth of 5 cm below the zone of sludge application (0–20 cm) for Pb, Cd, and Cu. In this study, sewage sludge was added annually for six years to field plots cropped to barley at rates varying from 0 to 225 tonnes/ha per year.

4.4 Forms and Speciation of Lead in Soils

Sposito et al (1982) fractionated Pb in arid-zone field soils amended with sewage sludge. They did the fractionation by sequential extraction of the soil samples with KNO_3, H_2O, NaOH, EDTA, and HNO_3 to estimate exchangeable, sorbed, organic, carbonate, and sulfide fractions of Pb, respectively. For the Greenfield soil treated with 45 tonnes/ha per year of composted sludge for four years, the Pb distribution (in ppm) was as fol-

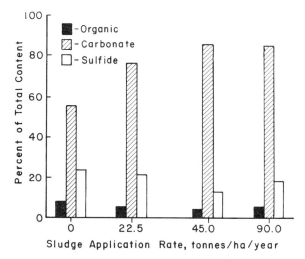

FIGURE 7.4. Distribution of lead fractions in an arid-zone soil amended with composted sewage sludge over a period of 6 years. The extractants used were 0.5 M NaOH, 0.05 M EDTA, and 4 M HNO_3 for the organic, carbonate, and sulfide fractions, respectively. *Source:* Sposito et al, 1982, with permission of the authors and the Soil Sci Soc of America.

lows: total Pb, 70 ± 2; exchangeable + sorbed, 1.6 ± 0.6; organic, 3.9 ± 0.1; carbonate, 50.8 ± 0.3; and sulfide, 17.6 ± 0.0. Thus, the dominant fractions of Pb are: carbonate >> sulfide > organic, shown in Figure 7.4. These data indicate that the most dominant fraction is the "carbonate" regardless of the sludge application rate. The "organic" fraction undergoes no significant change with application rate, whereas the "sulfide" fraction may decrease slightly. The low percentages of Pb in the "exchangeable" and "sorbed" forms could signify a low availability of this metal to plants, since the readily soluble form is often regarded as the most bioavailable. Chang et al (1984a) observed that the forms of Pb in soils were not substantially affected by applications of municipal sewage sludge.

In heavily Pb-polluted soils in Norway and Wales, Alloway et al (1979) gave the averages for the following fractions, as percent of total content: water soluble, <1; exchangeable, 7; organic, 61.7; and hydrous-oxide bound, 24.8.

Jurinak and Santillan-Medrano (1974) showed that the basic soils of the arid and semi-arid regions are an excellent sink for Pb. The movement of Pb in the percolating water is essentially nil, even in the presence of large amounts of excess salts. Cadmium, however, is less effectively removed from percolating water by these soils. The solubility of Pb in soil is about 100 times less than that of Cd in the pH range of 5 to 9.

Lead undergoes hydrolysis at low pH values and displays multiple hydrolysis at pH values encountered in the environment. The Pb-hydroxy

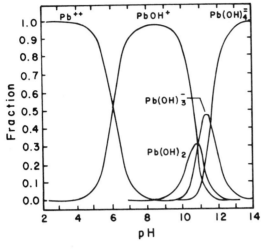

FIGURE 7.5. Simulations of the effects of pH (above) and chloride concentration (below) on the molecular and ionic species of lead. *Source:* Hahne and Kroontje, 1973, with permission of the authors and the Am Soc of Agronomy, Crop Sci Soc of America, Soil Sci Soc of America.

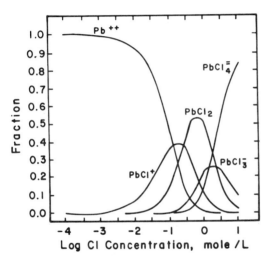

species indicate that $Pb(OH)_2$ formation is important above pH 9, while $PbOH^+$ is predominant between pH 6 and 10 (Hahne and Kroontje, 1973). Computer simulations for non-soil systems indicate that, at pH 6 to 7, the distribution of Pb^{2+} will be determined by the solubility of $PbOH^+$, with the latter predominating with increasing pH (Fig. 7.5, upper). In arid and semi-arid areas, soils are not only basic and saline in nature, but high chloride content is often associated with high salinity. Furthermore, chloride is ubiquitous in the natural soil environment and can be regarded as a very mobile and persistent complexing agent for the heavy metals. The chloride ion concentration determines the degree of complexation (Fig. 7.5, lower). The inorganic chloride complexes will always compete with sparingly soluble precipitates and organic complexes. However, because of their inorganic nature, chlorides are persistent unlike the organic complexes.

Under certain conditions, Pb can be transformed by microorganisms, as in the formation of a volatile Pb compound, tetramethyl Pb, by sediment microorganisms (Summers and Silver, 1978).

4.5 Adsorption, Fixation, and Complexation of Lead in Soils

As with other trace elements, the chemistry of Pb in soil can be qualitatively described as affected by: (1) the specific adsorption or exchange adsorption at a given mineral matrix; (2) the precipitation of sparingly soluble compounds of which it is a constituent, and (3) the formation of relatively stable complex ions or chelates that result from the interaction with organic matter.

Using equilibrium batch studies and deducing forms from equilibrium solubility diagrams, Santillan-Medrano and Jurinak (1975) presented data suggestive that, in noncalcareous soils, the solubility of Pb appears to be regulated by $Pb(OH)_2$, $Pb_3(PO_4)_2$, $Pb_4O(PO_4)_2$, or $Pb_5(PO_4)_3OH$, depending on the pH. In calcareous soils, $PbCO_3$ could assume importance. This tends to agree with the findings of Sposito et al (1982), which suggest that the carbonate form of Pb is a significant fraction of Pb in arid-zone soils (see section 4.4). Santillan-Medrano and Jurinak (1975) inferred from their studies that $Pb(OH)_2$ appears to regulate Pb^{2+} activity when the solution pH is less than about 6.6. As the pH increases, the formation of lead orthophosphate, lead hydroxypyromorphite, and also tetraplumbite phosphate becomes a distinct possibility. The precipitates formed under their experimental conditions may be the product of the coprecipitation of several compounds, each of which could exist alone in principle. These results tend to agree with Nriagu's (1972, 1973) contention that Pb phosphate formation can serve as a significant sink for Pb in the environment. This situation would be particularly important in areas contaminated with high levels of phosphate or in systems where high levels of indigenous phosphate exist.

Adsorption of Pb by soils and clay minerals has also been studied for a wide variety of reasons, including factors influencing the adsorption maximum. For all practical purposes, the Langmuir and Freundlich isotherms can describe the Pb adsorption by soil materials (Griffin and Au, 1977; Soldatini et al, 1976; Riffaldi et al, 1976). Depending on the mass of clay used, Griffin and Au (1977) were able to utilize the simple Langmuir equation over the entire concentration range of 0 to 1,200 ppm Pb. Soldatini et al (1976) found that Pb adsorption by soils conformed to both the Langmuir and Freundlich isotherms over a wide range of concentrations. Statistical analyses of their results indicate that organic matter and clay were the dominant constituents contributing to Pb adsorption, while the influence of other adsorbing surfaces, such as Mn oxides, was negligible. The former soil properties were also attributed by Hassett (1974) and Salim and Cooksey (1980) as primarily responsible for Pb adsorption by soils or sediments.

However, in view of the recent findings of McKenzie (1980), the role of the oxides of Mn, Fe, and possibly Al on Pb adsorption by soil can not be ignored. In measuring the adsorption of several trace elements (Co, Cu, Mn, Ni, Pb, and Zn) by synthetic Mn and Fe oxides, he found that Pb adsorption by Mn oxides was up to 40 times greater than that by the Fe oxides, and that Pb was adsorbed more strongly than any other metals studied. He considered this as proof for the accumulation of Pb in the Mn oxides of soils, observed previously by several investigators (Taylor and McKenzie, 1966; Norrish, 1975). McKenzie (1980) advanced three possible mechanisms to account for the binding of Pb by oxides of Mn: (1) strong specific adsorption; (2) a special affinity for oxides of Mn, as found for Co (McKenzie 1970, 1975) with the possibility of oxidation of the Pb; or (3) the formation of some specific Pb-Mn mineral, such as coronadite. The importance of hydrous ferric oxide on Pb adsorption has also been elaborated by Swallow et al (1980) and Gadde and Laitinen (1973).

Other investigators showed that the fixation of Pb by soil is principally caused by reactions involving essentially insoluble organic materials (Zimdahl and Skogerboe, 1977). In this case, precipitation by carbonate and sorption by hydrous oxides appear to be of secondary importance. The apparent involvement of hydrous oxides may be through their collection of or by organic matter. On the other hand, solubilization of Pb has been attributed to the formation of soluble chelate complexes with organic matter (Stevenson and Welch, 1979; Saar and Weber, 1980).

Exchange adsorption data by Bittel and Miller (1974), using pure clay systems (montmorillonite, illite, and kaolinite) show that Pb^{2+} was preferentially adsorbed over Ca^{2+}. The importance of competing ions on Pb adsorption was reconfirmed later by Griffin and Au (1977), who showed that the adsorption of Pb from $Pb(NO_3)_2$ solutions by a Ca-montmorillonite was dependent on the Pb/Ca ratios in solution. The importance of ion competition with Ca^{2+} was also shown by Riffaldi et al (1976). In studying the exchange reactions of Pb^{2+} with Al^{3+}, Ca^{2+}, and Na^+ ions in three soils, Lagerwerff and Brower (1973) found that a Gapon-type equation generally described the reactions. In the Na-systems, Pb^{2+} was found to precipitate, presumably as $Pb(OH)_2$ and other coprecipitates. The solubility of Pb-phase formed was found to be inversely related to pH and the concentration of salt (NaCl).

5. Lead in Plants

5.1 Root Uptake and Translocation of Lead by Plants

According to Wallace and Romney (1977), the distribution of trace elements in plants can be categorized into three groups: (1) somewhat uniformly distributed between roots and shoots—e.g., Zn, Mn, Ni, and B; (2) usually more in roots than in shoots, with moderate to sometimes large quantities in shoots—e.g., Cu, Cd, Co, and Mo; and (3) mostly in roots with very little in shoots—e.g., Pb, Sn, Ti, Ag, Cr, and V. This grouping, however, may change with some plant species and with very high soil levels of the elements.

Many researchers have shown that plants could accumulate Pb from the soil, stem, or foliar application. However, the results are rather conflicting as to the amounts accumulated and the amounts that can be translocated. Keaton (1937) found <3 ppm Pb in foliage of barley plants but up to 800 ppm in the roots of some plants when grown in soil containing up to 800 ppm Pb. Marten and Hammond (1966) studied Pb uptake by bromegrass from sandy loam soils with a range in Pb content of 12 to 680 ppm. Only plants grown in the soil with the 680-ppm level accumulated a significant amount of Pb. The maximum accumulation was only 34 ppm, supposedly enhanced by the addition of a chelate. Motto et al (1970) grew several crop species in contaminated soil and in acid-washed sand to which soluble Pb was added in low concentrations. Their data showed that Pb can be absorbed through the root system and that limited translocation to other plant parts occurred; most Pb taken up by the plants remains in the root system. Jones and Hatch (1945) compared various plant species grown on soils that had been treated with lead arsenate with those grown on untreated soils. Their mean values for the whole plants were 7.3 ppm Pb in the untreated soils and 11.2 ppm in the treated soils. The smallest amount of Pb taken up was in the edible portion of plants, and the greatest increase due to Pb in the soil was in the roots. In assessing the relative importance of air, water, and soil as sources of Pb for radish and ryegrass, Dedolph et al (1970) showed that only air and soil were significant sources and that both plant species derived only 2 to 3 ppm of Pb from soil sources. Jones et al (1973a) confirmed that plant roots restrict Pb movement into shoots.

In a field experiment (Baumhardt and Welch, 1972) where Pb acetate has been soil-applied at eight rates ranging from 0 to 3,200 kg Pb/ha, Pb concentration in corn foliage was significantly increased by Pb application (highest level was 27.6 ppm in leaves at tasselling at the 3,200-kg Pb/ha level vs. 3.6 ppm in the control). However, the Pb content of corn grain was not affected by added Pb. Their results indicate that Pb was absorbed by the roots and translocated to the plant tops, but it was not translocated from the stover to the grain. Miller and Koeppe (1970) reported that corn grown in sand culture absorbed considerable Pb, and that growth was retarded at high levels of added Pb. Rolfe (1973) showed that Pb uptake by eight tree species, grown on soil treated with five soil Pb levels (0 to 600 ppm), was significantly affected by soil Pb concentration with higher uptakes associated with higher soil Pb levels. Most of the Pb seems to accumulate in the roots, with significantly lower levels in the stem and leaves. Uptake studies of forage crops indicate that only a very small fraction (0.3% to 0.50%) of added Pb was utilized by the plants (Karamanos et al, 1976).

Apparently, Pb uptake by plants from soil is a passive process that proceeds rapidly until exchange sites in the root-free space are equilibrated with the solution concentration (Zimdahl and Arvik, 1973). As movement is mainly in the apoplast, translocation to the shoot is restricted by the Casparian band in the walls of the endodermal cells. Employing light and electron microscope studies of corn plants exposed to Pb in hydroponic solution, Malone et al (1974) showed that the roots generally accumulated a surface Pb precipitate and slowly accumulated Pb crystals in the cell walls. The root-surface precipitate formed without the apparent influence of cell organelles, whereas, Pb taken up by the roots was concentrated in dictyosome vesicles, fused with one another to encase the Pb deposit. Eventually, the Pb deposits concentrate in the cell wall outside the plasmalemma, representing the final stage of entry of Pb. They also obtained evidence that Pb was transported to the stems and leaves in a similar manner. Using an electron probe x-ray microanalyzer to study the mode of entry and localization of Cd and Pb in rice roots, Biddappa and Chino (1981) found that, initially, a small amount of Pb was found scattered broadly within the root tissues. After 180 minutes, Pb was concentrated more at the root surface than in the cortex. At 240 minutes, concentrations were sharply increased both at the surface and in the cortex, but the concentration was still higher at the surface. They concluded that Pb was

TABLE 7.4. Translocation of lead from roots, as indicated by concentrations (in ppm) in various plant parts, in selected crop species.[a]

Crop	Plant part	Plant Pb concentration		
		Control	200 ppm Pb in soil	1,000 ppm Pb in soil
Leaf lettuce	Leaf	2.5	3.0	54.2
	Root	5.8	84.5	867.7
Spinach	Leaf	0.7	7.9	39.2
	Root	4.7	73.3	—
Broccoli	Leaf	7.2	8.4	18.4
	Root	6.5	83.0	745.6
Cauliflower	Leaf	5.3	6.3	11.8
	Root	2.5	55.1	532.2
Oats	Grain	3.2	4.4	4.9
	Husk	11.1	11.8	16.4
	Leaf	6.0	6.8	20.1
	Stalk	1.6	2.5	9.2
	Root	4.5	82.0	396.6
Radish	Tops	3.7	9.9	14.3
	Tuber	6.3	7.0	44.6
Carrot	Tops	2.3	8.0	17.6
	Tuber	1.9	5.3	41.0
	Root	8.9	241.7	561.4

[a] *Source:* John and Van Laerhoven, 1972a, with permission of the Department of Agriculture, Government of Canada.

more slowly absorbed into the roots than Cd and subsequently translocated less to the cytoplasm.

Lead also may form insoluble complexes with cell wall constituents. Rains (1975) found little translocation in wild oat plants grown in nutrient solution in spite of Pb addition in the chelated form. He attributed Pb exclusion from the shoots (3,500 ppm in roots vs. 65 ppm in shoots) to precipitation or insoluble complexation of Pb on xylem elements and cell walls, thereby immobilizing it and greatly reducing transfer from the roots to the shoots. The immobile nature of Pb in plant roots is illustrated in Table 7.4, where very little translocation occurred from roots to the vegetative parts of crops.

Arvik and Zimdahl (1974b) suggest that a passive mode of entrance can account for the majority of Pb entering the plant root system when they found that Pb uptake follows a rapid influx and attainment of a concentration plateau; uptake is increased by increasing concentration, exhibits pH-dependency, is relatively insensitive to metabolic inhibitors, and does not exhibit unusual temperature effects.

5.2 Foliar Deposition and Uptake of Lead

In his review, Roberts (1975) listed at least three factors that can affect Pb retention by vegetation: (1) aerosol properties (particle size, chemical composition), (2) leaf surface characteristics (roughness, pubescence, moisture, and stickiness), and (3) meteorological factors (relative humidity, cloud density, and wind speed). These are some of the factors that account for variations observed in Pb concentrations for vegetation.

It is generally considered that atmospheric Pb increases the Pb content of vegetation primarily by particulate deposition. Although soluble materials applied to plants as foliar sprays enter the leaves in significant amounts (Brandt, 1970), since particulate Pb is quite insoluble, it is not expected to enter the leaf surfaces in substantial amounts. In studying plant contamination by Pb from mining and smelting operations, Palmer and Kucera (1980) concluded that the chemical form, level, and particulate size of atmospheric Pb, as well as the nature of the soil Pb, play an important role in the uptake of Pb by the plant tissue. Koslow et al (1977) found that Pb particle size was variable, with maximum dimensions ranging from 1 to 25 μm. While Pb sulfate, phosphate, and oxide have been detected in motor vehicle exhaust, Pb chlorobromide is generally assumed to be the primary Pb salt introduced into the atmosphere from the tailpipe (Ter Haar and Bayard, 1971). Lead particles leaving the exhaust system are generally in the range of <1 to 5 μm size.

The fact that generally high percentages of the Pb deposited on vegetation surfaces can be removed by water wash indicate that Pb is externally located. Kloke and Riebartsch (1964) reported that a very large fraction (80% and more) of Pb was washed off from grasses by water. Similarly, Crump and Barlow (1980) showed that between 45% and 80% of the total

Pb concentration in grasses originated from airborne deposition. Carlson et al (1976) obtained an unusually high (95%) removal of topically applied Pb from soybean leaves using simulated rainfall. The leaves were experimentally fumigated with $PbCl_2$ aerosol particulate. Other investigators (Rains, 1975; Lagerwerff et al, 1973; Davies and Holmes, 1972; Page et al, 1971) showed the importance of atmospheric Pb in the Pb burden of vegetation and its potential effect on quality of the vegetation for animal feeding.

In their review of Pb in plants, Höll and Hampp (1975) concluded that Pb is partially deposited on the surface and partially incorporated into the tissue because the total Pb content showed a correlation with the leaf area, whereas after a washing period of 24 hours, the Pb content was still a function of the leaf weight.

Large differences in aerosol deposition may occur between plant species because of differences in their surface structures (Page et al, 1971; Wedding et al, 1975). Plant tissue with a rough pubescent surface structure can accumulate considerably more—up to at least seven times more —Pb than tissues with a smooth surface. Thus, it can be reasonably expected that Pb accruing on rough pubescent tissues is not as easily washable.

Since it has been shown that Pb applied or deposited on the foliage remained as a topical coating on the foliage (Arvik and Zimdahl, 1974a; Carlson et al, 1976; Lagerwerff, 1972; Schuck and Locke, 1970), the barrier mechanism of foliar uptake of Pb should be elaborated. These barriers are the epicuticular waxes and the cuticular membrane (Arvik and Zimdahl, 1974a). The epicuticular waxes are extruded through the developing cuticle, and form structures that may be typical of a given species or set of environmental conditions. The cuticular membrane, which must be penetrated for entry to the foliage, is a noncellular structure covering the epidermal layers of leaves, stems, and fruits of higher plants, and is composed primarily of cutin, within the matrix of which can be found varying amounts and kinds of waxes, fatty acids, alcohols, and sugars. The waxes are external to the cuticle as well as being interspersed within the cutin structure and serve to make wetting of the surface more difficult. The chemical and physical nature of the waxes may act as a barrier to the entry of Pb into the leaf—i.e., particulates may be trapped within the wax matrix. Arvik and Zimdahl (1974a) found that only extremely small amounts of Pb penetrated the cuticle, even after extended exposure and exceptional conditions of Pb solubility and pH.

The Pb particulates that accumulate on leaf surfaces may cause adverse physiological effects, either by blocking the stomates and thus interfering with the normal exchange of gases between the leaf and its surrounding air, or by disrupting metabolic pathways after entry into the leaf. However, Carlson et al (1976) did not find any evidence that Pb particulates applied on soybean leaves interfered with photosynthesis.

5.3 Interactions of Lead with Other Ions

As discussed in section 4.5, the complexation of Pb with phosphate is a possible mechanism of immobilizing Pb in the environment. Consequently, less uptake of Pb by plants from soils having high levels of phosphate has been demonstrated (MacLean et al, 1969; Miller and Koeppe, 1970; Rolfe, 1973; Sheaffer et al, 1979). Miller and Koeppe (1970), using sand culture to grow corn plants, demonstrated that no Pb would be taken up by the plants as long as there is enough phosphate in the solution to precipitate Pb. Rolfe (1973) observed about 50% reduction in uptake of Pb by several tree species when high levels of phosphate were present.

Of the several soil factors studied to affect Pb uptake, soil phosphate content was the most dominant. Under high levels of Pb, but somewhat marginal levels of soil phosphate, phosphate deficiency may be exhibited by the plants because of the formation of insoluble Pb phosphate precipitates in the soil-plant systems (Koeppe and Miller, 1970; Lee et al, 1976).

Possible interaction of Pb with sulfate ions on the surface or in the plant roots is also possible (Jones et al, 1973b; Karamanos et al, 1976). The latter investigators found that the ratios, on a per pot basis, of total uptake of Pb by roots to total uptake of Pb by tops of alfalfa and bromegrass, were higher in the sulfate than in the nonsulfate-treated soils. They attributed the immobilization to complex formation of Pb with S-containing amino acids or proteins in the plant roots.

Farrah and Pickering (1977a), in studying the adsorption of various clay minerals, found that increasing the concentrations of sulfate and phosphate anions reduced the Pb^{2+} concentration in the residual solution. This indicates possible formation of sparingly soluble salts between anions and Pb.

5.4 Phytotoxicity of Lead

The phytotoxicity of Pb is relatively low compared with other trace elements. In rice plants, the order of phytotoxicity (Chino, 1981) in decreasing order is:

$$Cd > Cu > Co \approx Ni > As \approx Cr > Zn > Mn \approx Fe \geqslant Pb$$

In sugar beets grown in sand culture, Hewitt (1953) found that Cu, Co and Cd were highly and about equally active in causing chlorosis due to toxicity, followed by $Cr(VI) > Zn(II) \geqslant V(II) \geqslant Cr(III) > Mn(II) \geqslant Pb(II)$. Crop susceptibility to metal toxicity varied greatly. For example, Cu readily caused chlorosis in beet, tomato, and potato, but not in oats and kale. For oat and tomato plants, Berry (1924) found that a concentration of Pb, as $Pb(NO_3)_2$, of 25 mg/liter was required to cause toxicity. At a concentration of 50 mg/ liter, plant death occurred. Similarly, Davis et al

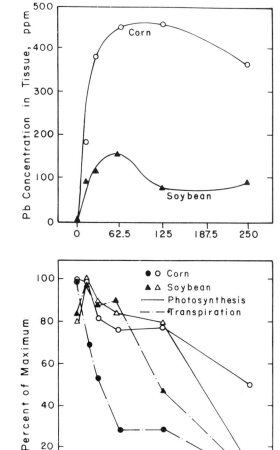

FIGURE 7.6. Lead accumulation in leaves of corn and soybeans and its effect on photosynthesis and transpiration. *Source:* Bazzaz et al, 1974, with permission of the authors and the Am Soc of Agronomy, Crop Sci Soc of America, Soil Sci Soc of America.

(1978) found that 25 mg/liter of Pb in solution used in sand culture was the critical level for barley growth. This concentration in the growth medium corresponds to an average concentration of 35 ppm (dry wt) in the leaves and shoots of these plants. Hopper (1937) found that 30 mg/liter of Pb in nutrient solutions was toxic to bean plants. Dilling (1926) reported that 2,000 mg/liter Pb (as nitrate and acetate) in aqueous solution prevented germination of vegetable (cress and mustard) seeds. His soil studies with Pb and germination were variable, and he attributed the variability to adsorption and precipitation of Pb in soils. In rice plants, Chino (1981) found the following toxic levels for Pb: 50 to 2,000 ppm in tops and 300 to 3,000 ppm in roots. Total Pb amounting to 400 to 500 ppm in the soil in a polluted area in Japan was found to be toxic to the plants. In radish plants grown on soil, Pb toxicity was manifested as stunted growth, which is more pronounced in roots than in shoots (Khan and Frankland, 1983). On a molarity basis, Cd was twenty times more toxic than Pb.

Excess Pb in plants alters several physiological and biochemical processes within the plants. It has been reported that Pb treatment affects mitochondrial respiration in corn (Koeppe and Miller, 1970); alters photosynthesis in soybean (Bazzaz et al, 1974; Huang et al, 1974); and inhibits photosynthetic electron transport in isolated spinach chloroplasts (Miles et al, 1972). These findings indicate that Pb affects energy-producing systems in plants. Growing corn and soybean seedlings in media treated with Pb (0–250 mg Pb, as $PbCl_2$, per plant, or 0–4,000 ppm Pb in solution), Bazzaz et al (1974) showed decreased net photosynthesis and transpiration with increasing Pb treatment levels (Fig. 7.6). The general simultaneous rise and decline in photosynthesis and transpiration of the two species in response to increased Pb level suggest that the rate changes of the two processes are related to changes in leaf stomatal resistance to CO_2 and water vapor diffusion. They suggested, however, that indirect rather than direct Pb toxicity effects can not be discounted. These include interference with ion uptake and translocation, growth retardation due to inhibition of mitochondrial respiration, and inhibition of chloroplast activity. With soybean seedlings grown in culture solution spiked with $Pb(NO_3)_2$ (0–100 mg/liter of Pb), Lee et al (1976) found increased respiration rate, increased activities of the enzymes acid phosphatase, peroxidase, and alpha-amylase, and increased levels of soluble protein and ammonia with Pb treatment. They found practically no change in malic dehydrogenase and total free amino acids. However, a decrease was observed for glutamine synthetase activity and nitrate. Increased activities of the hydrolytic enzymes and peroxidase indicates that the Pb treatment enhances senescense.

6. Lead in Natural Ecosystems

Lead is naturally present in small amounts in practically all environmental matrices. Since its use as a gasoline additive, Pb concentration in the atmosphere has increased. Thus, even remote terrestrial ecosystems are subject to Pb enrichment. Studies in many parts of the world, although limited, indicate considerable Pb enrichment of various forest ecosystem components from atmospheric input (Siccama and Smith, 1978; Van Hook et al, 1977; Schinner, 1980; Heinrichs and Mayer, 1980; Stahr et al, 1980; Reiners et al, 1975; Tyler, 1972).

Several investigators have conducted Pb budget studies in forest ecosystems. At the Hubbard Brook Experimental Forest in central New Hampshire, a remote northern hardwood forest ecosystem, Siccama and Smith (1978) found that the total Pb input to the ecosystem was 317 g/ha

TABLE 7.5. Lead concentrations (in ppm) in unwashed tree parts collected from the Hubbard Brook Experimental Forest, New Hampshire.[a]

Tree species	Leaves	Twigs	Bark	Wood	Roots[b]
Acer saccharum (sugar maple)	3.7	12.2	37.9	0.39	16.6
Betula alleghaniensis (yellow birch)	6.5	28.0	13.2	1.1	32.6
Fagus grandifolia (American beech)	8.0	13.0	19.8	0.72	10.2
Picea rubens (red spruce)	13.8	68.8	4.6	0.35	21.0
Abies balsamea (balsam fir)	7.4	38.2	8.7	0.55	—

[a] Source: Smith and Siccama, 1981, with permission of the authors and the Am Soc of Agronomy, Crop Sci Soc of America, Soil Sci Soc of America.
[b] Washed root material.

per year. The output of Pb from this ecosystem was approximately 12.1 g/ha per year, split as 11.4 g in the dissolved form and 0.7 g as coarse particulate matter in stream water draining from the ecosystem. Thus, Pb from the atmosphere is accumulating in the soil of this ecosystem at a rate of 305 g/ha per year, which could double the current Pb concentration in the forest floor humus (167 μg/g) in about 50 years. Lead concentration in the vegetation is very low, especially in the wood that makes up approximately 86% of the total living aboveground biomass. Their estimate of total Pb in the living biomass is <1 kg Pb/ha, whereas the forest floor has about 14.6 kg Pb/ha. In addition, Smith and Siccama (1981) found the following Pb concentrations in the biota: lichens (213 ppm) > mosses (190 ppm) > tree twigs (26 ppm) > roots (20 ppm) > bark (19 ppm) > leaves (7 ppm) = bracket fungi (7 ppm) > wood (0.7 ppm). The levels of Pb associated with tree parts are shown in Table 7.5.

In a regional (Connecticut, Massachusetts, and New York) study, Johnson et al (1982) found that the current average Pb concentration of the forest floor is 148 ± 5 ppm. These investigators observed increasing Pb levels with ages of forest floor up to 80 to 100 years (Fig. 7.7), indicating that nearly all of the Pb has accumulated during the last century. They found that Pb level in the forest floor is influenced by the following factors: precipitation rate, distance from the urban area, elevation, forest floor age, nature of underlying mineral horizon, and distance to Pb-emitting industries. Deposition on Camels Hump Mountain, Vermont increased Pb, Cu, and Zn in the forest floor by as much as 148% during a 14-year period (Friedland et al, 1984). At the Walker Branch forest watershed, in eastern Tennessee, the Pb budget (Fig. 7.8) indicates that the stream output was only 6 g/ha per year compared to the total atmospheric input of 286 g/ha per year. The Tennessee study also indicates that the litter and soils

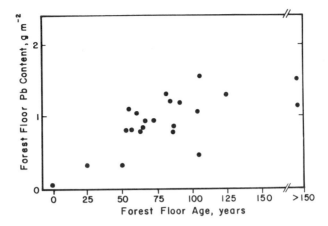

FIGURE 7.7. Relationships of forest floor lead content and forest floor age. Points represent site means of white pine plantations and pine/hemlock/hardwood stands in Massachusetts, New York, and Connecticut. All sites lie outside the urban corridor and below 500 m elevation. *Source:* Johnson et al, 1982, with permission of the authors and the Am Soc of Agronomy, Crop Sci Soc of America, Soil Sci Soc of America.

are the major sinks for Pb (Fig. 7.9). The major sites of Pb accumulation were the O1 (L layer) and O2 (F and H layers) litter horizons (O2 > O1) and small lateral roots, while the soils are the long-term sink. Estimated turnover time for Pb in vegetation, including root pools, is three years.

The distribution of Pb in the major vegetative components and soils at the Walker Branch forest ecosystem (Fig. 7.9) indicates that most of the Pb in the living vegetation is located in the bole (30%) and lateral roots (40%). The average Pb concentrations (in ppm) in the biomass of 11 tree species were: foliage, 4.7; branch, 3.2; and bole, 1.5. Thus, the concentration of Pb in vegetation followed the pattern: roots > foliage > branch > bole.

Forest ecosystems of New England, which are several hundred kilometers from major population centers, are subject to higher atmospheric Pb deposition rates than other rural regions of North America because of patterns of air movement and locations of sources (Reiners et al, 1975; Schlesinger et al, 1974; Siccama and Smith, 1978). Even within an ecosystem, altitudinal differences in Pb deposition rates can be expected due to differential exposure to the wind stream. Siccama and Smith (1978) noted higher Pb concentrations in the forest floor on the upper portion of a Hubbard Brook forest ecosystem than on the lower slope. In another forest ecosystem in New Hampshire, Reiners et al (1975) found that total Pb concentration in the forest floor increased with increasing altitude up to the fir forest, then declined in krummholz and alpine tundra. The decline in the latter ecosystems was attributed to the decline in interceptive veg-

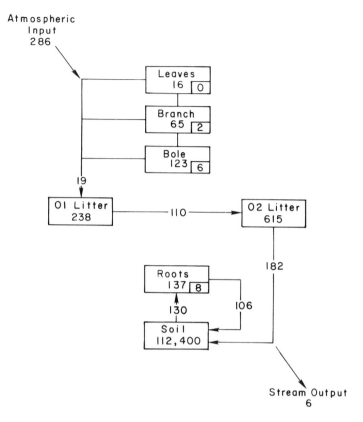

FIGURE 7.8. Lead cycle for an eastern Tennessee mixed deciduous forest. Standing pools are in units of g/ha (values in center of boxes), annual fluxes in g/ha per year (values in arrows), and annual increments in g/ha (values in lower right corner of boxes). O1 corresponds to L layer; O2 corresponds to F and H layers. *Source:* Van Hook et al, 1977, with permission of the authors and Pergamon Press, New York.

etative surfaces. The greater deposition of atmospheric Pb in the New England forest ecosystems and the effect of altitude and canopy surface area are illustrated with the Pb data in various ground components in Table 7.6. Lead values appear to be much higher in the New Hampshire forest ecosystem than in other unpolluted rural ecosystems elsewhere, with the contents rising with the altitude until the vegetative interceptive capacity limits accumulation.

In relatively rural forest (white pine) ecosystems in central Massachusetts, Siccama et al (1980) found that Pb has been accumulating, as indicated by significant increases in Pb content in the forest floor, during a 16-year period. The net average annual gain in the forest floor was about 300 g/ha per year of Pb, equivalent to approximately 80% of the estimated

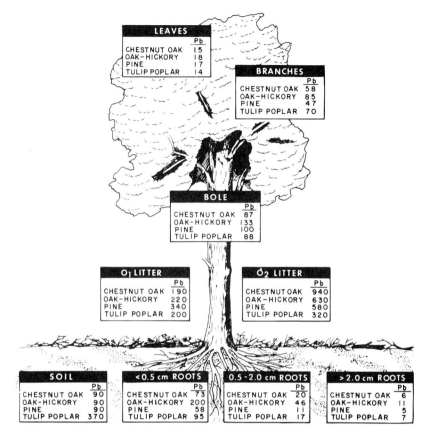

FIGURE 7.9. Distribution of lead in the major vegetative components (g/ha) and soils (kg/ha) in an eastern Tennessee mixed deciduous forest. *Source:* Van Hook et al, 1977, with permission of the authors and Pergamon Press, New York.

total annual deposition from precipitation and dry fallout in the region. The total amount of Pb in the forest floor in these pine forests (12 kg/ha) is greater than the amount found in the forest floor of the hardwood forest in northern New Hampshire (8 kg/ha).

In Europe, heavy metal concentrations in the Solling forest ecosystems in central Germany often exceed the values reported from other regions that are not influenced by local pollution (Heinrichs and Mayer, 1977, 1980). For example, concentrations of Pb in the forest litter are higher in the Solling than the mixed deciduous forest in eastern Tennessee (Van Hook et al, 1977), and in subalpine forests in New Hampshire (Reiners et al, 1975), but similar to a Missouri oak forest presumably free from local pollution (Jackson and Watson, 1977). In addition, the Pb concentrations in beech and spruce wood in the Solling are similar to the wood

TABLE 7.6. Lead concentrations (in ppm) in various ground components of relatively unpolluted forest ecosystems.

Location[a]	ppm Pb	Source
New Hampshire, USA		Reiners et al (1975)
Hardwoods		
O1	22	
O2 (F)	57	
(H)	30	
Spruce fir		
O1	110	
O2 (F)	114	
(H)	60	
Fir		
O1	274	
O2 (F)	220	
(H)	71	
Fir-krummholz		
O1	212	
O2 (F)	173	
(H)	73	
Alpine tundra		
(O1 + O2 + H)	38	
Tennessee, USA		Van Hook et al (1977)
Yellow poplar		
O1	31	
O2	42	
Chestnut oak		
O1	27	
O2	51	
Oak-hickory		
O1	25	
O2	35	
Pine		
O1	31	
O2	37	
Heath, Sweden		Tyler et al (1973)
(Aboveground litter)	66, 80	
(Below ground litter)	40, 45	
Spruce needle litter, Sweden		Nilsson (1972)
Least decomposed	27, 34, 78	
Intermediate	49, 61, 102	
Most decomposed	66, 105	

[a] O1 Horizon corresponds to L (litter) layer; O2 Horizon corresponds to F (fermentation) and H (humic) layers.

concentrations in oak and ash from a roadside woodland in England (Hall et al, 1975). Heinrichs and Mayer (1980) also observed that Pb is the only element that shows a clear concentration increase with the aging of spruce needles and a further drastic increase in the freshly fallen litter and in the forest floor organic matter. This was consistent with Tyler's (1972) and Nilsson's (1972) observations in Swedish spruce needles. Lead is strongly accumulated in the surface soil organic layer, which retained about 80 to 100 times the amount of Pb found in annual litterfall. The Pb concentrations decreased from the surface to the lower mineral soil. The pool of Pb in vegetation is about six times higher, and that in the forest floor layers about 20 times higher, in the Solling beech forest ecosystem than the ones reported for a Tennessee forest ecosystem (Van Hook et al, 1977).

7. Factors Affecting the Mobility and Availability of Lead

7.1 Soil Texture and Clay Mineralogy

Soil texture, in particular the soil's clay content, has been shown to influence plant uptake of Pb as well as sorption of Pb by soil materials. Soils with higher clay contents, assuming other soil constituents are constant, generally have higher CEC and therefore, higher binding capacity for cations.

In a growth chamber experiment, Karamanos et al (1976) reported that Pb concentrations were higher in the soil solutions obtained from the soil with the lowest clay and organic matter contents, causing usually higher Pb concentrations in the tops of alfalfa and bromegrass. In soils treated with sewage sludge, metal extractability by DTPA solution and plant uptake also correlated with soil texture (Gaynor and Halstead, 1976). Extractable Pb and Pb uptake by lettuce and tomato plants were higher in Granby sandy loam and Fox sandy loam than in Rideau clay soils. In investigating various soil properties influencing plant (oats and lettuce) uptake of Pb, John (1972) found that soil texture, in addition to soil pH, were significantly related to the amounts of Pb in plant parts. Organic matter in this case did not influence Pb phytoavailability.

Sorption studies of Pb by soils indicate that clay, in addition to organic matter, CEC, and soil pH are important soil parameters determining the fixation capacity of soils (Zimdahl and Skogerboe, 1977). Similarly, several investigators (Hassett, 1974; Soldatini et al, 1976; Riffaldi et al, 1976) have reported the dominant roles of clay and organic matter contents in soil sorption of Pb. The strong affinity of Pb and other metals to the clay fraction, compared to that of the sand and silt fractions, is demonstrated by Le Riche and Weir (1963) and Andersson (1979), who ranked them as: clay > silt > sand.

Clay mineralogy can also exert an influence on soil sorption of Pb. In an exchange adsorption study, Bittell and Miller (1974) reported that the average selectivity coefficients for montmorillonite, illite, and kaolinite were 0.60, 0.44, and 0.34, respectively, for Pb-Ca exchange and 0.58, 0.56, and 0.31 for Pb-Cd exchange. This indicates that Pb^{2+} adsorption was favored over Ca^{2+} on all three clay minerals, with the most preference

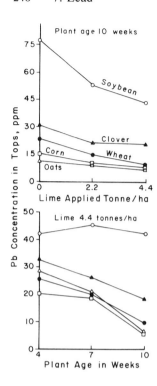

FIGURE 7.10. Effect of lime treatments in a lead-contaminated soil (top) and plant age (bottom) on lead concentration in tops of various crop species. *Source:* Cox and Rains, 1972, with permission of the authors and the Am Soc of Agronomy, Crop Sci Soc of America, Soil Sci Soc of America.

being on kaolinite. Using similar techniques, Lagerwerff and Brower (1973) found that in the presence of Al^{3+}, Pb was adsorbed more strongly by Cecil (kaolinitic) and Yolo (montmorillonitic) soils than by the Winsum (illitic) soil, whereas in the presence of Ca^{2+}, Pb adsorption is stronger with the Yolo than with the Cecil and Winsum soils. Harter (1979) found that sorption of Pb by soils sampled from the B horizon in eight northeastern states was significantly related to the vermiculite content.

7.2 pH

Soil reaction probably is the single most important factor affecting the solubility, mobility, and phytoavailability of Pb. Early to recognize this was Griffeth (1919), who noticed that Pb toxicity was inversely proportional to soil pH. Increasing soil pH decreased Pb in plant roots, although no consistent trends occurred in the foliage. MacLean et al (1969) suggested that lime reduced the uptake of Pb by oat and alfalfa plants via repression of solubility of Pb because of the greater capacity of organic matter to complex Pb with increasing pH. Cox and Rains (1972) and John and Van Laerhoven (1972b) reported that application of lime to Pb-contaminated soils reduced the foliar Pb content of the plants treated (Fig. 7.10) but

FIGURE 7.11. Effect of pH and redox potential on uptake of lead by rice shoots and roots and on the amount of water-soluble lead in soils. *Source:* Reddy and Patrick, 1977, with permission of the authors and the Am Soc of Agronomy, Crop Sci Soc of America, Soil Sci Soc of America.

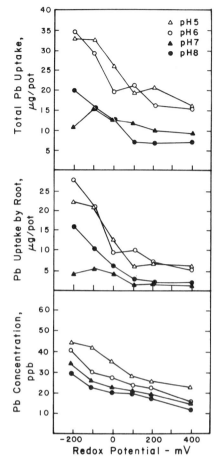

had little effect on Pb in roots. Lagerwerff (1972) rationalized that the occurrence of normal plants having high Pb concentrations in spite of lime addition may indicate a counteracting effect of lime, in the soil, on mobility and uptake of Pb, and also on the antidotal effect of intracellular Pb. Zimdahl and Foster (1976) found that liming the soil did not have a consistent effect on uptake by corn plants, but Pb translocation in the tops appeared to have been restricted by liming. Based on the comparable availability of insoluble $PbCO_3$ with $PbCl_2$ and $Pb(NO_3)_2$ to plants, John and Van Laerhoven (1972b) suggested that formation of $PbCO_3$ as a result of liming can not be considered a sole factor affecting Pb availability.

In rice, Reddy and Patrick (1977) found that Pb uptake, including root uptake, decreased with an increase in pH and suspension redox potential (Fig. 7.11). Redox potential and pH had a pronounced effect on the uptake of Pb by the root. Accumulation of Pb in the shoot was also influenced by pH, increasing with a decrease in pH from 8 to 5. There was a highly significant correlation of water-soluble Pb with both total Pb uptake (r = 0.83, significant at the 1% level) and Pb uptake by the root (r = 0.86, singificant at the 1% level). At a given pH, Pb translocation from root to the shoot increased with increasing redox potential. At low redox potential and at high pH levels, more Pb accumulated in the roots. They attributed this to precipitation of Pb on the root surface. Based on this common observation, that Pb may accumulate in roots but not in the tops, it appears that the primary effect of liming acidic soils may be on translocation of Pb rather than on its uptake by plants. The effect on uptake may be partly due to competition between Pb and other cations for available exchange sites within the soil matrix. The translocation effect conforms to that of Lagerwerff's (1972) hypothesis and is also explained in section 5.1. In case of flooded soils, especially in slightly acidic to acidic soils, the interaction between Pb and P, and Pb and Fe is a possibility. Increased uptake by the rice plants of P and Fe, and other trace elements such as Mn under reducing soil conditions, as compared to oxidized conditions, may result in the formation of Pb complexes with these elements in/on roots, retarding Pb translocation and resulting in more Pb accumulation in roots (Reddy and Patrick, 1977).

Also influenced by pH are the stability of molecular and ionic species of Pb (Fig. 7.5), forms of Pb (Gambrell et al, 1980), and the intensity of fixation by soil materials (Misra and Pandey, 1976; Farrah and Pickering, 1977b).

7.3 Organic Matter and Cation Exchange Capacity

It was indicated in section 4.3 that in the soil profile, Pb is often considerably higher in the surface than in the subsurface horizons. Lead accumulation in the surface horizon is associated with high organic matter content, partly due to reactions involving insoluble organic matter (Zimdahl and Skogerboe, 1977). These investigators concluded that Pb fixation by organic matter is more universally important than precipitation by carbonate or sorption by hydrous oxides and that the majority of Pb immobilized by soil is associated with organic matter.

Apparently, one mechanism of fixation by organic matter for metal cations is by ion exchange. Bunzl (1974, 1976) suggested that adsorption of Pb^{2+}, Cd^{2+}, Zn^{2+} or Ca^{2+} ions onto humic acid and peat occurs through an ion exchange process between the ion adsorbed and the H^+ ions from the organic matter—i.e., the adsorption of one divalent metal ion by peat was coupled with the release of two H^+ ions. Because the CEC of soils is often related to the organic matter content—i.e., they are positively correlated—an increase in CEC is often associated with reduced plant uptake but increased soil sorption of Pb (Hassett, 1974; Miller et al,

1975a,b). Organic matter may also serve as a trap for the hydrous oxides of Fe, Mn, and Al, which are known important soil constituents for sorption of trace elements.

The fixation capacity of organic matter for Pb is also reflected in several plant uptake studies. Additions of organic amendments to soil, such as animal wastes or peat, reduced plant uptake of Pb (Zimdahl and Foster, 1976; MacLean et al, 1969; Scialdone et al, 1980). Several investigators claimed that adsorption maxima for Pb by soils was primarily due to their organic matter and clay contents (Soldatini et al, 1976; Riffaldi et al, 1976). In contrast, Harter (1979) attached no significance to the role of organic matter to the adsortpion of Pb by soils from surface horizons of several northeastern states. Neither was organic matter found to be important in the availability of Pb to plants (John, 1972).

While organic matter may serve as a fixation medium for Pb, it can also have an important role in the solubility of this element via the formation of soluble chelate complexes with organic matter. Stevenson and Welch (1979) attributed some downward movement of Pb in the soil profile to this phenomenon (section 4.3).

7.4 Plant Factors

In section 5.2, it was indicated that foliar retention of Pb by plants partly depends on leaf surface characteristics, such as roughness and pubescence. Since atmospheric Pb is known to be primarily in a particulate form, plant tissue with a rough pubescent surface can intercept and accumulate considerably more Pb than tissue with smooth surfaces.

Interspecific variations in plant uptake of Pb should also be expected, as indicated in Figure 7.10, where Pb uptake from Pb-contaminated soil by crop species was soybeans >> clover > wheat ≥ corn ≥ oats. Other definite interspecific differences in Pb uptake were shown again in crop species by Sheaffer et al (1979), Czuba and Hutchinson (1980), John and Van Laerhoven (1972a), and also in tree species (Rolfe, 1973).

Differences in Pb concentration in various plant parts are also not uncommon, partly because of the unique translocation characteristics of Pb once it is taken up by the roots (see also section 5.1). Lead primarily concentrates in the roots and is generally poorly translocated to the vegetative parts and particularly to the reproductive organs (Table 7.4).

Plant age apparently influences Pb concentration also. As indicated in Figure 7.10, Pb in plant tops decreased as a function of plant age in four of five species. Similar observations were observed in young spring vegetables (higher) than in the mature fall tissues (Czuba and Hutchinson, 1980). Lead tends to decline during the rapid stage of growth, partly because of dilution of Pb by increased biomass (Cox and Rains, 1972). However, plant tissues may accumulate Pb again when they reach the senescent or dormant stage of growth (Rains, 1975; Mitchell and Reith, 1966) which can be partly due to the loss of plant matter (10% to 20% loss), thereby concentrating Pb in the remaining tissue.

7.5 Other Factors

In section 4.5, it was pointed out that the hydrous oxides of Fe, Mn, and Al can play an important role in the sorption of Pb and other trace elements by soils. The clay fractions of soil and sediments are known to contain appreciable amounts of these materials (Deshpande et al, 1968). These hydrous oxides have very high sorption capacity for certain trace elements compared to that of clays (Stumm and Morgan, 1970). Their roles in the environmental behavior of metals have been reviewed in detail by Jenne (1968). However, their influence on sorption can sometimes be obscured by other soil properties, such as organic matter and clay (Zimdahl and Skogerboe, 1977; Soldatini et al, 1976).

Temperature variation appears not to influence Pb uptake by plants (Arvik and Zimdahl, 1974b; Karamanos et al, 1976). However, its indirect effect on Pb concentration in plant tissue occurs when plants are in their senescent or dormant stage, resulting in elevation of Pb levels in plant tissue (see section 7.4).

Soil management and fertilization practices can affect Pb mobility and availability to plants. Impeded drainage could increase the mobilization of Pb in soils (Swaine and Mitchell, 1960; Reaves and Berrow, 1984b). As indicated in section 5.3, elevating phosphate levels through fertilization can immobilize Pb in soils. Because phosphate is a major nutrient applied to land before the growing season, low availability of Pb to plants is almost always ensured.

8. Lead in Drinking Water and Food

Lead in water originates from a variety of sources: atmosphere, geological formation, and—in the case of drinking water—plumbing fixtures. The US-EPA's ambient water quality criterion for Pb is identical to the existing drinking water standard, which is 50 μg Pb/liter (Strain et al, 1975). In isolated cases, tap-water in households can have considerable amounts of Pb, up to more than three times the EPA standard, caused by the household distribution system. Very high Pb concentrations can be expected in households with Pb plumbing (Moore, 1977).

United States surface waters, except when subject to special contamination, seldom contain Pb in excess of 50 μg/liter (Goyer and Chisolm, 1972). Only 5 of 876 samples examined in 1962–64 had higher levels than the standard. Approximately 14% of representative drinking water supplies contained more than 10 μg Pb/liter, and only 1% exceeded 30 μg Pb/liter. Both the US-EPA and the WHO have set a limit of 50 μg Pb/liter for drinking water (US-EPA, 1976; WHO, 1984).

The joint FAO/WHO Expert Committees on Food Additives have established provisional daily intakes of 430 μg of Pb, based on a 60-kg man (Lu, 1973; FAO/WHO, 1972). Intake values of Pb for most countries are considerably lower than that of the FAO/WHO. For example, in the United Kingdom, the daily intake value was about 200 μg Pb/person (Somers, 1974), higher than the Canadian value of about 130 μg/person, and much higher than for the United States value of 60 μg/person (Table 7.7). The values (Pfannhauser and Woidich, 1980) for other countries (in μg of Pb/person) are: Federal Republic of Germany, 113; Japan, 230 to 320; the Netherlands, 106; Rumania, 700 to 1,000; New Zealand, 413; Austria, 62; and German Democratic Republic, 42 to 47. In Denmark, the values are rather low, 23 to 43 μg/person, estimated from the Pb content in total diet of 8 to 15 μg per kg fresh weight and an assumed per capita daily intake of 2.9 kg (Solgaard et al, 1978). It should be noted that intake values for a given country are extremely variable, depending on the type and year of survey. For example, in their review, Pfannhauser and Woidich (1980) gave values of 60 to 300 μg Pb/person for the United States and 186 to 220 μg/ person for the United Kingdom.

Most human foods contain <1 ppm Pb when uncontaminated (Underwood, 1973; Warren and Delavault, 1962). The average concentration of Pb in the United States diet is roughly 0.2 ppm (Mahaffey, 1978), whereas the maximum safe intake level recommended by WHO is roughly 0.3 ppm in the adult diet (FAO/WHO, 1972). Very little is known of the natural levels of Pb in food of plant origin except that leafy and legume vegetables (Table 7.7) would tend to accumulate higher levels of airborne Pb. Tea plants are included in this category, although commercial leaf teas normally have low Pb contents (Michie and Dixon, 1977; Tsushida and Takeo, 1977; Seth et al, 1973). Also included are the salad plants, spinach, and also spices, such as parsley and thyme. Lead in parts of plants that are not in direct contact with the atmosphere, such as root, pod, and husk crops, should be expected to be low (Page et al, 1971). These include subterranean crops such as carrots, turnips, potatoes, etc.; husk crops such as corn; and crops with pods, such as beans, soybeans, etc. Canned food may contain significantly more Pb than frozen or fresh food (Pfannhauser and Woidich, 1980; Russell, 1979).

Duke (1970) reported data for food plants found in Central America and presumed to be relatively free of industrial and vehicular contamination to contain (ppm in dry matter): avocado, 1.0; breadfruit, 0.1; banana, 0.02; cassava, 0.1; coconut, 0.015; corn, <0.02; dry beans, 0.02; plantain fruits, 0.1; rice, 0.02; taro, 0.25; and yam, 0.03.

TABLE 7.7. Estimated dietary intakes and concentrations of lead in various food classes.

Class	Food Intake[a]		Pb Intake, USA[b]		Pb Intake, Canada[c]		Pb concentration, ppm[c]
	g food/day	% of total diet	μg Pb/day	% of total diet	μg Pb/day	% of total diet	Fresh weight basis
Dairy products	769	22.6	0.0	0.0	22	17	0.04
Meat, fish, and poultry	273	8.0	4.00	6.6	22	17	0.08
Grain and cereal	417	12.3	4.16	6.9	25	19	0.13
Potatoes	200	5.9	0.70	1.2	12	9	0.06
Leafy vegetables	63	1.9	3.03	5.0	2	—	0.05
Legume vegetables	72	2.1	18.80	31.1	6	4.7	0.17
Root vegetables	34	1.0	3.83	6.4	1	—	0.03
Garden fruits	89	2.6	11.36	18.8	11	8.6	0.14
Fruits	222	6.5	9.49	15.7	19	14.9	0.10
Oil and fats	51	1.5	0.67	1.1	1	—	0.06
Sugars and adjuncts	82	2.4	0.55	0.9	6	4.7	0.05
Beverages	1,130	33.2	3.81	6.3	<1	—	0.01
Total	3,402		60.4		127		

[a] Based on 1972–73 total diet survey by the US Food and Drug Administration for an adult person (Tanner and Friedman, 1977).
[b] Based on Mahaffey et al (1975).
[c] Average of 4 quarters in Vancouver, Canada (Kirkpatrick and Coffin, 1974).

9. Sources of Lead in the Terrestrial Environment

Lead is omnipresent in various environmental matrices, including air, water, soil, and plants. It is widely distributed in the environment, primarily because of its emission from combustion of Pb-containing fuel.

9.1 Global Contamination

The use of Pb as an anti-knock additive in gasoline constitutes its second largest use (Varma and Doty, 1980). In the United States, roughly 180,000 tonnes/year are released into the environment from this source and are believed to have contributed about 98% of all listed Pb emissions (Petrie, 1974). Lead in the form of very small particles may be airborne and disseminated over long distances. Those in larger particulate form fall near roadways. Detection of acid rain in the remote Amazon rainforest of southwestern Venezuela indicates the possibility of long-range transport of industrial pollutants, including heavy metals, to that region (Montagnini et al, 1984). An estimated 120,000 tonnes of fine Pb aerosols were dispersed in the northern hemisphere in 1966. In addition, it has been estimated that Pb falls on the surface of the earth at the rate of about 1.2 $\mu g/cm^2$ per year (Petrie, 1974). Lead pollution, derived mainly from the combustion of Pb additives in gasoline, is also being reflected in increased Pb fluxes into the sediments in the Southern California Coastal Basin (Chow et al, 1973). The average fluxes, in μg Pb/cm^2 per year of sea bottom, were: Santa Monica, 0.9; San Pedro, 1.7; Santa Barbara, 2.1; versus pre-pollution rates of 0.24, 0.26, and 1.0, respectively. Williams (1974) reported an increase ranging from 17% to 46% of the Pb content of the soils at the Rothamsted Experiment Station in the United Kingdom from 1883 to 1972. Cawse (1974), based on deposition data for Pb at several sites in the United Kingdom, proposed an approximate 5% increase per year in the Pb content of the top 1 cm of the soil.

Global Pb emissions from natural and anthropogenic sources are shown in Table 7.8. Of the latter sources, Pb alkyls have the highest emission factor (700 g/kg) and emission output (280 × 10^6 kg/year), constituting about 70% of the total from these sources. Recently in the United States, the mandatory use of unleaded gasoline in new automobiles has decreased the Pb concentrations in soils near highways (Byrd et al, 1983).

9.2 Near Roadways

Numerous studies in several countries, particularly in the United States and the United Kingdom, have been conducted on environmental contamination by Pb from automobile emission. United States studies almost invariably showed that most of the impacted area lies within 50 m from roadways; that most of the Pb on the surfaces of vegetation is suspendible

TABLE 7.8. Global lead emissions from natural and anthropogenic sources.[a]

Source	Production (10^9 kg/year)	Emission factor (g/kg)	Lead emission[b] (10^3 kg/year)
Natural			
Wind-blown and volcanic dust	200	1×10^{-2}	2,000
Sea spray	1000	$<1 \times 10^{-7}$	$<1,000$
Forest foliage	100	$<1 \times 10^{-5}$	<100
Volcanic sulfur	6	2×10^{-4}	1
Total			2,000
Anthropogenic			
Lead alkyls	0.4	700	280,000
Iron smelting	780	0.06	47,000
Lead smelting	4	6	24,000
Zinc and copper smelting	15	2.8	42,000
Coal burning	3300	4.5×10^{-3}	15,000
Total			400,000

[a] *Source:* Settle and Patterson, 1980, with permission of the authors and the Am Assoc for Advancement of Science.
[b] Estimated as the product of production (in 10^9 kg/year) and emission factor (in g/kg).

particulates; and that Pb in soil is primarily concentrated in the surface (0–5 cm) (Creason et al, 1971; Page and Ganje, 1970; Ganje and Page, 1972; Singer and Hanson, 1969; Milberg et al, 1980; Motto et al, 1970; Quarles et al, 1974; Cannon and Bowles, 1962; Lagerwerff and Specht, 1970). Similar observations were obtained in the United Kingdom (Davies and Holmes, 1972; Flanagan et al, 1980) and in Australia (David and Williams, 1975). The distribution of Pb in roadside soil and vegetation was reported to follow a double exponential function (Wheeler and Rolfe, 1979). The first exponent is associated with large particulates that settled out rapidly, usually within 5 m of the highway, and the second with smaller particles that settled out more slowly, within about 100 m of the source.

9.3 In Urban Areas

Environmental contamination in metropolitan areas with Pb arises primarily from automobile emission and from industry. Environmental enrichment in these areas has been reported for soil (Purves, 1972; Beavington, 1973; Rolfe and Jennett, 1975) edible plants (Preer et al, 1980), and woody plants (Smith, 1972). Based on an extensive survey in British Columbia, Canada, John (1971) reported that Pb contamination of agricultural soils in the outlying area was related to proximity to industrial and population centers.

9.4 Smelting

Collectively, smelting of Fe, Pb, Zn and Cu, contributes about 113×10^6 kg of Pb emitted globally per year (Table 7.8). Apparently, areal contamination from smelting can extend to several kilometers from the source (Kerin, 1975; Crecelius et al, 1974; Wixson et al, 1977; Kobayashi, 1971; Hutchinson and Whitby, 1974; Bolter et al, 1972; Little and Martin, 1972; Ragaini et al, 1977; Erviö and Lakanen, 1973). Because of the way smelters are sited, areas impacted are mostly natural ecosystems, primarily woodlands. Within 1 km of the smelter, Pb concentrations in the surface soil can range up to several thousand ppm (Palmer and Kucera, 1980; Ragaini et al, 1977; Nwankwo and Elinder, 1979; Bolter et al, 1972). Soils and/or vegetation of farmlands that are close to smelters can likewise be considerably enriched (Kobayashi, 1971; Dorn et al, 1975; Hemphill et al, 1973). Similarly, tissues of vegetation within this distance from the smelter can have several thousand ppm of Pb as indicated by the data of Bolter et al (1972) and Palmer and Kucera (1980) for trees and by Ragaini et al (1977) for grasses. Employing a Unified Transport Model at the Crooked Creek Watershed of the New Lead Belt in northeast Missouri, based on good data base, Patterson et al (1975) estimated that atmospheric transport deposits Pb an average of about 50 $\mu g/cm^2$ per month, surface water flow and erosion remove about 2.5 $\mu g/cm^2$ per month, and the soil adsorbs about 23 $\mu g/cm^2$ per month. The net accumulation of Pb on the watershed is about 24.5 $\mu g/cm^2$ per month, primarily occurring in the litter on the forest floor.

In a boreal forest in Manitoba, Canada, high levels of Pb, Zn, and Cu were detected in the surface soils close to a Cu-Zn smelter (<10 km) and declined rapidly with distance (Hogan and Wotton, 1984). Significant deposition of metals has been occurring at sites up to 35 km from the emission stacks.

9.5 Sewage Sludges and Effluents

Sewage sludges, especially those from metropolitan areas, usually have high levels of trace elements, including Pb (Sommers et al, 1976; Page, 1974). The review by Page (1974) indicates that Pb in sewage sludge ranges from <100 ppm to well over 1,000 ppm. Because of these high levels of Pb in sewage sludges, addition of large quantities of sludge to land can enrich the soil environment and possibly also increase the Pb contents of plants. Page (1974), however, indicated that enrichment of plant tissues by sludge-borne Pb was unlikely. Since Page's review, additional studies overwhelmingly indicate that, while the soil might be enriched by sludge-borne Pb, its uptake by crops would seldom be of major concern as far as the food chain is concerned (Kirkham, 1975; Dowdy and Larson, 1975; Zwarich and Mills, 1979; Soon et al, 1980). Of the critical trace elements in sludge, i.e., Cd, Ni, Cu, Zn, and Pb, Pb can be applied on agricultural lands in the largest amounts (Table 1.3, Chapter 1). The North Central Regional Technical Committee (NC-118), in cooperation with the Western Regional Technical Committee (W-124), published suggested limits for the total amount of sludge metals allowed on agricultural lands. The guidelines are based upon the CEC (meq/100 g) of the soil having pH of 6.5 or more. For Pb, the limits in kg Pb/ha are as follows: for soils with CEC of <5— 560; for CEC of 5 to 15—1,120; and for CEC > 15—2,240.

It has been concluded that, since Pb added to soils is usually not readily available to crops, accumulation of this element in crops would not limit the application of sludge to land (CAST, 1976).

9.6 Other Sources

Coal combustion can also emit Pb into the atmosphere (Table 7.8). Since fly ash-borne Pb is not readily available to plants, environmental contamination from coal combustion and land disposal of coal residues should not be of concern (Adriano et al, 1980).

Commercial fertilizers also contain from trace to several ppm and up to well over 150 ppm Pb (Stenström and Vahter, 1974; Mortvedt and Giordano, 1977; Williams, 1977). However, its effect on the food chain is much less than that for Cd in the fertilizers.

References

Adriano, D. C., A. L. Page, A. A. Elseewi, A. C. Chang, and I. Straughan. 1980. *J Environ Qual* 9:333–344.

Alloway, B. J., M. Gregson, S. K. Gregson, R. Tanner, and A. Tills. 1979. In *Management and control of heavy metals in the environment*, 545–548, CEP Consultants Ltd, Edinburgh, UK.

Andersson, A. 1977. *Swedish J Agric Res* 7:7–20.

——— 1979. *Swedish J Agric Res* 9:7–13.

Archer, F. C. 1980. *Minist Agric Fish Food* (Great Britain) 326:184–190.

Arvik, J. H., and R. L. Zimdahl. 1974a. *J Environ Qual* 3:369–373.

——— 1974b. *J Environ Qual* 3:374–376.

Aubert, H., and M. Pinta. 1977. *Trace elements in soils*. Elsevier, New York.

Baumhardt, G. R., and L. F. Welch. 1972. *J Environ Qual* 1:92–94.

Bazzaz, F. A., G. L. Rolfe, and P. Windle. 1974. *J Environ Qual* 3:156–158.

Beavington, F. 1973. *Aust J Soil Res* 11:27–31.

Berrow, M. L., and R. L. Mitchell. 1980. *Trans Roy Soc Edinburgh Earth Sci* 71:103–121.

Berrow, M. L., and G. A. Reaves. 1984. In *Proc Intl Conf Environ Contamination*, 333–340. CEP Consultants Ltd, Edinburgh, UK.

Berry, R. A. 1924. *J Agric Sci* 14:58–65.

Biddappa, C. C., and M. Chino. 1981. *Soil Sci Plant Nutr* 27:93–103.

Bittel, J. E., and R. J. Miller. 1974. *J Environ Qual* 3:250–253.

Bolter, E., D. Hemphill, B. Wixson, D. Butherus, and R. Chen. 1972. *Trace Subs Environ Health* 6:79–86.

Bowen, H. J. M. 1979. *Environmental chemistry of the elements.* Academic Press, New York. 333 pp.

Bradley, R. I., C. C. Rudeforth, and C. Wilkins. 1978. *J Soil Sci* 29:258–270.

Brandt, C. S. 1970. *Environ Sci Technol* 4:224.

Bunzl, K. 1974. *J Soil Sci* 25:517–534.

Bunzl, K., W. Schmidt, and B. Sansoni. 1976. *J Soil Sci* 27:32–41.

Butler, J. R. 1954. *J Soil Sci* 5:156–166.

Byrd, D. S., J. T. Gilmore, and R. H. Lee. 1983. *Environ Sci Technol* 17:121–123.

Cannon, H. L. 1974. In P. L. White and D. Robbins, eds. *Environmental quality and food supply.* Futura Publ Co, Mt Kisco, NY.

Cannon, H. L., and J. M. Bowles. 1962. *Science* 137:765–766.

Carlson, R. W., F. A. Bazzaz, J. J. Stukel, and J. B. Wedding. 1976. *Environ Sci Technol* 10:1139–1142.

Cawse, P. A. 1974. In *A survey of atmospheric trace elements in the UK (1972–73).* Harwell, UK Atom Energy Authority. Cited by Wilkins (1978).

Chang, A. C., A. L. Page, J. E. Warneke, and E. Grgurevic. 1984a. *J Environ Qual* 13:33–38.

Chang, A. C., J. E. Warneke, A. L. Page, and L. J. Lund. 1984b. *J Environ Qual* 13:87–91.

Chino, M. 1981. In K. Kitagishi and I. Yamane, eds. *Heavy metal pollution in soils of Japan,* 65–80. Japan Sci Soc Press, Tokyo.

Chisholm, D., and R. F. Bishop. 1967. *Phytoprotection* 48:78–81.

Chow, T. J., K. W. Bruland, K. Bertine, A. Soutar, M. Koide, and E. D. Goldberg. 1973. *Science* 181:551–552.

Council for Agricultural Science and Technology (CAST). 1976. *Application of sewage sludge to cropland: appraisal of potential hazards of the heavy metals to plants and animals.* EPA-430/9-76-013. General Serv Admin, Denver. 63 pp.

Cox, W. J., and D. W. Rains. 1972. *J Environ Qual* 1:167–169.

Creason, J. P., O. McNulty, L. T. Heiderscheit, D. H. Swanson, and R. W. Buechley. 1971. *Trace Subs Environ Health* 5:129–142.

Crecelius, E. A., C. J. Johnson, and G. C. Hofer. 1974. *Water, Air, Soil Pollut* 3:337–342.

Crump, D. R., and P. J. Barlow. 1980. *Sci Total Environ* 15:269–274.

Czuba, M., and T. C. Hutchinson. 1980. *J Environ Qual* 9:566–575.

David, D. J., and C. H. Williams. 1975. *Aust J Exp Agric Animal Husb* 15:414–418.

Davies, B. E. 1983. *Geoderma.* 29.67–75.

Davies, B. E., and P. L. Holmes. 1972. *J Agric Sci (Camb)* 79:479–484.

Davis, R. D., P. H. T. Beckett, and E. Wollan. 1978. *Plant Soil* 49:395–408.

Dedolph, R., G. Ter Haar, R. Holtzman, and H. Lucas, Jr. 1970. *Environ Sci Technol* 4:217–223.

Deshpande, T. L., D. J. Greenland, and J. P. Quirk. 1968. *J Soil Sci* 19:108–122.

Dilling, J. W. 1926. *Ann Appl Biol* 13:160–167.

Dorn, C. R., J. O. Pierce, G. R. Chase, and P. E. Phillips. 1975. *Environ Res* 9:159–172.

Dowdy, R. H., and W. E. Larson. 1975. *J Environ Qual* 4:278–282.

Duke, J. A. 1970. *Econ Bot* 24:344–366.

Erviö, R., and E. Lakanen. 1973. *Ann Agric Fenn* 12:200–206.

FAO/WHO Join Expert Comm. Food Additives. 1972. WHO Tech Rep Ser no 505, Geneva, Switzerland.

Farrah, H., and W. F. Pickering. 1977a. *Water, Air, Soil Pollut* 8:189–197.

——— 1977b. *Aust J Chem* 30:1417–1422.

Fisenne, I. M., G. A. Welford, P. Perry, R. Baird, and H. W. Keller. 1978. *Environ Inter* 1:245–246.

Flanagan, J. T., K. J. Wade, A. Currie, and D. J. Curtis. 1980. *Environ Pollut* 1:71–78.

Frank, R., K. Ishida, and P. Suda. 1976. *Can J Soil Sci* 56:181–196.

Friedland, A. J., A. H. Johnson, and T. G. Siccama. 1984. *Water, Air, Soil Pollut* 21:161–170.

Gadde, R. R., and H. A. Laitinen. 1973. *Environ Letters* 5:223–235.

Gambrell, R. P., R. A. Khalid, and W. H. Patrick, Jr. 1980. *Environ Sci Technol* 14:431–436.

Ganje, T. J., and A. L. Page. 1972. *Calif Agric* 26(4):7–9.

Gaynor, J. D., and R. L. Halstead. 1976. *Can J Soil Sci* 56:1–8.

Goyer, R. A., and J. J. Chisolm. 1972. In D. H. K. Lee, ed. *Metallic contaminants and human health*, 57–95. Acad Press, New York.

Griffeth, J. J. 1919. *J Agric Sci* 9:366–395.

Griffin, R. A., and A. K. Au. 1977. *Soil Sci Soc Am J* 41:880–882.

Hahne, H. C. H., and W. Kroontje. 1973. *J Environ Qual* 2:444–450.

Hall, C., M. K. Hughes, N. W. Lepp, and G. J. Dollard. 1975. In *Proc Intl Conf on Heavy Metals in the Environment*, 227–245. Toronto, Ontario, Canada.

Harter, R. D. 1979. *Soil Sci Soc Am J* 43:679–683.

Hassett, J. J. 1974. *Commun Soil Sci Plant Anal* 5:499–505.

Heinrichs, H., and R. Mayer. 1977. *J Environ Qual* 6:402–407.

——— 1980. *J Environ Qual* 9:111–118.

Heinrichs, H., B. Schulz-Dobrick, and K. H. Wedepohl. 1980. *Geochim Cosmochim Acta* 44:1519–1532.

Hemphill, D. D., C. J. Marienfeld, R. S. Reddy, W. D. Heidlage, and J. O. Pierce. 1973. *AOAC* 56:994–998.

Hewitt, E. J. 1953. *J Exp Bot* 4:59–64.

Hogan, G. D., and D. L. Wotton. 1984. *J Environ Qual* 13:377–382.

Höll, W., and R. Hampp. 1975. *Residue Rev* 54:79–111.

Hopper, M. C. 1937. *Ann Appl Biol* 24:690–695.

Huang, C. Y., F. A. Bazzaz, and L. N. Vanderhoef. 1974. *Plant Physiol* 54: 122–124.

Hutchinson, T. C., and L. M. Whitby. 1974. *Environ Conserv* 1:123–132.

Jackson, D. R., and A. P. Watson. 1977. *J Environ Qual* 6:331–338.

Jenne, E. A. 1968. *Adv Chem Series* 73:337–387.

John, M. K. 1971. *Environ Sci Technol* 12:1199–1203.

——— 1972. *J Environ Qual* 1:295–298.

John, M. K., and C. J. Van Laerhoven. 1972a. *Environ Letters* 3:111–116.

——— 1972b. *J Environ Qual* 1:169–171.

Johnson, A. H., T. G. Siccama, and A. J. Friedland. 1982. *J Environ Qual* 11:577–580.

Jones, J. S., and H. B. Hatch. 1945. *Soil Sci* 60:277–288.

Jones, L. H. P., C. R. Clement, and M. J. Hopper. 1973a. *Plant Soil* 38:403–414.

Jones, L. H. P., S. C. Jarvis, and D. W. Cowling. 1973b. *Plant Soil* 38:605–619.

Jurinak, J. J., and J. Santillan-Medrano. 1974. *The chemistry and transport of lead and cadmium in soils.* Res Rep 18, Utah State Univ Agric Exp Sta, Logan. 110 pp.

Karamanos, R. E., J. R. Bettany, and J. W. B. Stewart. 1976. *Can J Soil Sci* 56:485–494.

Keaton, C. M. 1937. *Soil Sci* 43:401–411.

Kerin, Z. 1975. In *Proc Intl Conf on Heavy Metals in the Environment*, 487–502. Toronto, Ontario, Canada.

Khan, D. H., and B. Frankland. 1983. *Plant Soil* 70:335–345.

Kirkham, M. B. 1975. *Environ Sci Technol* 9:765–768.

Kirkpatrick, D. C., and D. E. Coffin. 1974. *J Inst Can Sci Technol Aliment* 7:56–58.

Kloke, A., and K. Riebartsch. 1964. *Naturwissenschaften* 51:367–368.

Kobayashi, J. 1971. *Trace Subs Environ Health* 5:117–128.

Koeppe, D. E., and R. J. Miller. 1970. *Science* 167:1376–1378.

Koslow, E. E., W. H. Smith, and B. J. Staskawicz. 1977. *Environ Sci Technol* 11:1019–1021.

Krauskopf, K. B. 1979. *Introduction to geochemistry.* 2nd edition. McGraw-Hill, New York. 617 pp.

Lagerwerff, J. V. 1972. In J. J. Mortvedt, P. M. Giordano, and W. L. Lindsay, eds. *Micronutrients in agriculture,* 593–636. Soil Sci Soc Am Inc, Madison, WI.

Lagerwerff, J. V., and A. W. Specht. 1970. *Environ Sci Technol* 4:583–586.

Lagerwerff, J. V., and D. L. Brower. 1973. *Soil Sci Soc Am Proc* 37:11–13.

Lagerwerff, J. V., W. H. Armiger, and A. W. Specht. 1973. *Soil Sci* 115:455–460.

Lee, K. C., B. A. Cunningham, K. H. Chung, G. M. Paulsen, and G. H. Liang. 1976. *J Environ Qual* 5:357–359.

Le Riche, H. H., and A. H. Weir. 1963. *J Soil Sci* 14:225–235.

Little, P., and M. H. Martin. 1972. *Environ Pollut* 3:241–254.

Lu, F. 1973. WHO Chronicle 27:245.

MacLean, A. J., R. L. Halstead, and B. J. Finn. 1969. *Can J Soil Sci* 49:327–334.

Mahaffey, K. R. 1978. In J. O. Nriagu, ed. *The biogeochemistry of lead in the environment.* Part B. Elsevier, Amsterdam.

Mahaffey, K. R., P. E. Corneliussen, C. F. Jelinek, and T. A. Fiorino. 1975. *Environ Health Perspec* 12:63–69.

Malone, C., D. E. Koeppe, and R. J. Miller. 1974. *Plant Physiol* 53:388–394.

Marten, G. C., and P. B. Hammond. 1966. *Agron J* 58:553–554.

McKeague, J. A., and M. S. Wolynetz. 1980. *Geoderma* 24:299–307.

McKenzie, R. M. 1970. *Aust J Soil Res* 8:97–106.

――― 1975. *Aust J Soil Res* 13:177–188.

――― 1980. *Aust J Soil Res* 18:61–73.

Michie, N. D., and E. J. Dixon. 1977. *J Sci Food Agric* 28:215–224.

Milberg, R. P., J. V. Lagerwerff, D. L. Brower, and G. T. Biersdorf. 1980. *J Environ Qual* 9:6–8.

Miles, C. D., J. R. Brandle, D. J. Daniel, O. Chu-Der, P. D. Schnare, and D. J. Uhlik. 1972. *Plant Physiol* 49:820–825.

Miller, J. E., J. J. Hassett, and D. E. Koeppe. 1975a. *Commun Soil Sci Plant Anal* 6:349–358.

Miller, J. E., J. J. Hassett, and D. E. Koeppe. 1975b. *Commun Soil Sci Plant Anal* 6:339–347.

Miller, R. J., and D. E. Koeppe. 1970. *Trace Subs Environ Health* 4:186–193.

Misra, S. G., and G. Pandey. 1976. *Plant Soil* 45:693–696.

Mitchell, R. L. 1971. *Tech Bull no. 21,* Her Majesty's Sta Office London.

Mitchell, R. L., and J. W. S. Reith. 1966. *J Sci Food Agric* 17:437–440.

Montagnini, F., H. S. Neufeld, and C. Uhl. 1984. *Water, Air, Soil Pollut* 21:317–321.

Moore, M. R. 1977. *Sci Total Environ* 7:109–115.

Mortvedt, J. J., and P. M. Giordano. 1977. In H. Drucker and R. E. Wildung, eds. *Biological implications of metals in the environment*, 402–416. CONF-750-929. NTIS, Springfield, VA.

Motto, H. S., R. H. Daines, D. M. Chilko, and C. K. Motto. 1970. *Environ Sci Technol* 4:231–237.

National Research Council of Canada (NRCC). 1973. *Lead in the Canadian environment*. Ottawa, Canada. 116 pp.

Newland, L. W., and K. A. Daum. 1982. In O. Hutzinger, ed. *Handbook of environmental chemistry*. Vol. 3, Part B—*Anthropogenic compounds,* 1–26. Springer-Verlag, Berlin.

Nilsson, I. 1972. *Oikos* 23:132–136.

Norrish, K. 1975. In D. J. D. Nicholas and A. R. Egan, eds. *Trace elements in soil-plant-animal systems,* 55–81. Academic Press, New York.

Nriagu, J. O. 1972. *Inorg Chem* 11:2499–2503.

—— 1973. *Geochim Cosmochim Acta* 37:1735–1745.

—— 1978. In J. O. Nriagu, ed. *The biogeochemistry of lead in the environment*. Part A, 15–72. Elsevier, Amsterdam.

Nwankwo, J. N., and C. G. Elinder. 1979. *Bull Environ Contamin Toxicol* 22:625–631.

Page, A. L. 1974. *Fate and effects of trace elements in sewage sludge when applied to agricultural lands*. EPA-670/2-74-005. US-EPA Cincinnati, OH, 98 pp.

Page, A. L., and T. J. Ganje. 1970. *Environ Sci Technol* 4:140–142.

Page, A. L., T. J. Ganje, and M. S. Joshi. 1971. *Hilgardia* 41:1–31.

Page, A. L., A. A. Elseewi, and J. P. Martin. 1978. In *Proc Acceptable sludge disposal techniques*, 97–105. Info Transfer, Inc, Rockville, MD.

Palmer, K. T., and C. L. Kucera. 1980. *J Environ Qual* 9:106–110.

Patterson, M. R., J. K. Munro, and R. J. Luxmoore. 1975. *Trace Subs Environ Health* 9:217–225.

Petrie, W. L., ed. 1974. In *Geochemistry and the environment*, 43–56. Vol. 1. NAS, Washington, DC.

Pfannhauser, W., and H. Woidich. 1980. *Toxic Environ Chem Rev* 3:131–144.

Preer, J. R., H. S. Sekhon, B. R. Stephens, and M. S. Collins. 1980. *Environ Pollut* (Series B) 1:95–104.

Purves, D. 1972. *Environ Pollut* 3:17–24.

Quarles, H. D., R. B. Hanawalt, and W. E. Odum. 1974. *J App Ecology* 11:937–949.

Ragaini, R. C., H. R. Ralston, and N. Roberts. 1977. *Environ Sci Technol* 11:773–781.

Rains, D. W. 1975. *J Environ Qual* 4:532–536.

Rathjen, J. A., and T. J. Rowland. 1980. In US Bureau of Mines *Minerals Yearbook, 1978–79*, 507–538. US Dept of Interior, Washington, DC.

Reaves, G. A., and M. L. Berrow. 1984a. *Geoderma* 32:1–8.

——— 1984b. *Geoderma* 32:117–129.

Reddy, C. N., and W. H. Patrick, Jr. 1977. *J Environ Qual* 6:259–262.

Reiners, W. A., R. H. Marks, and P. M. Vitousek. 1975. *Oikos* 26:264–275.

Riffaldi, R., R. Levi-Minzi, and G. F. Soldatini. 1976. *Water, Air, Soil Pollut* 6:119–128.

Roberts, T. M. 1975. In *Proc of Intl Conf on Heavy Metals in the Environment*, 503–532. Toronto, Ontario, Canada.

Robinson, I. M. 1978. In J. O. Nriagu, ed. *The biogeochemistry of lead in the environment*. 99–118. Elsevier, Amsterdam.

Rolfe, G. L. 1973. *J Environ Qual* 2:153–157.

Rolfe, G. L., and J. C. Jennett. 1975. In P. A. Krenkel, ed. *Heavy metals in the aquatic environment*. Pergamon Press, New York. 852 pp.

Russell, L. H. 1979. In F. W. Oehme, ed. *Toxicity of heavy metals in the environment*, 3–23. Dekker Inc, New York.

Saar, R. A., and J. H. Weber. 1980. *Environ Sci Technol* 14:877–880.

Salim, R., and B. G. Cooksey. 1980. *Plant Soil* 54:399–417.

Santillan-Medrano, J., and J. J. Jurinak. 1975. *Soil Sci Soc Am Proc* 39:851–856.

Schinner, M. 1980. *Environ Pollut* 22:247–258.

Schlesinger, W. H., W. A. Reiners, and D. S. Knopman. 1974. *Environ Pollut* 6:39–47.

Schuck, E. A., and J. K. Locke. 1970. *Environ Sci Technol* 4:324–330.

Scialdone, R., D. Scognamiglio, and A. U. Ramunni. 1980. *Water, Air, Soil Pollut* 13:267–274.

Settle, D. M., and C. C. Patterson. 1980. *Science* 207:1167–1176.

Seth, T. D., M. Z. Hasan, and S. Sircar. 1973. *Bull Environ Contam Toxicol* 9:124–128.

Sheaffer, C. C., A. M. Decker, R. L. Chaney, and L. W. Douglas. 1979. *J Environ Qual* 8:455–459.

Siccama, T. G., and W. H. Smith. 1978. *Environ Sci Technol* 12:593–594.

Siccama, T. G., W. H. Smith, and D. L. Mader. 1980. *Environ Sci Technol* 14:54–56.

Singer, M. J., and L. Hanson. 1969. *Soil Sci Soc Am Proc* 33:152–153.

Smith, W. H. 1972. *Science* 176:1237–1239.

Smith, W. H., and T. G. Siccama. 1981. *J Environ Qual* 10:323–33.

Soldatini, G. F., R. Riffaldi, and R. Levi-Minzi. 1976. *Water, Air, Soil Pollut* 6:111–118.

Solgaard, P., A. Aarkrog, J. Fenger, H. Flyger, and A. M. Graabaek. 1978. *Nature* 272:346–347.

Somers, E. 1974. *J Food Sci* 39:215–217.

Sommers, L. E., D. W. Nelson, and K. J. Yost. 1976. *J Environ Qual* 5: 303–306.

Soon, Y. K., T. E. Bates, and J. R. Moyer. 1980. *J Environ Qual* 9:497–504.

Sposito, G., L. J. Lund, and A. C. Chang. 1982. *Soil Sci Soc Am J* 46: 260–264.

Stahr, K., H. W. Zöttl, and Fr. Hädrich. 1980. *Soil Sci* 130:217–224.

Stenström, T., and M. Vahter. 1974. *Ambio* 3:91–92.

Stevenson, F. J., and L. F. Welch. 1979. *Environ Sci Technol* 13:1255–1259.

Strain, W. H., A. Flynn, E. G. Mansour, F. R. Plecha, W. J. Pories, and O. A. Hill. 1975. In *Proc Intl Conf on Heavy Metals in the Environment*, 1003–1011. Toronto, Ontario, Canada.

Stumm, W., and J. J. Morgan. 1970. *Aquatic chemistry*. Wiley, New York.

Summers, A. O., and S. Silver. 1978. *Ann Rev Microbiol* 32:637–672.

Swaine, D. J. 1955. Commonwealth Bureau of Soil Science, Tech Commun no. 48, Harpenden.

Swaine, D. J. 1978. *J Royal Soc New South Wales* 111:41–47.

Swaine, D. J., and R. L. Mitchell. 1960. *J Soil Sci* 11:347–368.

Swallow, K. C., D. N. Hume, and F. M. M. Morel. 1980. *Environ Sci Technol* 11:1326–1331.

Tanner, J. T., and M. H. Friedman. 1977. *J Radioanal Chem* 37:529–538.

Taylor, R. M., and R. M. McKenzie. 1966. *Aust J Soil Res* 4:29–39.

Ter Haar, G., and M. A. Bayard. 1971. *Nature* 232:533–534.

Tsushida, T., and T. Takeo. 1977. *J Sci Food Agric* 28:255–258.

Tyler, G. 1972. *Ambio* 1:52–59.

Tyler, G., G. Christina, K. Holmquist, and A. Kjellstrand. 1973. *J Ecol* 61:251–268.

Underwood, E. J. 1973. In *Toxicants occurring naturally in foods*, 43–87. NAS, Washington, DC.

US Environmental Protection Agency (US-EPA, Washington, D.C.). 1976. *Quality criteria for water*.

Van Hook, R. I., W. F. Harris, and G. S. Henderson. 1977. *Ambio* 6:281–286.

Varma, M. M., and K. T. Doty. 1980. *J Environ Health* 42:68–71.

Vinogradov, A. P. 1959. *The geochemistry of rare and dispersed chemical elements in soils*. Consultants Bureau, Inc, New York. 209 pp.

Waldron, H. A. 1980. In H. A. Waldron, ed. *Metals in the environment*, 155–197. Academic Press, New York.

Wallace, A., and E. M. Romney. 1977. In H. Drucker and R. E. Wildung, eds. *Biological implications of metals in the environment*, 370–379. CONF-750929. NTIS, Springfield, VA.

Warren, H. V., and R. E. Delavault. 1962. *J Sci Food Agric* 13:96–98.

Warren, H. V., R. E. Delavault, and C. H. Cross. 1969. *Trace Subs Environ Health* 3:9–19.

Wedding, J. B., R. W. Carlson, J. J. Stukel, and F. A. Bazzaz. 1975. *Environ Sci Technol* 9:151–153.

Wheeler, G. L. and G. L. Rolfe. 1979. *Environ Pollut* 18:265–274.

Whitby, L. M., J. Gaynor, and A. J. MacLean. 1978. *Can J Soil Sci* 58: 325–330.

Williams, C. 1974. *J Agric Sci (Camb)* 82:189–192.

Williams, C. H. 1977. *J Aust Inst Agric Sci* (Sept–Dec):99–109.

Williams, D. E., J. Vlamis, A. H. Pukite, and J. E. Corey. 1984. *Soil Sci* 137:351–359.

Wixson, B. G., N. L. Gale, and K. Downey. 1977. *Trace Subs Environ Health* 11:455–461.

World Health Organization (WHO). 1984. *Guidelines for drinking-water quality*. Vol 1—Recommendations. WHO, Geneva, Switzerland. 130 pp.

Zimdahl, R. L., and J. H. Arvik. 1973. *CRC Critical Rev Environ Control* 3(2):213.

Zimdahl, R. L., and J. M. Foster. 1976. *J Environ Qual* 5:31–34.

Zimdahl., R. L., and R. K. Skogerboe. 1977. *Environ Sci Technol* 11:1202–1207.

Zwarich, M. A., and J. G. Mills. 1979. *Can J Soil Sci* 59:231–239.

8
Manganese

1. General Properties of Manganese

Manganese (atomic no. 25; atomic wt. 54.938; melting pt. 1,244 ± 3°C, specific gravity 7.21 to 7.22, depending on the allotropic form) is a member of Group VII-A of the periodic table. It is next to Fe in the atomic series, is similar to it in chemical behavior, and is often closely associated with it in its natural occurrence. It can exist in its compounds in the oxidation states of I, II, III, IV, VI, and VII. Its most stable salts are those of oxidation states II, IV, VI, and VII. The lower oxides, MnO and Mn_2O_3, are basic; the higher oxides are acidic. Manganese is a whitish-gray metal, harder than iron but quite brittle. Manganese metal oxidizes superficially in air and rusts in moist air.

Manganese is a ubiquitous metal in the earth's crust, the 12th most abundant element, and makes up about 0.10% of the earth's crust (Krauskopf, 1979). It is the principal metallic component of nodules deposited on the ocean floor. Manganese minerals are widely distributed, the most common being the oxides, carbonates, and silicates. The main ores are pyrolusite (MnO_2), rhodochrosite ($MnCO_3$), manganite ($Mn_2O_3 \cdot H_2O$), hausmannite (Mn_3O_4), braunite ($3Mn_2O_3 \cdot MnSiO_3$), and rhodonite ($MnSiO_3$).

2. Production and Uses of Manganese

Total worldwide production of Mn ores in 1980 was $26,780 \times 10^3$ tonnes, with the USSR, Republic of South Africa, Brazil, Gabon, Australia, India, and the Peoples Republic of China as the leading producers (Table 8.1). Although not a producer, the United States is a big consumer of imported Mn. For example $1,248 \times 10^3$ tonnes of Mn ore (35% or more Mn) and 888×10^3 tonnes of ferro-manganese were consumed in 1979.

Manganese ore deposits are widespread throughout the tropical, subtropical, and warmer temperate zones of the earth (NAS, 1973). The largest

TABLE 8.1. World production, in 10^3 tonnes, of manganese ore.[a]

Country	Percent Mn	1976	1978	1980
North America				
Mexico	35+	454	525	449
South America				
Bolivia	28–54	12.7	1.3	4.6
Brazil	38–50	1,701	1,923	2,184
Chile	36–40	23.6	23.3	22.6
Peru	26	0.6	—	—
Europe				
Bulgaria	30–	40	40	40
Greece	48–50	8.2	7.0	6.0
Hungary	30–33	125	115	88
Italy	22+	4.4	9.7	9.1
USSR	35	8,663	9,085	10,283
Yugoslavia	30+	19.1	27.4	30.0
Africa				
Egypt	28+	4.2	0.2	—
Gabon	50–53	2,224	1,666	2,153
Ghana	30–50	312	317	252
Morocco	43–50	117	126	150
South Africa	30–48+	5,469	4,330	5,713
Sudan	48	0.5	0.5	0.4
Zaire	30–57	182	—	—
Asia				
China, P. R. of	20+	1,001	1,274	1,592
India	10–54	1,840	1,623	1,650
Indonesia	47–56	10	6	7
Iran	33+	40	30	20
Japan	26–28	142	105	78
Korea, R. of	23–40	1.4	0.6	—
Pakistan	35–	—	0.3	0.1
Philippines	35–45	11	3.9	5.0
Thailand	46–50	50	73	49
Turkey	35–46	17	19	24
Oceania				
Australia	37–53	2,160	1,253	1,967
Others	40–44	35	21	—
TOTAL		24,673	22,606	26,780

[a] *Source:* DeHuff and Jones, 1980, with permission of the Bureau of Mines Publication Distribution Branch, Pittsburgh, PA.

deposits are found in the USSR, Peoples Republic of China, Brazil, India, Australia, Republic of South Africa, Gabon, Zaire, Ghana, and Morocco.

Since 1939, Mn has been used primarily in the metallurgical industry, and its use increased markedly after the introduction of the Bessemer process (Mena, 1980). As an essential ingredient of steel, it neutralizes the harmful effects of sulfur, serves as an anti-oxidant, and provides strength, toughness, and hardness. For these reasons, Mn is used also in the production of alloys of steel, aluminum, and copper. These alloys are used in the electrical industry and for ship propellers. Manganese, or its compounds, are used quite extensively in alkaline batteries, electrical coils, ceramics, matches, welding rods, glass, dyes, paints, and drying industries. Other compounds of Mn are used as driers for paints, varnishes and oils, fertilizers, disinfectants, and animal food additives. It is used also as a component of anti-knock compounds for internal combustion engines.

3. Natural Occurrence of Manganese

The average content of Mn in the lithosphere is about 1,000 ppm. The lithosphere contains about 50 times as much Fe, one-fifth as much Ni, and one-tenth as much Cu (NAS, 1973). Manganese is widely distributed in metamorphic, sedimentary, and igneous rocks. One reason for its wide distribution in different types of rocks is its similar ionic size to Mg and Ca, enabling it to replace the two elements in silicate structures (Stahlberg and Sombatpanit, 1974). Manganese also replaces ferrous Fe in magnetite. There are at least 100 minerals (e.g., sulfides, oxides, carbonates, silicates, phosphates, arsenates, tungstates, and borates) that contain Mn as an es-

TABLE 8.2. Commonly observed manganese concentrations (ppm) in various environmental matrices.

Material	Average concentration	Range
Igneous rocks[a]	—	390–1,620
Limestone	620	—
Sandstone[b]	460	—
Shale	850	—
Manganese nodules[b]	160,000	—
Coal[c]	100	—
Fly ash[c]		
Bituminous	145	—
Sub-bituminous	309	—
Lignite	543	—
Sewage sludges[d]	—	60–3,900
Normal soil[d]	850	100–4,000
Soils[b]	1,000	20–10,000
Common crops[d]	—	15–100
Herbaceous vegetables[b]	—	0.3–1,000
Ferns[b]	700	—
Fungi[b]	—	10–400
Lichens[b]	—	3–190
Woody angiosperm[b]	—	17–600
Woody gymnosperm[b]	—	300–650
Freshwater[b]*	8	0.02–130
Sea water[b]*	0.2	0.03–21

* μg/liter
Sources: Extracted from
[a] NAS (1973).
[b] Bowen (1979).
[c] Adriano et al (1980).
[d] Freedman and Hutchinson (1981).

sential element, and it is an accessory element in perhaps more than 200 others (NAS, 1973).

In igneous rocks, Mn is usually present in silicates in the II oxidation state. During weathering, Mn oxidizes and separates out as Mn oxides in the soil (Norrish, 1975). In soils, it can occur as dark nodules and coatings.

Normal soils contain an average Mn concentration of 850 ppm (range of 100–4,000 ppm); recently, world soils were reported to contain about 1,000 ppm (range of 20–10,000 ppm) (Table 8.2).

In nature, Mn chemically behaves similarly to Fe. In its naturally occurring compounds, Mn shows three oxidation states (II, III, and IV) in contrast to only two exhibited by Fe. Oxidation of Mn(II) compounds requires higher redox potentials than does oxidation of Fe(II) compounds; MnS is more soluble than FeS, and MnS_2 is far less stable than FeS_2 (pyrite) (Krauskopf, 1979). In reducing environments, the Mn(II) species are most stable, while in oxidizing environments the most stable compound is the dioxide, MnO_2.

4. Manganese in Soils

4.1 Total Soil Manganese

The U.S. Geological Survey reported that surficial soils in the conterminous United States have an arithmetic mean of 560 ppm Mn, with a range of <1 to 7,000 ppm ($n = 863$) (Shacklette et al, 1971). McKeague and Wolynetz (1980) reported a mean of 520 ppm (range of 100–1,200 ppm) for Canadian soils ($n = 173$ from 53 soils). Similarly, for Ontario (Canada) soils, Frank et al (1976) obtained a mean of 530 ppm Mn, with a range of 90–3,000 ppm. For soils collected from the plow layer in the central Russian province, an average Mn content of 740 ppm (range of 472–1,250 ppm) was obtained by Krupskiy et al (1978). The total Mn content of the soils of China ranges from 10 to 5,000 ppm, with an average of 710 ppm (Liu et al, 1983). Vinogradov (1959) reported 850 ppm as the mean Mn content and 100 to 4,000 ppm as the range for world soils; Aubert and Pinta (1977) reported a range of 500 to 1,000 ppm for most world soils. Swaine (1955) reported a range of 200 to 3,000 ppm for total Mn content in most soils; Bowen (1979) more recently reported a mean value of 1,000 ppm for world soils, with a range of 20 to 10,000 ppm (Table 8.2). Most recently, Berrow and Reaves (1984) reported a mean of 450 ppm Mn for world soils. Normal soils can be expected to have a maximum Mn content of 4,000 ppm.

4.2 Extractable (Available) Manganese in Soils

As with the other micronutrients, there is no universal method for determining extractable (sometimes called "available") Mn in soils that can accurately predict plant response to this element. However, it is universally agreed that total soil Mn is an inadequate predictor of Mn availability. Methods used in various regions of the United States and in various other countries largely depend on the properties of the soil.

Various extractants have been used to estimate water-soluble, exchangeable, acid-soluble, easily reducible, and complexable forms of soil Mn. Water-soluble and exchangeable Mn could be used to predict toxic conditions (Adams and Wear, 1957; Hoyt and Nyborg, 1971; Anderson and Boswell, 1964; Morris, 1948). Exchangeable Mn has been shown to relate closely to Mn uptake (Heintze, 1946; Steenbjerg, 1935; Shuman and Anderson, 1974; Walker and Barber, 1960; Sharpe and Parks, 1982; Fergus, 1953) and, in general, has correlated closely with Mn-deficiency symptoms in plants (Sherman and Harmer, 1942; Boken, 1968). The level of easily reducible Mn in soil [can be determined with 0.2% hydroquinone in 1 N NH_4OAc (pH 7)] has been used to evaluate available Mn for a number of years (Jones and Leeper, 1951; Leeper, 1947; Sherman et al, 1942). Several chelating agents have recently become more popular, especially DTPA as used in the method of Lindsay and Norvell (1978).

Relationships between soil Mn extracted by the two most commonly used methods and plant tissue Mn are exemplified in Figure 8.1. Relationships between extractable fractions and plant response have been extremely variable. For example, of the eight methods (two exchangeable extractants—1 N KCl and 1 N NH_4OAc at pH 3; reducible extractant—0.2% hydroquinone in 1 N NH_4OAc at pH 7; four soluble extractants–0.01 M $CaCl_2$, 0.1 N HOAc, 0.1 N H_3PO_4, and 0.002 N HCl; total Mn extraction), a 16-hour extraction with 0.01 M $CaCl_2$ gave the best estimate of plant-available Mn with barley ($r = 0.76$), rape ($r = 0.73$), and alfalfa ($r = 0.86$) as plant indicators (Hoyt and Nyborg, 1971). Later, Hoyt and Webber (1974) found that 0.01 M $CaCl_2$-soluble Mn was better correlated with the crop data than was 0.1 N H_3PO_4-soluble, water-soluble, reducible, or exchangeable Mn. Similarly, Hoff and Mederski (1958) found that, of the nine Mn extraction methods they employed, hydroquinone, 3 N $NH_4H_2PO_4$, and 0.1 N H_3PO_4 gave the highest correlation coefficients with Mn in soybean plants. Hammes and Berger (1960) found that Mn extracted by 0.1 N H_3PO_4 correlated better with Mn levels in oats than did Mn extracted by 1.5 M $NH_4H_2PO_4$ or 1 N H_3PO_4. Salcedo et al (1979b) noted that the correlation between extractable Mn and uptake of Mn by soybean plants decreased in the following order: 0.1 N H_3PO_4 > steam/1 N NH_4OAc > 1.5 M $NH_4H_2PO_4$ > 0.1 N HCl > 1 N NH_4OAc > 0.005 M DTPA.

In China, the following soil concentrations of extractable Mn (in ppm) were established as index values for plant growth (Liu et al, 1983): exchangeable Mn—<1.0 (very low), 1.0–2.0 (low), 2.1–3.0 (medium), 3.1–5.0 (high), and >5.0 (very high); easily reducible Mn—<50 (very low), 50–100 (low), 101–200 (medium), 201–300 (high), and >300 (very high). The exchangeable Mn was extracted with a neutral 1 N HOAc and NH_4OAc solution, and easily reducible Mn with the same solution containing 0.2% hydroquinone.

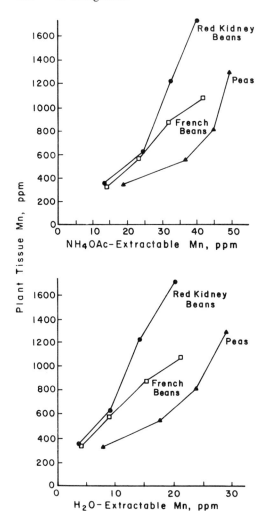

FIGURE 8.1. Relationships between plant tissue manganese levels and soil manganese extracted by ammonium acetate (above) and water (below). *Source:* White, 1970, with permission of the author and Soil Sci Soc of America.

Soil pH, to be discussed later in more detail, plays an important role in the selection of a suitable extractant. For example, Shuman and Anderson (1974) found that, for acid southeastern US soils, DTPA gives the best measure of plant-available Mn at pH 5.8 to 6.8; whereas for soils at pH 4.8, water-soluble Mn value would be the best measure. Because the correlations between amounts of Mn extracted from soils and the amounts found in plants are dependent on soil pH, to enhance their accuracy, prediction equations should include soil pH (Salcedo et al, 1979b; Randall et al, 1976).

Cox (1968) found that the degree of response of soybeans to Mn fertilization could be predicted from the soil pH and the level of dilute-acid-extractable ($0.050 \ N$ HCl + $0.025 \ N$ H_2SO_4 mixture) Mn in the soil. He developed a predictive model for yield response of soybeans based on

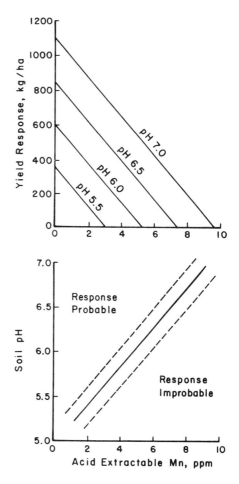

FIGURE 8.2. Yield response (above) of soybeans to fertilizer manganese application as influenced by soil pH and level of double acid-extractable soil manganese; the probability (below) of a yield response for this crop to manganese fertilization in relation to soil pH and acid-extractable manganese. Area between dashed lines indicates uncertainty. *Source:* Cox, 1968, with permission of the author and Am Soc of Agronomy.

soil tests over a soil pH range of 5.2 to 7.1 and a range of 0 to 10 ppm extractable soil Mn. The results of his predictions are displayed in Figure 8.2. As an example, Mn fertilizations could produce a yield response of at least 400 kg/ha in soils having low levels of extractable Mn (near 2 ppm) and pH values >6.0. Cox's (1968) results were then extended to develop a probability test for yield response (Fig. 8.2). For example, if lines are drawn from the pH and extractable Mn axes and they intersect above the main line, a response is probable; if below the line, a response is improbable. The area between the dashed lines represents the range of uncertainty, or where response is doubtful. Cox concluded that yield responses to fertilization occurred when soybeans contained <20 ppm in the uppermost developed leaf or in the mature seed.

4.3 Manganese in the Soil Profile

Leeper (1947) listed four possible distribution patterns of total Mn in the soil profile:

1. Surface accumulation, minimum in subsurface, then increase with depth. Examples are highly leached soils.
2. Steady decrease with depth. This is also found in highly leached soils.
3. Uniform distribution throughout the profile. Most unleached soils belong here.
4. Accumulation in the subsoil, particularly in horizons occurring above a calcareous layer.

Robinson (1929) stated that, as a general rule, Mn is highest in the surface soil, reaches a minimum in the B horizon, and then generally increases in the C horizon. The accumulation of Mn in the surface horizon could be attributed to plant root uptake of Mn and deposition on the surface upon decay of foliage. In major soil groups of Rajasthan (India), Mn (also Fe and Cu) was uniformly distributed in the profiles of relatively less weathered desert and old alluvial soils (Lal and Biswas, 1974). Accumulation of $CaCO_3$ in the lowest horizons of desert, gray, brown, and black soils substantially reduced the contents of total Mn. Extractable micronutrients were concentrated in surface horizons of well-drained soils but were concentrated in the lower horizons in poorly drained soils.

In other Indian soils, both total and extractable Mn generally decreased with depth (Agrawal and Reddy, 1972; Patel et al, 1972). Similarly, the concentration of DTPA-extractable Mn (and Zn, Fe, and Cu) decreased with depth in the profiles of Colorado soils (Follett and Lindsay, 1970). A similar pattern was observed for extractable Mn in Scottish soil profiles (Berrow and Mitchell, 1980). In contrast, water-soluble and exchangeable Mn showed little variation within profiles or between soils of Fraser Valley, British Columbia (Safo and Lowe, 1973). These soils were derived from alluvial and marine-deposited parent materials.

4.4 Forms and Transformations of Manganese in Soils

In their review of soil Mn, Stahlberg and Sombatpanit (1974) stated that the chemistry and biochemistry of soil Mn is very complex for the following reasons:

1. Mn can exist in several oxidation states;
2. It forms nonstoichiometric oxides in mixed oxidation states;
3. The higher oxides exist in several crystalline or pseudocrystalline states;
4. The oxides form coprecipitates, solid solutions, and perhaps super-structures, with iron oxides;
5. Solid Fe and Mn hydroxides exhibit amphoteric behavior and show a strong tendency to interact specifically with anions as well as with cations; and
6. Both oxidation and reduction processes involving Mn are probably influenced by both chemical and microbiological factors.

The forms of soil Mn are separated and defined by chemical extractants and may be classified as water-soluble; exchangeable; adsorbed; chelated or complexed; secondary clay minerals and insoluble metal oxides; and primary minerals. Such forms can be determined by sequential fractionation, as outlined in section 4.4 of the chapter on zinc. Results of a similar fractionation are presented in Table 8.3 for eight diverse Kentucky soils. In these soils, the reducible fraction accounted for an average of nearly 45% of the total Mn. The exchangeable, organic, and Fe-bound fractions each contained intermediate amounts (13%–19%), and the water-soluble had <2% of the total. However, the distribution varied among the soils and generally was related to soil pH. Soils of low pH generally had the highest amounts of water-soluble, exchangeable, and organic Mn, whereas soils with high pH contained large amounts of reducible Mn. Using a similar extraction scheme, Goldberg and Smith (1984) found that in selected UK soils, about 75% of the total indigenous soil Mn is found in the oxide-bound (easily reducible and resistant) fractions (Table 8.4). Less than 3% of the total is found in the water-soluble plus exchangeable and EDTA (organic-bound) fractions. The fate of soil-applied Mn fertilizers can be deduced from the results Goldberg and Smith obtained on the distribution of ^{54}Mn tracer between the different fractions. Most of the applied isotope, well over 70% in some cases, was recovered from the first three extractants (i.e., water-soluble plus exchangeable, EDTA-extractable, and easily reducible) in most of the soils, but the distribution is variable among soil samples. In general, the easily reducible fraction contained the most ^{54}Mn.

Using a similar fractionation scheme for a flooded soil, Sims and Patrick (1978) found that water-soluble Mn was inversely related to soil pH (4.5 to 7.5) and Eh (-150 to $+500$ mV). In general, greater amounts of Mn (also Fe, Cu, and Zn) were found in either the exchangeable or organic fractions at low pH and Eh than at high pH or Eh, indicating that under acidic and reducing conditions, micronutrients that were precipitated and occluded as oxides and hydroxides were solubilized. Grass et al (1973 a, b) found that conditions around tile drains became more favorable for solubilization of Mn at Eh <400 mV. In displaced soil solutions obtained from A horizons from several areas of the United States, about 84% to 99% of the total Mn was in the complex form (Geering et al, 1969). The Mn in the soil solution occurred in the II oxidation state. In contrast, Sanders (1983) found that free Mn^{2+} ions comprised the major Mn species in soil solution at pH 5, and the proportions of Mn in this form decreased as the pH increased. In general, Mn occurs primarily as the divalent ion in waters and soil solutions [Mn(II) $>>$ Mn(IV) $>$ Mn(III)] (NAS, 1977). The levels of Mn(IV) are very low because of the low solubility of manganese dioxide minerals, whereas the concentrations of Mn(III) are low because they readily reduce to Mn(II).

In flooded soils, the main transformations of Mn are the reduction of Mn(IV) to Mn(II), an increase in the concentration of water-soluble Mn^{2+} ions, precipitation of manganous carbonate, and reoxidation of Mn^{2+} ions diffusing or moving by mass flow to aerobic interfaces in the soil (Ponnamperuma, 1972). Ponnamperuma further reported that, within one to

TABLE 8.3. Distribution of indigenous manganese in several Kentucky soils.[a]

Mn fraction[b]	Soil series								
	Culleoka	Shelbyville	Maury	Huntington	Grenada	Eden	Zanesville	Russellville	Average
	Mn concentration, µg/g								
H$_2$O-soluble	28	35	40	7	4	1	48	2	21
Exchangeable	336	190	416	112	3	8	219	4	161
Organic	218	123	279	247	96	81	135	78	170
Reducible	297	1500	418	411	1507	377	165	371	631
Fe-bound	138	358	412	258	197	325	107	103	237
Residual	80	115	149	141	83	179	46	80	109
Soil pH[c]	4.8	5.2	6.2	6.3	6.4	6.5	6.5	7.0	

[a] *Source:* Adapted from Sims et al, 1979, with permission of the authors and Williams and Wilkens Co.
[b] The following extractants were used: exchangeable—neutral 1 N NH$_4$OAc; organic—K$_4$P$_2$O$_7$ solution; reducible—0.1 M NH$_2$OH·HCl in 0.01 M HNO$_3$ at pH 2; and Fe-bound—oxalate (0.1 M oxalic acid in 0.175 M ammonium oxalate, at pH 3.25) solution.
[c] After DTPA extraction.

TABLE 8.4. Manganese concentrations in soil fractions and distribution of ^{54}Mn tracer among soil fractions.[a,b]

Soil series	pH	Water-soluble + exch.	EDTA-extractable	Easily reducible	Resistant	Residual	Total (sum of fractions)	Total (single determination)
				Mn in soil fractions, ppm				
Darvel	5.55	4.4	1.3	142	210	55	412	388
Denchworth	5.85	11.4	6.1	84	235	160	496	438
Dreghorn	6.8	11.2	10.8	436	616	196	1270	1275
Dreghorn	5.9	14.2	3.7	105	250	51	424	462
Giffnock	5.5	0.7	0.1	0.4	11	42	54	60
Hexpath	3.7	0.3	<0.1	0.4	5	32	38	49
Hexpath	6.7	ND[c]	3.4[c]	49.2	151	186	389	400
Macmerry	6.25	4.1	1.9	126	328	106	566	525
Stirling	5.9	3.9	0.1	89	168	102	363	322
Worcester	6.3	6.4	16.5	270	245	180	718	712

Soil series	Remaining in soil after equilibration	Water-soluble + exch.	EDTA-extractable	Easily reducible	Resistant	Residual	Total recovered
			% of added ^{54}Mn				
			In individual soil fractions				
Darvel	28.2	6.2	2.0	14.7	3.5	3.5	26.7
Denchworth	29.0	9.1	5.6	7.1	3.5	0.7	26.0
Dreghorn	77.5	6.5	8.5	42.9	14.9	2.3	75.1
Dreghorn	25.4	8.1	2.8	9.4	4.1	0.6	25.0
Giffnock	3.0	2.3	0.3	0.08	0.16	0.04	2.9
Hexpath	13.3	11.7	<0.01	<0.01	<0.01	<0.01	11.7
Hexpath	61.0	—	28.0[d]	16.7	7.4	3.5	55.6
Macmerry	49.5	9.4	5.3	26.5	6.9	1.3	49.4
Stirling	4.0	1.2	0.2	1.6	0.5	0.1	3.6
Worcester	74.8	4.7	11.0	40.8	12.8	2.2	71.5

[a] Source: Adapted from Goldberg and Smith, 1984, with permission of the authors and Soil Sci Soc of America.
[b] Extractants used for the following fractions were: water-soluble and exchangeable—0.05 M $CaCl_2$; EDTA-extractable—0.03 M EDTA solution; easily reducible—neutral 1 M NH_4OAc in 0.2% hydroquinone; resistant—0.1 M oxalic acid plus 0.175 M ammonium oxalate at pH 3.25.
[c] Water-soluble + exchangeable fraction not determined; assumed to be included in value obtained for EDTA-extractable fraction.
[d] Sum of water-soluble + exchangeable and EDTA extractable fractions.

three weeks of flooding, almost all EDTA-dithionate-extractable Mn present in most soils is reduced. The reduction is both chemical and biological and precedes the reduction of Fe. Acid soils high in Mn and in organic matter build up water-soluble Mn^{2+} levels as high as 90 ppm within about two weeks of flooding, then exhibit a rapid decline to a fairly stable level of about 10 ppm. On the other hand, alkaline soils rarely contain as much as 10 ppm water-soluble Mn^{2+} at any period of flooding. Manganese is present in reducing soil solutions as Mn^{2+}, $MnHCO_3^+$, and as organic complexes. Most flooded soils contain sufficient water-soluble Mn for the growth of the rice plants, and Mn toxicity is not known to occur in flooded soils.

Secondary Mn oxides (usually along with Fe oxides) can occur in soils in several forms, including concretions, pans, coatings, and mottles. Their occurrence is fairly common in soils throughout the world, and in some cases, they constitute a considerable part of the soil. Some soils could contain upwards of 29% to 40% (by weight) of Fe-Mn concretions (Childs, 1975). Iron-Mn concretions are of interest in geochemical cycling of trace elements because of their ability to influence the distribution of metal ions in soils (Taylor and McKenzie, 1966; Jenne, 1968).

Mandal and Mitra (1982) noted that under flooded conditions, there was a sharp increase in the level of water-soluble plus exchangeable Mn accompanied by significant decrease in the level of reducible Mn. Manganous ions have been oxidized and precipitated (along with Fe) in soil tile drains (Grass et al, 1973 a, b). The black precipitates in the drains are primarily Mn oxide compounds. Meek et al (1973) concluded that this oxidation was microbiological in nature but that it was increased with increasing levels of bicarbonate and O_2. In Imperial Valley (California), tile lines, oxidation of Mn occurs down to a pH of 6. Tyler and Marshall (1967) concluded that the oxidation was primarily microbial in nature after demonstrating the absence of oxidation in autoclaved or azide-treated systems. Ehrlich (1968) reported that the increase in oxidation of Mn by bacteria may be the result of enzymatic catalysis, biodegradation of an Mn chelate followed by auto-oxidation of the freed Mn^{2+} ions, or auto-oxidation of Mn^{2+}.

However, the role of microorganisms in the transformation of Mn has not been fully elucidated. In their review of soil Mn, Mulder and Gerretsen (1952) indicated that bacteria and fungi are capable of oxidizing Mn(II). Organisms capable of oxidizing Mn(II) include strains of the bacterial genera *Arthrobacter, Bacillus,* etc., and of the fungal genera *Cladosporium, Curvularia,* etc. (Alexander, 1977). In tile lines having waters containing Fe(II) and Mn(II), oxidizing organisms colonizing on the inner surfaces and joints of the lines may cause the gradual accumulation of Fe and Mn deposits (Peterson, 1966; Spencer et al, 1963), resulting in sealing of tile drainage joints. Some of these microorganisms are heterotrophs, which grow on organic compounds and accumulate Fe(III) and Mn(III) salts

(Alexander, 1977). Others derive all or part of their energy from the oxidation of soluble inorganic substances such as Mn and Fe. In some cases, Mn and Fe may be precipitated in tile lines as a result of chemical oxidation due to increases in pH or Eh (Meek et al, 1968). Knezek and Ellis (1980) indicated that, although microorganisms are implicated in the oxidation of Mn(II), much of the effect may be indirect, such as a change in pH. Similarly, microorganisms may indirectly affect the reduction of Mn. For example, aerobic microbes may deplete the soil of O_2, causing a reduction in redox potential and the solubilization of Mn. However, Ghiorse and Ehrlich (1976) have shown that certain microbes may be directly involved in the reduction of MnO_2, through an enzymatic reaction.

4.5 Adsorption, Fixation, and Complexation of Manganese in Soils

Among the micronutrients, Mn adsorption is more complicated since it forms relatively insoluble oxides in response to pH-Eh conditions. This results in meager information on adsorption reactions and instead, data were accumulated on the influence of pH-Eh on Mn availability and plant uptake.

Data obtained by Shuman (1977) from equilibration studies showed that the adsorption of Mn by soils conformed to the Langmuir isotherm over the entire Mn concentration range tested (2–64 μg Mn/ml) except where there was very low adsorption by sandy, low-organic matter soils.

When adsorption data for 20 calcareous soils were plotted according to the conventional Langmuir equation, the isotherms were curvilinear at low Mn concentrations in the equilibration solution and linear at higher concentrations (Curtin et al, 1980). These isotherms were again found to be nonlinear when the data were fitted to a "competitive" Langmuir equation, indicating that the soils contain more than one population of adsorption sites.

Adsorption-desorption data (Table 8.5) indicate that the amounts of exchangeable Mn ranged from 6% to 76% of adsorbed Mn and the exchangeable fraction increased as the amount of adsorbed Mn increased. The DTPA data also indicate that part of the adsorbed Mn may be due to complexation with organic matter. Manganese not extracted by $CaCl_2$ and DTPA (i.e., fixed Mn) varied from 12% to 93% of total adsorbed Mn, indicating that quantities of this tightly bound form can be significant.

In high-organic matter soils, most of the retention of Mn in an unavailable form was attributed to organic matter complexation, particularly to humic acid (Pavanasasivam, 1973). Andersson (1977) found that sorption of heavy metals by organic soil is less pH-dependent than sorption by mineral soil. However, organic soils were more effective sorbents than mineral soils in acid environments; above neutrality, some of the organic matter could dissolve together with the heavy metal, increasing the concentration of

TABLE 8.5. Adsorbed, exchangeable, DTPA-extractable, and fixed manganese in soils of varying properties.[a]

Soil no.	CEC	pH	Org. C	Active CaCO$_3$	Free Fe (as Fe$_2$O$_3$)	Added Mn	Adsorbed[b] Mn	Exchangeable Mn in 0.1 N CaCl$_2$	DTPA-extractable Mn	Fixed[c] Mn
	meq/100 g			%			µg/g soil			
1	19.5	8.2	1.5	18.6	0.61	0	—	—	3.6	—
						125	51.3	3.0	4.0	47.9
						250	80.0	12.0	4.0	67.6
						500	112.5	33.0	4.0	79.1
2	41.9	8.0	0.86	4.3	8.6	0	—	—	38.0	—
						125	97.5	8.0	56.0	71.5
						250	171.3	56.0	68.0	85.3
						500	284.3	121.0	79.0	122.3
4	15.1	8.1	0.10	9.0	4.7	0	—	—	10.0	—
						125	52.5	34.0	21.0	7.5
						250	86.5	63.0	22.0	11.3
						500	134.5	102.0	27.0	15.5
10	36.4	7.7	0.56	0.8	6.3	0	—	—	55.0	—
						125	108.8	42.0	67.0	54.8
						250	161.3	87.0	70.0	59.3
						500	268.8	169.0	78.0	76.8

[a] Source: Adapted from Curtin et al, 1980, with permission of the authors and Soil Sci Soc of America.
[b] "Adsorbed" Mn was estimated as the difference between the amount of Mn added initially and that in solution after equilibration.
[c] "Fixed" Mn was the amount of Mn that was not extracted with CaCl$_2$ and DTPA solutions.

the metal in the equilibrium solutions. Page (1962) concluded that changes in the availability of Mn with pH were due neither to formation of higher oxides of Mn nor to biological oxidation of Mn, but to complexation of Mn by organic matter. Indeed, Geering et al (1969) found that the extent of complexation of Mn in soil solutions was intermediate between Zn and Cu. Bloom (1981) reported the following preference series for divalent ions for humic acids and peat: Cu > Pb >> Fe > Ni = Co = Zn > Mn = Ca. The complexing ability of organic matter is discussed in section 4.6 of the Cu chapter.

In flooded soils, Sims and Patrick (1978) found that the amounts of elements complexed with organic matter were in the order: Mn < Fe < Cu < Zn. Reddy and Perkins (1976) found that Mn could be strongly adsorbed by clay minerals and that the amounts held in nonexchangeable form by kaolinite, illite, and bentonite increased with increasing pH. Illite and bentonite fixed significant amounts of Mn under wetting and drying conditions. Soils kept at moisture saturation resulted in considerably less Mn fixed as when subjected to repeated wetting and drying. Reddy and Perkins attributed fixation of Mn to one or more of the following mechanisms: oxidation of Mn to higher valence oxides and/or precipitation of insoluble compounds in soils subjected to drying and wetting; physical entrapment in clay lattice wedge zones; and adsorption on the exchange sites. The term "fixed" Mn refers to the Mn that could not be extracted by neutral $1 N$ NH_4OAc solution.

In calcareous soils, chemisorption on $CaCO_3$ surfaces may be important in fixing Mn (McBride, 1979); this can be followed by precipitation of $MnCO_3$. Norvell and Lindsay (1972) indicated that soluble Mn could be lost from solution following additions of MnDTPA to soils. At low pH, Fe dissolves from the soils and displaces Mn from MnDTPA. In calcareous soil, Mn can easily be displaced by Ca. At neutral pH, both Fe and Ca can displace Mn. In the case of MnEDTA, the loss of Mn can be rapid in soils and essentially complete in less than one day from soil suspensions of pH 6.1 to 7.85 (Norvell and Lindsay, 1969). This instability would seriously limit the usefulness of this chelate as an Mn fertilizer. Norvell (1972) reported that none of the commonly used chelating agents—EDTA, DTPA, EDDHA, etc.—could form stable Mn chelates in soils, because either Fe or Ca, or both, can substitute for Mn. The free Mn can then be complexed by organic matter or precipitated as MnO_2.

5. Manganese in Plants

5.1 Essentiality of Manganese in Plants

Manganese has several functions in plants. It activates many enzyme reactions involved in the metabolism of organic acids, phosphorus, and nitrogen. In higher plants, Mn activates the reduction of nitrite and hydroxylamine to ammonia. It is also involved in photosynthesis. It is a constituent of some respiratory enzymes and of some enzymes responsible for protein synthesis. As an activator in enzymes involved in carboxylic acid cycle and carbohydrate metabolism, it may be replaced by Mg. Manganese functions along with Fe in the formation of chlorophyll.

5.2 Deficiency of Manganese in Plants

Manganese deficiencies have been observed in various crops throughout the world. Manganese deficiency is otherwise known as "gray speck" in oats, "yellow disease" in spinach, "speckled yellows" in sugar beets, "marsh spot" in peas, "crinkle leaf" in cotton, "stem streak necrosis" in potato, "streak disease" in sugar cane, "mouse ear" in pecan, and "internal bark necrosis" in apple. Deficiencies are common in cereal grains, beans, corn, potatoes, sugar beets, soybeans, and many vegetable crops (Rumpel et al, 1967). Manganese is reportedly the most common micronutrient deficiency in soybeans (Scott and Aldrich, 1970), which has been observed in the southeastern US region (Georgia, Florida, Kentucky, North Carolina, South Carolina, and Virginia), predominately in poorly drained soils with pH levels above 6.0 (Anderson and Mortvedt, 1982).

Labanauskas (1965) listed the following soil conditions as conducive to Mn deficiency:

1. Thin, peaty soils overlying calcareous subsoils
2. Alluvial soils and marsh soils derived from calcareous materials, such as calcareous silts and clays
3. Poorly drained calcareous soils with a high content of organic matter
4. Calcareous black sands and reclaimed acid heath soils
5. Calcareous soils freshly broken up from old grassland
6. Old black garden soils where manure and lime have been applied regularly for many years
7. Very sandy acid mineral soils that are low in native Mn content

In Michigan, the deficiency is closely related to soil reaction in mineral soils whose pH is >6.5. Response to Mn may occur in organic soils with pH >5.8 (Rumpel et al, 1967). In the United States, in general, Mn deficiency is most often associated with naturally wet areas that have been drained and brought into cultivation (NAS, 1977). This is apparently associated with the leveling losses of available soil Mn and the presence of calcareous subsoils.

The sensitivity of selected crops to Mn deficiency is indicated in Table 8.6. In general, members of the bean family are very sensitive to Mn deficiency. Among grain crops, oats is the most sensitive to Mn deficiency.

Data in Table 8.7 indicate the deficiency, sufficiency, and toxicity range of Mn in crops grown in various growth media. Jones (1972) reported the following ranges of Mn (in ppm, dry wt, of mature leaf tissue) for most

TABLE 8.6. Relative sensitivity of
various crops to manganese
deficiency.[a]

High sensitivity		
Beans	Peach	Soybean
Citrus	Pecan	Spinach
Lettuce	Potato	Sudangrass
Oats	Radish	Table beet
Onion	Sorghum	Wheat
Peas		
Medium sensitivity		
Alfalfa	Celery	Peppermint
Barley	Clover	Sorghum
Broccoli	Cucumber	Spearmint
Cabbage	Corn	Sugar beet
Carrot	Grass	Tomato
Cauliflower	Parsnips	Turnip
Low sensitivity		
Asparagus	Blueberry	Rye
Cotton		

[a] *Source:* Follett et al, 1981, with permission of the
authors and Prentice-Hall.

agricultural crops: deficient, <20; sufficient, 20–500; and toxic, >500. For
the following crops, Melsted et al (1969) found the following critical levels
of Mn (in ppm): corn, 15; soybeans, 20; wheat, 30; and alfalfa, 25. The
plant tissues analyzed were: corn—leaf at, or opposite and below ear level
at tassel stage; soybeans—the youngest mature leaves and petioles, on
the plant after pod formation; wheat—the whole plant at the boot stage;
and alfalfa—upper stem cuttings in early flower stage.

Manganese deficiency symptoms typically include interveinal chlorosis
(yellowish to olive-green) with dark-green veins. Under severe deficiency,
leaves develop brown speckling and bronzing in addition to interveinal
chlorosis, with abscission of developing leaves (Anderson and Ohki, 1977).
The patterns of chlorosis can be confused with those of Fe, Mg, or N
deficiency. Wheat and barley often exhibit colorless spots. In corn, deficient leaves are manifested by lighter green leaves with parallel, yellowish
stripes.

The following are some soil conditions, soil management, and other
practices that may aggravate or alleviate Mn deficiency (Labanauskas,
1965):

1. Soils with pH above 6.5 favor the oxidation of Mn(II) to Mn(IV), which
limits the solubility of Mn and consequently its availability to plants.
2. Manganese may be leached from strongly acidic soils, resulting in deficiency.
3. Excessive liming of soils to neutralize acidity is a common cause of
Mn deficiency.

TABLE 8.7. Deficient, sufficient, and toxic manganese concentrations in plants.[a,b]

Plant	Type of culture	Tissue sampled	Other information	Mn concentration, ppm (dry wt)		
				Deficient	Sufficient	Toxic
Horticultural crops						
Apple	Field	Leaves		15	30	—
Apple	Soil	Leaves	Interveinal bark necrosis	—	—	>400
Apricot	Field	Leaves		10	86–94	—
Avocado	Solution	Leaves		—	1,300	4,300–6,000
Avocado	Field	Leaves	September	—	366–655	—
Banana	Field	Leaves		<10	—	—
Bean	Field soil	Tops	pH, 4.7	—	40–940	1,104–4,201
Brussels sprouts	Field	Leaves plus petioles	Youngest fully expanded	—	78–148	760–2,035
Carrot	Solution	Tops	Reduced yields	—	—	7,100–9,600
Lima beans	Field	Tops		32–68	207–1,340	—
Onion	Limed organic soil	Tops	Maturity	—	34	—
Orange	Field	Leaves	7-month bloom cycle; leaves from non-fruiting terminals.	<19	20–90	>100
Peas	Podzol soil	Leaves		—	—	550
Potato	Field	Leaves	pH, 4.7	7	40	—
Spinach	Field	Plant		23	34–60	—
Tomato	Solution	Leaves		5–6	70–398	—
Turnip	Field	Leaves		—	75	—
Cereals						
Barley	Solution	Old leaves	Moderate to severe necrosis	—	—	305–410
Barley	Soil	Tops		—	14–76	—
Oats	Soil	Tops		8–12	30–43	—
Oats	Acid soil	Tops		—	301–370	—
Rice	Solution	Single leaf		—	—	4,000–8,000
Rye	Soil	Mature tops		—	10–50	—
Rye	Solution	Old leaves		—	—	1,400

Plant	Growth medium	Plant part	Remarks			
Wheat	Soil (pots)	Tops	Plants grown at pH 4.9 and 6.9	—	108–113	356–432
Wheat	Solution	Tops	Plants grown with different Mn contents	—	181–621	396–2,561
Corn	Field	Ear leaf	Single cross hybrids	—	116–214	—
Corn	Soil	Whole leaf	Moderately fertile soil	—	76–213	—
Cotton	Solution	Tops	var. Pima S-2	—	196–924	1,740–8,570
Cotton	Field	Leaves and petioles	16 varieties on 3 soils	—	58–238	—
Cowpeas	Solution	Tops	Toxicity symptoms	—	—	1,224
Peanut	Sand	Leaves		—	110–440	890–10,900
Soybean	Soil	Uppermost fully developed leaf		<20	—	—
Forage crops—temperate group						
Soybean	Solution	Leaves	30 days	2–3	14–102	173–199
Sugar beet	Field	Leaves		5–30	7–1,700	1,250–3,020
Tobacco	Soil	Upper leaf	var. Burley 21	10	45	—
Alfalfa	Field	Tops	Several soils of southeastern USA	—	—	477–1,083
Alfalfa	Soil (pots)	Tops		—	65–240	651–1,970
Lespedeza	Soil	Tops		—	—	>570
Ryegrass	Solution	Old leaves		—	—	800
Sweet clover	Solution	Tops	Toxic symptoms	—	—	321–754
Vetch	Field	Tops		—	—	500–1,117
White clover	Soil	Tops	var. New Zealand	—	—	650
Forage crops—tropical group						
Trifolium subterraneum	Soil (pots)	Tops		4–25	30–300	—
Trifolium subterraneum	Solution	Tops		—	200	—
Phaseolus lathyroides	Solution	Tops	Cultivar Murray	—	—	840
Glycine javanica	Solution	Tops	Cultivar Jineroo	—	—	560

[a] Source: From various sources, summarized by NAS (1973).
[b] Toxicity threshold levels are here defined as Mn concentrations found when yields were 5% below the maximum.

4. Burning of organic soils, especially those well supplied with lime, produces an alkaline condition and may cause Mn deficiency.
5. Some soils become more alkaline with irrigation, and in this case, Mn deficiency commonly develops.
6. Manganese deficiency is often produced in areas with fluctuating redox conditions. Waterlogging solubilizes Mn, and subsequent percolation removes the soluble Mn^{2+} ions. After the soils dry, conditions for rapid oxidation of Mn may ensue and Mn deficiency may result.
7. Liming of poorly buffered sandy soils often causes Mn deficiency.
8. Manganese deficiency in apple trees can be greater in dry summers than in ones with ample rainfall.
9. Manganese deficiency in peach trees (first noted in mid-June) can be aggravated by unusually dry weather in the late spring and early summer.

The relationships of relative yields and leaf Mn to an Mn deficiency index (MDI) are shown in Figure 8.3 for soybeans grown in the southeastern United States. An 8% yield reduction was observed for each unit increase in the MDI.

The level of Mn in wheat plants is believed to influence its tolerance to certain plant diseases. More recently, Graham and Rovira (1984) ob-

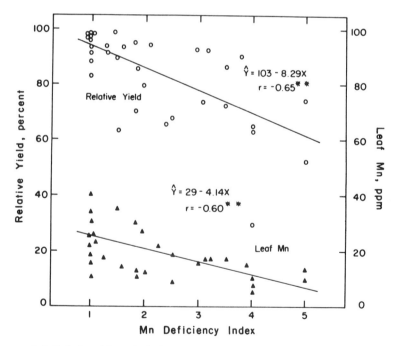

FIGURE 8.3. Relationships of relative yield and leaf manganese to an Mn-deficiency index (MDI) in leaf trifoliolates of soybeans grown in the southeastern United States. *Source:* Anderson and Mortvedt, 1982, with permission of the editors.

tained further evidence that wheat plants growing in Mn-deficient soils are more susceptible to infection by "take-all" *(Gaeumannomyces graminis* var. *tritici),* a worldwide fungal disease of wheat that at times has been devastating. They summarized their results in a hypothesis: the susceptibility of wheat to the disease depends in part on the availability of Mn in the soil and is inversely related to the concentration of Mn in host tissues. They offered one or more mechanisms by which Mn may exert its influence on the severity of "take-all": (1) Mn(II) may be directly toxic to the free inoculum of the fungus in the soil as shown for *Streptomyces scabies* in potato; (2) through Mn metabolism in the plants—i.e., Mn nutrition affects photosynthesis, which in turn controls the rate of exudation of soluble organic compounds by roots. These exudates affect the rhizosphere microflora, including the growth of the "take-all" fungus; and (3) through lignin production, which is controlled by Mn-activated enzyme systems. Lignin materials can provide partial defense against "take-all" and these materials may be more poorly developed in Mn-deficient plants.

5.3 Toxicity of Manganese in Plants

Under field conditions, Mn toxicity in plants could occur in acid soils. Manganese toxicity occurs frequently with Al toxicity. Aluminum and Mn toxicities are the most important growth-limiting factors in many acid soils (Foy and Campbell, 1984). In well-drained soils, Mn toxicity is generally found in soils having pH below 5.5 (NAS, 1973); in flooded soils, the reducing condition can produce levels of Mn^{2+} ions approaching toxic levels at much higher pH. Manganese toxicities in soybeans have been reported in Georgia and Mississippi, mainly on well-drained soils with pH levels <5.5 (Anderson and Mortvedt, 1982).

Manganese toxicity can be associated with Mn concentrations in plant tissues exceeding 500 ppm (dry wt basis) (Table 8.7). However, most crop species appear to tolerate as much as 200 ppm Mn in their tissues without showing toxicity effects. Levels of Mn in potato foliage in excess of 250 ppm Mn may cause toxicity (White et al, 1970). In rice plants, the toxicity levels are reported as 300 to 1,000 ppm in tops and 200 to 600 ppm in roots (Chino, 1981). In soybeans, the critical Mn deficiency (causing a 10% reduction in growth from maximum due to deficiency) and toxicity (causing a 10% reduction in growth from maximum due to toxicity) levels for leaf Mn are 20 and 250 ppm, respectively, at the 90% relative yield level (Fig. 8.4). A considerable number of yield depressions were obtained as leaf Mn excelled about 150 ppm, but near-maximum yields also occurred with leaf Mn up to 600 ppm. Hence, the critical toxicity level for soybean leaf Mn varies considerably among locations and among years.

On red basaltic soils in New South Wales, Australia, Mn toxicity is common in French beans and lettuce on soils with pH below 4.5 (Siman et al, 1974). In southern New South Wales, the establishment of pasture legumes, mainly lucerne, is restricted on some soils by the existence of

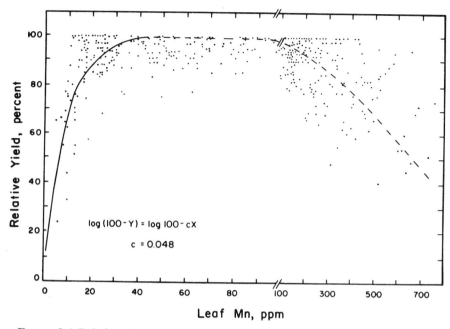

$$\log(100 - Y) = \log 100 - cX$$

$$c = 0.048$$

FIGURE 8.4. Relationships between relative yields and manganese concentrations of leaves of soybeans grown in Georgia. *Source:* Anderson and Mortvedt, 1982, with permission of the editors.

toxic levels of available Mn after heavy rain on slightly acid (pH 4.7–5.5) soils, or dry, hot conditions. In the Goulburn Valley in Victoria, water logging-induced Mn toxicity has caused substantial damage to young apple trees.

In potato growing areas, the soil is maintained at a pH of 5.4 or below to minimize the incidence of common scab. However, under soil conditions of low pH concomitant with high fertility, concentrations of soluble Mn and Al may be sufficiently high to depressed potato yields (Langille and Batteese, 1974). In some cases, soil pH has been allowed to drop as low as 4.5 to effectively control scab (Berger and Gerloff, 1948). These low pHs often produce stem streak necrosis in the potato plants, which causes premature death of the vines, thereby reducing yields. However, this condition could easily be corrected by lime application.

Several extraction procedures (see section 4.2) have been tested to diagnose soil toxic conditions, with inconsistent results. As pointed out earlier, no single extractant can be universally adopted. The most dependable approach is to perform both plant and soil tests to diagnose Mn status.

Manganese toxicity symptoms in plants are characterized by marginal chlorosis and necrosis of leaves (alfalfa, kale, lettuce, and rape), leaf puckering (cotton, snap beans, soybeans), and necrotic spots on leaves (barley, lettuce, and soybeans). In general, plants have deformed leaves, chlorotic areas, dead spots, stunted growth, and depressed yield.

In summary, the following soil conditions and soil management can be associated with excess Mn in soils:

1. Mn is frequently present in toxic concentrations in strongly acid soils. Soil pH is a controlling factor in the availability of Mn. The more acid the soil, the greater the solubility of Mn.
2. Waterlogging of soil can increase soluble Mn^{2+} ions caused by reduction of Mn(III) and Mn(IV) to Mn(II).
3. Fertilizers that lower the pH of the soil may increase the severity of Mn toxicity.
4. Liming of acid soils to raise the pH to 5.5 and above should alleviate Mn toxicity conditions.

6. Factors Affecting the Mobility and Availability of Manganese

6.1 pH

Manganese is generally considered to be one of the most important toxic metals in acid soils, Al being the most important. Several investigators found that pH had the greatest effect on the availability of Mn, followed in order by organic matter and moisture (Christensen et al, 1951; Sanchez and Kamprath, 1959). Marked increases in the availability of Mn should be expected when soil pH decreases below 5.5 (Lucas and Knezek, 1972). Morris (1949) found that water-soluble Mn in 25 naturally acid soils averaged 2.1 ppm in soils with pH below 5.2, 1.0 ppm in soils with pH of 5.2 to 5.4, and 0.5 ppm in soils having pH of 5.4. Gupta et al (1970) found that the highest content of Mn in carrot tops (grown in acid sphagnum peat soil under greenhouse conditions) occured at pH 4.4 to 5.0, and the lowest at pH 6.2 to 6.4. But the Mn content was lower at pH 4.0 to 4.1 than at pH 4.4 to 5.0. In general, the amount of extractable Mn is inversely related to soil pH (Hoyt and Nyborg, 1971; Sharpe and Parks, 1982; Sims et al, 1979).

Perhaps the influence of pH on plant growth and extractability of Mn can best be depicted as in Figure 8.5. In soils, both plant uptake and level of extractable Mn increase with acidity. However, in nutrient culture, Mn uptake tends to be maximal at pH 6.5 and decreases as the pH changes in either direction. Godo and Reisenauer (1980) concluded that reduction in Mn uptake at low pH may be caused by the competition for absorption between Mn, H, and Fe ions, the latter two being more available at lower solution pH (Vlamis and Williams, 1962). These data indicate that soil Mn availability is not controlled by either soil or plant characteristics *per se,* but by the combined effects of plant characteristics, soil properties, and the interactions of plant roots and the surrounding soil. Godo and Reisenauer (1980) found that root exudates made an important contribution

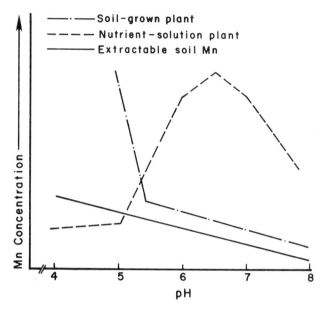

FIGURE 8.5. Generalized relationships between the concentrations of manganese in soil-grown plants and soil pH, between nutrient-solution grown plants and solution pH, and between manganese in soil extracts and soil pH. *Source:* Godo and Reisenauer, 1980, with permission of the authors and Soil Sci Soc of America.

to plant uptake of soil Mn, in that exudate compounds, such as hydroxy-carboxylates, increase soil Mn solubility through reducing MnO_2 and complexing the Mn^{2+} ions. These effects can be more significant in systems more acidic than pH 5.5 (Fig. 8.5).

Results from field experiments conducted in the southeastern United States indicate that relative yields of soybeans consistently decreased with increasing soil pH above pH 6.0 (Fig. 8.6) (Anderson and Mortvedt, 1982). These results further indicate that very little response to Mn applications can be expected at lower soil pH levels and emphasize the increasing need for Mn at higher soil pH levels, especially those soils with pH above 6.0.

Manganese toxicity in acidic soils can easily be alleviated or completely eliminated by application of lime at rates sufficient to raise the soil pH to about 6.5 (White et al, 1970; Snider, 1943). Calcium carbonate can have a strong negative effect on the uptake of Mn through adsorption and precipitation reactions, or in the formation of manganocalcite (Jauregui and Reisenauer, 1982).

6.2 Manganese Interactions with Other Ions

The antagonism between Mn and Fe is a well-documented interaction in higher plants. Ohki (1975) noted that, as Mn concentration in cotton tissue increased from the critical Mn level to a high but nontoxic level, Fe con-

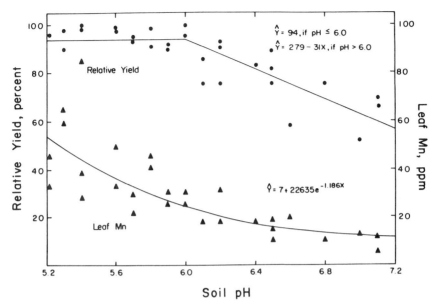

FIGURE 8.6. Influence of soil pH on relative yield and leaf manganese of soybeans grown in the southeastern United States without applied manganese. *Source:* Anderson and Mortvedt, 1982, with permission of the editors.

centrations in the plant decreased from a high to moderate level. In severe cases, high levels of Mn supply can reduce the Fe concentrations and induce Fe deficiency, as shown in pineapple culture (Sideris and Young, 1949). Others have also reported reciprocal relationships of Fe and Mn in rice and barley (Vlamis and Williams, 1964), soybeans (Somers and Shive, 1942) and other plants (Dokiya et al, 1968; Gerald et al, 1959). In nutrition of most plant species, the ratio of Fe to Mn in the nutrient medium must be maintained between 1.5 and 2.5 in order to get healthy plants (Mulder and Gerretsen, 1952). If the ratio is above 2.5, symptoms of Fe toxicity would occur; if it is below 1.5, the plant would suffer from Mn toxicity.

Conversely, addition of Fe to the growing medium may reduce Mn toxicity to plants (Hiatt and Ragland, 1963; Warington, 1954; Heenan and Campbell, 1983). Additions of Fe chelates to soil have induced Mn deficiency in bush beans (Holmes and Brown, 1955), reduced the Mn contents of plants (Wallace, 1958; Wallace et al, 1957), and have alleviated Mn toxicity (Shannon and Mohl, 1956; Wallace, 1962). Knezek and Greinert (1971) observed that the addition of MnEDTA or FeEDTA to the soil intensified Mn deficiency symptoms, depressed growth, and reduced Mn uptake, while the Fe concentration remained relatively constant, resulting in a wide Fe:Mn ratio in plants. The ineffectiveness of MnEDTA in overcoming plant Mn deficiency was due to a rapid substitution of Fe for Mn in the EDTA molecule, with the released Mn probably complexing with organic matter or precipitating as MnO_2. Mortvedt (1980) found that Mn in the MnEDTA added to limed or calcareous soils may be replaced by Ca^{2+} ions.

Other reported Mn interactions include P-Mn (Murphy et al, 1981), Mn-B (Ohki, 1973), and Mn-Ca (LeMare, 1977; Shuman and Anderson, 1976) couples.

6.3 Redox Potential

The redox status of soils influences the solubility and availability of Mn. Manganese in the III and IV oxidation states occurs as precipitates in oxidized environments, whereas Mn in the II oxidation state is dominant in solution and solid phases under reducing conditions. The chemistry of flooded soils has been extensively studied by the group of Dr. Ponnamperuma at the International Rice Research Institute in the Philippines and of Dr. Patrick and colleagues at Louisiana State University. In general, flooding a soil enhances Mn solubility and its availability to the rice plants (Clark et al, 1957; Chaudhry and McLean, 1963; Jugsujinda and Patrick, 1977; Gambrell and Patrick, 1978). However, in some instances flooding may decrease Mn uptake by plants due to increased solubility of Fe and its competition with Mn for plant absorption (see section 4).

In flooded soils, Gotoh and Patrick (1972) concluded that both pH and Eh largely control Mn behavior. Soluble and exchangeable Mn increased with decreases in both pH and Eh. Between pH 6 and 8, the conversion of insoluble soil Mn to the water-soluble and exchangeable forms was dependent on both pH and Eh, while at pH 5, Eh had little effect due to overriding effect of acidity.

6.4 Organic Matter

The role of organic matter in complexing Mn has already been discussed (section 4.5). Additions of organic matter amendments to soils can alleviate Mn toxicity symptoms (Cheng and Ouellette, 1971). In addition to its complexing ability, organic matter can affect the redox status of soils. In flooded conditions, microbial decomposition of organic matter present in soil as plant debris, humus, animal wastes, etc., leads to reducing conditions by utilizing free oxygen in the soil atmosphere and produces CO_2 and organic acids. Subsequently, microorganisms utilize oxygen associated with oxidized forms of Mn and Fe to form soluble forms of Mn and Fe. This effect of organic matter in lowering the redox potential of soil and enhancing the solubility of Mn has been demonstrated by Meek et al (1968) and Mandal and Mitra (1982).

6.5 Management and Fertilization Practices

Type of Mn carriers and method of fertilizer placement can affect Mn availability to plants. For example, soil applications of MnEDTA can cause higher incidence of Mn deficiency, as reported by Knezek and Greinert (1971) and Mortvedt (1980). Besides, it is an impractical Mn fertilizer since it is highly unstable in soils having pH above 6 (Norvell and Lindsay, 1969).

The effect of the amounts of Mn applied to soil on yield and Mn concentration of soybean leaves in instances where yields were increased by Mn fertilization is shown in Figure 8.7. Leaf and soil Mn increased with each increment of applied Mn, and near-maximum yields were obtained with the 20 kg/ha Mn rate. Leaf Mn and soil Mn associated with near-maximum yields were 19 ppm and 3.6 ppm, respectively.

Of the various application methods employed by farmers, banding and foliar applications of Mn may provide higher efficiency of Mn utilization (Randall et al, 1975). Foliar application is a common fertilizer practice for nut and fruit trees.

The use of certain fertilizers that create acidic soil reactions, such as $(NH_4)_2SO_4$, can cause reduced pH and subsequently, Mn toxicity to plants. Heavy and prolonged applications of NH_4NO_3 or $(NH_4)_2SO_4$ fertilizers can substantially lower soil pH and enhance Mn uptake by plants (Siman et al, 1971; Sillanpaa, 1972). In addition, Petrie and Jackson (1984b) found that application of an acid-forming N fertilizer, such as $(NH_4)_2SO_4$ or NH_4Cl, was more effective than other N sources (e.g., urea and various urea-phosphate formulations) in increasing the Mn concentration and alleviating Mn deficiencies in barley and oats. In laboratory experiments, Petrie and Jackson (1984a) demonstrated that application of either $(NH_4)_2SO_4$ or NH_4Cl alone or with $Ca(H_2PO_4)_2$ decreased the soil solution pH. They attributed the pH depression to nitrification of NH_4^+ ions and solubilization products of $Ca(H_2PO_4)_2$. Higher increases in soil solution

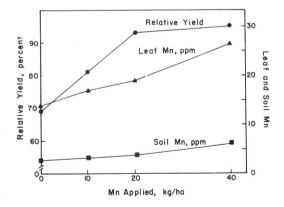

FIGURE 8.7. Influence of manganese application rates on relative yield and leaf manganese of soybeans and double acid-extractable soil manganese on Mn-deficient soils in the southeastern United States. *Source:* Anderson and Mortvedt, 1982, with permission of the editors.

Mn concentration were obtained from application of $(NH_4)_2SO_4$ or NH_4Cl in combination with $Ca(H_2PO_4)_2$ than from single application of $(NH_4)_2SO_4$, NH_4Cl, or $Ca(H_2PO_4)_2$.

6.6 Other Factors

Plant species and cultivars within species differ widely in their susceptibility to Mn deficiency and likewise in their tolerance to excess available Mn (Foy et al, 1978, 1981; Foy and Campbell, 1984; Horst, 1983; Nelson, 1983; Parker et al, 1981; Ohki et al, 1981).

7. Manganese in Drinking Water and Food

The US-EPA proposed a maximum level of 0.05 mg Mn/liter in drinking water (US-EPA, 1976). The maximum Mn content of tap waters collected from 50 homes in four urban areas of the United States was 17 μg/liter (Strain et al, 1975). Waters collected from 15 large rivers of the United States and Canada had a median concentration of 20 μg Mn/liter ($n = 52$); the range was 0 to 185 μg/liter (NAS, 1973). The mean for the rivers of the USSR is 11.9 μg/liter, but the global average is probably higher.

The ordinary foods consumed by man differ considerably in Mn content (Table 8.8). On a dry-weight basis, leafy vegetables contain higher concentrations of Mn than other foodstuffs. Total dietary Mn intakes by adults vary from 2 to 8 mg/day, depending upon the amounts and proportions of cereals, nuts, green leafy vegetables, and tea consumed (Underwood, 1973). Tea and cloves are exceptionaly rich in Mn; one cup of tea has

TABLE 8.8. Manganese contents of groups of principal foodstuffs.[a]

Class of food	Number of samples	Mn concentrations, ppm (fresh weight basis)		
		Minimum	Maximum	Average
Nuts	10	6.3	41.7	22.7
Cereal products	23	0.5	91.1	20.2
Dried legume seeds	4	10.7	27.7	20.0
Green leafy vegetables	18	0.8	12.6	4.5
Dried fruits	7	1.5	6.7	3.3
Roots, tubers, and stalks	12	0.4	9.2	2.1
Fresh fruits (including blueberries)	26	0.2	44.4	3.7
Fresh fruits (excluding blueberries)	25	0.2	10.8	2.0
Nonleafy vegetables	5	0.8	2.4	1.5
Animal tissues	13	0.08	3.8	1.0
Poultry and by-products	6	0.30	1.1	0.5
Dairy products	7	0.03	1.6	0.5
Fish and seafoods (excluding oysters)	7	0.12	2.2	0.5
Fish and seafoods (including oysters)	6	0.12	0.4	0.25

[a] *Source:* NAS (1973).

been reported to contain 0.30 to 1.3 mg Mn, compared with a very much lower amount in a cup of coffee. However, diets high in milk, sugar, and refined cereals might provide inadequate Mn, particularly in growing children, pregnant women, and persons with diabetes and rheumatoid arthritis. These diets may possess a slower turnover rate of Mn (NAS, 1973).

In Japan, the daily intake by adults is about 2.8 mg Mn/day, with more than 70% of Mn supplied by plant products (Murakami et al, 1965). In Canada, daily intake for Mn was reported at 2.9 to 3.6 mg, with most of the Mn being supplied by cereals, potatoes, and fruits (Kirkpatrick and Coffin, 1974). Other intake values reported are: the United Kingdom, 2.7 ± 0.8 mg/day (Hamilton, 1979); India, 8.3 mg/day (Soman et al, 1969); and the International Commission on Radiological Protection, 3.1 mg/day.

8. Sources of Manganese in the Terrestrial Environment

Anthropogenic sources of Mn contamination in terrestrial environments are primarily associated with certain industrial activities, such as metal smelting and refining. Other sources come from agricultural practices (fertilizer use, sewage sludge, and animal waste disposal), and atmospheric deposition from fossil fuel combustion and municipal incinerators.

8.1 Manganese Fertilizers

Manganese is available as inorganic and organic fertilizers for agricultural use. The most commonly used Mn fertilizer is $MnSO_4 \cdot 4H_2O$ (24% Mn), which has a high water solubility in both acid and alkaline conditions. Manganese oxide (MnO, 78% Mn) is also widely used. Organic carriers of Mn are seldom used because of their high costs and instability of certain products, such as MnEDTA. The rates of application depend on the soil type, soil pH, and type of crops. Application rates may range from about 7 kg Mn/ha for low to medium responsive crops growing in mineral soils, with pH of 6.5 to 7.2, to as high as 45 kg Mn/ha to high responsive crops growing in peats and mucks with pH of 7.3 to 8.5 (Lucas, 1967). In foliar applications, only 2 to 4.5 kg/ha may be required to correct Mn deficiency.

8.2 Sewage Sludges

A recent report (CAST, 1976) identified Cd, Cu, Ni, Zn, and Mo in sludge as the elements with the greatest potential to accumulate in plants and pose a threat to animals and man. Chaney (1983) considers Cd, Se, and Mo as elements posing the greatest risks to the plant-animal-man system. Therefore, in spite of the high levels of Mn in certain sludges (Appendix Table 1.6) and reported enrichment of some soils and crops from sludge application (Cunningham et al, 1975; Andersson and Nilsson, 1972), Mn in sludge is generally not a concern in the soil-plant-man system.

8.3 Fossil Fuel Combustion

Although fly ash is enriched in Mn compared to coal, it is considered a nonthreat to the soil-plant-animal pathway (Adriano et al, 1980), while, Se and Mo are identified as risks to the food-chain pathway. Burning of coal and fuel oil contributes to the Mn burden of the atmosphere (Smith, 1972). Other sources of Mn emission into urban air include municipal incinerators and diverse industrial manufacturing processes.

8.4 Iron and Steel Industry

Since most of the Mn is used in the iron and steel industry, this element is almost always found in the particulate matter of urban air.

8.5 Other Sources

Of localized importance is Mn being released from mining and smelting activities. Usually a majority of the Mn particulates suspended from the ground or emitted to the atmosphere settle within a few kilometers of the originating source (Hutchinson and Whitby, 1974).

The use of animal wastes in agriculture can provide substantial amounts of Mn (Atkinson et al, 1954; Capar et al, 1978).

The global cycle of Mn, which reflects man's impact on the fluxes and reservoir of this element, has been discussed by Garrels et al (1973). They noted that the major exchange of Mn between the atmosphere and the pre-human earth surface was due to continental dust being swept into the atmosphere by winds and then falling back onto the earth's surface. Today this dust flux is complemented by Mn emitted to the atmosphere in particulate form by industrial activities (e.g., iron and steel industry). The total river flux of Mn to the ocean today is nearly three times the pre-human flux. This increase represents principally an increase in the denudation rate of the land's surface from about 100×10^{14} g/year pre-human to today's rate of about 225×10^{14} g/year (Garrels et al, 1973). Because this increase in denudation reflects an increase in the suspended load of rivers from deforestation and agricultural activities, and because Mn is concentrated in the ferric oxide coatings on suspended material and in the suspended particles, the land to ocean flux is higher today than in the past. Manganese in particulate emissions from industrial activities rivals the natural input of continental dust to the atmosphere (30×10^{10} g vs 43×10^{10} g). Most particulate Mn falls out of the atmosphere near industrial sources. The mining of Mn ore has resulted in a net gain for the land reservoir and a net loss from the sediment reservoir. There is no evidence for change with time of dissolved Mn in the oceanic reservoir.

References

Adams, F., and J.I. Wear. 1957. *Soil Sci Soc Am Proc* 21:305–308.

Adriano, D. C., A. L. Page, A. A. Elseewi, A. C. Chang, and I. Straughan. 1980. *J Environ Qual* 9:333–344.

Agrawal, H. P., and C. J. Reddy. 1972. *J Indian Soc Soil Sci* 20:241–247.

Alexander, M. 1977. *Introduction to soil microbiology.* Wiley, New York. 467 pp.

Anderson, O. E., and F. C. Boswell. 1968. *Agron J* 60:488–493.

Anderson, O. E., and K. Ohki. 1977. *J Fert Solutions* 21(6):30.

Anderson, O. E., and J. J. Mortvedt, eds. 1982. In *Soybeans: diagnosis and correction of manganese and molybdenum problems.* Southern Coop Series Bull 281. Univ Georgia, Athens. 98 pp.

Andersson, A. 1977. *Swedish J Agric Res* 7:7–20.

Andersson, A., and K. O. Nilsson. 1972. *Ambio* 1:176–179.

Atkinson, H. J., G. R. Giles, and J. G. Desjardins. 1954. *Can J Agric Sci* 34:76–80.

Aubert, H., and M. Pinta. 1977. *Trace elements in soils.* Elsevier, New York.

Berger, K. C., and G. C. Gerloff. 1948. *Soil Sci Soc Am Proc* 12:310–314.

Berrow, M. L., and R. L. Mitchell. 1980. *Trans Royal Soc* (Edinburgh): *Earth Sci* 71:103–121.

Berrow, M. L., and G. A. Reaves. 1984. In Proc Intl Conf Environ Contamination, 333–340. CEP Consultants Ltd, Edinburgh, UK.

Bloom, P. R. 1981. In R. H. Dowdy, J. A. Ryan, V. V. Volk, and D. E. Baker, eds. *Chemistry in the soil environment,* 129–150. ASA Spec Publ. no. 40, Madison, WI.

Boken, E. 1968. *Plant Soil* 9:269–285.

Bowen, H. J. M. 1979. *Environmental chemistry of the elements.* Academic Press, New York. 333 pp.

Boxma, R., and A. J. de Groot. 1971. *Plant Soil* 34:741–749.

Capar, S. G., J. T. Tanner, M. H. Friedman, and K. W. Boyer. 1978. *Environ Sci Technol* 7:785–790.

Chaney, R. L. 1983. In J. F. Parr, P. B. Marsh, and J. M. Kla, eds. *Land treatment of hazardous wastes.* Noyes Data Corp, Park Ridge, NJ.

Chaudhry, M. S., and E. O. McLean. 1963. *Agron J* 55:565–567.

Cheng, B. T., and G. J. Ouellette. 1971. *Plant Soil* 34:165–181.

Childs, C. W. 1975. *Geoderma* 13:141–152.

Chino, M. 1981. In K. Kitagishi and I. Yamane, eds. *Heavy metal pollution in soils of Japan.* Japan Sci Soc Press, Tokyo. pp. 65–80.

Christensen, P. D., S. J. Toth, and F. E. Bear. 1951. *Soil Sci Soc Am Proc* 15:279–282.

Clark, F., D. C. Nearpass, and A. W. Specht. 1957. *Agron J* 49:586–589.

Council for Agricultural Science and Technology (CAST). 1976. *Application of sewage sludge to cropland: appraisal of potential hazards of the heavy metals to plants and animals.* EPA-430-9-76-013. General Serv Admin, Denver. 63 pp.

Cox, F. R. 1968. *Agron J* 60:521–524.

Cunningham, J. D., D. R. Keeney, and J. A. Ryan. 1975. *J Environ Qual* 4:448–454.

Curtin, D., J. Ryan, and R. A. Chaudhary. 1980. *Soil Sci Soc Am J* 44:947–950.

DeHuff, G. L., and T. S. Jones. 1980. In *Minerals yearbook,* vol. 1, *Metals and minerals,* 543–553. US Bureau of Mines, US Dept of Interior, Washington, DC.

Dokiya, Y., N. Owa, and S. Mitsui. 1968. *Soil Sci Plant Nutr* 14:169–174.

Ehrlich, H. L. 1968. *Appl Microbiol* 16:197–202.

Fergus, I. F. 1953. *Queensland J Agric Sci* 10:15–27.

Follett, R. H., and W. L. Lindsay. 1970. In *Profile distribution of zinc, iron, manganese, and copper in Colorado soils.* Colorado Agric Exp Sta Tech Bull 110:1–78.

Follett, R. H., L. S. Murphy, and R. L. Donahue. 1981. *Fertilizers and soil amendments.* Prentice-Hall, Englewood Cliffs, NJ. 557 pp.

Foy, C. D., R. L. Chaney, and M. C. White. 1978. *Ann Rev Plant Physiol* 29:511–566.

Foy, C. D., H. W. Webb, and J. E. Jones. 1981. *Agron J* 73:107–111.

Foy, C. D., and T. A. Campbell. 1984. *J Plant Nutr* 7:1365–1388.

Frank, R., K. Ishida, and P. Suda. 1976. *Can J Soil Sci* 56:181–196.

Freedman, B., and T. C. Hutchinson. 1981. In N. W. Lepp, ed. *Effect of heavy metal pollution in plants,* Vol. 2, *Metals in the environment,* 35–94. Applied Science Publishers, London.

Gambrell, R. P., and W. H. Patrick. 1978. In D. D. Hook and R. M. M. Crawford, eds. *Plant life in anaerobic environments,* 375–423. Ann Arbor Sci Publ, Ann Arbor, MI.

Garrels, R. M., F. T. MacKenzie, and C. Hunt, eds. 1973. *Chemical cycles and the global environment: assessing human influences.* W. Kaufmann Inc, Los Angeles, CA 206 pp.

Geering, H. R., J. F. Hodgson, and C. Sdano. 1969. *Soil Sci Soc Am Proc* 33:81–85.

Gerald, C. G., P. R. Stout, and L. H. Jones. 1959. *Plant Physiol* 34:608–613.

Ghiorse, W. C., and H. L. Ehrlich. 1976. *Appl Environ Microbiol* 31:977–985.

Godo, G. H., and H. M. Reisenauer. 1980. *Soil Sci Soc Am J* 44:993–995.

Goldberg, S. P., and K. A. Smith. 1984. *Soil Sci Soc Am J* 48:559–564.

Gotoh, S., and W. H. Patrick, Jr. 1972. *Soil Sci Soc Am Proc* 36:738–742.

Graham, R. D., and A. D. Rovira. 1984. *Plant Soil* 78:441–444.

Grass, L. B., A. J. MacKenzie, B. D. Meek, and W. F. Spencer. 1973a. *Soil Sci Soc Am Proc* 37:14–17.

——— 1973b. *Soil Sci Soc Am Proc* 37:17–21.

Gupta, U. C., E. W. Chipman, and D. C. MacKay. 1970. *Soil Sci Soc Am Proc* 34:762–764.

Hamilton, E. I. 1979. *Trace Subs Environ Health* 13:3–15.

Hammes, J. K., and K. C. Berger. 1960. *Soil Sci Soc Am Proc* 24:361–364.

Heenan, D. P., and L. C. Campbell. 1983. *Plant Soil* 70:317–326.

Heintze, S. G. 1946. *J Agric Sci* 36:227–238.

Hiatt, A. J., and J. L. Ragland. 1963. *Agron J* 55:47–49.

Hoff, D. J., and H. J. Mederski. 1958. *Soil Sci Soc Am Proc* 22:129–132.

Holmes, R. S., and J. C. Brown. 1955. *Soil Sci* 80:167–179.

Horst, W. J. 1983. *Plant Soil* 72:213–218.

Hoyt, P. B., and M. Nyborg. 1971. *Soil Sci Soc Am Proc* 35:241–244.

Hoyt, P. B., and M. D. Webber. 1974. *Can J Soil Sci* 54:53–61.

Hutchinson, T. C., and L. M. Whitby. 1974. *Environ Conserv* 1:123–132.

Jauregui, M. A., and H. M. Reisenauer. 1982. *Soil Sci* 134:105–110.

Jenne, E. A. 1968. *Adv Chem* 73:337–387.

Jones, J. B., Jr., 1972. In J. J. Mortvedt, P. M. Giordano, and W. L. Lindsay eds. *Micronutrients in agriculture,* 319–346, Soil Sci Soc Am Inc, Madison, WI.

Jones, L. H. P., and G. W. Leeper. 1951. *Plant Soil* 3:141–153.

Jugsujinda, A., and W. H. Patrick, Jr. 1977. *Agron J* 69:705–710.

Kirkpatrick, D. C., and D. E. Coffin. 1974. *J Inst Can Sci Technol Aliment* 7:56–58.

Knezek, B. D., and H. Greinert. 1971. *Agron J* 63:617–619.

Knezek, B. D., and B. G. Ellis. 1980. In B. E. Davis, ed. *Applied soil trace elements*, 259–286. Wiley, New York.

Krauskopf, K. B. 1972. In J. J. Mortvedt, P. M. Giordano, and W. L. Lindsay, eds. *Micronutrients in agriculture*, 7–40. Soil Sci Soc Am, Inc, Madison, WI.

——— 1979. *Introduction to geochemistry*. 2nd ed. McGraw-Hill, New York. 617 pp.

Krupskiy, N. K., L. P. Golovina, A. M. Aleksandrova, and T. I. Kisel. 1978. *Sov Soil Sci* 10:670–675.

Labanauskas, C. K. 1965. In H. D. Chapman, ed. *Diagnostic criteria for plants and soils*, 264–285. Quality Print Co, Abilene, TX.

Lal, F., and T. D. Biswas. 1974. *J Indian Soc Soil Sci* 22:333–346.

Langille, A. R., and R. I. Batteese. 1974. *Can J Plant Sci* 54:375–381.

Leeper, G. W. 1947. *Soil Sci* 63:79–94.

LeMare, P. H. 1977. *Plant Soil* 47:607–620.

Lindsay, W. L., and W. A. Norvell. 1978. *Soil Sci Soc Am J* 42:421–428.

Liu, Z., Q. Q. Zhu, and L. H. Tang. 1983. *Soil Sci* 135:40–46.

Lucas, R. E. 1967. In *Micronutrients for vegetables and field crops*. Ext Bull E-486. Mich State Univ, East Lansing, MI.

Lucas, R. E., and B. D. Knezek. 1972. In J. J. Mortvedt, P. M. Giordano, and W. L. Lindsay, eds. *Micronutrients in agriculture*, 265–288. Soil Sci Soc Am Inc, Madison, WI.

Mandal, L. N., and R. R. Mitra. 1982. *Plant Soil* 69:45–56.

Marrack, D. 1980. In *Environment and health*, 184–185. Ann Arbor Science, Ann Arbor, MI.

McBride, M. B. 1979. *Soil Sci Soc Am J* 43:693–698.

McKeague, J. A., and M. S. Wolynetz. 1980. *Geoderma* 24:299–307.

Meek, B. D., A. J. MacKenzie, and L. B. Grass. 1968. *Soil Sci Soc Am Proc* 32:634–638.

Meek, B. D., A. L. Page, and J. P. Martin. 1973. *Soil Sci Soc Am Proc* 37:542–548.

Melsted, S. W., H. L. Motto, and T. R. Peck. 1969. *Agron J* 61:17–20.

Mena, I. 1980. In H. A. Waldron, ed. *Metals in the environment*, 199–220. Academic Press, New York.

Morris, H. D. 1949. *Soil Sci Soc Am Proc* 13:362–371.

Mortvedt, J. J. 1980. *Soil Sci Soc Am J* 44:621–626.

Mulder, E. G., and F. C. Gerretsen. 1952. *Adv Agron* 4:221–277.

Murakami, Y., Y. Suzuki, T. Yamagata, and N. Yamagata. 1965. *J Rad Res* 6:105–110.

Murphy, L. S., R. Ellis, Jr., and D. C. Adriano. 1981. *J Plant Nutri* 3:593–613.

National Academy of Sciences (NAS). 1973. *Manganese*. NAS, Washington, DC. 191 pp.

——— 1977. In *Geochemistry and the environment*, 29–39. NAS, Washington, DC.

Nelson, L. E. 1983. *Agron J* 75:134–138.

Norrish, K. 1975. In D. J. D. Nicholas and A. R. Egan, eds. *Trace elements in soil-plant-animal systems,* 55–81. Academic Press, New York.

Norvell, W. A. 1972. In J. J. Mortvedt, P. M. Giordano, and W. L. Lindsay, eds. *Micronutrients in agriculture,* 115–138. Soil Sci Soc Am Inc, Madison, WI.

Norvell, W. A., and W. L. Lindsay. 1969. *Soil Sci Soc Am Proc* 33:86–91.

——— 1972. *Soil Sci Soc Am Proc* 36:778–783.

Ohki, K. 1973. *Agron J* 65:482–485.

——— 1975. *Agron J* 67:204–207.

Ohki, K., D. O. Wilson, and O. E. Anderson. 1980. *Agron J* 72:713–716.

Page, E. R. 1962. *Plant Soil* 16:247–257.

Parker, M. B., F. C. Boswell, K. Ohki, L. M. Shuman, and D. O. Wilson. 1981. *Agron J* 73:643–646.

Patel, M. S., P. M. Mehta, and H. G. Pandya. 1972. *J Indian Soc Soil Sci* 20:79–90.

Pavanasasivam, V. 1973. *Plant Soil* 38:245–255.

Peterson, L. 1966. *Acta Agric Scan* 16:120–128.

Petrie, S. E., and T. L. Jackson. 1984a. *Soil Sci Soc Am J* 48:315–318.

——— 1984b. *Soil Sci Soc Am J* 48:319–322.

Ponnamperuma, F. N. 1972. *Adv Agron* 24:29–96.

Randall, G. W., E. E. Schulte, and R. B. Corey. 1975. *Agron J* 67:502–507.

——— 1976. *Soil Sci Soc Am J* 40:282–287.

Reddy, M. R., and H. F. Perkins. 1976. *Soil Sci* 121:21–24.

Robinson, W. O. 1929. *Soil Sci* 27:335–350.

Rumpel, J., A. Kozakiewicz, B. Ellis, G. Lessman, and J. Davis. 1967. Mich Agric Exp Sta Quart Bull 50(1):4–11.

Safo, E. Y., and L. E. Lowe. 1973. *Can J Soil Sci* 53:95–101.

Salcedo, I. H., B. G. Ellis, and R. E. Lucas. 1979b. *Soil Sci Soc Am J* 43:138–141.

Sanchez, C., and E. J. Kamprath. 1959. *Soil Sci Soc Am Proc* 23:302–304.

Sanders, J. R. 1983. *J Soil Sci* 34:315–323.

Scott, W. O., and S. R. Aldrich. 1970. *The Farm Quarterly,* Cincinnati, OH. pp.84–85.

Shacklette, H. T., J. C. Hamilton, J. C. Boerngen, and J. M. Bowles. 1971. USGS Prof Paper 574-D, Washington, DC. 71 pp.

Shannon, L. M., and J. S. Mohl. 1956. In A. Wallace, ed. *Sympos on the use of metal chelates in plant nutrition,* 50–54. Los Angeles, CA.

Sharpe, R. R., and W. L. Parks. 1982. *Agron J* 74:785–788.

Sherman, G. D., and P. M. Harmer. 1942. *Soil Sci Soc Am Proc* 7:398–405.

Sherman, G. D., J. S. McHargue, and W. S. Hodgkiss. 1942. *Soil Sci* 54:253–257.

Shuman, L. M., and O. E. Anderson. 1974. *Soil Sci Soc Am Proc* 38:788–791.

——— 1976. *Commun Soil Sci Plant Anal* 7:542–557.

Sideris, C. P., and H. Y. Young. 1949. *Plant Physiol* 24:416–440.

Sillanpaa, M. 1972. *Trace elements in soils and agriculture.* Soils Bull no. 17. FAO of the United Nations, Rome. 67 pp.

Siman, A., F. W. Cradock, P. J. Nicholls, and H. C. Kirton. 1971. *Aust J Agric Res* 22:201–214.

Siman, A., F. W. Cradock, and A. W. Hudson. 1974. *Plant Soil* 41:129–140.

Sims, J. L., and W. H. Patrick, Jr. 1978. *Soil Sci Soc Am J* 42:258–262.

Sims, J. L., P. Duangpatra, J. H. Ellis, and R. E. Phillips. 1979. *Soil Sci* 127:270–274.

Smith, R. G. 1972. In D. H. K. Lee, ed. *Metallic contaminants and human health,* 139–162. Academic Press, New York.

Soman, S. D., V. K. Panday, K. T. Joseph, and S. J. Raut. 1969. *Health Phys* 17:35–40.

Somers, I. I., and J. W. Shive. *Plant Physiol* 17:582–602.

Snider, H. J. 1943. *Soil Sci* 56:187–195.

Spencer, W. F., R. Patrick, and H. W. Ford. 1963. *Soil Sci Soc Am Proc* 27:134–141.

Stahlberg, S., and S. Sombatpanit. 1974. *Acta Agric Scan* 24:179–194.

Steenbjerg, F. 1935. *Third Intl Congr Soil Sci Trans* (Oxford) 1:198–201.

Strain, W. H., A. Flynn, E. G. Mansour, F. R. Plecha, W. J. Pories, and O. A. Hill. 1975. In Proc Intl Conf Heavy Metals in the Environ, 1003–1011. Toronto, Ontario, Canada.

Swaine, D. J. 1955. *The trace element content of soils.* Commonwealth Bur Soil Sci (Great Britain) Tech Commun 48. Herald Printing Works, York, England. 157 pp.

Taylor, R. M., and R. M. McKenzie. 1966. *Aust J Soil Res* 4:29–39.

Tyler, P. A., and K. C. Marshall. 1967. *Antonie van Leeuwenhoek* 33:171–183.

Underwood, E. J. 1973. In *Toxicants occurring naturally in foods,* 43–87. NAS, Washington, DC.

US Environmental Protection Agency (US-EPA). 1976. *Quality criteria for water.* US-EPA, Washington, DC.

Vinogradov, A. P. 1959. The *geochemistry of rare and dispersed chemical elements in soils.* Consultants Bureau Inc, New York. 209 pp.

Vlamis, J., and D. E. Williams. 1962. *Plant Physiol* 37:650–655.

⸺ 1964. *Plant Soil* 20:221–231.

Walker, J. M., and S. A. Barber. 1960. *Soil Sci Soc Am Proc* 24:485–488.

Wallace, A. 1958. *Agron J* 50:285–286.

⸺ ed. 1962. *Decade of synthetic chelating agents in inorganic plant nutrition.* Los Angeles, CA.

Wallace, A., L. M. Shannon, O. R. Lunt, and R. L. Impley. 1957. *Soil Sci* 84:27–41.

Warington, K. 1954. *Ann Appl Biol* 41:1–22.

White, R. P. 1970. *Soil Sci Soc Am Proc* 34:625–629.

White, R. P., E. C. Doll, and J. R. Melton. 1970. *Soil Sci Soc Am Proc* 34:268–271.

9
Mercury

1. General Properties of Mercury

Mercury, also called liquid silver, has the atomic number 80, an atomic weight of 200.59, boiling point of 356.6°C, melting point of −38.9°C, specific gravity of 13.55, vapor pressure of 1.22×10^{-3} mm at 20°C (2.8×10^{-3} mm at 30°C), and solubility in water of 6×10^{-6} g per 100 ml (25°C). It is a heavy, glistening, silvery-white metal, a rather poor conductor of heat but a fair conductor of electricity. It has seven stable isotopes with the following percent abundances: ^{195}Hg (0.15); ^{198}Hg (10.1); ^{199}Hg (17.0); ^{200}Hg (23.3); ^{201}Hg (13.2); ^{202}Hg (29.6); and ^{204}Hg (6.7). There are many minerals of Hg; the commonest are the sulfides cinnabar and metacinnabar. Mercury is recovered almost entirely from cinnabar (α-HgS, 86.2% Hg); less important sources are livingstonite (HgS·2Sb$_2$S$_3$), metacinnabar (β-HgS), and about 25 other Hg-containing minerals. Its unusual high volatility, which increases with increasing temperature, accounts for its presence in the atmosphere in appreciable amounts.

There are three stable oxidation states of Hg: 0, I, and II. Native elemental Hg exists in nature but generally is rare; however, other forms of Hg may be transformed to elemental Hg. Mercury also occurs either in the form of inorganic or organic complexes, or as stable mercuric sulfides and sulfosalts.

2. Production and Uses of Mercury

World Hg production was fairly constant, about 3.6×10^3 tonnes/year, from 1900 to 1939 (Gavis and Ferguson, 1972). Since the 1960s, however, production has more than doubled. The leading producers, in descending order, are: USSR, Spain, Algeria, United States, China, and Czechoslovakia (Drake, 1980). The principal deposits are found at Almaden in Spain, Idria in Yugoslavia, and Monte Amiita in Italy. On the American continent, significant Hg deposits are found in the Pacific Coast area, in particular

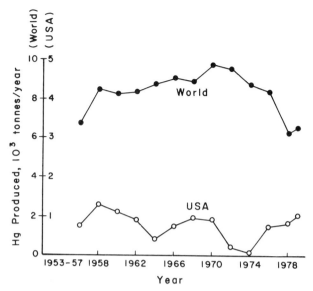

FIGURE 9.1. Trends in mercury production in the United States and worldwide, 1953–1979. *Source:* US Bureau of Mines, *Minerals yearbook,* several issues.

in Peru, Mexico, and California. World production in 1979 was about 6.7 × 10^3 tonnes, considerably lower than the rates in the early 1970s (Fig. 9.1). Estimated worldwide production, by decade, since the 18th century, is presented in Table 9.1. The recent surge in US production has been inspired by higher prices.

The US consumption of Hg is about double of its production capacity, resulting in a net import of the element roughly equivalent to that produced domestically (Drake, 1980). Consumption in the states and abroad has been declining during the last few years (Table 9.2). In the United States

TABLE 9.1. Estimated worldwide mercury production and release to the environment.[a]

Period	Mercury produced, × 10^6 kg	Mercury release, 10^6 kg		
		Air	Water	Soil
Pre-1900	200	232	37.0	252
1900–1909	34.3	39.8	6.3	43.3
1910–1919	37.6	43.6	6.9	47.5
1920–1929	37.8	43.8	7.0	47.8
1930–1939	35.9	41.6	6.6	45.3
1940–1949	61.5	71.3	11.4	77.8
1950–1959	65.0	75.4	12.0	82.2
1960–1969	78.5	91.1	14.5	99.2
1970–1979	87.6	102	16.2	111
Total (all time)	638	741	118	806

[a] *Source:* Andren and Nriagu, 1979, with permission of the authors and Elsevier, Amsterdam.

TABLE 9.2. Estimated consumption of mercury in the United States and abroad.[a]

User	Primary function	EEC[b] 1973	USA 1973	USA 1975	FRG 1973	FRG 1975
Electrolysis	Cathode electrode	1,030	486	520	372	137
Electrotechnique and instruments	Batteries, lamps, arc rectifiers, valves, meters, etc.	280	868	690	60.1	61.1
Paints	Antifouling and mildew proofing	70	262	250	18.7	5.6
Catalysis	Catalysts	60	23	25	42	12
Agriculture	Fungicide	95	63	30	50.3	31.9
Dental use	Amalgams for fillings	135	92	60	31	25
Pharmaceutical products	Preservative, patented medicines	40	21	20	0.5	0.5
Laboratory products	Thermometers, reagents, etc.	170	23	15	40	61.3
Others and stock	Slimicide	690	35	160	190	44.2
Total		2,570	1,873	1,770	808	377.6

[a] Source: Modified from Kaiser and Tölg, 1980, with permission of the authors and Springer-Verlag, New York.
[b] EEC: European Economic Community.

the overall decline was led by reduced demand for Hg use in catalysts, paints, dental fillings, electrical equipment, and for several laboratory purposes. However, there was increased use for electrolysis.

Mercury has wide applications in science, industry, and agriculture (Table 9.2). Its biggest consumers are the chlor-alkali industry (electrolysis), electrical and control instrument industry, laboratory products, dentistry, and agriculture. It is also widely used in paints, catalysis, pharmaceutical products, and in the paper and pulp industry (as slimicides).

The chlor-alkali industry, which manufactures chlorine and caustic soda, has been one of the biggest Hg polluters. Its use in agriculture also poses a threat to the quality of the food chain because of its use as a seed dressing in grains, potatoes, flower bulbs, sugar cane, cotton, and others and as a foliar spray against plant diseases. Thus, contamination of foodstuff and wildlife is often inevitable (Smart, 1968; Fimreite, 1970; Leong et al, 1973). As of 1968 (Smart, 1968), the biggest users of Hg (in tonnes of Hg compounds) in decreasing order were: Japan ($\approx$1,600), United States (400), Federal Republic of Germany (41), Italy (26), Turkey (22.5), and the United Kingdom (20).

3. Natural Occurrence of Mercury

The earth's crust contains approximately 50 ppb Hg, mainly as sulfide. Mercury occurs in all types of rocks (igneous, sedimentary, and metamorphic). The Hg content of most igneous rocks is generally <200 ppb

TABLE 9.3. Commonly observed mercury concentrations (ppb) in various environmental matrices.

Material	Average concentration	Range
Igneous rocks[a]	—	5–250
Limestone[a]	40	40–220
Sandstone[a]	55	<10–300
Shale[a]	—	5–3,250
Petroleum[a]	—	20–2,000
Coal[a]	—	10–8,530
Fly ash[b]		
Bituminous	100	—
Sub-bituminous	40	—
Lignite	100	—
Rock phosphate[a]	120	—
Peat[a]	—	60–300
Soils (normal)[a]	70	20–150
Soils near Hg deposits[a]	—	Up to 250 ppm
Soil horizons (normal)[a]		
A	161	60–200
B	89	30–140
C	96	25–150
Terrestrial plants (dry wt)[a]	—	30–700
Marine plants (fresh wt)[a]	—	0.01–37
Normal ground waters[a]	0.05	0.01–0.10
Rain water, snow[a]	0.2	0.01–0.48
Freshwater[a]	0.03	0.01–0.1
Sea water[a]	0.1	0.005–5.0

Sources: Extracted from
[a] NRCC (1979).
[b] Adriano et al (1980).

and probably averages <100 ppb (Fleischer, 1970). Most sedimentary rocks have Hg contents of <200 ppb, except for shales. Shales high in organic matter are particularly enriched in Hg. Sedimentary rocks, in general, tend to contain more Hg than igneous rocks (Table 9.3). Mercury minerals consist essentially of cinnabar, metacinnabar, or both (polymorphs of HgS), and one or more of native Hg, stibnite, native S, quartz, fluorite, and carbonates (NRCC, 1979). Mercury is present naturally in soils at concentrations ranging from a few ppb to a few hundred ppb. Soils may be considered normal if their Hg contents fall within the <100 ppb level. In the vicinity of gold, molybdenum, and base-metal deposits, soils may contain from 50 to 250 ppb Hg, and in some instances, as much as 2,000 ppb Hg (Warren et al, 1966). In the vicinity of Hg deposits, soil levels of Hg can be expected to be considerably higher, in the range of 1,000 to 10,000 ppb Hg, and in some cases at the 100,000-ppb level. Some environments, like that of Hawaii, are naturally high in Hg. Measurements of Hg contents of air, water, rainfall, soils, rocks, and biological materials from the islands of Hawaii, Maui, and Oahu indicate that the source of environmental Hg in Hawaii is geothermal (Siegel et al, 1973).

Most natural waters (ground water, river water, sea water) contain <2 ppb Hg (Gavis and Ferguson, 1972).

4. Mercury in Soils

4.1 Total Soil Mercury

Normal soils are expected to have Hg levels not exceeding 100 ppb. Although Warren and Delavault (1969) reported very high concentrations of Hg in certain UK soils, ranging as high as 15,000 ppb, they estimated that the levels of Hg in most UK agricultural soils would probably range between 10 and 60 ppb. Shacklette et al (1971), based on 912 surface soil samples collected in the continental United States, reported an arithmetic mean of 112 ppb (geometric mean = 71 ppb). The arithmetic mean for western states was 83 ppb (n = 492) and for eastern states, 147 ppb (n = 420). Andersson (1967) reported that Swedish soils had Hg contents ranging from 20 to 920 ppb, with an average of 70 ppb. For Canadian soils, McKeague and Wolynetz (1980) reported a mean of 59 ppb (n = 173), with values ranging from 5 to 100 ppb. Vinogradov (1959) gave a mean value of 10 ppb Hg for world soils. More recently, Berrow and Reaves (1984) reported that the typical concentration of Hg for world soils is about 60 ppb. Other values for soil Hg are indicated in Appendix Table 1.17.

In the North Ossetian Plains in the USSR, the background level of Hg in surface soils averaged 60 ppb (Zyrin et al, 1981). These investigators found that >50% of the total Hg content of soils was found in soil particles <0.005 mm in size. Some uncontaminated virgin soils in Kyushu, Japan, had an average Hg concentration of 40 ppb (range of 3–245 ppb) (Gotoh et al, 1979). Paddy soils however, showed surface enrichment of Hg up to 245 ppb due to past mercurial applications. Soil near the Nifu Hg mine in Japan had Hg concentrations as high as 100 ppm (Morishita et al, 1982). The area in the vicinity of this mine is reportedly one of the most heavily polluted areas in the world by a Hg mining activity.

For agricultural soils, Archer (1980) reported an average value of 40 ppb (n = 53) for the United Kingdom, with a range of 8 to 190 ppb. Except in rare instances, agricultural practices apparently do not substantially enrich the soils and edible portions of crops with Hg (Table 9.4). From a survey of farm soils and grain in 16 major wheat-growing states in the United States, Gowen et al (1976) concluded that the Hg levels in samples collected from sites where Hg compounds had been used were not significantly different from those where no Hg was applied (Table 9.4). Mercury compounds, used in amounts ranging from 0.005 to 0.01 kg/ha, were used for seed treatment. Similar results were obtained by Wiersma and Tai (1974) for surface agricultural soils collected from 29 eastern US states, where there were no statistical differences between the Hg levels in cropland ($\bar{x}$ = 80 ppb, n = 275) and noncropland ($\bar{x}$ = 70 ppb, n = 104) soils. For soils collected from north central states, the results of analysis for Hg residues indicate that for all practical purposes, Hg levels in soils were comparable for sites where Hg compounds were used and not used (Sand et al, 1971).

TABLE 9.4. Mercury levels (ppm dry wt) in soil and wheat grain of 16 states in the United States in 1969.[a]

Sample	Arithmetic mean	Geometric mean	95% confidence limits		Extremes	Number of samples
			Lower	Upper		
Soil						
Mercury compounds used	0.12	0.098	0.080	0.119	0.05–0.29	24
Mercury compounds not used	0.13	0.105	0.079	0.139	0.05–0.36	24
All soil samples	0.12	0.101	0.086	0.120	0.05–0.36	48
Wheat grain						
Mercury compounds used	0.27	0.247	0.204	0.300	0.07–0.59	24
Mercury compounds not used	0.31	0.266	0.212	0.332	0.11–1.06	25
All wheat samples	0.29	0.257	0.222	0.296	0.07–1.06	49

[a] Source: Gowen et al (1976).

In Manitoba (Canada), Mills and Zwarich (1975) found similar levels of Hg in A horizon soils of cultivated ($\bar{x}$ = 36 ppb) and noncultivated areas ($\bar{x}$ = 35 ppb). However, in certain agricultural areas in Ontario (Canada), soils from apple orchards where phenylmercuric acetate was used for foliar treatment, had Hg levels of 290 ± 290 ppb (Frank et al, 1976). This value is much higher than the overall mean value of 110 ± 180 ppb Hg (n = 296) for agricultural soils, with extreme values of 10 to 1,140 ppb. Total Hg contents of surface soils from experimental plots, which the Agronomy Department at the University of Illinois have maintained for about 63 years, are rather unusually high and variable, ranging from 100 to 3,920 ppb (Jones and Hinesly, 1972). They attributed the high variation among samples to changes in management strategies effecting drainage differences between plots and possible sample contamination.

4.2 Extractable (Available) Mercury in Soils

There are only scanty data on the extractable forms of Hg in soil and their relationships to plant uptake. Of the various extractants used ($CaCl_2$, NH_4OAc, EDTA, and various HCl solutions), Gracey and Stewart (1974a) suggested that the amount of Hg available for root uptake can be estimated from the amount of Hg in the 0.01 M $CaCl_2$ extracts of soil samples taken after a lapse of about five months of Hg application. The soils were treated with 5% mercuric fungicide in the field experiment and samples taken at 4, 34, 49, and 111 days after treatment. Results showed an increase in the amount of Hg detectable in the $CaCl_2$ extract between days 4 and 34, and a decrease by the next sampling date. The $CaCl_2$ extractant does not include organically-bound Hg. Using similar extractants, Hogg et al (1978a) found that extraction of the 0- to 10-cm soil layer by either $CaCl_2$, NH_4OAc, EDTA, or DTPA-TEA removed <1.0% of the applied Hg. However, larger amounts of Hg were removed by acid extraction (0.5 N HCl solution), which amounted to as much as 11% of total Hg applied. This indicates the strong binding of Hg to soil components. This phenomenon was apparently associated with the organic fraction of the soil, as indicated by the data of Lindberg et al (1979) for the surface soils collected near an Hg mining/refining operation in Almaden, Spain. Their conclusion was based on two observations: (1) the 0.5 M $NaHCO_3$-extractable Hg, which is the organic associated fraction, accounted for ≈200 times more Hg than the 1 N NH_4OAc-extractable Hg, which is the exchangeable fraction; and (2) soil density gradient fractionation experiments indicate ≈2 times more of the total Hg to be associated with the light-density organo-clay fraction in the Almaden soil than in the control soil.

4.3 Mercury in the Soil Profile

Two generalizations can be made regarding profile distribution of Hg: (1) below the surface layer, Hg is fairly mobile in the soil profile (Aomine et al, 1967; Hogg et al, 1978b); and (2) Hg tends to accumulate in the surface horizons (Whitby et al, 1978; Poelstra et al, 1974; Andersson, 1967; Jonasson and Boyle, 1971). The Hg content of a soil horizon has been reported to be related to the organic matter and/or clay content of that horizon. Goldschmidt (1954) attributed the higher levels of Hg in A horizons to accumulation of decayed plant material. Warren et al (1966) have reported that soil samples taken from horizons with either a high organic matter or clay content may contain substantially higher amounts of Hg than the average for the whole profile. Although Hg levels are expected to accumulate primarily in surface horizons, there are exceptions. Mills and Zwarich (1975) reported that contents of Hg in A and C horizons from soils in Manitoba were not systematically different. However, 10 of the 16 soils reported in their study contained noticeably lower levels of Hg in A horizons in comparison to amounts in the respective C horizons. On the other hand, in a study of Hg contents of some Saskatchewan soils, Gracey and Stewart (1974b) found no significant differences in levels in A and C horizons. Gotoh et al (1978) found that in northern Kyushu, Japan, the Hg in the soil profiles was generally higher than in the underlying parent material (igneous rock, $\bar{x} \leqslant 10$ ppb Hg). They observed a generally steep concentration gradient of Hg down the profiles, especially in paddy soils, and found that surface soils had a mean Hg level of 197 ppb (range of 64–459 ppb).

In chernozemic soils from Alberta, Dudas and Pawluk (1976) found that contents of Hg in surface horizons were considerably lower than the contents in the respective C horizons. For eluviated soils, highest contents of Hg were found in B horizons, with lowest levels in A horizons. They gave two possibilities that may have caused decreased content of Hg in the surface horizons: (1) Hg may have been translocated from the surface horizons to C horizons due to leaching by precipitation, and (2) portions of Hg in surface horizons may have been mobilized and lost to the atmosphere. In a very extensive soil survey, virgin soils sampled from all provinces of Canada and from the Northwest Territories and representing 234 horizons from 65 soils, McKeague and Kloosterman (1974) observed that in more than half of the samples, particularly podzolic and gleysolic soils, the highest Hg concentrations were in the horizons of organic matter accumulation. However, for a few profiles, the highest Hg levels occurred in the C horizon, and in some others, the Hg concentrations were similar in all horizons.

The behavior of Hg in the soil profile is typified in Figure 9.2 in diverse European soils and demonstrates that numerous factors influence Hg levels and mobility in the profile. Soils 1, 2, 4, 5, 6, and 7 represent the "reference soils" where no Hg has been introduced and no disturbances in the profile have occurred during the last 20 years. The Hg values in the profiles of these soils seem to represent the natural background level. Soil 11 represents soils from the bulb-growing area in the Netherlands, where application of Hg as a fungicide has been a common practice for about 50

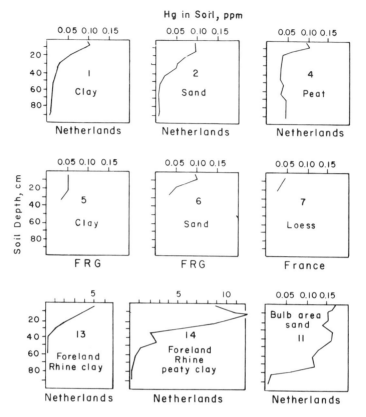

FIGURE 9.2. Mercury concentrations (in ppm) in some western European soil profiles. *Source:* Adapted from Poelstra et al, 1974, with permission of the authors.

years, with its use being intensified about 20 years ago. Soils 13 and 14 are foreland soils from River Rhine that are frequently flooded, usually during the winter. The Rhine is heavily polluted and discharges some 70 tonnes of Hg annually into the North Sea.

Andersson (1979) indicated that different distribution patterns of Hg can be expected for different soil types and horizons (Fig. 9.3). Part A of Figure 9.3 shows the pattern of Hg and organic matter in a podzolic soil developed from glacial till under a beech forest. This soil type is usually acidic (pH 3.9–4.7), thereby rendering the organic matter more effective in retaining Hg; hence, a similar pattern for Hg and organic matter. Part B of the figure illustrates the distribution of Hg, organic matter, and Fe in a brown earth soil developed on a heavy illitic clay, calcareous below the 35-cm depth. The pH in the profile ranges from 6.2 to 8.2; thus the iron oxides and clay minerals can be expected to effectively retain Hg. In this case, the distribution for Hg is more like the one for Fe than for organic matter.

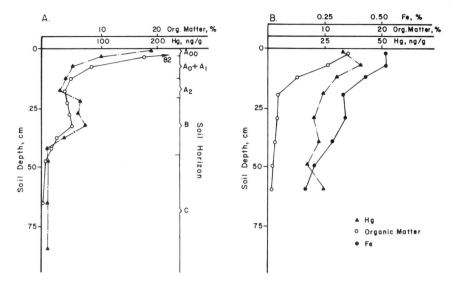

FIGURE 9.3. Distribution of mercury, organic matter, and oxalate-soluble iron in a podzol (A) and a brown earth (B) soil. *Source:* Andersson, 1979, with permission of the author and Elsevier, Amsterdam.

4.4 Sorption of Mercury in Soils

When an Hg compound is added to the soil, the adsorption process is probably dominant initially in determining the fate of this element. The remaining Hg in soil solution eventually will be volatilized, precipitated, leached, or taken up by the plants. Adsorption of Hg is apparently dependent on several factors, among them the chemical form of the Hg applied, the amount and chemical nature of inorganic and organic soil colloids, the soil pH, the type of cations on the exchange complex, and the redox potential (Hogg et al, 1978a; Aomine and Inoue, 1967; Inoue and Aomine, 1969). Sorption of Hg(II) has been demonstrated to occur on alumino-silicate clays (Farrah and Pickering, 1978; Reimers and Krenkel, 1974), oxides of iron and manganese (Kinniburg and Jackson, 1978; Lockwood and Chen, 1973), and analogs of organic matter (Reimers and Krenkel, 1974).

The retention of Hg in soil is due not only to valence-type ionic adsorption by organic and inorganic materials and the formation of covalent bonds with organic compounds, but also to low solubilities of Hg as carbonate, phosphate, and especially sulfide (Lagerwerff, 1972). Thus, the Hg contents of soil leachates are usually low. The sorption of Hg by soils follows both the Langmuir (Hogg et al, 1978a) and the Freundlich-type adsorption isotherms (Fang, 1978). Adsorption of Hg by synthetic $Fe(OH)_3$ also follows the Freundlich isotherm (Lockwood and Chen, 1974).

Hogg et al (1978a) found that sorption of Hg by soils depends not only on the soil properties but also on the chemical form of Hg. They observed that the highest adsorption maxima for all Hg compounds tested were found for soils that had the most organic matter and clay contents. Adsorption maxima increased in the order, methylmercuric chloride (MMC) < phenylmercuric acetate (PMA) < mercuric chloride ($HgCl_2$). Fang (1978) observed that, in addition to organic matter, clay mineralogy also plays an important role in the sorption of Hg by soil. Among the clay minerals, he found that illite had the highest sorption capacity and kaolinite the lowest.

4.5 Transformations of Mercury in Soils

Mercury in nature and that from industrial discharges is largely in inorganic form, although organic (aryl) mercurials are used in agriculture. In general, Hg is very unstable in the environment because it is subject to chemical, biological, and photochemical reactions (Kaiser and Tölg, 1980). Many mercurial compounds, both organic and inorganic, decompose to yield elemental Hg, which may volatilize, or be converted to HgS, or complexed with inorganic ligands. The general scheme for the environmental/biological interconversion of the forms of Hg is depicted in Figure 9.4.

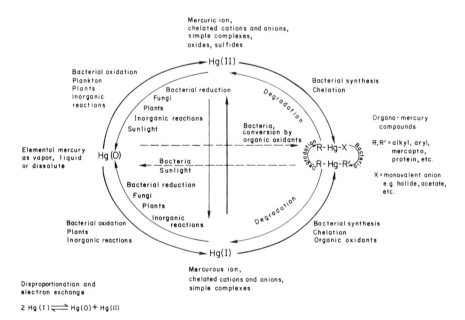

FIGURE 9.4. Environmental/ biological interconversion of Hg(0), Hg(I), and Hg(II). *Source:* Carty and Malone, 1979, with permission of the authors and Elsevier, Amsterdam.

Wood (1974) has proposed a scheme whereby various forms of Hg entering the aquatic environment are transformed into mercuric ion and then methylated, resulting in the formation of toxic and volatile monomethyl or dimethylmercury (Fig. 9.5). These transformations depend largely on biological processes. Consequently, a generalized scheme of the reactions involved in the formation of methylmercury compounds has been proposed (NRCC, 1979): (1) the precursor Hg^{2+} can be methylated by both aerobic and anaerobic bacteria, although methylation is usually enhanced under anaerobic (but not completely anaerobic) conditions; (2) $(CH_3)Hg^+$ can be formed from Hg^{2+} by methane-producing bacteria; (3) in sediments and soils, like paddy soils, reducing conditions produce sulfide ions, forming HgS, which is quite insoluble and hence resistant to methylation; (4) under aerobic conditions, HgS may be oxidized to the sulfate form, which can then undergo methylation; (5) dimethylmercury, $(CH_3)_2Hg$, may be transformed into monomethylmercury at low pH and in the presence of Hg^{2+}, or in the presence of ultraviolet light; (6) monomethylmercury may attach to -SH groups of organic matter and may then decompose photolytically to Hg(0) and CH_4; (7) analogous reactions may occur with other alkyl, alkoxyalkyl, and aryl mercury compounds; (8) methylmercury salts may be volatilized into the air by the action of H_2S; and (9) bacterial action can also cause demethylation of methylmercury compounds. These reactions are discussed in more detail by Kaiser and Tölg (1980), by Löfroth (1970), and by Summers and Silver (1978).

In the aquatic environment, Jensen and Jernelöv (1967) found that microorganisms in lake sediments could methylate inorganic Hg, while Wood

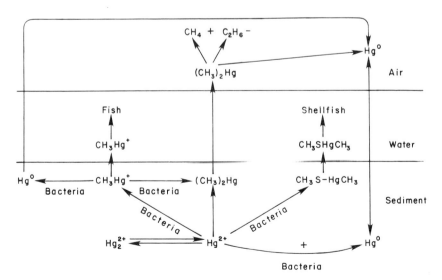

FIGURE 9.5. Pathways of mercury breakdown and methylation in sediments and its potential bioaccumulation. *Source:* Wood, 1974, with permission of the author.

et al (1968) found that cell-free extracts from methanogenic bacteria could also methylate Hg. Thus, methylmercury could be formed by both enzymatic and non-enzymatic reactions, which are discussed in more detail by Bisogni (1979). The processes by which Hg can be methylated chemically or biologically are discussed by Beijer and Jernelöv (1979).

Beckert et al (1974) discovered that methylation of inorganic Hg occurs also in terrestrial environments and found that methylmercury was present even in desert soils that had been amended with mercuric nitrate. This form of Hg has also been detected in the atmosphere above a soil amended with mercuric chloride (Braman and Johnson, 1974). Lexmond et al (1976) indicated that methylmercury may be formed from mercuric Hg by both anaerobic and aerobic microorganisms and that dimethylmercury may be formed by further methylation of methylmercury or by methylation of Hg^0. Rogers subsequently found that inorganic Hg was methylated under a variety of conditions in alkaline agricultural soils and that it is possible that the methylation process could, in part, be abiotic in nature (Rogers, 1977). Rogers (1976) found that methylation was directly proportional to the clay content, moisture content, temperature, and Hg concentration. However, van Faasen (1975) found only little methylation of inorganic Hg in soils. Others (Imura et al, 1971; Bertilsson and Neujahr, 1971; DeSimone, 1972) have also shown that mercuric ions can be abiotically methylated by a variety of substances. It was later suggested by Rogers (1977) that the methylation reaction in soil is abiological and that the still-unclear methylating substances in the soil are possible to extract. It appears from his studies that the site and nature of methylmercury synthesis in soil is largely unknown.

In terrestrial environments, Hg can be mobilized by volatilization, leaching, and/or plant uptake. It has been shown that volatilization is a result of chemical reaction (Booer, 1944; Frear and Dills, 1967; Toribara et al, 1970) and/or microbial activity (Rogers and McFarlane, 1979; Corner and Rigler, 1957; Landa, 1978; Magos et al, 1964). Both appear to be operative in natural systems. Toribara et al (1970) proposed that the mechanism of Hg loss follows the chemical reduction of Hg(II) to Hg(I), which disproportionates spontaneously to Hg(II) and elemental Hg.

Volatilization of Hg from turfgrass and soil has been considered as the dominant pathway in the loss of Hg, more important than leaching or plant uptake. It accounted for as much as 56% loss of the total Hg added over a period of 57 days (Gilmour and Miller, 1973) and, in another case, for as much as 45% of total Hg added to a soil in one week (Rogers and McFarlane, 1979). They, along with Landa (1978), Rogers (1976), and Johnson and Braman (1974) indicated that Hg(0) is likely to be the predominant Hg species to be evolved.

4.6 Complexation and Speciation of Mercury in Soils

In natural systems, heavy metals such as Zn, Hg, Pb, Cd, etc. undergo complexation with organic and inorganic ligands. In environments where organic matter is present, organo-metallic complexes may occur in soluble or colloidal form, which differ in mobility. There is evidence that Hg could be strongly chelated by soil organic matter (Andersson and Wiklander, 1965). Humic substances containing S (e.g., in cysteine) are believed to keep Hg in soluble form. Added Hg(II) in soils may be immobilized by its incorporation with the sulfhydryl groups of the organic matter, or depending on the redox potential of the soil, may precipitate as HgS (Landa, 1978). Gambrell et al (1980) indicated that most dissolved Hg in sediments was probably bound as inorganic, non-ionic, or negatively charged complexes, or complexed with soluble organics. Lindberg and Harriss (1974) suggested that relatively soluble polysulfide complexes may form under some conditions, but that under reducing conditions, very insoluble mercuric sulfide (cinnabar) may form.

Gilmour (1971) discussed the formation of inorganic complexes of Hg(II), with special reference to chloride and hydroxyl complexes and showed how stability constants could be used to predict the distribution of various complexes in aqueous systems. Furthermore, Hahne and Kroontje (1973a) have calculated the theoretical complexes of Hg(II) with hydroxyl and chloride ions; Hg(II) was calculated to hydrolyze in the pH range of 2 to 6, with the final species being the soluble $Hg(OH)_2$ at about pH 6 (Fig. 9.6). The precipitation of $Hg(OH)_2$ would occur only if the concentration of Hg(II), as $Hg(OH)_2$, exceeds 107 ppm, the intrinsic solubility of $Hg(OH)_2$ being 5.37×10^{-4} M. The monohydroxy species $HgOH^+$ predominates between pH 2.2 and 3.8. Thus, at pH 6 to 7, the distribution of Hg(II) would be dependent on the solubility of $Hg(OH)_2$.

Chlorides, like H^+ and OH^- ions, occur in all natural soil and water systems and may be regarded as one of the most mobile and persistent complexing agents for heavy metals (Hahne and Kroontje, 1973a). High Cl^- concentrations in soil solution can be expected in saline and saline-sodic soils. Chlorides complex with Hg(II) at Cl^- concentrations above 10^{-9} M (35×10^{-6} ppm); $HgCl_2$ forms above $10^{-7.5}$ M Cl^- (1.1×10^{-3} ppm), peaking at about 10^{-4} M (≈ 3.5 ppm Cl^-) and $HgCl_3^-$ and $HgCl_4^{2-}$ form above 10^{-2} M Cl^- (350 ppm) (Fig. 9.6); $HgCl^+$ peaks at Cl^- concentration of 10^{-7} M (≈ 0.0035 ppm). At Cl^- concentrations above 3,550 ppm, $HgCl_4^{2-}$ predominates.

Hahne and Kroontje (1973b) further predicted that at pH 4 and 5, Cl^- concentrations of 3.5 and 14 ppm, respectively, are sufficient for all Hg(II) to be in the $HgCl_2$ form. At higher pH values, partial to complete mobilization is possible depending on the prevailing Cl^- and Hg(II) concentrations. For example, the range of Cl^- concentrations where partial to complete mobilization of Hg(II) in the form of Cl complexes will occur at pH 6, 7, 8, and 9 are: pCl 5 to 2 (0.35–354 ppm), pCl 4 to 1 (3.54–3,540 ppm), pCl 2.8 to 0.5 (56–11,213 ppm), and pCl 2 to 0.1 (354–28,166 ppm), respectively.

The effects of Hg(II) complexation with Cl^- on Hg mobility are reflected in several adsorption studies. Andersson (1970) showed a drastic reduction in Hg(II) adsorption by inorganic colloids at Cl^- concentrations of 10^{-3}

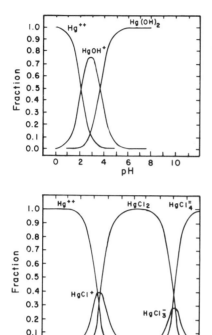

FIGURE 9.6. Predicted speciation of mercury as influenced by pH (above) and chloride ions (below). *Source:* Hahne and Kroontje, 1973a, with permission of the authors and Am Soc of Agronomy, Crop Sci Soc of America, Soil Sci Soc of America.

M or higher. Similarly, Aomine and Inoue (1967) found that $HgCl_2$ salts were adsorbed only slightly by soils and clays. Newton et al (1976) found that Cl^- sharply reduced Hg(II) sorption by bentonite clay, especially at low pH. At pH 6 or lower, increasing the $CaCl_2$ concentration from 10^{-5} to 10^{-4} M depressed adsorption; higher $CaCl_2$ levels were required to decrease adsorption at neutral pH. Others found similar inhibitory effects of Cl^- on Hg(II) adsorption by hydrous iron oxide gel (Kinniburgh and Jackson, 1978) and by precipitated iron(III) oxide (Lockwood and Chen, 1974).

5. Mercury in Plants

5.1 Uptake and Translocation of Mercury in Plants

Mercury and its compounds are absorbed by the roots and translocated, although to a limited extent, to other plant parts. Uptake has been demonstrated for both economic crops and trees. However, Smart (1968) has summarized evidence to show that Hg compounds, applied to some parts of plants and trees, can be readily translocated to other parts. Movement within foliage, stems, fruits, and tubers is greater than that from the root upwards. In general, the availability of soil Hg to plants is low, and there is a general tendency for Hg to accumulate in the roots—i.e., the roots serve as a barrier to Hg uptake (Hogg et al, 1978b; Gracey and Stewart, 1974a; Beauford et al, 1977; Fang, 1978). For example, Gracey and Stewart (1974a) found very high concentrations of Hg in alfalfa roots (up to 133 ppm in dry matter) from several forms of Hg compared to <1 ppm (as low as 0.07 ppm) in alfalfa foliage.

Using solution culture, Beauford et al (1977) found that uptake of inorganic Hg (as $HgCl_2$) by higher plants, such as peas, was a function of external concentration. Over the concentration range (up to 10 ppm Hg) tested, they observed the accumulation of Hg in the roots to be linear on a log-log scale. The translocation of the element into the shoots appeared to be two-phased—i.e., up to 0.01 ppm of external Hg, a low threshhold level of the element seemed to be maintained, and between 0.1 ppm and 10 ppm, the transport of Hg to the shoots was, like the roots, showing a linear trend on a log-log scale. In addition, they found that the proportion of Hg retained in the roots, relative to the shoots, remained fairly constant (about 95%). Lindberg et al (1979) found a dual mechanism of Hg uptake by alfalfa; roots accumulated Hg in proportion to the soil levels, while aerial plant material absorbed Hg vapor directly from the atmosphere. Volatilization of Hg from soils contributed to foliar uptake of Hg, which may occur in plants inhabiting environments near smelters.

The form of Hg could influence the absorption of Hg and its subsequent transport in plants. Huckabee and Blaylock (1973) reported that [203]Hg from methylmercury accumulated more in parts of snap beans than [203]Hg from the inorganic form. In the bean, the actual Hg concentration was 890 ppb from the organic versus only 80 ppb from the inorganic form. Furthermore, they reported that inorganic Hg is not readily translocated in cedar trees. However, Gracey and Stewart (1974a) did not find any effect of the form of Hg on its uptake by alfalfa.

Ross and Stewart (1960, 1962) have shown that Hg compounds, such as phenylmercury acetate, can be translocated and redistributed in plants when applied as fungicides to the foliage of apple trees. Stewart and Ross (1960) further found that, following single cover sprays of phenylmercury acetate, applied from June to July, Hg residues in apple fruits declined until early in August and then increased until at harvest the residues were greater than the initial deposits. Thus, when apple trees are sprayed, some of the mercury fungicide is absorbed by the leaves, as shown in Figure 9.7, and may appear later in growing fruit or to new foliage. The fruit of the apple seems to accumulate more of the Hg than other parts of the plants. Similarly, when the foliage of potatoes is sprayed, the Hg can move downward and appear in the tuber (Smart, 1964).

5.2 Mercury in Economic Crops and Other Plant Species

Smart (1968) has suggested that Hg in vegetable crops can be expected to be at the parts per billion (ppb) level. In general, total Hg concentrations in common agronomic plants and products derived from these plants range

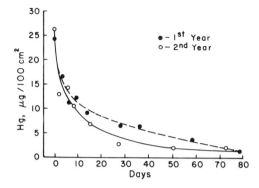

FIGURE 9.7. Decline of mercury residues (as phenylmercury acetate) on apple leaves caused mainly by translocation to other plant parts, such as fruit or new foliage. *Source:* Ross and Stewart, 1960, with permission of the authors.

from <1 to 300 ppb, with the higher levels caused by natural Hg deposits in soil (NAS, 1978). The actual level depends on the location, plant species, and other factors. The results (Table 9.5) of Gracey and Stewart (1974b) for crops grown in Saskatchewan field plots with normal soils containing <40 ppb Hg, indicate that plant tissues contain Hg usually in the <50 ppb level. Higher Hg levels were recorded for the straw than for the grain or seed of the cereal and oil seed crops. Mercury was not detected in the seed of flax and rape. The levels in the straw were approximately equal to those in the soil, with the exception of barley, in which levels in straw were higher than the level found in soil. Later analysis by Gowen et al (1976) for wheat grain sampled at harvest in 16 major wheat-producing

TABLE 9.5. Mercury contents of some crops grown in soil in field plots containing less than 40 ppb of mercury.[a]

Crop	Number of varieties analyzed	Crop part	Hg content, ppb (dry wt basis) Range	Hg content, ppb (dry wt basis) Mean
Alfalfa	10	Foliage	15–57	39
Barley	7	Straw	67–89	80
		Grain	5–17	12
Wheat	6	Straw	27–47	36
		Grain	7–15	11
Oats	6	Straw	27–42	33
		Grain	4–19	9
Flax	4	Straw	9–33	19
		Seed	ND[b]	ND[b]
Rape	1	Straw	24	24
		Seed	ND[b]	ND[b]
Rutabagas	1	Tops	51	51
		Tubers	40	40

[a] *Source:* Gracey and Stewart, 1974b, with permission of the authors and Can Soc Soil Science.
[b] No detectable mercury.

states, however, indicate much higher levels of Hg ($\bar{x}$ = 290 ppb, with the range of 70–1,060 ppb; n = 49). These rather high concentrations in the wheat grain are not surprising in view of the findings of Saha et al (1970), which indicate that Hg compounds can be absorbed by wheat plants and translocated into the grain, producing levels of Hg similar to those found by Gowen et al (1976). In addition, Selikoff (1971) reported that grain seeds treated with 20,000 ppb methylmercury produced a crop containing 100 ppb in the grain and 70 to 80 ppb in the leaves.

Although Hg compounds have been demonstrated to be taken up by the roots of vegetable crops (e.g., broccoli, bean, lettuce, and carrot) from nutrient solution or soil, and it has been shown that there is little translocation to the aerial parts (Pickard and Martin, 1959), there are exceptions. Gravelly loam soil treated with up to 10 ppm Hg, as methylmercury dicyandiamide, caused very high concentrations in plant stems and leaves of potatoes (1,045 ppb) and tomatoes (341 ppb) (Bache et al, 1973). Onion bulbs absorbed up to about 1,087 ppb Hg from this soil. Other edible parts usually contained <100 ppb. Similarly, Byrne and Kosta (1969) showed elevated levels of Hg in carrot (up to 603 ppb) grown in soils contaminated by Hg mining and processing in Yugoslavia.

5.3 Sensitivity to and Phytotoxicity of Mercury

Since Hg compounds are used in agriculture and other industry for the control of plant diseases (e.g., the use of mercuric oxide, mercuric chloride, or mercurous chloride for the control of common scab of potatoes, club root in species in the family *Brassica*, white rot of onions, and brown patch, dollar spot, and snow mold in turf and others) there is a concern not only about the transfer of Hg to the food chain but also about the direct effect of Hg on plant growth.

In his review, Booer (1951) reported that metallic Hg and mercury compounds in soil can retard the growth of plants. In his earlier work, Booer conducted emergence tests on numerous species to determine their relative susceptibility to Hg. He found that the growth of monocotyledons was only slightly retarded. Oats and barley behaved similarly to wheat, and lawn grasses were only slightly retarded. The percentage emergence of onions was unaffected by amounts of Hg of up to 100 ppm Hg in soil. In the pre-emergence stage, most of the dicotyledons appear to be more sensitive, but more variable in their reactions, to Hg. He stated that severe pre-emergence losses may be caused by as little as 50 ppm Hg in the soil, particularly to sensitive species, like lettuce and carrot. However, Gray and Fuller (1942) observed that the percentage germination of numerous vegetable seeds was not severely affected by the presence of Hg vapor, with subsequent growth only slightly retarded.

In nutrient culture containing 0.06 to 0.28 ppm phenylmercuric acetate (PMA), roots of dwarf bean seedlings accumulated about 10 ppm Hg and in terminal tissues, from 0.09 to 0.90 ppm in 21 days (Pickard and Martin,

1959). The seedling biomass was reduced 50% by 0.11 ppm PMA in the solution. Beauford et al (1977) observed that 5 ppm of Hg, as $HgCl_2$, in solution culture inhibited growth of higher plants (*Pisum sativum* and *Mentha spicata*) and affected both physiological and biochemical processes in the plants. Puerner and Siegel (1972), using petri dish (agar gel) culture with Hg concentrations of up to 1,000 ppm, found that $HgCl_2$ alone, in admixture with fluorescein, or chemically combined as mercurochrome, inhibited the growth of cucumber seedlings and induced disorientation of root and shoot. About 400 ppm of Hg (as $HgCl_2$), which was found to be more toxic than mercurochrome, completely inhibited cucumber growth. Siegel et al (1984) observed that seed germination and early growth of several crop species and cultivars were not significantly affected by their exposure to Hg vapor at air saturation levels (14 ppb). They suggested that the general toxic effects of Hg vapor on growing plants is premature senescense, which may not be associated with leaf necrosis or chlorosis. Young plants grown on nutrient culture spiked with $HgCl_2$ were sensitive to Hg, exhibiting toxicity symptoms at as low as 10 ppb Hg in the nutrient culture (Mhatre and Chaphekar, 1984). Toxic effects included reduced yield and reduced chlorophyll contents.

In sand culture experiments, Davis et al (1978) found that the critical level of Hg in dry matter tissue of barley was about 3 ppm. The critical concentration of Hg in the solution was 4 ppm Hg(II) as $HgCl_2$. Visual symptoms of Hg toxicity in barley include yellowing of leaves and the presence of reddish stems. Chino (1981) reported that, for rice, the critical level of Hg was 0.5 ppm in stem and leaf and 1,000 ppm in roots. In soils in the area near the Nifu mine in Japan, roots of rice plants contained 0.1 to 7.23 ppm Hg (dry wt basis) (Morishita et al, 1982). The leaves and stems of these plants contained 0.031 to 0.113 ppm, while the rice grain contain 0.010 to 0.060 ppm. The relationships between Hg in highly polluted soils and Hg in the roots of rice plants growing on them are depicted in Figure 9.8.

Other species are more tolerant to Hg. For example, Shacklette (1970) reported that as much as 3.5 ppm Hg in dry matter of labrador tea was not injurious to the plants, whose roots were above cinnabar mineral deposits. Velvet bentgrass showed no effect when grown on soils with Hg levels of 450 ppm, which accumulated from organo-mercurial fungicides applied annually over a period of 15 years (Estes et al, 1973). These turf plants had an average Hg concentration of 1.68 ppm.

Growth of common bermudagrass was reduced by 50 ppm of Hg in soil, applied as $HgCl_2$ (Weaver et al, 1984). Apparently, excess levels of Hg in plants inactivate both biochemical and physiological processes. Significant portions of Hg in plants are associated with the cell (nuclear) wall (Beauford et al, 1977; Rao et al, 1966). Beauford et al (1977) have shown that Hg inhibits the synthesis of proteins in leaves of plants. They also suggested that, as a heavy metal, Hg could affect the water-absorbing and

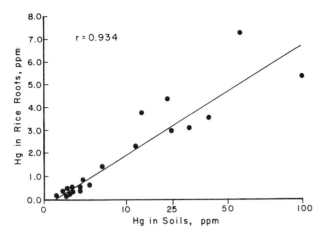

FIGURE 9.8. Relationships between mercury levels in soils and in rice roots (dry weight) where the value of soil mercury was converted to square root. *Source:* Morishita et al, 1982, with permission of the authors.

transporting mechanisms in plants. Bolli (1947) found that the photosynthetic activity in several plant species was reduced by treatment of $HgCl_2$. Similarly, Harriss et al (1970) noted that Hg compounds, such as organomercurial fungicides, reduced photosynthesis in plankton. Because Hg has a strong affinity for sulfhydryl or thiol groups involved in enzyme reactions (Passow et al, 1961), this element presumably can affect metabolic activity by producing structural changes in enzymes following mercaptide formation with -SH groups (Kuramitsu, 1968).

6. Factors Affecting the Mobility and Availability of Mercury

6.1 pH

Undoubtedly, pH can affect the speciation and stability of Hg in the soil environment, as discussed in section 4.6 and shown in Figure 9.6. In addition, the nature of the species that will predominate in a solution depends upon the ambient pH and redox potential and upon the nature of the anions and other groups, such as ligands, present. These can be easily described by means of an Eh-pH diagram for both undissolved and soluble species, as demonstrated by Gavis and Ferguson (1972) and by Hem (1970).

   Gilmour (1971) predicted that as the pH or Cl^- concentration increased, $Hg(OH)_2$ or $HgCl_2$ becomes dominant, while HgClOH serves as a transitional species. He calculated that divalent Hg in four Wisconsin rivers was present as $Hg(OH)_2$ (63%–92%), HgClOH (8%–35%), and $HgCl_2$

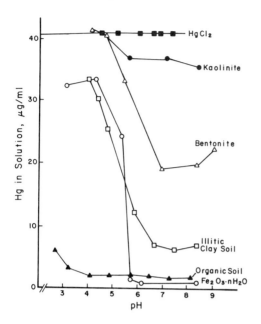

FIGURE 9.9. Retention of mercury by soils and soil components as influenced by pH. *Source:* Andersson, 1979, with permission of the author and Elsevier, Amsterdam.

(<1%–2%). The stability of Hg in soil can also be influenced by pH. From a series of limed loamy soils treated with HgCl₂, as soil pH increased from 5.3 to 6.4, Frear and Dills (1967) showed increased volatile losses, or reduction of mercury salts to metallic Hg. Similarly, Landa (1978) found increasing losses of Hg with increasing pH of Montana soils (pH 6.6–8.3) treated with Hg(NO₃)₂. In contrast, loss of Hg(II) via chemical reduction from dilute solutions of 0.01 M CaCl₂ and Ca(NO₃)₂ was reduced when pH was increased in the pH range of 4 to 8 (Newton and Ellis, 1974). Alberts et al (1974) found that the loss of Hg, via chemical reduction of mercuric ion from Hg(NO₃)₂-humic acid suspension, was reduced by about 9% over a 290-hour period when the pH was increased from 6.5 to 8.2.

Rogers (1977) found that the ability of 0.5 N NaOH extracts of soils to methylate Hg is also influenced by solution pH. In all extracts, there was a decrease in the occurrence of methylmercury when the solution pH was above 4.5, particularly in the clayey and loamy soil extracts as compared to sandy soil extracts. The lowest amount of methylmercury was detected at pH 9. In general, a high pH (neutral to alkaline) tends to favor the microorganisms that produce dimethyl Hg, while an acid medium tends to favor monomethyl Hg formation, or conversion of dimethyl to monomethyl Hg (D'Itri, 1972). The former form of Hg is more volatile than the latter.

The pH can also affect the sorption of Hg by clay minerals and soil materials. Farrah and Pickering (1978) found that, in the absence of ligands, sorption of Hg by illite and kaolinite changed little with pH; with montmorillonite, however, Hg sorption decreased with increasing pH. The influence of pH on the sorption of Hg from solutions by soils and soil constituents is shown in Figure 9.9. The samples were adjusted to the desired pH by adding $Ca(OH)_2$. Then 10^{-5} M $HgCl_2$ was added and suspensions equilibrated for 2 hours. Andersson's data (Fig. 9.9) indicate that the effective sorbent for Hg in acid soils (pH <4) is the organic matter, whereas at higher pH (pH >5.5), iron oxides and clay minerals may become the more effective sorbents.

6.2 Carrier of Mercury

Undoubtedly, the carrier of Hg can also influence the stability and mobility of this element in soils. Rogers (1979) found that over a 144-hour period, Hg compounds most soluble in water [i.e., $Hg(NO_3)_2$, $HgCl_2$, and $Hg(C_2H_3O_2)_2$] had a greater loss of Hg from three soils tested (12%–38%), followed by the less soluble HgO (6%–19%), with the insoluble HgS losing only a minor amount of applied Hg (0.2%–0.3%). These results indicate that the form of Hg added to the soil has an effect on the initial volatilization of applied Hg. Kimura and Miller (1964) reported differences in the tendencies of degradation of organo-mercurials in soil and their modes of loss from soil. The nonvolatile phenylmercury acetate was degraded to Hg(0) and was lost as Hg vapor. The volatile ethylmercury acetate was also degraded in soil to Hg(0), but its loss from the soil occurred both as organo-mercury vapor and as Hg vapor. Methylmercury dicyandiamide and chloride gave no significant Hg vapor, and their loss from soil was entirely due to the volatilization of the organo-mercury compound. They concluded that the chloride form was about twice as volatile as the dicyandiamide form. Apparently, the more volatile the compound the more phytotoxic Hg is as indicated by Byford (1975).

Hogg et al (1978b) found that the movement of Hg in soil columns was generally greater for the methylmercury chloride (MMC) than either the $HgCl_2$ or phenylmercuric acetate (PMA). In addition, they found that Hg levels in bromegrass foliage were significantly higher from the MMC-treated soil than from either the PMA- or $HgCl_2$-treated soils. This is expected since the adsorption maxima for soils found by Hogg et al (1978a) increased in the order MMC < PMA < $HgCl_2$. Similarly, Dolar et al (1971) found that watermilfoil rooted in sediment had much greater uptake of Hg from organic- than from inorganic Hg carriers. They hypothesized that, due to relatively high sulfide-S content of the sediment, inorganic Hg was precipitated largely as unavailable HgS. However, there are instances when plant uptake of Hg may not be influenced by the carrier (Gracey and Stewart, 1974a).

6.3 Organic Matter and Soil Texture

Several investigators have demonstrated the importance of organic matter in the retention of Hg against leaching, volatilization, or plant uptake. MacLean (1974) noted that, in an incubation experiment, where PMA was added to various soils, the volatilization of Hg varied from 59% of that applied in a sandy soil to 3% of that applied in a peat. Landa (1978) found the same preventive effect of organic matter from volatile loss of Hg from soils. This finding was further corroborated by Hogg et al (1978a) and Fang (1978), who attributed this effect to the higher adsorption maxima of Hg by soils to both organic matter and clay contents. Thus, in the soil profile, the highest Hg accumulations are usually in horizons characterized by high organic matter contents (section 4.3). Gotoh et al (1978) found that total Hg and organic matter contents of forest soils were very highly correlated (Fig. 9.10). However, Gracey and Stewart (1974b) found that Hg in soil horizons was significantly correlated with clay content only and not with organic C content. The poor correlation of Hg with organic C content might be due to complexation of Hg by organic matter, which then facilitated its mobility (Niebla et al, 1976).

Soil type or clay content also influences the transformation and mobility of Hg in soils. In general, volatile loss of Hg can be expected to occur higher in coarse-textured than in fine-textured soils (Rogers, 1979; Rogers and McFarlane, 1979). For example, during the first week, soils amended with 1 ppm Hg, as $Hg(NO_3)_2$, 20% of the applied Hg was lost from a silty clay loam soil compared with 45% loss from a loamy sand soil. In addition, Dudas and Pawluk (1976) observed that, for any given soil sample, the highest quantity of Hg was associated with the clay fraction compared with other particle size separates. However, the texture effect might be in conjunction with organic matter effect, since clay and organic matter content of soil can increase or decrease in the same order (MacLean et

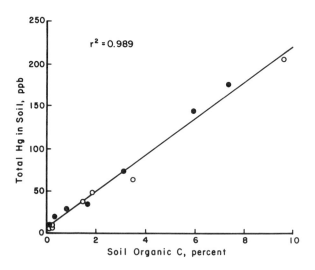

FIGURE 9.10. Relationships between total mercury and organic matter content in two forest soil profiles (solid and open circles) developed on granite. *Source:* Gotoh et al, 1978, with permission of the authors.

al, 1973; Rogers, 1976). Because of the usual parallel occurrence of clay and organic matter, Rogers (1976) found methylmercury in a decreasing order: clay > loam > sand. He proposed that increased methylation was caused by either the increased organic matter with increasing clay content, or increased microbial population with increasing clay and organic matter contents, or both.

6.4 Redox Potential

Soil moisture content also influences the stability of Hg in soil. In sediments, a common hypothesis has been that reducing conditions are necessary for the maximum formation of methylmercury. However, other evidence is available indicating that methylmercury formation occurs at a higher rate in oxic aqueous systems than in reducing ones. Losses of methylmercury could be greater under aerobic than anaerobic conditions, as indicated by the findings of Rogers (1976) for soils and Olson and Cooper (1976) for ocean sediments. Their results seemed contradictory to the trend of increased methylmercury loss with increasing soil moisture. However, a greater rate of methylmercury production than the rate of loss under reducing conditions cannot be ruled out. Soil moisture can also influence the ability of soil to sorb Hg vapor. In general, sorption of Hg vapor increases as the initial soil moisture content increases (Kimura and Miller, 1964; Fang, 1981). Fang (1981) found that, in all soil samples tested, the sorption of Hg vapor increased almost linearly as the soil moisture content increased from about 2% to 20%. The moisture contents that gave maximum Hg vapor sorption coincide closely with the soil's water-holding capacity values at 1/3 bar. Further increase in moisture content could decrease the Hg vapor sorption due to decrease of soil pore volume, which is decreased by the filling of soil spaces with water.

Redox potential also influences the stability of solid phases of Hg. Wiklander (1969) reported that insoluble HgS produced under reducing conditions can be gradually converted to the more soluble $HgSO_4$ when exposed to oxidizing conditions. Fagerström and Jernelöv (1970) showed that methylmercury was formed from HgS in aerobic organic sediments, although the process was substantially slower than when $HgCl_2$ was the Hg source. The solid phase HgS is stable in flooded soils, as in paddy soils, but can become unstable when the soil undergoes aerobic conditions (Aomine et al, 1967; Engler and Patrick, 1975).

7. Mercury in Drinking Water and Food

The occurrence of "Minamata disease"—a form of alkylmercury poisoning, which affected at least 111 persons, causing 41 deaths, in Minamata, Japan, in the early 1950s and in Niigata, Japan in the early 1960s (Irukayama, 1967; Ui, 1967)—spurred the recent interest in the environmental hazards of Hg, particularly in aquatic systems. Drinking waters and foods of terrestrial origin are insignificant sources of alkylmercury in the diet. However, fish, as demonstrated by "Minamata disease," is recognized as a major source of alkylmercury in the diet (Simpson et al, 1974; D'Itri, 1972). Thus, based on available data, it appears that almost all the alkylmercury in the human diet comes from seafoods, primarily fish (NAS, 1978). The amounts of methylmercury in edible terrestrial plants and plant products are generally extremely low.

Gavis and Ferguson (1972) reported that most river and lake waters from many parts of the world have Hg concentrations in the range of <0.1 to 6.0 ppb. In central Illinois public sewer systems, total Hg ranged from 1.3 to 1.8 ppb (Evans et al, 1973). Nearly all public water supplies in the United States contain <2 ppb of Hg (NAS, 1977). Ground waters in Sweden ($n = 34$) showed Hg levels in the range of 0.020 to 0.070 ppb, with a mean of 0.050 ppb (Wicklander, 1969). The WHO has an upper guideline limit of 1 ppb for the total Hg content of water for human consumption (WHO, 1984). For drinking water in the United States, the proposed limit is 2 ppb, as set by the US Environmental Protection Agency (US-EPA, 1976).

Mercury can enter the terrestrial food chain mainly through atmospheric deposition (coal combustion, smelting, volcanic activity, etc.) and the use of agricultural pesticides. Recent surveys of common foods in the United Kingdom (Cross et al, 1979) and the United States (Tanner and Friedman, 1977) showed levels generally lower than the former US-FDA guideline of 0.5 ppm in fish. Gomez and Markakis (1974) found that Hg concentrations in land-produced foods (dairy and poultry, cereals, meat, fruits, vegetables, etc.) are generally low, in the range of <0.01 to 0.03 ppm. Of the

TABLE 9.6. Estimated dietary intakes of mercury in various food classes.

Class	Food intake[a]		Hg intake[b]		Hg concentration[a]
	g/food/day	% of total diet	μg Hg/day	% of total diet	ppm wet wt
Dairy products	769	22.6	0.0	0.0	<0.001
Meat, fish, and poultry	273	8.0	2.89	100.0	0.017
Grain and cereal	417	12.3	0.0	0.0	<0.003
Potatoes	200	5.9	0.0	0.0	<0.002
Leafy vegetables	63	1.9	0.0	0.0	<0.001
Legume vegetables	72	2.1	0.0	0.0	<0.001
Root vegetables	34	1.0	0.0	0.0	<0.001
Garden fruits	89	2.6	0.0	0.0	<0.001
Fruits	222	6.5	0.0	0.0	—
Oils and fats	51	1.5	0.0	0.0	—
Sugars and adjuncts	82	2.4	0.0	0.0	—
Beverages	1,130	33.2	0.0	0.0	—
Total	3,402	100.0	2.89		

[a] Based on 1972–73 total diet survey by the US Food and Drug Administration for an adult person (Tanner and Friedman, 1977).
[b] Based on 1973 total diet survey by the US Food and Drug Administration (Russell, 1979).

various food classes, meat, fish, and poultry accounted for practically the entire source of Hg in diets (Table 9.6).

The Joint FAO/WHO Expert Committee on Food Additives has established a provisional tolerable weekly intake for Hg (Lu, 1973; FAO/WHO, 1972). For a 60-kg man on a daily basis, the intake limit is 43 μg total Hg ($\approx$ 0.3 mg Hg/person/week). For methylmercury, expressed as Hg, the daily intake limit is 29 μg Hg (0.2 mg/person/week). Recent surveys on dietary intakes of Hg (μg Hg/day) have been conducted: 2.9 in the United States (Russell, 1979); 10 to 16 in Canada (Somers, 1974; Kirkpatrick and Coffin, 1974; Meranger and Smith, 1972); <5 to 25 in Italy (Clemente et al, 1977), and <16 in the United Kingdom (Hamilton and Minski, 1972).

8. Sources of Mercury in the Terrestrial Environment

Estimated releases of Hg to air, soil, and water since the 18th century are presented in Table 9.1.

There are several sources of Hg in the environment, both natural and anthropogenic. Agricultural and industrial uses of Hg and its compounds have caused a widespread occurrence of this metal in the environment and the presence of Hg residues in foods. The former uses of Hg, primarily as fungicides and pesticides, constitute a low hazard (Smart, 1968; Novick, 1969). Amounts of Hg translocated from treated seed to edible portions of crops, in general, appear to be below significant levels (0.3 mg Hg/ person per week) (FAO/WHO, 1972). In addition, uses of ordinary sewage sludges (van Loon, 1974; Zwarich and Mills, 1979) and commercial fertilizers (Williams, 1977; Berrow and Burridge, 1979) do not pose hazards to the crop tissues. In contrast, Hg emanating from industrial releases, particularly from chemical and pulp mill industries, and ending in aquatic systems, is far more hazardous toxicologically. Some of these industrial sources known to enrich the environment and the biota include: chloralkali plants (Hildebrand et al, 1980; Temple and Linzon, 1977; Jernelöv and Wallin, 1973) and pulp and paper mills (Marton and Marton, 1972). It should be noted that mercury releases from chlor-alkali plants are rigidly controlled and mercurials are no longer being used in the United States pulp and paper industry.

Sources of airborne Hg include combustion of fossil fuels, chlorine manufacturing, smelting of Zn, Pb, Ni, and Cu ores; Hg mining and smelting, and incineration of Hg-containing products such as batteries, lamps, and others. Joensuu (1971) has estimated that coal combustion alone releases globally about 3,000 tonnes of Hg annually. The amount of Hg release annually from all fossil fuel burning, worldwide, may be as much as 5,000 tonnes; a similar amount of Hg is being discharged into the rivers yearly through multi-industrial and agricultural activities (Marton and Marton, 1972). In comparison, natural weathering of Hg from geological materials has been estimated to contribute annually from 230 (Joensuu,

1971) to about 5,000 tonnes (Goldberg, 1970). Mercury emissions from geothermal power plants, in such geothermal areas as Hawaii, New Zealand, and Iceland are comparable to releases from coal-burning power plants on a per megawatt basis (Robertson et al, 1977).

References

Adriano, D. C., A. L. Page, A. A. Elseewi, A. C. Chang, and I. Straughan. 1980. *J Environ Qual* 9:333–344.

Alberts, J. J., J. E. Schindler, R. W. Miller, and D. E. Nutter, Jr. 1974. *Science* 184:895–897.

Andersson, A. 1967. *Oikos Suppl* 9:13–15.

——— 1979. In J. O. Nriagu, ed. *The biogeochemistry of mercury in the environment,* 79–112. Elsevier, Amsterdam.

——— 1970. *Grundförbättring 23:31–40.*

Andersson, A., and L. Wiklander. 1965. *Grundförbättring* 18:171–177.

Andren, A. W., and J. O. Nriagu. 1979. In J. O. Nriagu, ed. *The biogeochemistry of mercury in the environment,* 1–21. Elsevier, Amsterdam.

Aomine, S., and K. Inoue. 1967. *Soil Sci Plant Nutr* 13:195–200.

Aomine, S., H. Kawasaki, and K. Inoue. 1967. *Soil Sci Plant Nutr* 13:186–188.

Archer, F. C. 1980. *Minist Agric Fish Food* (Great Britain) 326:184–190.

Bache, C. A., W. H. Gutenmann, L. E. St. John, Jr., R. D. Sweet, H. H. Hatfield, and D. J. Lisk. 1973. *J Agric Food Chem* 21:607–613.

Beauford, W., J. Barber, and A. R. Barringer. 1977. *Physiol Plant* 39:261–265.

Beckert, W. F., A. A. Moghissi, F. H. F. Au, E. W. Bretthauer, and J. C. McFarlane. 1974. *Nature* 249:674–675.

Beijer, K., and A. Jernelöv. 1979. In J. O. Nriagu, ed. *The biogeochemistry of mercury in the environment,* 203–210. Elsevier, Amsterdam.

Berrow, M. L., and J. C. Burridge. 1979. In Proc Intl Conf Heavy Metals in the Environment, 304–311. CEP Consultants Ltd, Edinburgh, UK.

Berrow, M. L., and G. A. Reaves. 1984. In Proc Intl Conf Environ Contamination, 333–340. CEP Consultants Ltd, Edinburgh, UK.

Bertilsson, L., and H. Y. Neujahr. 1971. *Biochemistry* 10:2805–2808.

Bisogni, J. J., Jr. 1979. In J. O. Nriagu, ed. *The biogeochemistry of mercury in the environment,* 211–230. Elsevier, Amsterdam.

Bolli, M. 1947. *Ann Fac Agrar Univ Stud Perugia* 4:180–186, cited by Beauford et al (1977).

Booer, J. R. 1944. *Ann App Biol* 31:340–359.

——— 1951. *Ann Appl Biol* 38:334–347.

Braman, R. S., and D. L. Johnson. 1974. In E. D. Copenhauer, ed. *Proc NSF-RANN Trace Contaminants in the Environment,* 75–78. ORNL, Oak Ridge, TN.

Byford, W. J. 1975. *Ann Appl Biol* 79:221–230.

Byrne, A. R., and L. Kosta. 1969. Inst Josef Stefan (Yugoslavia) Rep R-562.

Carty, A. J., and S. F. Malone. 1979. In J. O. Nriagu, ed. *The biogeochemistry of mercury in the environment,* 433–479. Elsevier, Amsterdam.

Chino, M. 1981. In K. Kitagishi and I. Yamane, eds. *Heavy metal pollution in soils of Japan,* 65–80. Japan Sci Soc Press, Tokyo.

Clemente, G. F., L. C. Rossi, and G. P. Santaroni. 1977. *J Radioanal Chem* 37:549–558.

Corner, E. D. S., and F. H. Rigler. 1957. *J Mar Biol Assoc* (UK) 36:449–458.

Cross, J. D., I. M. Dale, H. Smith, and L. B. Smith. 1979. *J Radioanal Chem* 48:159–167.

Davis, R. D., P. H. T. Beckett, and E. Wollan. 1978. *Plant Soil* 49:395–408.

DeSimone, R. E. 1972. *J Chem Commun* (Chem Soc London) 13:780–781.

D'Itri, F. M. 1972. *The environmental mercury problem*. CRC Press, Cleveland, OH.

Dolar, S. G., D. R. Keeney, and G. Chesters. 1971. *Environ Letters* 1:191–198.

Drake, H. J. 1980. In *Minerals yearbook,* 1978–79, 593–600. US Bureau of Mines, US Dept of Interior, Washington, DC.

Dudas, M. J., and S. Pawluk. 1976. *Can J Soil Sci* 56:413–423.

Engler, R. M., and W. H. Patrick, Jr. 1975. *Soil Sci* 119:217–221.

Estes, G. O., W. E. Knoop, and F. D. Houghton. 1973. *J Environ Qual* 2:451–452.

Evans, R. L., W. T. Sullivan, and S. Lin. 1973. *Water Sewage Works* 120:74–76.

Fagerström, T., and A. Jernelöv. 1970. *Water Res* 5:121–122.

Fang, S. C. 1978. *Environ Sci Technol* 12:285–288.

────── 1981. *Arch Environ Contam Toxicol* 10:193–201.

FAO/WHO. *See* World Health Organization.

Farrah, H., and W. F. Pickering. 1978. *Water, Air, Soil Pollut* 9:23–31.

Fimreite, N. 1970. *Environ Pollut* 1:119–131.

Fleischer, M. 1970. In *Mercury in the environment.*USGS Prof Paper 713, 6–13. USDI, Washington, DC.

Frank, R., K. Ishida, and P. Suda. 1976. *Can J Soil Sci* 56:181–196.

Frear, D. E. H., and L. E. Dills. 1967. *J Econ Entom* 60:970–974.

Gambrell, R. P., R. A. Khalid, and W. H. Patrick, Jr. 1980. *Environ Sci Technol* 14:431–436.

Gavis, J., and J. F. Ferguson. 1972. *Water Res* 6:989–1008.

Gilmour, J. T. 1971. *Environ Letters* 2:143–152.

Gilmour, J. T., and M. S. Miller. 1973. *J Environ Qual* 2:145–148.

Goldberg, E. D. 1970. In *Chemical invasion of the ocean by man. Yearbook of Sci and Technol* McGraw-Hill, New York.

Goldschmidt, V. M. 1954. *Geochemistry*. Oxford Press, London. 730 pp.

Gomez, M. I., and P. Markakis. 1974. *J Food Sci* 39:673–675.

Gotoh, S., H. Otsuka, and H. Koga. 1979. *Soil Sci Plant Nutr* 25:523–537.

Gotoh, S., S. Tokudome, and H. Koga. 1978. *Soil Sci Plant Nutr* 24:391–406.

Gowen, J. A., G. B. Wiersma, and H. Tai. 1976. *Pesticides Monit J* 10:111–113.

Gracey, H. I., and J. W. B. Stewart. 1974a. In *Proc Intl Conf on Land Waste Mgmt,* 97–103. Ottawa, Ontario, Canada.

────── 1974b. *Can J Soil Sci* 54:105–108.

Gray, N. E., and H. J. Fuller. 1942. *Am J Bot* 29:456–459.

Hahne, H. C. H., and W. Kroontje. 1973a. *J Environ Qual* 2:444–450.

────── 1973b. *Soil Sci Soc Am Proc* 37:838–843.

Hamilton, E. I., and M. J. Minski. 1972. *Sci Tot Environ* 1:375–394.

Harriss, R. C., D. B. White, and R. B. MacFarlane. 1970. *Science* 170:736–737.

Hem, J. D. 1970. In *Mercury in the environment*. USGS Prof Paper 713, 19–24.USDI, Washington, DC.

Hildebrand, S. G., R. H. Strand, and J. W. Huckabee. 1980. *J Environ Qual* 9:393–400.

Hogg, T. J., J. W. B. Stewart, and J. R. Bettany. 1978a. *J Environ Qual* 7:440–444.

———— 1978b. *J Environ Qual* 7:445–450.

Huckabee, J. W., and B. G. Blaylock. 1973. *Adv Exp Med Biol* 40:125–160.

Imura, N., E. Sukegawa, S. Pan, K. Nagae, J. Kim, T. Kwan, and T. Ukita. 1971. *Science* 172:1248–1249.

Inoue, K., and S. Aomine. 1969. *Soil Sci Plant Nutri* 15:86–91.

Irukayama, K. 1967. In *Proc 3rd Intl Conf Advances in Water Pollution Research,* vol. 3:153. Water Pollut Contr Fed, Washington, DC.

Jensen, S., and A. Jernelöv. 1967. As cited by Beijer and Jernelöv (1979).

Jernelöv, A. 1969. In M. W. Miller and G. C. Berg, eds. *Chemical fallout,* chap. 4, 68–74. Thomas Publ, Springfield, IL.

Jernelöv, A., and T. Wallin. 1973. *Atmos Environ* 7:209–214.

Joensuu, O. I. 1971. *Science* 172:1027–1028.

Johnson, D. L., and R. S. Braman. 1974. *Environ Sci Technol* 8:1003–1008.

Jonasson, I. R., and R. W. Boyle. 1971. In *Mercury in man's environment.* Proc R Soc Can Sympos, Ottawa, Ontario, Canada.

Jones, R. L., and T. D. Hinesly. 1972. *Soil Sci Soc Am Proc* 36:921–923.

Kaiser, G., and G. Tölg. 1980. In O. Hutzinger ed. *Handbook of environmental chemistry,* Vol. 3, part A. Springer-Verlag, New York.

Kimura, Y., and V. L. Miller. 1964. *J Agric Food Chem* 12:253–257.

Kinniburgh, D. G., and M. L. Jackson. 1978. *Soil Sci Soc Am J* 42:45–47.

Kirkpatrick, D. C., and D. E. Coffin. 1974. *J Inst Can Sci Technol Aliment* 7:56–58.

Kuramitsu, H. K. 1968. *J Biol Chem* 243:1061–1071.

Lagerwerff, J. V. 1972. In J. J. Mortvedt, P. M. Giordano, and W. L. Lindsay, eds. *Micronutrients in agriculture,* 593–636. Soil Sci Soc Am Inc, Madison, WI.

Landa, E. R. 1978. *J Environ Qual* 7:84–86.

Leong, L., B. Olson and R. Cooper. 1973. J. Environ. Health 35:436–442.

Lexmond, T. M., F. A. M. de Haan, and M. J. Frissel. 1976. *Netherlands J Agric Sci* 24:79–97.

Lindberg, S. E., and R. C. Harriss, 1974. *Environ Sci Technol* 8:459–462.

Lindberg, S. E., D. R. Jackson, J. W. Huckabee, S. A. Janzen, M. J. Levin, and J. R. Lund. 1979. *J Environ Qual* 8:572–578.

Lockwood, R. A., and K. Y. Chen. 1973. *Environ Sci Technol* 7:1028–1034.

———— 1974. *Environ Letters* 6:151–166.

Löfroth, G. 1970. In *Methylmercury: a review of health hazards and side effects associated with the emission of mercury compounds into natural systems.* Ecol Res Comm Bull no. 4. Swed Natl Sci Res Council, Stockholm, Sweden. 56 pp.

Lu, F. 1973. *WHO Chron* 27:245–253.

MacLean, A. J., B. Stone, and W. E. Cordukes. 1973. *Can J Soil Sci* 53:130–132.

MacLean, A. J. 1974. *Can J Soil Sci* 54:287–292.

Magos, L., A. A. Tuffery, and T. W. Clarkson. 1964. *Brit J Ind Med* 21:294–298.

Marton, J., and T. Marton. 1972. *Tappi* 55:1614–1618.

McKeague, J. A., and B. Kloosterman. 1974. *Can J Soil Sci* 54:503–507.

McKeague, J. A., and M. S. Wolynetz. 1980. *Geoderma* 24:299–307.

Meranger, J. C., and D. C. Smith. 1972. *Can J Public Health* 63:53–57.

Mhatre, G. N., and S. B. Chaphekar. 1984. *Water, Air, Soil Pollut* 21:1–8.

Mills, J. G., and M. A. Zwarich. 1975. *Can J Soil Sci* 55:295–300.

Morishita, T., K. Kishino, and S. Idaka. 1982. *Soil Sci Plant Nutr* 28:523–534.

National Academy Sciences (NAS). 1977. *Drinking water and health,* 250–488. NAS, Washington, DC.

———— 1978. *An assessment of mercury in the environment.* NAS, Washington, DC. 185 pp.

National Research Council Canada (NRCC). 1979. *Effects of mercury in the Canadian environment* NRCC no. 16739. Ottawa, Ontario, Canada. 290 pp.

Newton, D. W., and R. Ellis, Jr. 1974. *J Environ Qual* 3:20–23.

Newton, D. W., R. Ellis, Jr., and G. M. Paulsen. 1976. *J Environ Qual* 5:251–254.

Niebla, E. E., N. E. Korte, B. A. Alesii, and W. H. Fuller. 1976. *Water, Air, Soil Pollut* 5:399–401.

Novick, S. 1969. *Environment* 11:2–9.

Olson, B. H., and R. C. Cooper. 1976. *Water Res* 10:113–116.

Passow, H., A. Rothstein, and T. W. Clarkson. 1961. *Pharmacol Rev* 13:185–224.

Pickard, J. A., and J. T. Martin. 1959. In Ann Rep Long Ashton Res Sta, Bristol, UK. Pages 93–100, as cited by Smart (1964).

Poelstra, P., M. J. Frissel, N. van der Klugt, and W. Tap. 1974. In *Comparative studies of food and environmental contamination,* 281–292. Proc Series IAEA-SM-175/46, Vienna.

Puerner, N. J., and S. M. Siegel. 1972. *Physiol Plant* 26:310–312.

Rao, A. V., E. Fallin, and S. C. Fang. 1966. *Plant Physiol* 41:443–446.

Reimers, R. S., and P. A. Krenkel. 1974. *J Water Pollut Control Fed* 46:352–365.

Robertson, D. E., E. A. Crecelius, J. S. Fruchter, and J. D. Ludwick. 1977. *Science* 196:1094–1097.

Rogers, R. D. 1976. *J Environ Qual* 5:454–458.

———— 1977. *J Environ Qual* 6:463–467.

———— 1979. *Soil Sci Soc Am J* 43:289–291.

Rogers, R. D., and J. C. McFarlane. 1979. *J Environ Qual* 8:255–260.

Ross, R. G., and D. K. R. Stewart. 1960. *Can J Plant Sci* 40:117–122.

———— 1962. *Can J Plant Sci* 42:280–285.

Russell, L. H., Jr. 1979. In F. W. Oehme, ed. *Toxicity of heavy metals in the environment,* chap 2:3–23. Marcel Dekker Inc, New York.

Saha, J. G., Y. W. Lee, R. D. Tinline, S. H. F. Chinn, and H. M. Austenson. 1970. *Can J Plant Sci* 50:597–599.

Sand, P. F., G. B. Wiersma, H. Tai, and L. J. Stevens. 1971. *Pesticides Monit J* 5:32–33.

Selikoff, I. J. 1971. *Environ Res* 4:1–69.

Shacklette, H. T. 1970. In *Mercury in the environment.* USGS Prof Paper 713, 35–39. USDI, Washington, DC.

Shacklette, H. T., J. G. Boerngen, and R. L. Turner. 1971. Geol Survey Circ 644. USDI, Washington, DC.

Siegel, B. Z., M. Lasconia, E. Yaeger, and S. M. Siegel. 1984. *Water, Air, Soil Polut* 23:15–24.

Siegel, S. M., B. Z. Siegel, A. M. Eshleman, and K. Bachmann. 1973. *Environ Biol Med* 2:81–89.

Simpson, R. E., W. Horwitz, and C. A. Roy. 1974. *Pesticides Monit J* 7:127–138.

Smart, N. A. 1964. *J Sci Food Agric* 15:102–108.

——— 1968. *Residue Rev* 23:1–36.

Somers, E. 1974. *J Food Sci* 39:215–217.

Stewart, D. K. R., and R. G. Ross. 1960. *Can J Plant Sci* 40:659–665.

Summers, A. O., and S. Silver. 1978. *Ann Rev Microbiol* 32:637–672.

Tanner, J. T., and M. H. Friedman. 1977. *J Radioanal Chem* 37:529–538.

Temple, P. J., and S. N. Linzon. 1977. *Trace Subs Environ Health* 11:389–397.

Toribara, T. Y., C. P. Shields, and L. Koval. 1970. *Talanta* 17:1025–1028.

Ui, J. 1967. In Proc 3rd Intl Conf Advances in Water Pollution Research, Vol. 3. Water Pollut Cont Fed, Washington, DC.

US Environmental Protection Agency (US-EPA). 1976. In *Quality criteria for water*. US-EPA, Washington, DC.

Van Faasen, H. G. 1975. *Plant Soil* 44:505–509.

Van Loon, J. C. 1974. *Environ Letters* 6:211–218.

Vinogradov, A. P. 1959. *The geochemistry of rare and dispersed chemical elements in soils*. Consultants Bureau Inc, New York. 209 pp.

Warren, H. V., R. E. Delavault, and J. Barakso. 1966. *Econ Geol* 61:1010–1028.

Warren, H. V., and R. E. Delavault. 1969. *Oikos* 20:537–539.

Weaver, R. W., J. R. Melton, D. Wang, and R. L. Duble. 1984. *Environ Pollut* (A) 33:133–142.

Whitby, L. M., J. Gaynor, and A. J. MacLean. 1978. *Can J Soil Sci* 58:325–330.

Wiersma, G. B., and H. Tai. 1974. *Pesticides Monit J* 7:214–216.

Wiklander, L. 1969. *Geoderma* 3:75–79.

Williams, C. H. 1977. *J Aust Inst Agric Sci* (Sept/Dec.): 99–109.

Wood, J. M. 1974. *Science* 183:1049–1052.

Wood, J. M., F. S. Kennedy, and C. G. Rosen. 1968. *Nature* 220:173–174.

World Health Organization (WHO). 1984. *Guidelines for drinking-water quality*, Vol. 1. WHO, Geneva, Switzerland. 130 pp.

World Health Organization (WHO), and FAO. Joint FAO/WHO Expert Committee on Food Additives (FAO/WHO) 1972. WHO Tech Rep Ser no. 505, Geneva, Switzerland.

Zwarich, M. A., and J. G. Mills. 1979. *Can J Soil Sci* 59:231–239.

Zyrin, N. G., B. A. Zvonarev, L. K. Sadovnikova, and N. I. Voronova. 1981. *Sov Soil Sci* 13:44–52.

10
Molybdenum

1. General Properties of Molybdenum

Molybdenum (atomic wt. 95.95) is in the second row of the transition metal elements and occurs as five isotopes. It is in Group VI-B with Cr and W and shares some chemical properties with each of these elements. It has a density of 10.2, melting point of 2,620 ± 10°C, and boiling point of 4,800°C. Molybdenum has five possible oxidation states (II, III, IV, V, and VI), but in nature, the IV and VI oxidation states predominate, with the latter being the most stable (Krauskopf, 1972). The most important compound is the trioxide (MoO_3), from which most of the known Mo compounds can be prepared. Molybdenum is resistant to HCl, H_2SO_4, H_3PO_4, and HF solutions under many conditions of concentration and temperature. However, the metal is attacked by oxidizing acids and fused alkalis. It is rapidly oxidized in air at >500°C (Shamberger, 1979). At moderate to high concentrations in solution, molybdate readily polymerizes into polymolybdates with a wide variety of very complex structures. However, in dilute solutions, such as those found in soils or in most natural waters, the predominant form of soluble Mo is the molybdate anion, MoO_4^{2-}. Only under unusual conditions of very high enrichment will Mo be found as soluble polymolybdate in waters. In nature the only important ore is molybdenite (MoS_2); however, some powellite and deposits containing wulfenite also occur.

2. Production and Uses of Molybdenum

Economically recoverable resources of Mo in the world have been estimated at 26×10^6 tonnes, while total resources have been estimated at about 1×10^9 tonnes (King et al, 1973). About half of the world's identified resources are in the United States, with about 75% occurring in deposits where Mo is the principal metal mined. Most of the Mo deposits in the United States are found in Colorado, the Sangre de Cristo Range in New

Mexico, and the Basin and Range Province of the southwest. In other parts of the world, major deposits are found in Peru and Chile and in the Altai and Ural Mountains of the USSR.

Quantities of Mo produced annually have steadily increased from 1962 to 1979, with that of the United States more than doubling and that of the world about quadrupling (Fig. 10.1). The world's largest Mo mine is located at Climax, Colorado, which produces approximately 60% of the Mo mined in the United States and 40% of that mined in the western world (Chappell, 1973). World mine production of Mo increased from 95.4 × 10³ tonnes in 1977 to an average of 101.8 × 10³ tonnes in 1978 and 1979. The United States, Canada, Chile, and the USSR continued to supply nearly all of the world's output. The United States accounted for about 60% of the world's output in 1978 and 1979 and exported over half of its production primarily to Japan and western Europe. Indirect evidence indicates that world demand exceeded production in 1978 and 1979, with the demand particularly high in west European countries (Kummer, 1980).

One of the most important reasons for the increased demand for Mo is that it is considered nontoxic to humans and can be substituted for Cr or other toxic metals used in steel alloys, corrosion inhibitors, and pigments. The major uses of Mo in various end-use products are indicated in Figure 10.2. Its chief industrial use is in various alloy steels. Molybdenum uses can be categorized into chemical and metallurgical (Lander, 1977). There are three major types of Mo chemicals that are commercially utilized: molybdenum disulfide (MoS_2), molybdic oxide (MoO_3), and the various molybdates [$Na_2MoO_4 \cdot 2H_2O$, $(NH_4)_2Mo_2O_7$, and $CoMoO_4$]. These chemicals are used as catalysts, corrosion inhibitors, or as ingredients in pigments, dyes, plastic and rubber parts, industrial gear oils, and high-pres-

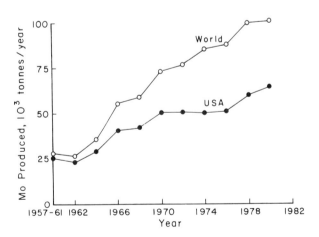

FIGURE 10.1. Trends in molybdenum production in the United States and worldwide, 1957–1979. *Source:* US Bureau of Mines, *Minerals yearbook,* several issues.

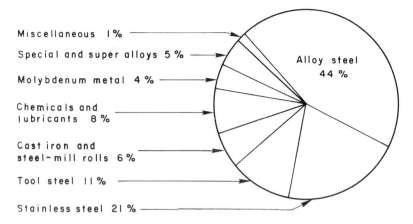

Miscellaneous 1%

Special and super alloys 5%

Molybdenum metal 4%

Chemicals and lubricants 8%

Cast iron and steel-mill rolls 6%

Tool steel 11%

Stainless steel 21%

Alloy steel 44%

FIGURE 10.2. World industrial uses of molybdenum, 1974 consumption. *Source:* Lander, 1977, with permission of the author and Dekker, Inc.

sure greases. With the exception of $CoMoO_4$, these chemicals are commonly used as sources of Mo fertilizers in agriculture (Murphy and Walsh, 1972). The metallurgical uses of Mo include production of stainless steels (1%–6% Mo), high-strength steels (0.25%–0.75% Mo), chrome-moly elevated temperature steels (0.5%–1.0% Mo), superalloys (1%–28% Mo), and molybdenum metal (99%–100% Mo).

The domestic use of Mo is approximately 70% for production of steels; 22% for production of cast irons, superalloys, other alloys, and as a refractory metal; and 8% for use in catalysts, lubricants, pigment, and other chemical uses (Kummer, 1980).

3. Natural Occurrence of Molybdenum

The average concentrations of Mo in the earth's crust vary from 1.0 to 2.3 ppm (Day, 1963; Sandell, 1946). Among elements, it ranks 53rd in crustal abundance, following As (Krauskopf, 1979). For different rock types, the following Mo concentrations (ppm) are commonly reported (Table 10.1): igneous rock, 0.9–7; phosphorite, 5–100; shale, 5–90; limestone and dolomite, <3–30; and sandstone, <3–30. Black shales are usually high in Mo, with values up to 300 ppm. In the eastern Yukon Peninsula, paleozoic bedrock (shale, limestone, siltstone) Mo concentrations have been found ranging from <2–100 ppm, producing stream sediments with 3 to 35 ppm Mo, and soils with 4 to 29 ppm Mo (Doyle et al, 1972). Proterozoic and cretaceous bedrocks had much lower concentrations (<2 ppm). In the United Kingdom, black shales had Mo concentrations of <2 to 40 ppm, much higher than for other rocks (Thornton, 1977). They caused corresponding high values for stream sediments (3–60 ppm) and subsoils

TABLE 10.1. Commonly observed molybdenum concentrations (ppm) in various environmental matrices.

Material	Average (or median)	Range
Basaltic igneous[a]	1.5	0.9–7
Granitic igneous[a]	1.6	1–6
Phosphorite[a]	30	5–100
Shale[b]	16	5–90
Black shale[c]	72	<22–290
Black shale (hi C)[a]	10	1–300
Limestone and dolomite[c]	0.79	<3–30
Siltstone[b]	4	<2–5
Sandstone[c]	~3	<3–30
Coal[d]	3	0.3–30
Fly ash[d]	—	8.4–33
Lime[e]	—	1.3–3.8
Fertilizers[e]	—	<0.05–1.7
Sewage sludge[f]	15.6	1–76
Soils[d]	—	0.2–5
Common crops[g]	—	0.03–5
Legumes[g]	—	0.3–5
Common plants[h] (mostly weeds)	0.74	0.17–2.0
Ferns[i]	—	0.8–2.5
Fungi[i]	1.4	—
Lichens[i]	—	0.03–3
Woody angiosperm[i]	—	0.06–3
Woody gymnosperm[i]	—	0.2–2.8
Freshwater*[i]	0.5	0.3–10
Sea water*[i]	10	4–10

* μg/liter
Sources: Extracted from
[a] Fleischer (1972).
[b] Doyle et al (1972).
[c] Connor and Shacklette (1975).
[d] Adriano et al (1980).
[e] Evans et al (1950a).
[f] Jarrell et al (1980).
[g] Gupta and Lipsett (1981).
[h] Robinson and Edgington (1954).
[i] Bowen (1979).

(<2–85 ppm). Others found Mo-rich soils that were derived at least partially from molybdeniferous shale bedrock (Doyle and Fletcher, 1977). Goldschmidt (1954) and Rankama and Sahama (1950) showed that shale and granite are the major rocks contributing Mo to soil parent materials. Goldschmidt (1954) suggested that the average Mo concentration in igneous rocks may be as high as 15 ppm, although North American granites average only 2 ppm. Manskaya and Drozdova (1968) stated that Mo can substitute for Fe, Ti, and Al in mineral crystal lattice structures, probably contributing to a somewhat uniform distribution of Mo in the environment. High Mo geochemical environments produce vegetation with correspondingly high concentrations of Mo (Thornton, 1977; Doyle et al, 1972).

Several important primary minerals contain Mo (Krauskopf, 1972;

Northcott, 1956). Among them, in approximate order of importance, are: molybdenite (MoS_2), powellite ($CaMoO_4$), ferrimolybdite [$Fe_2(MoO_4)_3$], wulfenite ($PbMoO_4$), ilsemanite (molybdenum oxysulfate), and jordisite (amorphous molybdenum disulfide). Molybdenite, ferrimolybdite, and jordisite currently are the most economically important of these minerals, because they are the most abundant and are found in deposits that can be mined solely for their Mo content.

Molybdenum concentrations in normal soils average about 1 to 2 ppm, with some unusual concentrations as high as 24 ppm (Allaway, 1968; Aubert and Pinta, 1977). Aubert and Pinta (1977) reported that ancient alluvial sands and sandy loams from Byelorussia (USSR) contained 0.28 to 0.90 ppm Mo, morainic clay loams and loessic loams had 0.9 to 4.9 ppm, and clays averaged 2.8 to 3.7 ppm.

The concentration of Mo in plants varies quite widely (Table 10.1). Levels as low as 0.01 ppm in common crops to several hundred ppm in legumes have been reported (Sauchelli, 1969).

4. Molybdenum in Soils

4.1 Total Soil Molybdenum

The Mo content of soils is very dependent on the nature of parent material, degree of weathering, and organic matter content. The soil levels may be greater than that of the parent material (Aubert and Pinta, 1977). Accumulation of Mo in the "humiferous" zone of the soil can be attributed to soil organic matter fixation. Another possibility is that the inorganic soil constituents in the surface are more weathered and adsorb more Mo. Most reports have associated high Mo levels in soils with sedimentary parent materials, especially shales (Swaine and Mitchell, 1960).

In general, levels of 0.5 to 5 ppm are considered normal (Robinson and Alexander, 1953; Williams, 1971) and are in agreement with the relative abundance of Mo in the earth's crust (1.0–2.3 ppm). Robinson and Alexander (1953) suggested a mean soil total Mo concentration of 2.5 ppm, based on nearly 500 United States soil samples and 237 soils from other parts of the world. However, current assessments, with inclusion of a broader range of soils, suggest that a median of 1.2 to 1.3 ppm Mo may be a more representative value, with the range from 0.1 to 40 ppm (Kubota, 1977). Levels of 0.5 ppm or less could be considered somewhat low (Williams, 1971). Berrow and Reaves (1984) recently estimated a mean of 1.5 ppm Mo for world soils. Shacklette and Boerngen (1984) gave a mean concentration of 0.97 ppm Mo (range of <3–15 ppm) for United States surface soils (Appendix Table 1.17). For tropical Asian paddy soils, a mean concentration of 2.9 ppm Mo (range of trace–8 ppm) was reported by Domingo and Kyuma (1983).

Barshad (1948) obtained total soil Mo values of 0.1 to 9.7 ppm from 20 different locations in California. MacLean and Langille (1973) reported that the Mo content of Nova Scotia (Canada) podzols ranged from 0.05 to 12.1 ppm. Russian and Argentine investigators gave 2.0 to 2.6 ppm (Sauchelli, 1969), whereas some soils in France had Mo concentrations varying from 4.3 to 6.9 ppm. The mean content of Mo in British surface soils has been reported as 1 ppm and in Scottish soils, originating from different rock types, as <1 ppm (andesite, granite, shale, sandstone), 1 ppm (serpentine, quartzite), 2 ppm (olivine gabro), 3 ppm (trachyte), and 5 ppm Mo (quartz mica schist) (Thornton, 1977). The average Mo content of surface soils in the European USSR was about 1.7 ppm, whereas the Russian Plain was reported to contain 2.6 ppm (Zborishchuk and Zyrin, 1974). The total content of Mo in the soils of China varies from 0.1 to 6 ppm, with an average of 1.7 ppm (Liu et al, 1983). Soils derived from loessial deposits in the loess plateau and along the Yellow River are usually Mo-deficient. Also, acid Chinese soils are often deficient in Mo.

In his review of Mo levels in soils and plants of the United States, Kubota (1977) found that soils from the western United States had a median of 6 ppm, while soils from the eastern side had a median of 0.5 ppm. United States soils have a median Mo concentration of slightly over 1 ppm, with a range of 0.08 to >30 ppm. The trend toward less Mo in the eastern states is attributed to the fact that soils in the eastern region are formed in sandy materials of glacial and marine origin (Table 10.2), while shale and granite are rock sources of Mo in many western US soils. Areas of high (toxic) accumulations are usually sparsely distributed, whereas Mo-deficient soils may extend over larger areas, particularly in acid soils. Because Mo accumulates in soils, it has been used as an indicator element for geochemical prospecting, both for Mo ores and for Cu minerals, with which it is sometimes associated (Jarrell et al, 1980).

4.2 Extractable (Available) Molybdenum in Soils

Total Mo content of soils, except possibly when occurring in very low levels, is generally not a good indicator of Mo phytoavailability. Extractable Mo may be not necessarily correlated to the total Mo content of soils (Stone and Jencks, 1963), although soils richest in extractable Mo are generally those having high total Mo contents. A number of extractants have been used to evaluate the "plant-available" Mo, ranging from mildly acid, neutral, and alkaline extractants including water, to strong acids and alkalis, but the method that has been tested against plant response and used in advisory work is that of Tamm, who used an acid oxalate extractant buffered to pH 3.3 (Williams, 1971). However, in some instances this type of extractable Mo has not been found to correlate well with Mo uptake by plants on various soils (Gupta and Lipsett, 1981). Grigg (1960) attributed this poor relationship to the strong action of acid ammonium oxalate in dissolving Fe and Al oxides, which brings into solution not only exchangeable Mo, but also ferric molybdate that is not available to plants.

TABLE 10.2. Molybdenum concentrations (ppm) in US soils derived from various rocks or parent materials.[a]

Rock or parent material	Number of samples	Mean	Range
Mancos shale (Colorado)	16	2.5	0.3–8.1
Colorado shale (Montana)	8	2.1	0.5–3.9
Marine and nonmarine shales (Oregon)	15	2.2	1.0–7.0
Shales, interbedded (Maryland)	10	2.3	1.5–3.0
Shales and limestone			
Virginia	12	2.9	1.4–31.5
W Virginia	80	1.8	0.8–3.3
Gneiss and schist (Connecticut)	30	1.2	0.8–2.1
Sandstone (Connecticut)	17	0.7	0.6–0.9
Glacial outwash			
New York	8	0.49	0.22–0.86
Michigan	10	0.96	0.16–1.76
Wisconsin	7	0.30	0.08–0.80
Wyoming	3	0.51	0.25–0.65
Montana	1	0.38	—
Glacial till			
New York	13	0.49	0.14–0.92
Michigan	6	1.81	0.16–4.68
Minnesota	13	0.66	0.48–0.88
Loess (Illinois)	15	2.53	0.75–6.40
Lacustrine (Ohio)	12	4.14	1.20–7.15
Granitic alluvial fan (Nevada)	35	11.5	4.4–42.5
Volcanic ash			
Oregon	33	1.2	0.4–1.8
Alaska	18	2.3	1.3–2.8

[a] Source: Adapted from Kubota, 1977, with permission of the author and Dekker, Inc.

Using the ammonium-oxalate extractant, Walsh et al (1952) found 0.04 to 0.12 ppm Mo in soils (pH 5–5.5) where Mo deficiency in plants occurred. However, soils of the same total Mo content (0.4–3.5 ppm) with higher pH (pH 7–8) gave extractable values of 0.2 to 0.7, and produced plants with levels of Mo potentially harmful to livestock. In New Zealand, Grigg (1953) obtained ammonium-oxalate extractable Mo of 0.03 to 0.22 ppm. The water-soluble Mo content of soils was generally low. Water extraction usually yields very low amounts of Mo, making this extractant analytically unattractive. However, in alkaline soils with fairly high Mo contents, water extraction may be considered a suitable soil test (Vlek and Lindsay, 1977). In general, concentrations of extractable Mo in soils are rarely higher than 1 ppm.

Soil Mo extraction with 1 M $(NH_4)_2CO_3$ at pH 9 has been tested by Vlek and Lindsay (1977) to represent the more Mo-labile fraction. The amount of Mo obtained using this extractant on soils having pH >7 amended with 0 to 2 ppm Mo showed a correlation of $r = 0.98$ with Mo uptake by alfalfa (Fig. 10.3). However, inclusion of soils having pH <7

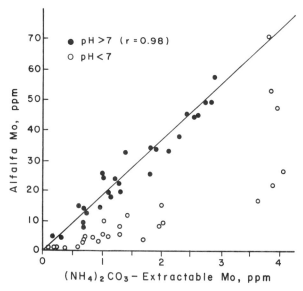

FIGURE 10.3. Relationships between ammonium carbonate-extractable molybdenum and molybdenum in alfalfa grown on Colorado soils in the greenhouse. *Source:* Vlek and Lindsay, 1977, with permission of the authors and Dekker, Inc.

caused poor correlation due to extraction of some soil Mo by $(NH_4)_2CO_3$ than was available to plants. This method is considered good for testing soil Mo in alkaline soils.

Williams and Thornton (1972) found that, when considered with soil pH and organic matter, Mo extracted by NH_4OAc and EDTA can provide useful information on the potential uptake of Mo by pasture plants.

Soils supplying excessive amounts of Mo (>10 ppm) usually show better correlation between extractable Mo and plant uptake than do soils having low extractable Mo. Barshad (1951) and Kubota et al (1963) obtained good correlations between water-extractable soil Mo and plant Mo concentrations of up to 40 ppm. Correlation coefficients between plant Mo and Mo extracted by acid oxalate and 0.1 M NaOH were 0.80 and 0.85, respectively, for plants having up to 11 ppm Mo (Haley and Melsted, 1957). Ahmed et al (1982) obtained a significant relationship between the Mo in soils extracted by the Grigg's reagent (acid ammonium oxalate) and the Mo concentration of the rice plant.

Molybdenum extracted by anion-exchange resin has been shown to satisfactorily correlate with plant Mo uptake over a wide range of soil pH (Bhella and Dawson, 1972; Jackson et al, 1975; Jarrell and Dawson, 1978).

4.3 Molybdenum in the Soil Profile

Little information is available on the distribution of Mo in the soil profile. In soils where leaching occurs, soluble Mo may be lost from the soil profile, especially in alkaline soils. In alkaline soils, Mo is more mobile, and if not leached from the profile by rain or irrigation water, the Mo may accumulate in plants (Johnson et al, 1952). As indicated by resin-extractable Mo, there was a net increase in Mo in subsoils due to irrigating a field planted to alfalfa (Jackson et al, 1975). Although the field was irrigated with water containing trace amounts of Mo, the increase of Mo in the subsoil (pH $\geqslant 8$) was probably caused by leaching of Mo from the surface soil. Utilizing soil columns under excessive precipitation, Jones and Bellings (1967) found that Mo leached through most soils, particularly soils with alkaline pH, but some accumulation of Mo occurred in the top 5 cm of the soil. Pratt and Bair (1964), using lysimeters in irrigated fields, found that in soils with leaching fractions of about 7% to 8%, Mo buildup in soils was about 75% of the amount added in the irrigation water. They calculated the soil buildup as the difference between the Mo input from irrigation water and Mo output through crop removal and leaching.

In general, Mo tends to accumulate in the A layer, plow layer, or the few top centimeters of the soil profile (Kubota, 1977; Barshad, 1948; Dobritskaya, 1964; Bradford et al, 1967; Swaine and Mitchell, 1960). Molybdenum tends to be high in layers of organic matter accumulation, usually corresponding to the A layer or plow layer.

4.4 Forms and Speciation of Molybdenum in Soils

Molybdenum may be present in soils in various forms (Barrow, 1977; Barshad, 1951; Davies, 1956; Ellis and Knezek, 1972; Williams and Thornton, 1973): (1) fixed within the crystal lattice of primary and secondary minerals and not available to plants; (2) adsorbed by soil materials as an anion; (3) bound with organic matter; (4) exchangeable; and (5) water-soluble. Molybdenum in the last two categories should be easily available to plants. The distribution of these various forms of Mo will vary between soils of differing characteristics. Williams and Thornton (1973) defined four forms of Mo for British soils (Table 10.3): (1) water-soluble, which includes readily soluble molybdates and water-soluble organic complexes; (2) exchangeable (extracted by NH_4OAc), which is part of the Mo adsorbed by ion exchange material; (3) complexed (extracted by EDTA), which includes Mo loosely bound in organic complexes and associated with organic compounds soluble in EDTA; and (4) nonextractable, which is that fraction of Mo not extracted by the alkaline extractants (NaOH or NH_4OH). Their data (Table 10.3) indicate that at least 45% of the total Mo in these soils is nonextractable and that soils with the lowest organic C contents (5.1%–7.2%) had the most nonextractable Mo (74%–79%). The nonextractable form may be considered fixed or occluded in the crystal lattice of soil minerals. Data in Table 10.3 also indicate that, except on the noncalcareous brown earth soil, between 5% and 7% of soil Mo was in the exchangeable form; EDTA extracted greater amounts of Mo from the organic (peaty) than from the inorganic soils; and the water-soluble fraction constitutes only a minor source of plant-available Mo, only around 1% of total Mo.

TABLE 10.3. Distribution of molybdenum fractions (as % of total Mo) in selected soils.[a]

Soil	Parent material	pH	Organic C, %	Total Fe, %	Total Mo, ppm	Forms of Mo (as % of total Mo)[b]			
						Nonextractable	Exchangeable	Complexable	Water-soluble
Peat	Lake alluvium of mixed limestone/shale	5.4	35.5	1.4	35	47	7	12	1
Peat	Lake alluvium of mixed limestone/shale	5.7	40.2	3.0	260	45	6	3	1
Peaty gley	Glacial drift of mixed limestone/shale	5.1	20.1	3.6	34	67	5	9	<1
Gley	Lower Namurian shale	5.8	7.2	4.2	13	74	7	8	<1
Calcareous	Lower Lias shale	7.1	6.9	5.1	39	79	6	1	3
Brown earth	Lower Lias shale	6.4	5.1	4.2	31	76	1	4	1

[a] Source: Adapted from Williams and Thornton, 1973, with permission of the authors.
[b] The exchangeable fraction was extracted by 1 M NH₄OAc, the complexable fraction by 0.05 M EDTA, and the nonextractable fraction was represented by that amount of soil Mo not extracted by NH₄OH and NaOH extractants.

Adsorption or precipitation of the molybdate anions by hydrous oxides of Fe and Al partly accounts for some unavailable Mo in soils (Davies, 1956). Water-soluble Mo would normally be very low, except possibly under alkaline conditions (Barshad, 1951).

The species of Mo comprising soluble Mo in soils are anionic, with MoO_4^{2-} as the predominant ion (Lindsay, 1972). The species of Mo in solution varies with pH. Above pH 4.2, MoO_4^{2-} is the major solution species (Lindsay, 1979). The solution species generally decrease in the order: $MoO_4^{2-} > HMoO_4^{-} > H_2MoO_4^{0} > MoO_2(OH)^{+} > MoO_2^{2+}$. The first three species predominate the total Mo in solution in the pH range of 3 to 5. The latter two ions can be ignored in soils. In very dilute solutions (10^{-5} to 10^{-8} M) the following Mo species were reported by Chojnacka (1963): above pH 5, MoO_4^{2-} predominates; from pH 2.5 to 4.5, MoO_4^{2-}, $HMoO_4^{-}$, and H_2MoO_4 predominate; from pH 1.0 to 2.5, H_2MoO_4 predominates; and below pH 1.0, H_2MoO_4 disappears and cationic forms appear. Since most soils have a pH above 5.0, the simple tetrahedral MoO_4^{2-} would be the predominant ion. Below this pH value, the MoO_4^{2-} ion becomes protonated.

In soil solution, these species are expected to exist in monomeric forms, since polymer formation of Mo does not occur in solutions having an Mo (VI) concentration of less than 10^{-4} M (Averton et al, 1963; Jenkins and Wain, 1963). Saturation extracts of some Indiana soils have shown Mo concentrations in the neighborhood of 2 to 8 $\times$ 10^{-8} M (Lavy and Barber, 1964).

4.5 Adsorption and Fixation of Molybdenum in Soils

Molybdenum sorption on soil has been used as an indicator of a soil's potential to lower the phytoavailability of Mo (Barrow and Spencer, 1971; Wells, 1956). Molybdate is sorbed strongly by Fe and Al oxides (Reyes and Jurinak, 1967; Jones, 1957; Reisenauer et al, 1962). Jones (1956, 1957) found that hydrous Fe oxide adsorbs Mo much more strongly than Al oxide, halloysite, nontronite, and kaolinite, listed in decreasing order of amounts adsorbed. Recent studies indicate the major role of amorphous Fe oxide (Jarrell and Dawson, 1978) or free Fe oxide (Karimian and Cox, 1978) on the Mo sorption by soils. A portion of the sorbed Mo becomes unavailable to the plant, while the remainder is in equilibrium with the soil solution Mo. As plant roots deplete Mo from solution, more Mo is desorbed into the solution by mass action.

The amount of Mo adsorbed by a soil or hydrous Fe or Al oxide is strongly pH-dependent. Reisenauer et al (1962) showed that Mo sorption increased with decreasing pH from 7.75 to 4.45. Possible explanations for this effect on adsorption in this pH range are that hydroxide and molybdate ions compete for adsorption sites, or Fe and Al oxides become more active as pH decreases. The adsorption was described by Freundlich's equation if pH was held constant over the Mo concentrations studied. Other adsorption isotherms also followed the Freundlich's equation (Karimian and Cox, 1978; Jarrell and Dawson, 1978).

The amount of Mo adsorbed is closely related to soil organic matter and Fe oxide contents (Fig. 10.4). In acid soils, Fe oxides carry positive charges and can react with molybdate. Adsorption isotherms for three histosols (organic soils) indicate that the level of Mo adsorbed increases with organic matter content; isotherms for four mineral soils show that the amount of adsorbed Mo increases with free Fe oxide content (Fig. 10.4). Since Mo in soil exists in anionic form, it is difficult to explain its adsorption by organic matter, especially in view of the findings by Bloomfield and Kelso (1973) that the anionic form of Mo persisted in solution even after a three-week incubation period with anaerobically decomposing material. However, based on the known reaction of organic matter and

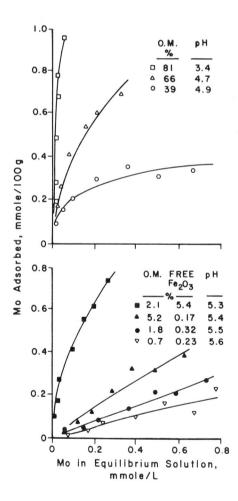

FIGURE 10.4. Molybdenum adsorption isotherms for three histosols (above) and four mineral soils (below). *Source:* Modified from Karimian and Cox, 1978, with permission of the authors and Soil Sci Soc of America.

Fe in soil, it has been suggested that the Fe oxide bound to organic matter is actually responsible for Mo adsorption (Karimian and Cox, 1978).

In volcanic ash-derived soils, Mo adsorption is due mainly to the presence of allophane and amorphous Al, Si, and Fe compounds and may not be influenced by the organic matter content of the soil (Gonzales et al, 1974).

5. Molybdenum in Plants

5.1 Essentiality of Molybdenum in Plants

The role of Mo in biological processes was first established by Bortels (1930), who showed its requirement by free-living *Azotobacter* to fix atmospheric N_2. Subsequently, Bortels (1937), Jensen (1941), and Anderson (1942) found Mo to be required for symbiotic N_2 fixation by legumes. Arnon and Stout (1939) and Piper (1940) showed it to be essential for the growth of higher plants. Anderson (1942) showed that Mo was essential in pasture production in the process of N_2 fixation by the *Rhizobium*-legume complex. Since that time, Mo has been demonstrated to be required by a variety of nonlegumes, including lettuce, spinach, beet, and species in the family *Brassica* (Johnson, 1966).

In biological systems, Mo is a component of at least five enzymes that catalyze unrelated reactions. Three of these enzymes (nitrate reductase, nitrogenase, and sulfite oxidase) are found in plants (Gupta and Lipsett, 1981). The most important functions of Mo in plants are associated with N metabolism. Its role is primarily involved in enzyme activation, primarily with nitrogenase and nitrate reductase enzymes. Nitrogenase catalyzes the reduction of atmospheric N_2 to NH_3, the reaction by which *Rhizobium* bacteria in root nodules supply N to the host plant. For this reason, Mo-deficient legumes often exhibit symptoms of N deficiency (Hagstrom, 1977). Nitrogenase contains Mo and Fe ions, both of which are required for activity of the enzyme (Allaway, 1977).

Molybdenum is needed by plants when N is absorbed in the nitrate (NO_3^-) form, because it is a critical constituent of the nitrate reductase enzyme. This enzyme catalyzes the biological reduction of NO_3^- to NO_2^-, which is the first step toward the incorporation of N, as NH_2, into proteins. The increased Mo requirement of most plants grown on NO_3^--N compared with NH_4^+-N attests to the catalytic role of Mo for nitrate reductase (Evans, 1956).

5.2 Deficiency of Molybdenum in Plants

Molybdenum deficiency in plants often occurs in acid and strongly acidic soils, partly due to the presence of hydrous oxides of Fe and Al. It can also occur in plants growing in coarse-textured soils and soils low in organic matter, like the oxisols of the tropics (Newton and Said, 1957).

Among the micronutrients that are essential for plant growth, Mo is required in the smallest amounts. Because of these low requirements, there is only a narrow range between the deficient and sufficient levels in most plants (Table 10.4). Different crop species have widely varying Mo requirements, as demonstrated by Johnson et al (1952) in growing 30 different species on an Mo-deficient soil. Healthy barley, maize, oats, and wheat contained 0.03 to 0.07 ppm Mo; tomatoes, sugar beets, squash, and spinach contained as much as 0.10 to 0.20 ppm Mo while still being moderately to severely Mo-deficient. Legumes appear to have a two to three times greater Mo requirement than that of nonlegumes (Jarrell et al, 1980). In general, plants that demand a high soil pH for satisfactory growth appear to suffer most from Mo deficiency (Williams, 1971). For example, cauliflower, broccoli, Brussels sprouts, spinach, lettuce, tomato, sugar beets, kale, and rape are relatively highly sensitive; celery, lucerne, carrot, flax, and the clovers are intermediate; while the cereals (including corn, rice, and small grains), grasses, cotton, and large-seeded legumes are the least sensitive (Hewitt, 1956; Lucas and Knezek, 1972; Williams, 1971). Among the orchard or fruit crops, citrus has shown a high requirement for Mo (Stewart and Leonard, 1952).

Overall, symptoms of Mo deficiency have been known to occur in more

TABLE 10.4. Deficient and sufficient levels of molybdenum in plants.[a]

Plant	Plant parts	Mo in dry matter (ppm)	
		Deficient	Sufficient
Alfalfa	Leaves at 10% bloom	0.26–0.38	0.34
	Whole plants at harvest	0.55–1.15	
	Top 15.2 cm of plant prior to bloom	<0.4	1.5
	Upper stem cutting at early flowering stage		0.5[b]
	Shortly before flowering (top ⅓ of plants)	<0.2	0.5–5.0
	Whole tops at 10% bloom		0.12–1.29
Barley	Blades 8 weeks old		0.03–0.07
	Whole tops at boot stage		0.09–0.18
	Grain		0.26–0.32
Beans	Tops 8 weeks old		0.4
Beets	Tops 8 weeks old	0.05	0.62
Broccoli	Tops 8 weeks old	0.04	
Brussels sprouts	Whole plants when sprouts began to form	<0.08	0.16
	Leaves	0.09	0.61
Cabbage	Leaves	0.09	0.42
	Aboveground plants at the first signs of curding	<0.26	0.68–1.49

TABLE 10.4 *Continued*

Plant	Plant parts	Mo in dry matter (ppm) Deficient	Mo in dry matter (ppm) Sufficient
Cauliflower	Whole plants before curding	<0.11	0.56
	Young leaves showing whiptail	0.07	
	Leaves	0.02–0.07	0.19–0.25
Corn	Roots	0.023–0.3	2.8–11.9
	Stems	0.013–0.11	1.4–7.0
	At tassel middle of the first leaf opposite and below the lower ear	<0.1	>0.2 0.2^b
Lettuce	Leaves	0.06	0.08–0.14
Pasture grass	First cut at first bloom		0.2–0.7
Red clover	Total aboveground plants at bloom	<0.15	0.3–1.59
	First cut at flowering		0.26
	Plants at 10% bloom	<0.22	
	Whole plants at the bud stage	0.1–0.2	0.45
	Whole plants at the bud stage		0.46–1.08
Soybeans	Plants 26–28 cm high	0.19	
Spinach	Leaves 8 weeks old	0.1	1.61
	Whole tops at normal maturity		0.15–1.09
Sugar beets	Blades shortly after symptoms appear	0.01–0.15	0.2–20.0
	Fully developed stemless leaf; late June or early July	<0.1	0.2–2.0 $>20^c$
Temperate pasture legumes	Plant shoots		>0.1
Timothy	Whole tops at prebloom; head out of the panicle	0.11	
Tobacco	Leaves 8 weeks old		1.08
Tomatoes	Leaves 8 weeks old	0.13	0.68
Tropical pasture legumes	Plant shoots		>0.02
Wheat	Whole tops; boot stage		0.09–0.18
	Grain		0.16–0.20
Winter wheat	Aboveground plants at ear emergence, 40 cm high		>0.3

[a] *Source:* From various sources, summarized by Gupta and Lipsett, 1981, with permission of the authors and Academic Press.
[b] Considered critical.
[c] Considered toxic.

than 40 higher plant species (Hewitt, 1956). Thus, more sensitive plants, like cauliflower, broccoli, cabbage, lettuce, and tomato may be used as indicator crops for diagnostic purposes. Citrus may be useful in the tropics.

Molybdenum deficiencies have been reported throughout the world, especially on acid soils in North America (Rubins, 1956), Australia (Anderson, 1956; Leeper, 1970), the United Kingdom (Plant, 1951; Williams, 1971), the Netherlands (Mulder, 1954), and New Zealand (Davies and Grigg, 1953). In the United Kingdom, Mo deficiency has been most commonly reported in cauliflower or broccoli, but it only occurs sporadically (Williams, 1971). The deficiency can often be corrected by liming the soil, as indicated by response lines for soybeans and clover-grass pastures in Figure 10.5. In soybeans, application of Mo resulted in an average yield increase of 4.2 hl/ha (hectoliter/ha) when the soil pH was 5.0. The response of soybeans to Mo application diminished as the pH increased up to 6.2, when no more response was obtained.

A value of about 0.1 ppm Mo in the dry matter of plant tops is regarded as the critical level below which Mo deficiency is likely to occur (Anderson, 1956; Jones, 1972). However, the value may vary widely (Table 10.4). For example, good responses by legumes to Mo application have been found where the tissues contained 0.5 ppm Mo; on the other hand, clover may not respond even if the tissue content is only about 0.1 ppm (Anderson and Oertel, 1946; Walker et al, 1955). In Illinois (Melsted et al, 1969), the following critical levels of Mo (in ppm) were used for diagnostic interpretation: corn, 0.2; soybeans, 0.5; wheat, 0.3; and alfalfa, 0.5. Growth reduction may be expected to occur if Mo levels in plants fall below these values. Grigg (1953) found that responses to applications of Mo were obtained when the ammonium oxalate-extractable Mo level of the soil was below 0.14 ppm, provided the soil pH was lower than 6.3.

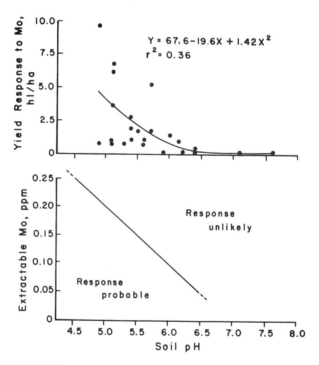

FIGURE 10.5. Molybdenum response lines for soybeans, in hectoliter per hectare (top) in Louisiana and clover-grass pastures (bottom) in New Zealand as affected by soil pH. *Sources:* Sedberry et al, 1973; Davies and Griggs, 1953.

Some of the most spectacular symptoms of Mo deficiency have been displayed in the *Brassica* family of plants as the "whiptail" disease (Sauchelli, 1969) and the "yellow spot" disease in citrus (Stewart and Leonard, 1952, 1953). The most general manifestation of Mo deficiency is leaf chlorosis that resembles N deficiency (Johnson et al, 1952). In "yellow spot" in citrus, the symptoms appear on the leaves as large interveinal chlorotic spots. In "whiptail," blade distortion is exhibited, with some leaves showing perforations or marginal scorching. The chlorosis is followed by marginal curling, wilting, and finally necrosis and withering of the leaf. Symptoms usually appear first in the older leaves, then show up in younger leaves until the growing point is killed (Sauchelli, 1969). In wheat, Mo-deficient plants exhibit golden-yellow coloration of older leaves along the apex and apical leaf margins. Plants showed reduced foliage and shorter internodes. Normal tillering may not occur, and young plants have whitish, necrotic areas extending back along the leaves from the tips (Lipsett and Simpson, 1971). In onions, it shows up as a drying of the leaf tips (Lucas, 1967). Below the dead tip, the leaf shows about 2.5 to 5.0 cm of wilting and flabby formation. As the deficiency progresses, the wilting and dying advances down the leaves. In severe cases, the plant dies.

The following methods have all been shown effective in correcting Mo deficiency in horticultural and field crops: (1) lime application; (2) lime application plus Mo; (3) foliar sprays of Mo; (4) soil application of Mo; (5) seed treatment with Mo; (6) mixing Mo with potting compost; and (7) treatment of seedbed with Mo (Murphy and Walsh, 1972; Williams, 1971).

5.3 Accumulation and Toxicity of Molybdenum in Plants

In normal crop plants, Mo usually ranges from 0.8 to 5.0 ppm in the tissue (Lucas, 1967). However, some plants have been found to contain >15 ppm. Deficient plants usually contain <0.5 ppm. Certain nonresponsive crops, such as the grasses and corn, may contain as low as 0.1 ppm.

Molybdenum toxicity in plants has not been observed under field conditions. However, toxicity has been induced under extreme experimental conditions. Using sand culture, Davis et al (1978) found that the upper critical level of Mo for barley was about 135 ppm. Using a similar technique, Hunter and Vergnano (1953) found 200 ppm Mo to be critical for oat plants. They found that Mo was less toxic than Ni, Co, Cu, Cr, and Zn, in decreasing order. Using solution culture, Wallace et al (1977) found that 710 ppm in bush bean leaves markedly reduced plant yield. This tissue level was caused by 96 ppm Mo in the culture solution. On the other hand, turnips with up to 1,800 ppm in the leaves showed normal growth pattern when grown in nutrient solution containing 20 ppm Mo (Lyon and Beeson, 1948). Soybeans, grown on pot culture, having 80 ppm Mo in leaves were normal (Golov and Kazakhkov, 1973), as were cotton plants containing 1,585 ppm in the leaves (Joham, 1953).

Applied Mo is readily available to plants, as indicated by 60% recovery by French beans grown in the greenhouse (Widdowson, 1966). When an acid soil (pH 5.6) was limed to pH 7.5, Mo recovery rose to 80%. Plants suffering from excess levels of Mo exhibit chlorosis and yellowing, apparently due to interference with Fe metabolism in plants (Warington, 1954; Millikan, 1949).

Under field conditions, plant uptake of Mo is well correlated with the amount of Mo in the soil solution, which can range from 10^{-8} M in low-Mo soils (Lavy and Barber, 1964) to 10^{-5} M in soils producing toxic herbage (Kubota et al, 1963). When livestock ingest forages containing relatively high Mo levels (10 to 20 ppm), they develop a condition known as "molybdenosis," which is often fatal. For this reason, high levels of Mo in plants are a concern to agriculturalists and animal nutritionists. Actually, "molybdenosis," also known as "teart" or "peat scours," is Mo-induced Cu deficiency in livestock, and can be rapidly corrected by injections of Cu. The symptoms of the disease are severe diarrhea, followed by roughening of the coat, change of color, and loss in weight (Barshad, 1948). Young animals are more prone to this disease, often showing stiffness in the hind legs. In severe cases, their hair sheds in spots, leaving reddened, dry, and cracked skin.

Most animals require <1.0 ppm Mo in their diet (Allaway, 1968). Neathery and Miller (1977) reported that 6 ppm dietary Mo is safe, with sheep tolerating slightly higher levels than cattle, and that animals could tolerate substantially higher levels if the diet is amply augmented with Cu. Where Cu is low (<4 ppm), 5 ppm Mo may adversely affect animals. Barshad (1948) found injury to cattle grazing herbage with 20 ppm, but no adverse effect with 10 ppm in the herbage. Marclise et al (1970) and Wynne and McClymont (1955) found 10 ppm to be toxic. In general, Mo toxicity is expressed much earlier in animals than in plants.

Kubota (1977) showed the generalized US regional pattern of Mo concentration in legumes. The principal legumes, collected over a period of 10 years, were alfalfa, clovers, and biennial sweet clovers. The general pattern is one of decreasing Mo content from the west (6 ppm) to the east (0.5 ppm) across the United States. The pattern roughly coincides with soil changes from neutral to alkaline (often calcareous) soils in the western region to predominantly acid soils in the east.

6. Factors Affecting the Mobility and Availability of Molybdenum

6.1 Soil Type and Parent Material

Mitchell (1955) stated that the trace element content of a soil is dependent almost entirely on that of the rocks from which the parent material was derived and on the weathering process to which the soil-forming materials have been subjected. The more mature and older the soil, the less may be the influence of the parent rock. This was shown by the work of Oertel (1961), who concluded that pedogenic factors and the nature of parent material play important roles in determining the Mo levels in the solum.

Soils formed from shale generally have more Mo than those from other kinds of rocks (Kubota, 1977). In general, shale and granite are the major rocks contributing Mo to soil parent materials, like the alluvium in the western United States. The Appalachian region in the eastern states has extensive areas of soils formed in shale or interbedded shale and sandstone. A similar pattern of high Mo soils might be expected in the Rockies, where shales are important rock sources of soil parent materials. Massey et al (1967) determined ammonium oxalate-extractable Mo on 73 diverse Kentucky soils. Soils formed from Devonian black fissile shale contained approximately 20 times as much Mo as any other soil. Among the other soils, the parent rock averages ranged from 0.44 ppm Mo for older Ordovician limestone (generally high phosphatic) to 0.08 ppm Mo for Pennsylvanian sandstones and shales.

Soils in the north-central and northeastern United States reflect amounts of Mo in glacial drift and related deposits, such as loess and lacustrine deposits (Table 10.2). Coarse-textured soils formed in glacial outwash of Wisconsin, Michigan, and New York have about 0.6 ppm Mo. Eight Canadian soils formed in glacial drift ranged from 0.20 to 0.82 ppm Mo (Wright et al, 1955).

In his review of factors affecting Mo availability, Williams (1971) summarized the following properties of acid soils where responses to Mo applications are likely to occur:

1. High in oxides of Fe (e.g., red loams and clays), eroded laterites, and ironstone soils
2. Highly podzolized, derived from sedimentary quartzitic material and often high in Fe
3. High in Mo-retention capacity and with low pH
4. Highly leached, or depleted of Mo supply through intensive cropping
5. High in available Mn
6. Heavily amended with sulfates, as from gypsum and $(NH_4)_2SO_4$

6.2 pH

Availability of Mo to plants is largely dependent on soil pH. Unlike most other micronutrients, availability of Mo in soils is greatest under alkaline conditions and less under acidic conditions. Thus, soil reaction is considered as one of the two most important factors affecting Mo availability, along with Fe and Al oxides in soils (Reisenauer et al, 1962). Liming acid Mo-deficient soils often corrects the deficiency. Liming acid soils may substitute for Mo fertilization by releasing Mo from soils into forms readily available for plant uptake (Evans et al, 1950, 1951; Gammons et al, 1954;

Gupta, 1972; John et al, 1972; Robinson et al, 1951). Conversely, plant response to application of Mo, under field conditions, is more effective on acid soils (Fig. 10.5). The amount of lime required for maximum production of Mo-responsive crops may be sharply reduced or eliminated by Mo fertilization (Ahlrichs et al, 1963; Giddens and Perkins, 1960; James et al, 1968). It is a common practice to control Mo deficiency by liming the soil to a pH of 6.0 to 6.5. Anderson (1956), however, cautioned that the effect of lime on Mo deficiency varies with conditions. It may partially or entirely correct the deficiency, it may have practically no effect, or it may increase the response to Mo.

The marked effect of soil reaction on the availability of Mo in soils, observed by numerous investigators (Oertel et al, 1946; Lewis, 1943a, b; Gupta et al, 1971; Gupta and Kunelius, 1980) can be explained by the proposed simplified reaction:

$$MoO_4^{2-} + Soil\begin{matrix} \nearrow^{OH} \\ \searrow_{OH} \end{matrix} \leftrightharpoons Soil\text{-}MoO_4 + 2\ OH^-$$

The equation shows the dependency of Mo solubility in soils on the hydroxyl ion concentration. Thus, the concentration of Mo in soil solution is inversely proportional to the square of the H^+ ion concentration; a one-unit increase in pH should increase Mo concentration a hundredfold. However, this relationship seldom if ever occurs in non-ideal soil systems because of the roles of Fe, Al, organic matter, and other factors on Mo solubility.

The importance of soil reaction on the availability of applied Mo to alfalfa and the anion extractability of soil Mo are demonstrated in Figure 10.6. The first relationship indicates a correlation between soil pH, available Mo, and plant uptake. The latter relationship suggests that the anion exchange resin method can detect increases in available soil Mo brought about by increases in soil pH from 4.95 to 7.10.

In the southeastern United States, yields of alfalfa and clover were significantly correlated with soil pH and soil Mo, with soil pH having the greatest effect (Mortvedt and Anderson, 1982). There was also a consistent significant correlation between plant Mo and soil pH (Fig. 10.7).

In tropical and subtropical regions, legumes are often grown in acidic soils and are dependent on N_2 fixation. Nodulation and N_2 fixation frequently are poor under these conditions. Franco and Munns (1981) reported that beans generally do not respond to Mo when grown in acidic soils (pH <5.2) that are Mo-deficient. However, they observed that six bean cultivars grown in solution culture had similar Mo concentrations in the pH range of 4.8 to 5.8.

6.3 Iron and Aluminum Oxides

Adsorption or precipitation of molybdate ions by hydrous oxides of Fe and Al are major processes in the formation of unavailable Mo in soils

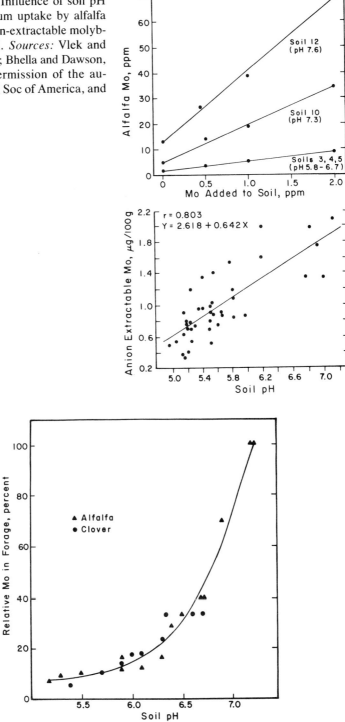

FIGURE 10.6. Influence of soil pH on molybdenum uptake by alfalfa (top) and anion-extractable molybdenum in soil. *Sources:* Vlek and Lindsay, 1977; Bhella and Dawson, 1972, with permission of the authors, Soil Sci Soc of America, and Dekker, Inc.

FIGURE 10.7. Effect of soil pH on the relative molybdenum concentrations in alfalfa and clover grown in the southeastern United States region. *Source:* Mortvedt and Anderson, 1982, with permission of the editors.

(Davies, 1956). The solubilities of Fe and Al molybdates increase with increasing pH values. Oxides of Fe have been shown to be the major soil component responsible for holding Mo (Reisenauer et al, 1962; Reyes and Jurinak, 1967). Jones (1956, 1957) reported that freshly prepared Fe oxides and a soil high in Fe oxide removed large amounts of Mo from aqueous solution. Removal of Fe oxide from the soil drastically reduced its effectiveness in adsorbing Mo. Jones (1956) suggested that in ferroginous soils, molybdate anions are held on the surface of colloidal Fe oxides, which are then replaceable by hydroxyl ions.

The Fe oxide content of soil has been found to be negatively correlated with Mo uptake by plants (Robinson and Edgington, 1954; Takahashi, 1972; Williams and Moore, 1952). More specifically, in analyzing eight western Oregon soils, Jarrell and Dawson (1978) found that the amorphous fraction of Fe oxides, extracted with ammonium oxalate, plays a major role in Mo sorption by these soils. In volcanic ash-derived soils, the presence of allophane and amorphous Al, Si, and Fe compounds contribute primarily to the Mo sorption by these soils (Gonzales et al, 1974).

6.4 Redox Potential and Organic Matter

Soils that produce high Mo plants are generally wet, poorly drained, and frequently neutral to alkaline in reaction, and high in organic matter (Allaway, 1975). Allaway attributed this phenomenon to three factors: (1) the Fe in these soils is likely to be in the reduced state (II), and ferrous molybdate is more soluble than ferric molybdates at pH values commonly found in soils; (2) loss of soluble Mo by leaching is restricted, causing accumulation of soluble Mo in the root zone; and (3) movement of Mo to the plant roots, via mass flow, is expected to be greater in wet than in dry soils.

The "teart" areas in the United Kingdom (Lewis, 1943) are reportedly poorly drained, as are the problem peat areas in New Zealand (Taylor et al, 1956) and Florida (Kretschmer et al, 1956). Molybdenum toxicity to animals is a problem in some isolated valleys with floodplains of Mo-rich alluvium (Kubota et al, 1961; Kubota, 1977). In Nevada, the highest Mo concentrations in forage samples (max. = 185 ppm) were from soils with poor drainage on granitic alluvial fans. Kubota et al (1961) related Mo toxicity in animals to two distinctly different kinds of soil. One soil group includes the imperfectly to poorly drained soils, such as those in California, Nevada, and the "teart" areas in the United Kingdom. The other group includes the organic soils. Levels of Mo in forages for the latter group are in the range of 2 to 7 ppm. Peats and mucks are formed in wet environments and are associated with Mo toxicity in the soils of the California delta, the Klamath area in Oregon, and the Everglades in Florida (Kubota, 1972). High organic matter apparently increases soluble Mo due to its effect on redox potential and more rapid reduction of Fe rather than through formation of soluble organic forms of Mo (Allaway, 1975).

In contrast, a sorption study by Karimian and Cox (1978) showed that the amounts of Mo sorbed by soils is related not only to the Fe oxide but also to organic matter content (Fig. 10.4). In this case, sorpton was attributed to the Fe oxide content of the organic matter.

6.5 Phosphorus, Sulfur, and Other Elements

Of the nutrient elements commonly used in agriculture, P and S can exert a major influence on the availability and uptake of Mo by plants. Several workers reported synergistic effects of soil application of P on Mo uptake by plants. Stout et al (1951) pointed out that the molybdates could be released from the anion-exchange complex by $H_2PO_4^-$ replacement and thus increase the concentration of Mo in the solution phase available for plant uptake. Barshad (1951) suggested that P stimulation of Mo uptake was due to the formation of a complex phosphomolybdate anion more readily adsorbed by plants. However, his data indicate depression of Mo uptake by P, when supplied as normal superphosphate that contains about 12% S as sulfate. Molybdenum uptake was enhanced by P supplied as orthophosphoric acid, a P source that contains no S. Further evidence of the beneficial effects of P on Mo uptake were obtained by several investigators (Gupta and Munro, 1969; Singh and Kumar; 1979). Bingham and Garber (1960) found that this relationship depends on soil pH. They observed synergistic effects of P fertilization on Mo uptake by orange seedlings in acid soils, whereas in alkaline soils, excessive P reduced Mo availability.

In tropical Asian countries, large-scale use of cultures of N_2-fixing blue-green algae or "Azolla" is being implemented for paddy rice production to supplement N fertilizers. Application of both P and Mo are known to enhance the growth and N_2 fixation by these organisms. Basak et al (1982 a, b) found that applied P had a synergistic effect on the Mo uptake by the rice plant.

Sulfur application to soil has been found to depress Mo uptake by plants (Gupta and Munro, 1969; Gupta and Mehla, 1980; Pasricha and Randhawa, 1972; Singh and Kumar, 1979; Stout et al, 1951). Since S is present in ordinary superphosphate fertilizers, application to soil of these fertilizers may decrease Mo uptake by plants. The inhibitory effects of SO_4^{2-} anions on Mo uptake have been suggested to occur primarily during the absorption process, with some antagonistic mechanism involved during translocation from roots to leaves. Since the MoO_4^{2-} anion is divalent and has the same size as the SO_4^{2-} anions, it is possible that these two anions may compete directly for absorption sites on the roots during the first step of absorption by plants (Stout et al, 1951).

Other antagonistic relationships of Mo with other elements in plant nutrition have been reported for Mn (Anderson and Arnot, 1953; Cheng and Ouelette, 1972; Mulder, 1954; Plant, 1950; Ralph, 1948), Fe (Gerloff et al, 1959; Hanger, 1965; Olsen and Watanabe, 1979), Zn (Singh and Steenberg, 1975), and Cu (MacKay et al, 1966).

6.6 Plant Factors

Plant species are known to accumulate different amounts of Mo from the soil. Legumes normally contain higher Mo than do grasses growing in the same area, and this has been used to define areas of potential Mo toxicity to livestock (Allaway, 1977). Members of the *Brassica* family (Table 10.5), such as cauliflower and broccoli, are highly sensitive to Mo deficiency; however, they generally contain higher Mo than other species growing on the same soil if the Mo level is sufficient. The cereal crops and large-seeded legumes also contain less Mo, in general. Others have shown significant plant interspecific variation in Mo contents (Barshad, 1951; Rinne et al, 1974).

Cultivars also vary in their uptake potential for Mo, as demonstrated in alfalfa (Younge and Takahashi, 1953), cauliflower (Gammon et al, 1954), and maize (Brown and Clark, 1974).

Although plants may tend to decrease in trace element content with maturity, Mo tends to increase with age of several plant species, both legumes and nonlegumes, especially during periods of slow growth (Barshad, 1948, 1951). In contrast, the Mo content of leaves and stems of soybeans was found to decrease with maturity (Singh and Kumar, 1979).

Once absorbed by the roots, Mo is differentially translocated to various plant parts, as indicated in Table 10.5. Molybdenum tends to accumulate, in decreasing order: leaves > stems > seed or fruit. In general, the greatest concentration of Mo was found in the blades of the leaves and in the

TABLE 10.5. Molybdenum concentrations of plant parts of various crop species (values in parentheses are the means).

Plant species	Leaves	Petioles	Stems	Seed or fruit
Brussels sprouts[a]	0.11–2.98	0.12–0.62		
	(0.90)	(0.36)		
Broccoli[a]	2.16–11.16	0.37–4.28		
	(3.76)	(1.76)		
Cauliflower[a]	0.54–3.72	0.37–1.10		
	(1.65)	(0.98)		
Rutabaga[a]	0.20–1.59	0.17–0.78		
	(0.65)	(0.32)		
Alfalfa[a]	0.10–0.72		0.08–0.28	
	(0.28)		(0.15)	
Red clover[a]	0.06–0.23		0.08–0.35	
	(0.12)		(0.15)	
Cotton[b]	21.6		2.7	1.7
Tomato[b]	34.0		4.3	3.1
Cowpeas[b]	141		281	70

Sources:
[a] Gupta and Lipsett (1981).
[b] Barshad (1951).

actively growing plant parts (Barshad, 1948). More specifically, the interveinal areas of leaves are found to preferentially accumulate Mo (Stout and Meagher, 1948).

7. Molybdenum in Drinking Water and Food

Molybdenum has been found in many surface and ground waters, usually in the range of a few ppb, but on numerous occasions, the levels were below the detection limits of the analytical method employed. Hern (1970) reported an average of 0.68 ppb Mo for surface waters of the United States, but only about 33% of the total number analyzed (1,577 samples) had levels of Mo greater than detection limits (Kopp and Kroner, 1970).

Based on very limited data, total dietary intake of Mo by adults is about 100 μg Mo/day, or about 0.13 ppm of the dry diet (Tipton et al, 1966). This value compares favorably with that for the United Kingdom of 128 ± 34 μg Mo/day and less than the International Commission on Radiological Protection average diet intake of 300 μg Mo/day (ICRP II = 450 μg Mo/day) (Hamilton, 1979; Hamilton and Minski, 1972). Underwood (1973) indicated that Mo toxicity in man is impossible from the Mo occurring naturally in ordinary foods. Molybdenum concentrations in vegetables commonly range from <5 to 70 ppm in ash (Connor and Shacklette, 1975). Duke (1970) reported the following Mo contents (ppm in dry matter): avocado, 0.1; dry beans, 0.3; corn grain, 0.1; cassava, 0.2; coconut, 0.06; rice, 0.4; taro, 0.5; and yam, 0.25. Cereal grains varied from 0.12 to 1.14 ppm and leguminous seeds from 0.2 to 4.7 ppm Mo (Westerfeld and Richert, 1953). In North American wheat and flours, the whole grain samples ranged from 0.30 to 0.66 ppm Mo, with a mean of 0.48 ppm (Czerniejewski et al, 1964). Milk is usually low in Mo but is susceptible to levels of dietary intake of the element, with concentrations of 20 to 30 ppb Mo reported from cows fed ordinary rations (Hart et al, 1967).

8. Sources of Molybdenum in the Terrestrial Environment

8.1 Molybdenum Fertilizers

Sodium molybdate ($Na_2MoO_4 \cdot 2H_2O$, 39% Mo) and [$(NH_4)_6Mo_7O_{24} \cdot 4H_2O$, 54% Mo] are very soluble and are the most widely used sources of Mo in agriculture (Murphy and Walsh, 1972). Molybdenum is present in the VI oxidation state as the MoO_4^{2-} ion. Less satisfactory carriers of Mo are MoO_3 (66% Mo), MoS_2 (60% Mo), and Mo frits (2%–3% Mo).

The amount of Mo required to correct a deficiency varies with the soil, plant species, source, and application method. The amount of Mo needed to produce optimum yield is very small. Consequently, Mo is not commonly added to mixed fertilizers. As summarized by Murphy and Walsh (1972), rates of field soil application producing optimum yields ranged from 28 g/ha in pasture to 444 g/ha of Mo for cauliflower.

In general, Mo deficiencies are corrected with a soil application or seed treatment. However, a foliar spray can be employed when the deficiency occurs during the growing season. Because seed treatment can be more uniformly applied and proved to be more effective, it has become probably the most commonly used technique of correcting Mo deficiency in crops. However, seed treatment tends to increase the Mo content of the plant much more than soil treatment (Reisenauer, 1963). In acid soils, soil, foliar, or seed treatment proved to be equally effective in correcting Mo deficiency (Sedberry et al, 1973). In this case, seed treatment was more economical, because of the smaller amount of Mo required. However, seed treatments can vary from about 50 g/ha to as much as 900 g/ha of Mo (Murphy and Walsh, 1972). Most work indicates that soil treatments with 50 to 100 g/ha of Mo can satisfactorily correct most Mo deficiencies. Certain species, especially members of the *Brassica* family, like cauliflower, may need as much as 450 g/ha of Mo.

Except in acid soils with large amounts of Fe and Al oxides, a single application of Mo at the rate of 143 g/ha can remain effective for most crops for several years (Anderson, 1956).

8.2 Sewage Sludges

Molybdenum is frequently found in municipal sewage sludges at levels above the natural soil concentrations. The mean Mo concentration in sewage sludges from 16 US cities was about 15 ppm, with the range of 1.0 to 40 ppm (Appendix Table 1.6). Similar data were obtained for Canadian cities, with an average of 16 ppm (van Loon and Lichwa, 1973). In a survey of nine states, Dowdy et al (1976) reported a median of 30 ppm, with a range of 5 to 39 ppm. Since the average concentration of Mo in sludges is much higher than that in normal soils, it can be expected that application of high rates of sludge can increase soil Mo levels, and hence plant uptake of Mo. In pot culture, adding sewage sludge up to 4% by weight of soil showed enrichment of plants in Mo (1.6 ppm in barley leaf with 4% addition vs. 0.5 ppm in control) (Jarrell et al, 1980). Several other investigators showed potential Mo enrichment of the soil and/or plants with sludge application (Andersson and Nilsson, 1972; Bradford et al, 1975; Lahann, 1976; Wells and Whitton, 1977).

8.3 Coal Combustion

Straughan et al (1978) calculated that about 1,000 tonnes of Mo/year can be potentially mobilized from the combustion of coal for power generation in the United States. It was calculated that 15 tonnes of Mo/year can be emitted into the surrounding environment for every 1,000 MW of power produced (Schwitzgebel et al, 1975). The amount of aerial release is largely determined by the type of emission control devices, with very little released if electrostatic precipitators are used.

The potential enrichment of the terrestrial environment with Mo arises primarily from coal combustion residues, now amounting to about 80×10^6 tonnes/year (Adriano et al, 1980). Approximately 65×10^6 tonnes of this amount is fly ash. Most of these residues are destined for landfill disposal, with a fraction used for agricultural production. Its potential use in agriculture is primarily due to its nutrient contents and its liming potential. However, fly ash is known to provide readily available forms of B, Mo, Se, and possibly As, which can result in toxic levels for plants or animals fed these plants (Adriano et al, 1980). Substantial enrichment of plant tissues grown on fly ash-treated soils with Mo was shown by Doran and Martens (1972), Elseewi et al (1980), and Straughan et al (1978). In addition, Elseewi et al (1980) obtained plants having low Cu:Mo ratios, suggesting problems for this type of plant when intended for animal use. A survey of 21 states revealed Mo average concentrations of 15.0 to 25.4 ppm in fly ash produced from various coal ranks (Furr et al, 1977). Fly ashes derived from bituminous coals tend to contain higher levels of Mo.

Elseewi and Page (1984) concluded that Mo-deficient soils may benefit from small applications ($\leqslant 40$ g fly ash per kg soil) of fly ash, but cautioned that Mo levels in plants growing in such soils should be continuously monitored.

8.4 Molybdenum Mining and Processing

Mining and milling of Mo is a potential source of Mo release into the environment. The source would very likely be primarily from the drainage system adjoining the mine (Chappell, 1975). Mining activities in the US Rocky Mountains have increased the Mo in some of the streams draining the mined areas. Concentrations of Mo in Colorado river waters higher than 5 ppb are generally associated with Mo mineralization, whereas levels above 60 ppb are associated with mining activities (Vlek and Lindsay, 1977). Use of high-Mo waters to irrigate pastures can adversely affect the quality of herbage produced, as shown in Colorado (Jackson et al, 1975; Vlek and Lindsay, 1977).

References

Adriano, D. C., A. L. Page, A. A. Elseewi, A. C. Chang, and I. Straughan. 1980. *J Environ Qual* 9:333–344.
Ahlrichs, L. E., R. G. Hanson, and J. M. MacGregor. 1963. *Agron J* 55:484–486.
Ahmed, M. S., and L. Rahman. 1982. *Plant Soil* 69:287–291.

Allaway, W. H. 1968. *Adv Agron* 20:235–274.

Allaway, W. H. 1975. In Proc Intl Conf on Heavy Metals in the Environment, 35–47. Toronto, Ontario, Canada.

—— 1977. In W. R. Chappell and K. K. Petersen, eds. *Molybdenum in the environment,* 317–339. Dekker, Inc, New York.

Anderson, A. J. 1942. *J Aust Inst Agric Sci* 8:73–75.

—— 1956. *Adv Agron* 8:163–202.

Anderson, A. J., and A. C. Oertel. 1946. *Council Sci Ind Res Bull* 198, Part 2:25–44.

Anderson, A. J., and R. H. Arnot. 1953. *Aust J Agric Res* 4:29–43.

Andersson, A., and K. O. Nilsson. 1972. *Ambio* 1:176–179.

Arnon, D. I., and P. R. Stout. 1939. *Plant Physiol* 14:599–602.

Aubert, H., and M. Pinta. 1977. *Trace elements in soils.* Elsevier, New York.

Averton, J., E. W. Anaeker, and J. S. Johnson. 1963. *Inorg Chem* 3:735–746.

Barrow, N. J. 1977. In W. R. Chappell and K. K. Petersen, eds. *Molybdenum in the environment,* 583–595. Dekker Inc, New York.

Barrow, N. J., and K. Spencer. 1971. *Aust J Exp Agric Anim Husb* 11:670–676.

Barshad, I. 1948. *Soil Sci* 66:187–195.

—— 1951. *Soil Sci* 71:297–313.

Basak, A., L. N. Mandal, and M. Haldar. 1982a. *Plant Soil* 68:261–269.

—— 1982b. *Plant Soil* 68:271–278.

Berrow, M. L., and G. A. Reaves. 1984. In Proc Intl Conf Environ Contamination, 333–340. CEP Consultants Ltd, Edinburgh, UK.

Bhella, H. S., and M. D. Dawson. 1972. *Soil Sci Soc Am Proc* 36:177–179.

Bingham, F. T., and M. J. Garber. 1960. *Soil Sci Soc Am Proc* 24:209–213.

Bloomfield, C., and W. I. Kelso. 1973. *J Soil Sci* 24:368–379.

Bortels, H. 1930. *Arch Mikrobiol* 1:333–342.

—— 1937. *Arch Mikrobiol* 8:13–26.

Bowen, H. J. M. 1979. *Environmental chemistry of the elements.* Academic Press, New York. 333 pp.

Bradford, G. R., R. J. Arkley, P. F. Pratt, and F. L. Bair. 1967. *Hilgardia* 38:541–556.

Bradford, G. R., A. L. Page, L. J. Lund, and W. Olmstead. 1975. *J Environ Qual* 4:123–127.

Brown, J. C., and R. B. Clark. 1974. *Soil Sci Soc Am Proc* 38:331–333.

Chappell, W. R. 1975. In P. A. Krenkel, ed. *Heavy metals in the aquatic environment,* 167–192. Pergamon Press, New York.

Cheng, B. T., and G. J. Ouelette. 1972. *Soil Sci* 114:55–60.

Chojnacka, J. 1963. *Rocz Chem* 37:259–272.

Connor, J. J., and H. T. Shacklette. 1975. USGS Prof Paper 574-F. USDI, Washington, DC. 168 pp.

Czerniejewski, C. P., C. W. Shank, W. G. Bechtel, and W. B. Bradley. 1964. *Cereal Chem* 41:65–72.

Davies, E. B. 1956. *Soil Sci* 81:209–221.

Davies, E. B., and J. L. Grigg. 1953. *N Z J Exp Agric* 87:561–567.

Davis, R. D., P. H. T. Beckett, and E. Wollan. 1978. *Plant Soil* 49:395–408.

Day, F. H. 1963. *The chemical elements in nature.* Reinhold, New York.

Dobritskaya, Y. I. 1964. *Sov Soil Sci* 5:517–526.

Domingo, L. E., and K. Kyuma. 1983. *Soil Sci Plant Nutr* 29:439–452.

Doran, J. W., and D. C. Martens. 1972. *J Environ Qual* 1:186–189.

Dowdy, R. H., W. E. Larson, and E. Epstein. 1976. In *Land application of waste materials*. Soil Conserv Soc Am, Ankeny, Iowa.

Doyle, P. J., and K. Fletcher. 1977. In W. R. Chappell and K. K. Petersen, eds. *Molybdenum in the environment*, 371–386. Dekker Inc, New York.

Doyle, P., K. Fletcher, and V. C. Brink. 1972. *Trace Subs Environ Health* 6:369–375.

Duke, J. A. 1970. *Econ Bot* 24:344–366.

Elseewi, A. A., I. R. Straughan, and A. L. Page. 1980. *Sci Total Environ* 15:247–259.

Elseewi, A. A., and A. L. Page. 1984. *J Environ Qual* 13:394–398.

Evans, H. J. 1956. *Soil Sci* 81:199–208.

Evans, H. J., E. R. Purvis, and F. E. Bear. 1950a. *Soil Sci* 71:117–124.

——— 1950b. *Plant Physiol* 25:555–566.

Fleischer, M. 1972. *Ann N Y Acad Sci* 199:6–16.

Follett, R. F., and S. A. Barber. 1967. *Soil Sci Soc Am Proc* 31:26–29.

Franco, A. A., and D. N. Munns. 1981. *Soil Sci Soc Am J* 45:1144–1148.

Furr, A. K., T. F. Parkinson, R. A. Hinrichs, D. R. van Campen, C. A. Bache, W. H. Gutenmann, L. E. St. John, Jr., I. S. Pakkala, and D. J. Lisk. 1977. *Environ Sci Technol* 11:1104–1112.

Gammons, N., Jr., G. M. Volk, E. N. McCubbin, and A. H. Eddins. 1954. *Soil Sci Soc Am Proc* 18:302–305.

Gerloff, G. C., P. R. Stout, and L. H. P. Jones. 1959. *Plant Physiol* 34: 608–613.

Giddens, J., and H. F. Perkins. 1960. *Soil Sci Soc Am Proc* 24:496–497.

Goldschmidt, V. M. 1954. *Geochemistry*. Oxford Press, London.

Golov, V. I., and Y. N. Kazakhkov. 1973. *Sov Soil Sci* 5:551–558.

Gonzales, R., H. Appelt, E. B. Schalscha, and F. T. Bingham. 1974. *Soil Sci Soc Am Proc* 38:903–906.

Grigg, J. L. 1953. *N Z J Sci Technol A* 34:405–414.

——— 1960. *N Z J Agric Res* 3:69–86.

Gupta, U. C. 1972. *Soil Sci* 114:131–136.

Gupta, U. C., and J. A. Cutcliffe. 1968. *Can J Soil Sci* 48:117–123.

Gupta, U. C., and D. C. Munro. 1969. *Soil Sci* 107:114–118.

Gupta, U. C., F. W. Calder, and L. B. MacLeod. 1971. *Plant Soil* 35:249–256.

Gupta, U. C., and H. T. Kunelius. 1980. *Can J Plant Sci* 60:113–120.

Gupta, U. C., and D. S. Mehla. 1980. *Plant Soil* 56:229–234.

Gupta, U. C., and J. Lipsett. 1981. *Adv Agron* 34:73–115.

Hagstrom, G. R. 1977. *Fert Solutions* (July–Aug):18–28.

Haley, L. E., and S. W. Melsted. 1957. *Soil Sci Soc Am Proc* 21:316–319.

Hamilton, E. I. 1979. *Trace Subs Environ Health* 13:3–15.

Hamilton, E. I., and M. J. Minski. 1972. *Sci Tot Environ* 1:375–394.

Hanger, B. C. 1965. *Aust Inst Agric Sci* 31:315–317.

Hart, L. I., E. C. Owen, and R. Proudfoot. 1967. *Br J Nutr* 21:617–630.

Hern, J. D. 1970. In *Study and interpretation of the chemical characteristics of natural water*. USDI-USGS Water Supply Paper 1473.

Hewitt, E. J. 1956. *Soil Sci* 81:159–171.

Hunter, J. G., and O. Vergnano. 1953. *Ann Appl Biol* 40:761–777.

Jackson, D. R., W. L. Lindsay, and R. D. Heil. 1975. *J Environ Qual* 4: 223–229.

James, D. W., T. L. Jackson, and M. E. Harward. 1968. *Soil Sci* 105:397–402.

Jarrell, W. M., and M. D. Dawson. 1978. *Soil Sci Soc Am J* 42:412–415.

Jarrell, W. M., A. L. Page, and A. A. Elseewi. 1980. *Residue Rev* 74:1–43.

Jenkins, I. L., and A. G. Wain. 1963. *J Appl Chem* (London) 13:561–564.

Jensen, H. L. 1941. *Aust J Sci* 3:98–99.

Joham, H. E. 1953. *Plant Physiol* 28:275–280.

John, M. K., G. W. Eaton, V. W. Case, and H. H. Chuah. 1972. *Plant Soil* 37:363–374.

Johnson, C. M. 1966. In H. D. Chapman, ed. *Diagnostic criteria for plants and soils,* 286–301. Univ Calif Div Agric Sci, Berkeley.

Johnson, C. M., G. A. Pearson, and P. R. Stout. 1952. Plant Soil 4:178–196.

Jones, G. B., and G. B. Belling. 1967. *Aust J Agric Res* 18:733–740.

Jones, J. B., Jr. 1972. In J. J. Mortvedt, P. M. Giordano, and W. L. Lindsay, eds. *Micronutrients in agriculture,* 319–346. Soil Sci Soc Am Inc, Madison, WI.

Jones, L. H. P. 1956. *Science* 123:1116.

——— 1957. *J Soil Sci* 8:313–327.

Karimian, N., and F. R. Cox. 1978. *Soil Sci Soc Am J* 42:757–761.

King, R. U., D. R. Shawe, and E. M. Mackevett, Jr. 1973. In D. A. Robst and W. P. Pratt, eds. *US mineral resources.* USGS Prof Paper 820. USDI, Washington, DC.

Kopp, J. F., and R. C. Kroner. 1970. In *Trace elements in waters of the United States.* USDI-FWPCA, Div Pollut Surv, Cincinnati, OH.

Krauskopf, K. B. 1979. *Introduction to geochemistry,* 2nd ed. McGraw-Hill, New York. 617 pp.

——— 1972. In J. J. Mortvedt, P. M. Giordano, and W. L. Lindsay, eds. *Micronutrients in agriculture,* 7–40. Soil Sci Soc Am Inc, Madison, WI.

Kretschmer, A. E., Jr., and R. J. Allen, Jr. 1956. *Soil Sci Soc Am Proc* 20:253–257.

Kubota, J., V. A. Lazar, L. N. Langan, and K. C. Beeson. 1961. *Soil Sci Soc Am Proc* 25:227–232.

Kubota, J., E. R. Lemon, and W. H. Allaway. 1963. *Soil Sci Soc Am Proc* 27:679–683.

Kubota, J. 1972. In H. C. Hopps and H. L. Cannon, eds. *Geochemical environment in relation to health and disease,* 105–115. Geol Soc Am, Boulder, CO.

Kubota, J. 1977. In W. R. Chappell and K. K. Petersen, eds. *Molybdenum in the environment,* 555–581. Dekker Inc, New York.

Kummer, J. T. 1980. In *Minerals yearbook,* Vol. 1—*Metals and minerals,* 615–628. US Bureau of Mines, US Dept of Interior, Washington, DC.

Lahann, R. W. 1976. *Water, Air, Soil Pollut.* 6:3–8.

Lander, H. N. 1977. In W. R. Chappell and K. K. Petersen, eds. *Molybdenum in the environment,* 773–806. Dekker Inc, New York.

Lavy, T. L., and S. A. Barber. 1964. *Soil Sci Soc Am Proc* 28:93–97.

Leeper, G. W. 1970. *Six trace elements in soils.* Melbourne Univ Press, Melbourne, Australia. 59 pp.

Lewis, A. H. 1943a. *J Agric Sci* 33:52–57.

——— 1943b. *J Agric Sci* 33:58–63.

Lindsay, W. L. 1972. In J. J. Mortvedt, P. M. Giordano, and W. L. Lindsay, eds. *Micronutrients in agriculture,* 41–57. Soil Sci Soc Am Inc, Madison, WI.

——— 1979. *Chemical equilibria in soils.* Wiley, New York. 449 pp.

Lipsett, J., and J. R. Simpson. 1971. *J Aust Inst Agric Sci* 37:348–351.

Liu, Z., Q. Q. Zhu, and L. H. Tang. 1983. *Soil Sci* 135:40–46.

Lucas, R. E. 1967. In *Micronutrients for vegetables and field crops.* Ext Bull E-486. Michigan State Univ, East Lansing. 13 pp.

Lucas, R. E., and B. D. Knezek. 1972. In J. J. Mortvedt, P. M. Giordano, and W. L. Lindsay, eds. *Micronutrients in agriculture,* 265–288. Soil Sci Soc Am Inc, Madison, WI.

Lyon, C. B., and K. C. Beeson. 1948. *Botan Gaz* 109:506–520.

MacKay, D. C., E. W. Chipman, and U. C. Gupta. 1966. *Soil Sci Soc Am Proc* 30:755–759.

MacLean, K. S., and W. M. Langille. 1973. *Commun Soil Sci Plant Anal* 4:495–505.

Manskaya, S. M., and T. V. Drozdova. 1968. *Geochemistry of organic substances.* Pergamon, New York.

Marclise, N. A., C. B. Ammerman, R. M. Valsecchi, D. G. Dunavant, and G. K. Davis. 1970. *J Nutr* 100:1399–1406.

Massey, H. F., R. H. Lowe, and H. H. Bailey. 1967. *Soil Sci Soc Am Proc* 31:200–202.

Melsted, S. W., H. L. Motto, and T. R. Peck. 1969. *Agron J* 61:17–20.

Millikan, C. R. 1949. *Proc R Soc Victoria* 61:25–42.

Mitchell, R. L. 1964. In F. E. Bear, ed. *Chemistry of the soil,* 320–368. Am Chem Soc Monog no 160. Reinhold, New York.

Mortvedt, J. J., and O. E. Anderson, eds. 1982. *Forage legumes: diagnosis and correction of molybdenum and manganese problems.* Southern Coop Ser Bull 278. Univ of Georgia, Athens. 68 pp.

Mulder, E. G. 1954. *Plant Soil* 4:368–415.

Murphy, L. S., and L. M. Walsh. 1972. In J. J. Mortvedt, P. M. Giordano, and W. L. Lindsay, eds. *Micronutrients in agriculture,* 347–387. Soil Sci Soc Am Inc, Madison, WI.

Neathery, M. W., and W. J. Miller. 1977. *Feedstuff* (Aug):18–20.

Newton, J. D., and A. Said. 1957. *Nature* 180:1485–1486.

Northcott, L. 1956. *Molybdenum.* Butterworth, London.

Oertel, A. C. 1961. *J Soil Sci* 12:119–128.

Oertel, A. C., J. A. Prescott, and C. S. Stephens. 1946. *Aust J Sci* 9:27–28.

Olsen, S. R., and F. S. Watanabe. 1979. *Soil Sci Soc Am J* 43:125–130.

Pasricha, N. S., and N. S. Randhawa. 1972. *Plant Soil* 37:270–274.

Piper, C. S. 1940. *J Aust Inst Agric Sci* 6:162–164.

Plant, W. 1950. *Nature 164:533–534.*

———— 1951. *J Hort Sci* 26:109–117.

Pratt, P. F., and F. L. Bair. 1964. *Agric Chem* 19:39.

Ralph, P. M. 1948. *Nature* 161:528.

Rankama, K. and T. G. Sahama. 1950. *Geochemistry.* Univ Chicago Press, Chicago.

Reisenauer, H. M. 1963. *Agron J* 55:459–460.

Reisenauer, H. M., A. A. Tabikh, and P. R. Stout. 1962. *Soil Sci Soc Am Proc* 26:23–27.

Reyes, E. D., and J. J. Jurinak. 1967. *Soil Sci Soc Am Proc* 31:637–641.

Rinne, S. L., M. Sillanpaa, E. Houkuna, and S. R. Hiivola. 1974. *Ann Agric Fenniae* 13:109–118.

Robinson, W. O., and L. T. Alexander. 1953. *Soil Sci* 75:287–291.

Robinson, W. O., G. Edgington, W. H. Armiger, and A. V. Breen. 1951. *Soil Sci* 72:267–274.

Robinson, W. O., and G. Edgington. 1954. *Soil Sci* 77:237–251.

Rubins, E. J. 1956. *Soil Sci* 81:191–197.

Sandell, E. B. 1946. *Am J Sci* 244:643.

Sauchelli, V. 1969. *Trace elements in agriculture.* Reinhold, New York. 248 pp.

Schwitzgebel, K., F. B. Meserole, R. G. Oldham, R. A. Magee, F. G. Mesich, and T. L. Thoem. 1975. In *Proc Intl Conf on Heavy Metals in the Environ,* 533–551. Toronto, Ontario, Canada.

Sedberry, J. E., T. S. Dharmaputra, R. H. Brupbacher, S. A. Phillips, J. G. Marshall, L. W. Sloane, D. R. Melville, J. L. Rabb, and J. H. Davis. 1973. *Louis Agric Exp Sta Bull* 670:3–39.

Shacklette, H. T., J. C. Hamilton, J. C. Boerngen, and J. M. Bowles. 1971. USGS Prof Paper 57-D. USDI, Washington, DC.

Shacklette, H. T., and J. G. Boerngen. 1984. *Element concentrations in soils and other surficial materials of the conterminous United States.* USGS Prof Paper 1270. USDI, Washington, DC. 105 pp.

Shamberger, R. J. 1979. In F. W. Oehme, ed. *Toxicity of heavy metals in the environment,* 689–796. Dekker Inc, New York.

Singh, M., and V. Kumar. 1979. *Soil Sci* 127:307–312.

Singh, B. R., and K. Steenberg. 1975. *Soil Sci Soc Am Proc* 39:674–679.

Stewart, I., and C. D. Leonard. 1952. *Nature* 170:714–715.

——— 1953. *Am Soc Hort Sci Proc* 62:111–115.

Stone, K. L., and E. M. Jencks. 1963. W Virginia Univ Agric Exp Sta Bull no. 484. 10 pp.

Stout, P. R., and W. R. Meagher. 1948. *Science* 108:471–473.

Stout, P. R., W. R. Meagher, G. A. Pearson, and C. M. Johnson. 1951. *Plant Soil* 3:51–87.

Straughan, I., A. A. Elseewi, and A. L. Page. 1978. *Trace Subs Environ Health* 12:389–402.

Swaine, D. J., and R. L. Mitchell. 1960. *J Soil Sci* 11:347–368.

Takahashi, T. 1972. *Plant Soil* 36:665–674.

Theisen, A. A., and A. Pinkerton. 1968. *Soil Sci Soc Am Proc* 32:440–441.

Thornton, I. 1977. In W. R. Chappell and K. K. Petersen, eds. *Molybdenum in the environment,* 341–369. Dekker Inc, New York.

Tipton, I. H., P. L. Stewart, and P. G. Martin. 1966. *Health Phys* 12:1683–1690.

Underwood, E. J. 1973. In *Toxicants occurring naturally in foods,* 43–87. NAS, Washington, DC.

Van Loon, J. C., and J. Lichwa. 1973. *Environ Letters* 4:1–8.

Vlek, P. L. G., and W. L. Lindsay. 1977. In W. R. Chappell and K. K. Petersen, eds. *Molybdenum in the environment,* 619–650. Dekker Inc., New York.

Walker, T. W., A. F. R. Adams, and H. D. Orchiston. 1955. *Plant Soil* 6:201–220.

Wallace, A., E. M. Romney, G. V. Alexander, and J. Kinnear. 1977. *Commun Soil Sci Plant Anal* 8:741–750.

Walsh, T., M. Neenan, and L. B. O'Moore. 1952. *J Dept Agric* (Ireland) 48:32–43.

Warington, K. 1954. *Ann Appl Biol* 41:1–22.

Westerfeld, W. W., and D. A. Richert. 1953. *J Nutr* 51:85–95.

Wells, N. 1956. N Z J Sci Technol B 37:482–502.

Wells, N., and J. S. Whitton. 1977. *N Z J Exp Agric* 5:363–369.

Widdowson, J. P. 1966. *N Z J Agric* 9:59–67.

Williams, C. H., and C. W. E. Moore. 1952. *Aust J Agric Res* 3:343–361.

Williams, C., and I. Thornton. 1972. *Plant Soil* 36:395–406.

———— 1973. *Plant Soil* 39:149–159.

Williams, J. H. 1971. In *Trace elements in soils and crops*. Minist Agric *Fish Food* (UK). Tech Bull no. 21, 119–136.

Wright, J. R., R. Levick, and H. J. Atkinson. 1955. *Soil Sci Soc Am Proc* 19:340–344.

Wynne, K. N., and G. L. McClymont. 1955. *Nature* 175:471–472.

Younge, O. R. and M. Takahashi. 1953. *Agron J* 45:420–428.

Zborishchuk, Y. N., and N. G. Zyrin. 1974. *Agrokhimiya* 3:88–93.

11
Nickel

1. General Properties of Nickel

Nickel (atomic no. 28) belongs in Group VIII of the periodic table, the so-called iron-cobalt group of metals, and has an atomic weight of 58.71, specific gravity of 8.902, and melting point of 1,453°C. It is silvery white, takes on a high polish, is hard, malleable, ductile, and a good conductor of heat and electricity. It is insoluble in water, soluble in dilute HNO_3, slightly soluble in HCl and H_2SO_4, and insoluble in NH_4OH.

Five stable isotopes, with their percent natural abundance, have been reported: $^{58}Ni(68.27)$; $^{60}Ni(26.10)$; $^{61}Ni(1.13)$; $^{62}Ni(3.59)$; and $^{64}Ni(0.91)$. Nickel forms stable complexes with many organic ligands; however, complexes with other naturally occurring inorganic ligands are formed only to a small degree and in the order $OH^- > SO_4^{2-} > Cl^- > NH_3$. Under anaerobic conditions, sulfide may control the solubility of Ni (Richter and Theis, 1980).

Normally, Ni occurs in the 0 and II oxidation states, although I, III, and IV states can exist under certain conditions. The latter ions are not stable in aqueous solution.

Nickel is closely related to Co in both its chemical and biochemical properties (Memon et al, 1980). Nickel can replace other heavy metals located at active sites, especially in metallo-enzymes, and thus inactivate the function of certain essential metallo-organic compounds.

2. Production and Uses of Nickel

United States production of Ni peaked in the late 1960s, leveled off until about 1976, and declined in 1978 (Fig. 11.1). World production continued to rise until 1976, then paralleled the United States and declined afterward. As of 1979, the primary mine producers of Ni in the world, in decreasing order were: the USSR, Canada, New Caledonia, Australia, Cuba, and the Philippines (Matthews, 1980). However, in smelter production, Japan

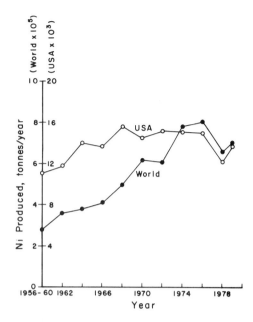

FIGURE 11.1. Nickel production in the United States and worldwide. *Source:* US Bureau of Mines, *Minerals yearbook,* several issues.

followed the USSR, with Canada, third. Traditionally, however, Canada has been the world's largest producer (Duke, 1980b). All Canadian, Zimbabweian, and South African Ni, and some Russian and Australian Ni are produced from sulfide ores. Others are produced from oxide ores.

The major uses of refined Ni in industry include electroplating, alloy production and fabrication, the manufacture of Ni-Cd batteries and electronic components, and the preparation of catalysts for hydrogenation of fats and methanation.

Most Ni is used in alloys that are strong and corrosion-resistant, such as stainless steel. Stainless steel production is the single largest application of Ni and has been the most rapidly growing use in recent years (Duke, 1980b). Thus, Ni is found in a wide variety of commodities, such as automobiles, batteries, coins, jewelry, surgical implants, kitchen appliances, and sinks and utensils. Nickel is also used in Ni-Fe (used for magnetic components in electrical equipment), Ni-Cu (noted for resistance to corrosion, especially in marine applications), Ni-Cr (used in heating elements for stoves and furnaces), and in Ni-Ag (used as a base for electroplated articles) alloys. Nickel-steel alloys are crucial in armour plating and in armaments. Other high-Ni alloys are used for such high-technology applications as turbine blades, jet engine components, and in nuclear reactors. In 1973, the worldwide consumption of Ni was reported to be about 660×10^3 tonnes (Sevin, 1980). Of that amount, about 35% was used in the United States in the following percentages: transportation, 20.9%; chemicals, 14.9%; electrical goods, 12.9%; fabricated metal products, 10%; petroleum, 8.9%; construction, 8.9%; machinery, 6.9%; household appliances, 6.9%; and miscellaneous uses, 9.7%.

3. Natural Occurrence of Nickel

Nickel constitutes about 80 ppm of the earth's crust (NAS, 1975) making it the 23rd most common element. Most of this is in igneous rocks, of which Ni comprises approximately 100 ppm. Hence, it is ubiquitous in the environment. Deposits that are commercially exploitable contain at least 10,000 ppm Ni. Nickel content varies greatly between rock types (Table 11.1). Sandstone has very low Ni content, whereas peridotite or serpentine (ultrabasic igneous rocks) averages as much as 2,000 ppm (Cannon, 1974). Serpentine rocks also contain high Cr and high Mg and Fe, but low Ca; serpentines have anomalously high Mg:Ca ratios (200,000:15) compared with the crustal average, and relatively low levels of silica (Cannon, 1978). Thus, serpentine-derived soils are usually sparse

TABLE 11.1. Commonly observed nickel concentrations (ppm) in various environmental matrices.

Material	Average concentration	Range
Igneous rocks[a]	75	2–3,600
Limestone[a]	20	—
Sandstone[a]	2	—
Shale[a]	68	20–250
Sedimentary rocks[b]	20	—
Coal[c,d]	15	3–50
Fly ash[d]		
Bituminous	11	—
Sub-bituminous	2	—
Lignite	13	—
Petroleum[e]	10	—
Soils (world)	40[f]	5–500[c]
Herbaceous vegetables[e]	—	0.02–4
Common crops[g]	—	<.1–4.0
Ferns[e]	1.5	—
Fungi[e]	1.8	—
Lichens[e]	—	1.5–3.8
Woody angiosperm[e]	1.4	—
Woody gymnosperm[e]	—	2.1–5.3
Freshwater*[e]	0.5	0.02–27
Sea water*[e]	0.56	0.13–43

* μg/liter
Sources: Extracted from
[a] Cannon (1978).
[b] Polemio et al (1982).
[c] Trudinger et al (1979).
[d] Adriano et al (1980).
[e] Bowen (1979).
[f] Vinogradov (1959).
[g] Vanselow (1965).

in vegetation because of nutrient imbalances and toxicities. In the tropics, Ni is concentrated with Fe in laterite, a hard red soil characterized by low silica but high Fe and Al contents. The Ni content of laterite may be as much as 60 times that of the parent rocks.

The primary category of Ni ore is laterite, an Ni oxide ore mined mostly by open-pit technique in Australia, Cuba, Indonesia, New Caledonia, the United States, and the USSR (Sevin, 1980). An Ni sulfide ore, pentlandite, is being mined underground in Canada and the USSR. There are almost 100 minerals of which Ni is an essential constituent (Duke, 1980a). The majority of Ni minerals are sulfides and arsenides. Pentlandite is the principal ore mineral of Ni and is a common accessory mineral in igneous rocks. The most important Ni ore deposits of commercial importance are the Ni sulfide minerals associated with mafic and ultramafic igneous rocks. The deposits of Sudbury in Ontario, Thompson in Manitoba, Norilsk and Pechenga in the USSR, and Kambalda in Australia are of this type (Duke, 1980a).

The soil content of Ni is also extremely variable (Table 11.1), with the world's average reported around 40 ppm; normal soils are reported to contain from 5 to 500 ppm (Swaine, 1955). Mitchell (1945) grouped soils into two categories: (1) those derived from sandstones, limestones, or acid igneous rocks, containing <50 ppm; and (2) those derived from argillaceous sediments or basic igneous rocks, containing from 5 to over 500 ppm of Ni. It is not uncommon for soils derived from ultrabasic igneous rocks to contain 5,000 ppm or more of total Ni.

4. Nickel in Soils

4.1 Total Soil Nickel

The values for total Ni contents of soils reported in the literature are quite variable. These variations are obviously caused by several factors, including soil parent materials. For world soils, Vinogradov (1959) reported a mean value of 40 ppm, while Aubert and Pinta (1977) reported a range of 20 to 30 ppm. More recently, Berrow and Reaves (1984) reported a value of 25 ppm Ni for soils from various parts of the world, while an average value of 22 ppm Ni was reported for tropical Asian paddy soils (Domingo and Kyuma, 1983). Values (in ppm) for several countries were reported as follows: United States—20; Canada—20 (range, 5–50); Italy—28 (range, 4.0–97.5); England and Wales—26 (range, 4.4–228); Sweden—8.7 (Archer, 1980; McKeague and Wolynetz, 1980; Polemio et al, 1982; Andersson, 1977).

Considerable variations in Ni contents especially occur in serpentine soils. Proctor (1971) reported values of 100 to 3,000 ppm for these soils in the United Kingdom and Scandinavia. Brooks et al (1974a) summarized average values (in ppm) for serpentine soils for other countries: Italy—2,600; the USSR—5,000; Portugal—4,000; Zimbabwe—7,000; New Caledonia—5,400; western Australia—700; and New Zealand—3,300.

Mitchell (1971) reported Ni contents (in ppm) for typical Scottish surface soils developed from different parent materials as: serpentine—800; olivine gabbro—50; andesite—10; granite—10; shale—40; and sandstone—15. Andersson (1977) reported 7.8 ppm total Ni for noncultivated Swedish soils and 9.5 ppm for cultivated soils.

In the Volgograd Oblast region of the USSR, the Ni content of the soils varied from about 5 to 40 ppm depending on the soil texture and soil group (Ogoleva and Cherdakova, 1980). The highest Ni concentrations are usually found in the illuvial horizons.

4.2 Extractable (Available) Nickel in Soils

Since total Ni in soil is not a reliable index of plant availability and mobility of Ni, the extractable fraction is commonly used for this purpose. As expected, there are considerable variations in the amount (as percentage of total concentration) of Ni removed by a particular extractant: 0.5–2.0% by DTPA extractant for some Canadian soils (Whitby et al, 1978); as much as 7.0%, also by DTPA extractant, for some Egyptian soils in the Nile Delta (Elrashidi et al, 1979); about 3% by NH₄OAc-EDTA extractant for some polluted Greek soils (Nakos, 1982); from 1.3% to 5.7% by DTPA-TEA extractant for some Southern California soils (Valdares et al, 1983);

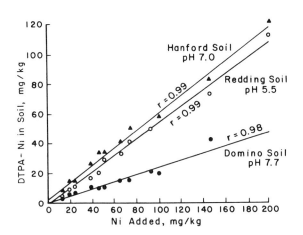

FIGURE 11.2. Relationships between DTPA-extractable nickel and total added nickel in soils. Nickel enrichment was via sewage sludge addition at various rates up to 4% (by wt). Soils were planted to successive crops of lettuce and Swiss chard, after which extractable nickel was determined. *Source:* Valdares et al, 1983, with permission of the authors and the Am Soc of Agronomy, Crop Sci Soc of America and Soil Sci Soc of America.

and about 3.8% by HOAc extractant for soils in England and Wales (Archer, 1980). Amounts of Ni extracted with water are usually very low, commonly <0.5%, even for contaminated soils (Haq et al, 1980) and serpentine soils (Shewry and Peterson, 1976).

Just like the other trace elements, several extractants to relate plant-available Ni to extractable Ni have been tested: mild acids (Crooke, 1956; Haq et al, 1980; Pierce et al, 1982; Shewry and Peterson, 1976); neutral salts (Halstead et al, 1969; Shewry and Peterson, 1976); and chelating agents (Valdares et al, 1983; Whitby et al, 1978; Haq et al, 1980; Shewry and Peterson, 1976). Although some investigators found that a suite of extractants was unsatisfactory in predicting the availability of Ni to plants (Shewry and Peterson, 1976), some extractants, like the DTPA solution could predict plant uptake quite well (Valdares et al, 1983). In addition, the latter investigators found an excellent correlation between DTPA-extractable Ni and the total quantity of this element added to the soil (Fig. 11.2). In one case, the extractability of Ni is similar in the acid (pH 5.5) and neutral soils (pH 7.0), but considerably lower in the calcareous soils (pH 7.7).

4.3 Nickel in the Soil Profile

Reports from several countries indicate that for nonpolluted soils, there is no discernible pattern in the distribution of Ni in the soil profile. In the United States, data for 20 agricultural soils in New Jersey (Painter et al, 1953) indicate about equal concentrations of total Ni between the A ($\bar{x}$ = 14.4 ppm) and B ($\bar{x}$ = 16.5 ppm) horizons. However, additional data from New Jersey (Connor et al, 1957) indicate that, for some podzols, total Ni tends to increase with depth, from the A, to B, then to C horizon. In examining 16 major soil series in Minnesota, Pierce et al (1982) found similar total Ni contents of the surface soils ($\bar{x}$ = 23 ± 4 ppm) and subsoils ($\bar{x}$ = 27 ± 11). In an extensive survey in California encompassing 49 soil profiles (Bradford et al, 1967), basically four patterns for Ni soil concentrations can be discerned: (1) uniform distribution throughout the profile (16 profiles); (2) increasing concentration with depth (20 profiles); (3) decreasing concentration with depth (6 profiles); and (4) increasing concentration, followed by a decreasing pattern (7 profiles). In Wales, similar total Ni contents were observed in the surface soils ($\bar{x}$ = 29 ppm) and subsoils ($\bar{x}$ = 33 ppm) (Bradley et al, 1978). Elsewhere in the United Kingdom, observations for several profiles (Butler, 1954; Swaine and Mitchell, 1960) revealed the same patterns observed by Bradford et al (1967) for California soils.

In agricultural soils in Canada, the distribution of total Ni in the profile was, for all practical purposes, fairly similar (Mills and Zwarich, 1975; Whitby et al, 1978). Even for Canadian serpentine soils, the distribution of total Ni in the profile was remarkably uniform (Roberts, 1980). However,

TABLE 11.2. Nickel concentrations (ppm) in litter and soil of the Solling Forests, Hartz Mountains, the Federal Republic of Germany.[a]

	Spruce forest	Beech forest
Organic top layer		
O_L-horizon[b]	12(0.1)[c]	15(1.7)
O_F-horizon	23(7.5)	—
O_H-horizon	22(6.4)	—
O_{F+H}-horizon	—	22(2.9)
Mineral soil		
0–10 cm	11	6
10–20 cm	16	7
20–30 cm	18	Not determined
30–40 cm	22	13
40–50 cm	27	14

[a] *Source:* Extracted from Heinrichs and Mayer, 1980, with permission of John Wiley & Sons, Inc, New York.
[b] O_L = litter layer, not or slightly decomposed; O_F = fermentation layer, partly decomposed; O_H = humic layer, completely decomposed and humified.
[c] Standard deviation in parentheses.

in agricultural muck soils in Canada, there was a declining trend in Ni concentration with depth (Hutchinson et al, 1974). The decline in concentration with depth was even more dramatic in virgin muck soils. In other parts of the world, the pattern may be uniform, as in the case of several profiles in New Zealand (Quin and Syers, 1978), or variable, as in Papua-New Guinea soils (Bleeker and Austin, 1970). Considerable leaching of Ni in the soil profile could occur in instances where the soil has been severely acidified, such as in soils close to smelters (Hazlett et al, 1984).

In forest ecosystems, the Ni content of soils may tend to be higher in the mineral A horizon ($\bar{x} = 0.52 \pm 0.13$) than in the underlying horizons ($\bar{x} \leq 0.10$ ppm) as indicated by data for extractable Ni in Nagoya University Forest (Memon et al, 1980). Conversely, Ni content tended to increase with mineral soil depth in the Solling Forest ecosystem in the Hartz Mountains of the Federal Republic of Germany (Table 11.2) (Heinrichs and Mayer, 1980). It should be noted that Ni tended to accumulate in the surface organic layer (Table 11.2), a trend also observed for other trace elements.

4.4 Forms of Nickel in Soils

As with the other trace elements, Ni in the solid phase of soils occurs in several chemical forms. These include occurrence in the usual exchange sites, specific adsorption sites, adsorbed or occluded into sesquioxides, fixed within the clay mineral lattice, or fixed in organic residues and microorganisms. In the aqueous phase (i.e., soil solution), Ni may occur in

TABLE 11.3. Distribution (as %) of chemical forms of nickel in sewage sludge and sludge-treated soils.[a]

			Treated soil		
Extractant	Chemical form	Sewage sludge	Holland (pH 5.8–6.4)	Ramona (pH 5.2–6.7)	Helendale (pH 7.4–8.0)
0.5 M KNO$_3$	Exchangeable	10.9	<0.1	<0.1	<0.1
H$_2$O	Adsorbed	0.5	<0.1	<0.1	<0.1
0.5 M NaOH	Organically bound	24.3	<0.1	<0.1	<0.1
0.05 M EDTA	Carbonate	31.9	3.4	3.9	8.1
4.0 M HNO$_3$	Sulfide/residual	26.4	79.8	93.5	87.1

[a] *Source:* Emmerich et al, 1982a, with permission of the authors and the Am Soc of Agronomy, Crop Sci Soc of America, Soil Sci Soc of America.

the ionic form and in forms complexed with either organic or inorganic ligands.

The most common technique of partitioning solid phase metals into chemically similar forms is by using sequential extraction with selective chemical reagents. Such fractionation scheme has been conducted by Emmerich and associates whose results (Table 11.3) indicate that, for three southern California soils treated with sewage sludge, most of the soil Ni exists in the residual form ($\bar{x}$ = 87%), followed by the carbonate form ($\bar{x}$ = 5.1%). The residual form is considered the most stable form of Ni because it requires the most drastic extracting reagent for its removal. Using the same sequential extraction, Chang et al (1984a) found that Ni in a sandy loam soil treated for seven years with composted sludge was mostly in the sulfide/residual (41%), carbonate (34%), and organic (23%) forms. Hickey and Kittrick (1974), using another sequential extraction scheme, found that for soils heavily enriched with heavy metals, Ni was distributed as follows: residual >> Fe-Mn oxide > organic ≈ carbonate > exchangeable form. However, with a polluted sediment, Ni was distributed as Fe-Mn oxide >> residual = carbonate > organic = exchangeable form. Cottenie et al (1979) reported that for contaminated soils and river sediments, Ni occurred primarily in the precipitated and 0.5 N HNO$_3$-extractable forms. Very small amounts were in the water-soluble form, while adsorbed or complexed form can be significant fractions of the total Ni.

Using the GEOCHEM computer program (Sposito and Mattigod, 1980), Emmerich et al (1982b) predicted the forms of Ni in the soil solutions for the Ramona soil treated with sewage sludge as follows: in the sludge layer—free ionic form, 52.5%; inorganic complex form, 22%; and organic complex form, 26.5%; in the mineral soil below the sludge layer—free ionic form, 68%; inorganic complex form, 26.5%; and organic complex form, 6.0%.

Using a different fractionation scheme, Lindau and Hossner (1982) found the following percentages for sediments in natural marsh ecosystems near Galveston, Texas: residual, 52.1%; moderately reducible, <7.8%; organic matter plus sulfide, 28.5%; easily reducible, 7.6%; and exchangeable plus water-soluble, <1.2%.

However, the exchangeable form of Ni may constitute a significant portion of the total Ni for certain soils, such as in serpentine soils (Crooke, 1956) and some soils of Japan (Biddappa et al, 1982).

4.5 Sorption and Complexation of Nickel in Soils

A number of soil factors would have a bearing on the sorption of trace elements including Ni. These include CEC, pH, texture, $CaCO_3$ content, organic matter, sesquioxides, chelating agents, and others. Since Ni(II) is stable over a wide range of pH and redox conditions (Cotton and Wilkinson, 1980), this is the species of Ni expected to occur in most soils. Nickel is known to complex readily with a variety of inorganic and organic ligands. Nickel-halides and salts of oxo-acids are generally soluble in water, while Ni carbonate is fairly insoluble.

In examining soils from agricultural and potential waste disposal sites in New Mexico, Bowman et al (1981) found that sorption of Ni from 0.01 N $CaCl_2$ was extensive for all samples; >99% of added Ni was sorbed in some cases. In spite of this almost complete sorption, Ni sorption was not completely irreversible, as some Ni was desorbed by the chelating agent, DTPA. However, in the presence of high levels of Ca^{2+} ions, as $CaCl_2$, and small amounts of the chelating agent EDTA in the equilibrating solution, the Ni sorption by the soils was minimized. In this case, Ca^{2+} ions competed with Ni^{2+} for sorption sites rather than Ni-Cl complex formation, while EDTA complexed with Ni, maintaining it in soluble form in the aqueous phase. Sommers and Lindsay (1979) indicated that Ni would bind with a variety of synthetic chelates to the virtual exclusion of competing cationic species at typical soil pHs. In some instances, inorganic complexation of Ni may be significant in affecting its mobility. Some investigators found increased Ni mobility (Doner, 1978), or decreased sorption of Ni (Bowman and O'Connor, 1982) in the presence of Cl^- ions, apparently due to Cl^- complex formation. Similarly, SO_4^{2-} or NO_3^- ions may also have a complexing effect on Ni (Mattigod et al, 1979; Bowman and O'Connor, 1982).

Nickel is retained in certain soils primarily by specific sorption mechanisms at low (≤ 10 ppm) concentrations (Bowman et al, 1981). The isotherm conforms to the Freundlich equation up to about the initial solution concentrations of 10 ppm Ni. Depending on sorption capacities, the Freundlich equation may describe the adsorption of Ni at concentrations well above 10 ppm Ni. However, Harter (1983) noted that Ni sorption by soil could display a complex (i.e., multiphasic) isotherm. Further elaboration on the Ni retention mechanism by soil is needed. In all soils (pH 4.3–5.6) used by Harter (1983), adsorption of metals was in the order Pb > Cu > Zn > Ni. Using soil columns, the mobility of metals in soils (pH

3.8–7.1) tended to be: Cd ≥ Ni ≥ Zn >> Cu (Tyler and McBride, 1982).
In general, differences in sorption or mobility of the metals can be a result
of the differences in soil chemical properties.

5. Nickel in Plants

5.1 Nickel in Plant Nutrition and Its Uptake and Translocation

The essentiality of Ni has not been established for higher plants. However,
some beneficial effects of Ni on plant growth have been reported. For
example, Ni seems to be required for optimum growth of certain pine tree
species and some Ni accumulator species of *Alyssum,* but its function is
not known (Welch, 1981). Nickel has also been shown to stimulate seed
germination and growth of a number of plant species. In a review, Hutchinson (1981) indicated that, although the essentiality of Ni to plants has
not yet been demonstrated conclusively, evidence for this continues to
accumulate. This includes growth stimulation, yield increases of some
plants, such as potatoes and grapes, growth stimulation of certain microorganisms, and functioning of certain enzymes. Hutchinson and his associates at the University of Toronto have demonstrated growth stimulation of the grass *Deschampsia cespitosa* to quite high Ni concentrations
in the soil. Similar response was obtained for *Phragmites communis* seedlings grown in nutrient culture spiked with 0.1 to 0.5 ppm Ni. Using soybean tissue culture experiments, Polacco (1976, 1977) showed that Ni^{2+}
is required in urea utilization and urease synthesis, indicating its possible
role on N metabolism by plants. More recently, Bick et al (1982) observed
that leaves of the wild plant *Alyssum bertolonii* showed a direct relationship
between their total free amino acid N and Ni contents. These plants thrive
on serpentine outcrops in northern Italy.

Since Ni is ubiquitous in the environment, it is a normal constituent of
plant tissues. Vanselow (1965) reported that the Ni content of field grown
crops and natural vegetation ranges from 0.05 to 5.0 ppm in the dry matter.
Subsequently, Connor et al (1975) have reported mean values of about
0.20 to 4.5 ppm for nearly 2,000 specimens of field crops and natural vegetation from the United States. More recently, Cottenie et al (1979) reported that, for normal plants, the range is from trace level up to about
8 ppm Ni.

Nickel is usually absorbed in the ionic form, Ni^{2+}, from the soil or
culture media. Mishra and Kar (1974) in their review, concluded that Ni
is more easily absorbed by plants when supplied in the ionic form than
when chelated. A possible explanation is that the charge on the chelated
molecule inhibits its absorption by the roots—i.e., a molecule with no
charge or a slight negative charge can be taken up, while a complexed
molecule with strong negative charge cannot. High Ni content of the soil,

particularly when the exchangeable form is also high, accelerates the absorption of this element by the plants when other conditions are also favorable (Mishra and Kar, 1974).

Using 21-day old soybean plants, Cataldo et al (1978a) demonstrated the presence of multiphasic absorption isotherms for Ni^{2+}. Each of the three isotherms conformed to Michaelis-Menten kinetics. Such multiple absorption isotherms are a common characteristic of ion uptake in plants. Subsequently, Cataldo et al (1978b) found that, following its absorption by the root, Ni, transported within the xylem as organic complexes, was highly mobile in plants, with leaves being the major sink in the shoots during vegetative growth. At senescence, >70% of the Ni present in the shoot was remobilized to seeds. This behavior of Ni is shared by a number of micronutrients, i.e., Fe, Cu, Mn, and Zn (Cataldo and Wildung, 1978). Nickel accumulated in the seeds is primarily associated with the cotyledons. Comparison of the translocation of Ni and Cr showed that Ni was translocated much more rapidly than Cr from roots to other parts where it was accumulated (Ishihara et al, 1968). Chromium was accumulated in fine roots, and its translocation was very slow compared to that of Ni. Thus, it can be generalized that since Ni occurs in plants as a chemically stable, soluble species, it can accumulate readily. In addition, Ni, unlike most nonessential elements, is fairly mobile in plants and readily accumulates in the seeds (Mishra and Kar, 1974). Cataldo et al (1978b) concluded that the behavior of Ni^{2+} in plants may be regulated by the same mechanisms utilized for nutrient ions such as Cu^{2+} and Zn^{2+}, since Ni^{2+} has been shown to be a competitor for root transport sites employed for both Cu^{2+} and Zn^{2+}.

5.2 Plant Accumulators and Hyperaccumulators of Nickel

Concentrations of Ni in plants growing in normal soils seldom exceed 5 ppm (Vanselow, 1965). However, this is not the case in environments with ultrabasic substrates, such as periodotite or serpentinite. The Ni content in dry matter of vegetation from serpentine areas is usually <50 ppm and seldom exceeds 100 ppm on dry weight basis (Brooks and Radford, 1978).

Brooks and his co-workers (Brooks, 1980; Brooks et al, 1977b; Kelly et al, 1975; Severne and Brooks, 1972; Brooks and Wither, 1977; Wither and Brooks, 1977) in New Zealand have worked extensively on Ni plant "indicators" or "accumulators." They defined the term "accumulator" as a plant whose mean content of a particular element (expressed on an ash weight basis) is greater than the content of the same element in the fine earth fraction of the substrate (Brooks et al, 1974a). Severne (1974) more broadly defines "accumulators" as plants that have dry-weight elemental concentrations greater than either the associated substrate or normal plants. Since there is a small number of plants whose Ni content

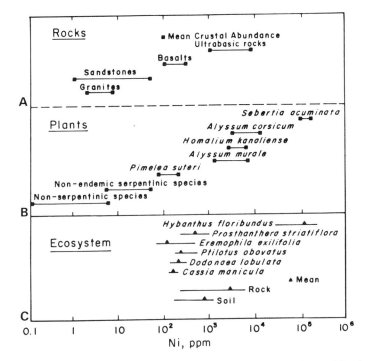

FIGURE 11.3. Range of nickel levels (ppm) in rocks (a), plants (b), and in shrubs, rocks and soils from the eastern Goldfields of western Australia (c). *Sources:* Adapted from Brooks, 1980; and Severne and Brooks, 1972, with permission of the authors.

can, under certain circumstances, exceed by a factor of 10 even the highest values observed for plants growing over ultramafic nickeliferous substrates, Brooks et al (1977a) ultimately suggested the term "hyperaccumulators" to denote such species. Leaves of such species should have at least 1,000 ppm Ni, on dry weight basis. Some of these species are indicated in Figure 11.3. "Hyperaccumulators" are not always restricted to Ni-rich ultrabasic substrates but are always associated with them when the 1,000 ppm level is exceeded.

Brooks (1980) listed three reasons for interest in high Ni values in plants: (1) the challenge of reclaiming serpentine soils, i.e. making fertile the appreciable area of the earth covered by ultrabasic rocks; (2) the potential usefulness of using Ni levels in vegetation as an indication of the Ni content of the soil, e.g., in biogeochemical prospecting; and (3) the phytochemical and geobotanical considerations. Serpentine soils are unsuitable to support normal vegetation, primarily because of their low nutrient contents, high Mg content ($\approx 20\%$) which can antagonize Ca uptake by plants, and direct toxicity of elements, such as Ni and Cr. Application of biogeochemical prospecting has been relatively successful for Ni in New Zealand, Australia, New Caledonia, and the USSR. It may also be applicable to U, Co, Cr, Pb, and platinum metal prospecting but tends to be unreliable for Cu and Zn, which are essential elements for plant nutrition. "Hyperaccumulators" are of special interest to geobotanists and phytochemists because of their adaptability to ultrabasic areas and the excessive levels of Ni in plant tissues. This phenomenon is usually considered a tolerance mechanism, the details of which are as yet still largely unknown. No one has claimed a universal mechanism for this phenomenon.

Brooks (1980) also indicated some interest in the geographical distribution of "hyperaccumulators" of Ni. Of the identified species, 12 are from New Caledonia, 14 are from southern Europe, 5 are from Southeast Asia, 2 are from central Africa, and 1 is from western Australia. However, he stated the possibility of a geographical bias for the distribution of such plants caused by strong selective pressures acting on isolated and endemic floras, such as the one in New Caledonia. Similar selection pressures may have acted on the serpentine floras of Czechoslovakia and the western United States (Kruckeberg, 1954; Walker, 1954).

5.3 Nickel Interactions with Other Elements

In plant nutrition, elements can interact either synergistically or antagonistically. Serpentine soils usually have elevated levels of Fe, Co, Cr, Ni, and Mg. Their high Mg contents can inhibit Ca uptake by plants (Kruckeberg, 1954). Therefore, plants growing on serpentine soils can be considered as unusually tolerant of low available Ca.

Since the uptake behavior of Ni by plants is characteristic of nutrient ions, Ni may serve as an analog of an essential element for which known transport mechanisms are operating. To assess whether Ni^{2+} uptake was the result of its behavior as an analog, Cataldo et al (1978a) measured the rate of Ni^{2+} uptake by soybean plants in the presence or absence of Co^{2+}, Cu^{2+}, Fe^{2+}, Mg^{2+}, Mn^{2+}, and Zn^{2+}. Of the ions investigated, Mg^{2+} and Mn^{2+} did not inhibit Ni^{2+} absorption, while Co^{2+}, Cu^{2+}, Fe^{2+}, and Zn^{2+} did. In the presence of the latter four elements, Ni^{2+} uptake was reduced by 25% to 42% of the rates found for control treatments. In addition, Cataldo et al showed that Cu^{2+} was a competitive inhibitor to Ni^{2+} uptake and that Cu^{2+} appears to be a better competitor of Ni^{2+} than is Zn^{2+}.

Mizuno (1968) noted that several crop species grown on serpentine soil did not exhibit Ni toxicity symptoms when the Cu:Ni ratio was equal to or greater than one or when the Fe:Ni ratio was equal to or greater than five. Crops, such as kidney beans, soybeans, cabbage, and alfalfa, which contained low ratios of Fe:Ni and Cu:Ni, were extremely susceptible to Ni toxicity, whereas crops such as potatoes and corn, which contained high ratios, were not affected. Iizuka (1975) alleviated the severity of Ni toxicity symptoms in mulberry plants grown on serpentine soil by either top-dressing the soil or foliar spraying with Fe and Zn compounds. In studying the interactions among Cu, Ni, and Zn in young barley, Beckett and Davis (1978) noted that the interactions were insignificant and thus may be ignored when dealing with subtoxic accumulations. However, at higher concentrations, the interactions became larger and more complex.

Because of the role of Fe in overcoming part of the phytotoxicity caused by some metals, it can be suggested that trace metal-induced chlorosis is a form of Fe deficiency. Hunter and Vergnano (1953) reported that excessive concentrations of heavy metals applied to plants produced Fe deficiency chlorosis, concomitant with the secondary symptoms specific to the metal in question, with Ni being more pronounced than the other metals. Later, Crooke and Knight (1955), using autoradiographic technique, found that the concentration of Fe in the necrotic areas of oat plants suffering from Ni toxicity was very low. More recently, Foy et al (1978) concluded in their review that Ni, along with Cu, could cause an actual Fe deficiency by inhibiting its translocation from roots to shoots.

Crooke et al (1954) found that an increase in Fe level in the culture solution produced a reduction in Ni toxicity symptoms. Furthermore, they reported a correlation of necrotic symptoms with the Ni content and with the Ni:Fe ratio in the plant, similar to that indicated in Figure 11.4. More recently, Khalid and Tinsley (1980) reported that the Ni:Fe ratio, rather than the Ni and Fe concentrations in plants, shows a better relationship with Ni toxicity effects. Crooke (1955) found that Ni consistently reduced Fe contents of the roots and tops of oats and tomato plants. In contrast, Agarwala et al (1977) found that while excess Co^{2+}, Cu^{2+}, Mn^{2+}, and Zn^{2+} reduced Fe translocation to shoots of barley, excess Ni^{2+}, which was most active in producing toxic effects, did not.

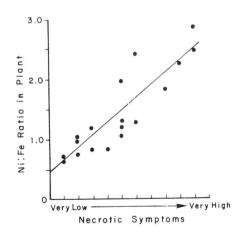

FIGURE 11.4. Relationships between Ni:Fe ratio in the plant and the degree of necrotic symptoms. *Source:* Crooke, 1955.

5.4 Toxicity of Nickel in Plants

In general, when Ni concentrations in vegetative tissues of plants exceed 50 ppm (dry wt basis), plants may suffer from excess Ni and exhibit toxicity symptoms. Exceptions are the endemic species of serpentine soils or the so-called Ni accumulators and hyperaccumulators, which may contain several thousands ppm of Ni (Fig. 11.3).

A wide range of levels of Ni in the growth media and in plant tissues has been reported toxic to plants. Mishra and Kar (1974) reported that in nutrient solution, Ni may be toxic to plants at levels as low as <1 ppm to levels as high as 300 ppm, depending on the plant species.

With trees, levels $\geqslant 10$ ppm Ni in the nutrient solution substantially reduced the growth of white pine and white spruce seedlings (Lozano and Morrison, 1982). In soil, Ni levels can be much higher before phytotoxicity occurs. For example, Khalid and Tinsley (1980) noted no yield depressions and the absence of chlorotic symptoms in ryegrass until at about the 90 ppm level in a noncalcareous (pH 4.7) soil.

In plants, the following toxic levels (ppm dry wt) in the foliage were found: rice, 20–50 (Chino, 1981); ryegrass, 154 (Khalid and Tinsley, 1980); barley, 26 (Davis et al, 1978); hardwood species, 100–150 (Lozano and Morrison, 1981); and citrus, 55–140 (Vanselow, 1951). These results indicate that plants are divergent in their sensitivity to Ni.

The toxicity symptoms produced by Ni are similar to those produced by several heavy metals and consist of (1) chlorosis caused by Fe-induced deficiency, and (2) specific effects of the metal itself. A typical toxic symptom produced by Ni is the chlorosis or yellowing of leaves followed by necrosis. Other symptoms include stunted growth of root and shoot, deformed plant parts, unusual spotting, and in severe cases, death of the whole plant. Clark et al (1981) found that, with sorghum, the symptoms of Ni excess are similar to those for excess Co, which in turn resembles Fe deficiency, except that the yellow streaks of the Co-induced symptoms did not extend as readily to the leaf tips as in Fe deficiency. Roots were stubby, short, thick, dark-colored, and brittle. In rice, interveinal chlorosis appears in new leaves, and the number of leaves and tillers are decreased (Chino, 1981). Root growth is severely depressed. Hewitt (1953) noted that Ni toxicity symptoms resemble those of Mn deficiency in potato and tomato. Excess Ni can cause mitotic disturbances in root tips of some plants (Mishra and Kar, 1974).

Several investigators have determined the phytotoxicity of several metals, relative to Ni. In barley, the effectiveness of metals in inducing visual toxicity symptoms was in the order Ni > Co > Cu > Mn > Zn (Agarwala et al, 1977). In rice, the order was Cu > Ni > Co > Zn > Mn and Hg > Cd > Zn (Chino, 1981). In oats, the order was Ni > Cu > Co > Cr > Zn > Mo > Mn (Hunter and Vergnano, 1953). In corn, however, Cd proved more toxic than Ni (Traynor and Knezek, 1974). In corn and sunflower, the toxic effect to net photosynthesis and growth was in the order Tl > Cd > Ni > Pb (Carlson et al, 1975).

6. Factors Affecting the Mobility and Availability of Nickel

6.1 pH

The solubility and plant availability of most heavy metals in soils are known to be inversely related to pH. The effects of pH on Ni chemistry in soils have been demonstrated in soil retention studies, sewage sludge application on land, reclamation of serpentine soils, plant uptake studies, and others. In soil sorption studies, the amount of Ni retained was dependent upon the pH of the soil, with retention increasing with increasing pH (Harter, 1983; Gerritse et al, 1982). Harter (1983) found that Ni extracted with 0.01 N HCl averaged about 70% to 80% of the amount retained by soils from pH <5.0 to ≥8.0. However, the percent extractability decreased to <75% of the Ni retained at pH 7 or above. Also pH influences the precipitation of Ni with other compounds, such as phosphates. Pratt et al (1964b) reported that formation of Ni-P complexes could occur at pH values ≥7.0 thereby reducing Ni toxicity in soils.

In lands receiving sewage sludge high in metal contents, liming of soils to elevate or maintain pH at 6.5 or above, Ni and other metals should not cause toxicity to crops or pose a hazard to the food chain (CAST, 1976). Liming sludged soils is important since sludge high in ammoniacal-N would tend to acidify the soil through the production of H^+ ions from the mineralization and nitrification of sludge N. John and Van Laerhoven (1976) found that differences in plant tissue Ni concentrations were associated with sludge-lime interaction.

Liming serpentine soils also has a dramatic effect on Ni uptake by plants (Crooke, 1956; Halstead, 1968; Hunter and Vergnano, 1953). Elevating the pH of these soils reduces the amount of exchangeable Ni and therefore, Ni toxicity to plants. The drastic effect of elevated soil pH on Ni uptake by plants growing on serpentine soils is demonstrated in Figure 11.5.

Other investigators (Painter et al, 1953; Halstead et al, 1969; Bingham et al, 1976; Wallace et al, 1977) have also demonstrated the effect of pH on Ni uptake by plants. Bingham et al (1976) indicated that the critical soil concentration of Ni was much lower for acid than for calcareous soil. For example, in wheat as plant indicator, the initial Ni level in soil that produced 50% yield decrement was 195 ppm in acid soil compared with 510 ppm in calcareous soil; with Romaine lettuce, the values were 110 and 440 ppm, respectively, for acid and calcareous soils. Painter et al (1953) found that approximately 20% more Ni was removed by alfalfa

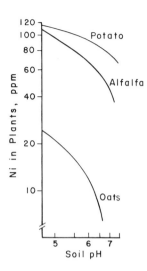

FIGURE 11.5. Relationships between soil pH and nickel content (ppm) in selected crops. *Source:* Mizuno, 1968. Reprinted by permission.

from plots having pH values lower than 6.8 than from plots having higher pH values. However, there might be instances where liming the soil at even close to pH 8.0 may not alleviate Ni toxicity to plants (Soane and Saunder, 1959).

6.2 Organic Matter

Organic matter, depending on its nature, can either immobilize or mobilize metals. Organic matter has been shown to fix Ni, thereby rendering it less available to plants (Halstead, 1968; Halstead et al, 1969). It is a common practice to reclaim serpentine soils by either liming or adding organic matter or both. This is because organic matter can add to the sorption capacity of soils for metals. For example, in comparing the additions of heavy metals to soils as inorganic salts with the addition of sludge spiked with metals, Cunningham et al (1975b) found that Cu, Cr, and Zn were less available to plants when applied in the latter form. The binding mechanism in sludge may involve weakly acidic functional groups (Gould and Genetelli, 1978). A large portion (50%–80%) of organic matter of stabilized sewage sludge is fairly resistant to decomposition in the soil and is similar to humified organic matter (Terry et al, 1979a, b). Latterell et al (1978) estimated that the average rate of decrease in soil organic matter from a sludge application ranged from 10% to 18% per year.

In contrast, metals can be complexed by dissolved organic matter and therefore rendered more mobile. Pratt et al (1964a) reported that the capacity for chelation of Ni by organic matter is approximately 2.2% by weight. For example, a soil with 1% organic matter could complex at least 220 μg Ni/g. Bloomfield et al (1976) found that the metals were mobilized partly in association with colloidal humified organic matter and partly in true solution as complexes that appeared to be anionic. Also, during the decomposition of organic matter, a number of organic acids is produced by microbial activity. These acids can act as chelating agents and form complexes with Ni. Misra and Pande (1974) noted that the organic acids most effective in complexing Ni are of the di- and tri-carboxylic acids. Thus, Jackson and Sherman (1953) inferred that the movement of metals to lower horizons or out of soil profiles is due to organic constituents having chelating abilities.

6.3 Soil Texture and Parent Material

Although data are meager, Ni content in the various soil fractions tends to increase with decreasing particle size and can be generalized as sand < silt < clay (LeRiche and Weir, 1963; Painter et al, 1953). The effects of parent material on Ni content of soils are discussed in sections 3 and 4.1.

6.4 Plant Factors

Like the other metals, uptake of Ni by plants varies according to species (Sheaffer et al, 1979; Giordano and Mays, 1977; Memon et al, 1980; Ogoleva and Cherdakova, 1981) and cultivars (Welch and Cary, 1975). In addition, Ni accumulates differentially between plant parts (Brooks et al, 1974a; Fleming, 1963) and tends to accumulate higher in younger leaf blades (Hunter and Vergnano, 1952).

7. Nickel in Drinking Water and Food

Most of the data on Ni obtained in the early 1960s and earlier probably are unreliable because of analytical unreliability, whereas later reports benefit from the recent development of precise analytical techniques. For example, the limit of detection for Ni by most modern instruments is about 5 ppb (Russell, 1979). Therefore, the data to be considered are those obtained only fairly recently.

The acceptable Ni level for drinking water in the United States has not been established (US-EPA, 1976). Clemente et al (1980) reported that the Ni content of most mineral and deep spring waters in Italy is below 10 ppb. A few water samples had about 100 ppb.

The Ni content of representative Canadian diets are presented in Table 11.4, which indicates that the primary sources of Ni in diets are meat, fish, and poultry, followed by grain and cereal. Because of the divergent nature of Ni concentrations in various food items and differences in eating habits, intake of Ni can vary considerably. For example, the levels of Ni in diets of institutionalized children, aged 9 to 12, from 28 United States cities ranged from 0.140–0.321 ppm (Murthy et al, 1973). Clemente et al (1980) reported that the intakes of Ni in Italy by adults range from 100 to 700 μg/day, comparable with range values of 150 to 800 μg/day in the

TABLE 11.4. Estimated dietary intakes and concentrations of nickel by food classes.[a]

Food class	Ni Intake		Ni concentration ppm (fresh wt basis)
	μg Ni/day	% of total diet	
Dairy products	44	9.5	0.09
Meat, fish, and poultry	98	21.1	0.36
Grain and cereal	77	16.7	0.41
Potatoes	50	10.8	0.26
Leafy vegetables	14	3.0	0.29
Legume vegetables	24	5.2	0.74
Root vegetables	8	1.7	0.18
Garden fruits	42	9.1	0.50
Fruits	30	6.5	0.16
Oils and fruits	16	3.5	0.61
Sugars and adjuncts	46	10.0	0.32
Beverages	13	2.8	0.22
Total	462		

[a] Based on the survey of Kirkpatrick and Coffin (1974) in Halifax and Vancouver, Canada.

United States. The Canadian dietary intake of Ni (Table 11.4) ranged from 347 to 576 μg/day for Halifax and Vancouver, respectively (Kirkpatrick and Coffin, 1974). Hamilton and Minski (1972) reported an intake of <300 μg/day for the United Kingdom; that of the International Commission on Radiological Protection is about 400 μg/day/person. Because of the usual low levels of Ni in drinking water, the contribution of drinking water in intake is usually <60 μg/day (Clemente, 1976).

8. Sources of Nickel in the Terrestrial Environment

8.1 Metal-Ore Smelting

Release of air pollutants from metal-ore smelting—chiefly SO_2 and heavy metals, primarily Ni, Cu, Co, and Fe—has been implicated as a cause of damage to terrestrial vegetation and soils in the Sudbury area in Ontario, Canada (Lozano and Morrison, 1981; Hutchinson and Whitby, 1974; Whitby and Hutchinson, 1974; Freedman and Hutchinson, 1980; Temple and Bisessar, 1981). Three major air pollution processes have been identified in the Sudbury region. In the outermost zone, pollution takes the form of acid deposition, merged with the larger background of atmospheric acidity over eastern North America. Closer to the smelters, acid deposition intensifies, and direct injury to vegetation may occur due to SO_2 fumes. Nearest to the smelters, still more intensified acid aerosol washout and SO_2 fumes occur, augmented by substantial deposition of particulates laden with Ni, Cu, and to a lesser degree, Fe and Co.

In the Coniston Ni smelter near Sudbury, Ontario, about 68 tonnes of Ni were released in the last quarter of 1971 (Rutherford and Bray, 1979), while the large smelter at Copper Cliff released 167 tonnes per month at

TABLE 11.5. Nickel concentrations (ppm, dry wt) of crops and feedstuffs with and without the influence of nickel emission.[a,b]

	Average	
Plant	Without emission	With emission
Onions	0.830	2.063
Potatoes	0.565	1.038
Carrots	0.504	1.410
Turnips	0.495	2.199
Dwarf beans	3.075	8.223
Parsley	1.724	11.146
Lettuce	1.403	10.695
Cabbage	0.842	4.652
Cucumbers	0.718	—
Kohlrabi	0.616	1.716
Tomatoes	0.575	1.914
Apples	0.429	1.268
Cauliflower	0.351	—
Oats	0.712	1.456
Winter wheat	0.301	1.259
Winter rye	0.263	1.044
Summer barley	0.246	0.395
Maize	0.197	—
Meadow red clover	1.848	—
Acre red clover	1.529	—
Lucerne	1.404	—
Turnip leaf	1.458	5.757
Pasture grass	1.002	8.996
Green maize	1.054	2.806
Green wheat	0.530	—
Green rye	0.482	—

[a] Source: Anke et al, 1983, with permission of the authors and Raven Press.
[b] Samples were taken from four different areas of the GDR without Ni emission and within the radius (20 km) of a Ni-processing plant that began operation 15 years prior to sampling.

the same time (Hutchinson and Whitby, 1974, 1977). The highest concentrations of metals in soils in the Sudbury region occurred in surface soils close to the smelters (levels up to 3,000–5,000 ppm Ni), but elevated levels of Ni and Cu in soils have been reported from as far away as 50 km from the smelters (Hutchinson and Whitby, 1974). The impacts of the smelting activity are also detectable in lakes within a 30-km radius of the smelting complex. In fact, the enrichment factors in surficial sediments and deposition rates for Ni and Cu are among the highest ever recorded anywhere in the world (Nriagu et al, 1982). In response to this exposure to high Ni over the past 60 years, both plants and animals have evolved some Ni-tolerant populations (Cox and Hutchinson, 1980; Stokes et al, 1973).

Nickel pollution from metal smelting has also been reported in Australia by Ashton (1972) and Beavington (1975). The effects of Ni emission on the concentration of Ni in food and foodstuff are indicated in Table 11.5.

TABLE 11.6. Estimated global emissions of nickel to the atmosphere.[a]

Natural source	Aerosol production (10^9 kg/year)	Nickel concentration in particles (ppm)	Nickel emissions (10^6 kg/year)
Soil suspension	120	40	4.8
Volcanoes	25	100	2.5
Vegetation	75	11	0.82
Forest fires	12	15	0.19
Meteoric dust	0.0036	50,000	0.18
Sea salt	1000	0.009	0.009
Total			8.5 (16%)

Anthropogenic source	World consumption (10^9 kg/year)	Nickel emission factor	Nickel emissions (10^6 kg/year)
Residual oil	578	0.03 kg/tonne burned	17
Fuel oil	323	0.03 kg/tonne burned	9.7
Nickel mining and refining	0.80	9 kg/tonne produced	7.2
Municipal incinerators	2550	0.002 kg/tonne burned	5.1
Steel production	0.24[b]	5 kg/tonne Ni charged	1.2
Transportation	—	—	0.9
Nickel alloy production	0.14	5 kg/tonne Ni charged	0.70
Coal burning	3300	0.0002 kg/tonne burned	0.66
Cast iron production	0.03[b]	10 kg/tonne Ni charged	0.30
Sewage sludge incineration	48	0.001 kg/tonne burned	0.048
Cu-Ni alloy production	0.04[b]	1 kg/tonne Ni charged	0.04
Total			43 (84%)
Total, all emissions			51

[a] Source: Schmidt and Andren, 1980, with permission of John Wiley & Sons, Inc, New York.
[b] Amount of nickel consumed in this activity.

In general, aside from smelters, Ni in air probably comes primarily from fuel oil and coal combustion, since petroleum contains Ni and there is a high correlation between V and Ni in air (Schroeder, 1971). Swaine (1980) reported that coal combustion does not contribute as much Ni to the atmosphere as oil combustion. The former emits only 0.66×10^6 kg Ni/year vs 26.7×10^6 kg/year for the latter (Table 11.6).

8.2 Sewage Sludges

Since sewage sludges can sometimes contain up to 0.5% total Ni (Berrow and Webber, 1972; Sommers et al, 1976), sewage sludge applied to land can cause soil enrichment of Ni. Of the metals in sludge, Cu, Ni, and Zn are considered the ones most likely to cause phytotoxicity, while Cd, Cr, Mn, Hg, and Pb are regarded as less of a potential problem in this regard (Berrow and Webber, 1972). Although a considerable amount of information has indicated increased uptake of metals due to sludge application (Hinesly et al, 1972; Dowdy and Larson, 1975; Cunningham et al, 1975a; Andersson and Nilsson, 1972; Sterrett et al, 1983), long-term studies show

TABLE 11.7. Nickel contents (ppm) in coal by rank and by location and fly ash from several countries.[a]

	Range and/or mean[b]
Coal	
Rank	
Brown	0.1–3 (1)
Sub-bituminous	0.7–15 (5)
Bituminous	1–70 (15)
Anthracite	3–70 (20)
Location	
Australia	
Victoria	0.1–3 (1)
New South Wales and Queensland	3–50 (15)
Europe	5–60 (15)
United States	
Appalachian region	2–70 (15)
Midwest	5–50 (18)
Rocky Mountains	0.70–10 (3)
Northern Great Plains	0.7–15 (2)
Fly ash	
Germany	270
Anthracite combustion, Belgium	900–960
United States (several values)	45–300
	143–151
	1.8–115 (33)
	10–200
Mojave Generating Station, USA	136–420 (200)
Western USA coal	207
Power plant, western USA coal	25–43
Colstrip Power Plant, USA	81
New South Wales bituminous coal	6–60 (25)
Queensland bituminous coals	25–30
South Australian hard-brown coal	40–50
Victorian brown coal	30

[a] *Source:* From several sources, summarized by Swaine, 1980, with permission of John Wiley & Sons, Inc, New York.
[b] Data in parenthesis are means.

that while Ni has accumulated substantially in soils, uptake by plants of this element has not been sufficient to be of concern in the food chain (Kirkham, 1975; Chang et al, 1982, 1984b; El-Bassam et al, 1979).

8.3 Other Sources

Since metals are found in phosphate rocks in varying amounts, application of phosphatic fertilizers in agricultural lands may elevate the levels of some metals in soils. In general, phosphate rocks from the western United States contain higher concentrations of most heavy metals than do rocks found in the eastern states (Mortvedt and Giordano, 1977). Selected phosphate rocks in the United States can have as much as 64 ppm Ni, whereas Williams (1977) gave a range of 1 to 10 ppm Ni for Australian super-phosphates. Although long-term use of phosphate fertilizers may raise the Cd levels in soils (Mulla et al, 1980), plant uptake of most metals in fertilizers should not cause concern in the food chain at the rates of phosphorus fertilizers usually applied under field conditions.

Coal contains varying amounts of Ni (<2–150 ppm in European and Canadian coals; <1–90 ppm in Australian bituminous coals; <2–>300 ppm in bituminous coals from the Appalachian region of the United States). Upon combustion, Ni may be concentrated in the fly ash (Swaine, 1980). Nickel is distributed between the fly ash and the bottom slag, with about 80% of the Ni being in the fly ash. Swaine (1980) reported that fly ashes from several countries can have Ni from <2 to 960 ppm (Table 11.7). In spite of possible high levels of Ni in fly ashes, Ni, unlike Se, Mo, and B, does not appear to be taken up by the plants substantially (Adriano et al, 1980).

Schmidt and Andren (1980) reported that Ni release to the atmosphere is mostly anthropogenic in nature, accounting for about 84% of the total emission.

References

Adriano, D. C., A. L. Page, A. A. Elseewi, A. C. Chang, and I. Straughan. 1980. *J Environ Qual* 9:333–344.

Agarwala, S. C., S. S. Bisht, and C. P. Sharma. 1977. *Can J Bot* 55:1299–1307.

Andersson, A. 1977. *Swed J Agric Res* 7:7–20.

Andersson, A., and K. O. Nilsson. 1972. *Ambio* 1:176–179.

Anke, M., M. Grün, B. Groppel, and H. Kronemann. 1983. In B. Sarkar, ed. *Biological aspects of metals and metal-related diseases,* 89–105. Raven Press, New York.

Archer, F. C. 1980. *Minist Agric Fish Food* (Great Britain) 326:184–190.

Ashton, W. M. 1972. *Nature* 237:46–47.

Aubert, H., and M. Pinta. 1977. *Trace elements in soils.* Elsevier, Amsterdam.

Beavington, F. 1975. *Environ Pollut* 9:211–217.

Beckett, P. H. T., and R. D. Davis. 1978. *New Phytol* 81:155–173.

Berrow, M. L., and G. A. Reaves. 1984. In Proc Intl Conf on Environ Contamination, 333–340. CEP Consultants Ltd, Edinburgh, UK.

Berrow, M. L., and J. Webber. 1972. *J Sci Fed Agric* 23:93–100.

Bick, W., P. C. DeKock, and O. V. Gambi. 1982. *Plant Soil* 66:117–119.

Biddappa, C. C., M. Chino, and K. Kumazawa. 1982. *Plant Soil* 66:299–316.

Bingham, F. T., G. A. Mitchell, R. J. Mahler, and A. L. Page. 1976. In Proc of Intl Conf on Environ Sensing and Assessment, Vol 2. Inst Elec and Electron Engrs Inc, New York.

Bleeker, P., and M. P. Austin. 1970. *Aust J Soil Res* 8:133–143.

Bloomfield, C., W. I. Kelso, and G. Pruden. 1976. *J Soil Sci* 27:16–31.

Bowen, H. J. M. 1979. *Environmental chemistry of the elements.* Academic Press, New York. 333 pp.

Bowman, R. S., and G. A. O'Connor. 1982. *Soil Sci Soc Am J* 46:933–936.

Bowman, R. S., M. E. Essington, and G. A. O'Connor. 1981. *Soil Sci Soc Am J* 45:860–865.

Bradford, G. R., R. J. Arkley, P. F. Pratt, and F. L. Bair. 1967. *Hilgardia* 38:541–556.

Bradley, R. I., C. C. Rudeforth, and C. Wilkins. 1978. *J Soil Sci* 29:258–270.

Brooks, R. R. 1980. *In* J. O. Nriagu, ed. *Nickel in the environment*, 407–430. Wiley, New York.

Brooks, R. R., J. Lee, and T. Jaffre. 1974a. *J Ecol* 62:493–499.

——— 1974b. *J Ecol* 62:434–439.

Brooks, R. R., and E. D. Wither. 1977. *J Geochem Explor* 7:295–300.

Brooks, R. R., J. Lee, R. D. Reeves, and T. Jaffre. 1977a. *J Geochem Explor* 7:49–57.

Brooks, R. R., E. D. Wither, and B. Zepernick. 1977b. *Plant Soil* 47:707–712.

Brooks, R. R., and C. C. Radford. 1978. *Proc Royal Soc London B* 200:217–224.

Butler, J. R. 1954. *J Soil Sci* 5:156–166.

Cannon, H. L. 1974. In P. L. White and D. Robbins, eds. *Environmental quality and food supply*, 143–164. Futura Publ Co, Mt Kisco, NY.

Cannon, H. L. 1978. *Geochem Environ* 3:17–31.

Carlson, R. W., F. A. Bazzaz, and G. L. Rolfe. 1975. *Environ Res* 10:113–120.

Cataldo, D. A., and R. E. Wildung. 1978. *Environ Health Perspect* 27:149–159.

Cataldo, D. A., T. R. Garland, and R. E. Wildung. 1978a. *Plant Physiol* 62: 563–565.

Cataldo, D. A., T. R. Garland, R. E. Wildung, and H. Drucker. 1978b. *Plant Physiol* 62:566–570.

Chang, A. C., A. L. Page, and F. T. Bingham. 1982. *J Environ Qual* 11:705–708.

Chang, A. C., A. L. Page, J. E. Warneke, and E. Grgurevic. 1984a. *J Environ Qual* 13:33–38.

Chang, A. C., J. E. Warneke, A. L. Page, and L. J. Lund. 1984b. *J Environ Qual* 13:87–91.

Chino, M. 1981. In K. Kitagishi and I. Yamane, eds. *Heavy metal pollution in soils of Japan*, 65–80. Japan Sci Soc Press, Tokyo.

Clark, R. B., P. A. Pier, D. Knudsen, and J. W. Maranville. 1981. *J Plant Nutr* 3:357–374.

Clemente, G. F. 1976. *J Radioanal Chem* 32:25–41.

Clemente, G. F., L. C. Rossi, and G. P. Santaroni. 1980. In J. O. Nriagu, ed. *Nickel in the environment*, 493–498. Wiley, New York.

Connor, J., N. F. Shimp, and J. C. F. Tedrow. 1957. *Soil Sci* 83:65–73.

Connor, J. J., H. T. Shacklette, R. J. Ebens, J. A. Erdman, A. T. Miesch, R. R. Tidball, and H. A. Tourtelot. 1975. In USGS Prof Paper 574-F: 106–111.

Cottenie, A., R. Camerlynck, M. Verloo, and A. Dhaese. 1979. *Pure Appl Chem* 52:45–53.

Cotton, F. A., and G. Wilkinson. 1980. *Advanced inorganic chemistry*, 4th ed. Wiley, New York.

Council for Agricultural Science and Technology (CAST). 1976. In *Application of sewage sludge to cropland*. EPA-430/9-76-013, US-EPA, Washington, DC. 63 pp.

Cox, R. M., and T. C. Hutchinson. 1980. *New Phytol* 84:631–647.

Crooke, W. M. 1955. *Ann Appl Biol* 43:465–476.

——— 1956. *Soil Sci* 81:269–276.

Crooke, W. M., J. G. Hunter, and O. Vergnano. 1954. *Ann Appl Biol* 41:311–324.

Crooke, W. M., and A. H. Knight. 1955. *Ann Appl Biol* 43:454–464.

Cunningham, J. D., D. R. Keeney, and J. A. Ryan. 1975a. *J Environ Qual* 4:448–454.

——— 1975b. *J Environ Qual* 4:455–459.

Davis, R. D., P. H. T. Beckett, and E. Wollan. 1978. *Plant Soil* 49:395–408.

Domingo, L. E., and K. Kyuma. 1983. *Soil Sci Plant Nutr* 29:439–452.

Doner, H. E. 1978. *Soil Sci Soc Am J* 42:882–885.

Dowdy, R. H., and W. E. Larson. 1975. *J Environ Qual* 4:278–282.

Duke, J. M. 1980a. In J. O. Nriagu, ed. *Nickel in the environment*, 27–50. Wiley, New York.

Duke, J. M. 1980b. In J. O. Nriagu, ed. *Nickel in the environment*, 51–65. Wiley, New York.

El-Bassam, N., C. Tietjen, and J. Esser. 1979. In *Management and control of heavy metals in the environment*, 521–524. CEP Consultants Ltd, Edinburgh, UK.

Elrashidi, M. A., A. Shehata, and M. Wahab. 1979. *Agrochimica* 23:245–253.

Emmerich, W. E., L. J. Lund, A. L. Page, and A. C. Chang. 1982a. *J Environ Qual* 11:178–181.

——— 1982b. *J Environ Qual* 11:182–186.

Fleming, G. A. 1963. *J Sci Fd Agric* 14:203–208.

Foy, C. D., R. L. Chaney, and M. C. White. 1978. *Ann Rev Plant Physiol* 29:511–566.

Freedman, B., and T. C. Hutchinson. 1980. *Can J Bot* 58:1722–1736.

Gerritse, R. G., R. Vriesema, J. W. Dalenberg, and H. P. de Roos. 1982. *J Environ Qual* 11:359–364.

Giordano, P. M., and D. A. Mays. 1977. In H. Drucker and R. E. Wildung, eds. *Biological implications of metals in the environment*. CONF-750–929. NTIS, Springfield, VA.

Gould, M. S., and E. J. Genetelli. 1978. *Water Res* 12:889–892.

Halstead, R. L. 1968. *Can J Soil Sci* 48:301–305.

Halstead, R. L., B. J. Finn, and A. J. MacLean. 1969. *Can J Soil Sci* 49:335–342.

Hamilton, E. I., and M. J. Minski. 1972. *Sci Total Environ* 1:375–394.

Haq, A. U., T. E. Bates, and Y. K. Soon. 1980. *Soil Sci Soc Am J* 44:772–777.

Harter, R. D. 1983. *Soil Sci Soc Am J* 47:47–51.

Hazlett, P. W., G. K. Rutherford, and G. W. van Loon. 1984. *Geoderma* 32:273–285.

Heinrichs, H., and R. Mayer. 1980. In J. O. Nriagu, ed. *Nickel in the environment*, 431–455. Wiley, New York.

Hewitt, E. J. 1953. *J Exp Bot* 4(10):59–64.

Hickey, M. G., and J. A. Kittrick. 1984. *J Environ Qual* 13:372–376.

Hinesly, T. D., R. L. Jones, and E. L. Ziegler. 1972. *Compost Sci* 13:26–30.

Hunter, J. G., and O. Vergnano. 1952. *Ann Appl Biol* 39:279–284.

——— 1953. *Ann Appl Biol* 40:761–777.

Hutchinson, T. C. 1981. In N. W. Lepp, ed. *Effects of heavy metal pollution on plants*, 171–211. Applied Science Publ, London.

Hutchinson, T.C., M. Czuba, and L. M. Cunningham. 1974. *Trace Subs Environ Health* 8:81–93.

Hutchinson, T. C., and L. M. Whitby. 1974. *Environ Conserv* 1:123–132.

———— 1977. *Water, Air, Soil Pollut* 7:421–428.

Iizuka, T. 1975. *Soil Sci Plant Nutr* 21:47–55.

Ishihara, M., Y. Hase, H. Yolomizo, S. Konno, and K. Sato. 1968. *Engei Shikenjo Hokoku* (ser. A, 1968) 7:39–54 (as cited by Mishra and Kar, 1974).

Jackson, M. L., and G. D. Sherman. 1953. *Adv Agron* 5:219–318.

John, M. K., and C. J. Van Laerhoven. 1976. *J Environ Qual* 5:246–251.

Kelly, P. C., R. R. Brooks, S. Dilli, and T. Jaffre. 1975. *Proc Royal Soc London B* 189:69–80.

Khalid, B. Y., and J. Tinsley. 1980. *Plant Soil* 55:139–144.

Kirkham, M. B. 1975. *Environ Sci Technol* 9:765–768.

Kirkpatrick, D. C., and D. E. Coffin. 1974. *J Inst Can Sci Technol Aliment* 7:56–58.

Kruckeberg, A. R. 1954. *Ecology* 35:267–274.

Latterell, J. J., R. H. Dowdy, and W. E. Larson. 1978. *J Environ Qual* 7:435–440.

LeRiche, H. H., and A. H. Weir. 1963. *J Soil Sci* 14:225–235.

Lindau, C. W., and L. R. Hossner. 1982. *J Environ Qual* 11:540–545.

Lindsay, W. L., and W. A. Norvell. 1978. *Soil Sci Soc Am J* 42:421–428.

Lozano, F. C., and I. K. Morrison. 1981. *J Environ Qual* 10:198–204.

———— 1982. *J Environ Qual* 11:437–441.

Lyon, G. L., P. J. Peterson, R. R. Brooks, and G. W. Butler. 1971. *J Ecol* 59:421–429.

Marrack, D. 1980. In M. M. Trieff, ed. *Environment and health,* 184–185. Ann Arbor Science, Ann Arbor, MI.

Matthews, N. A. 1980. In *Minerals yearbook,* 629–641. US Bureau of Mines, US Dept of Interior, Washington, DC.

Mattigod, S. V., A. S. Gibali, and A. L. Page. 1979. *Clays Clay Miner* 27:411–416.

McKeague, J. A., and M. S. Wolynetz. 1980. *Geoderma* 24:299–307.

Memon, A. R., S. Ito, and M. Yatazawa. 1980. *Soil Sci Plant Nutr* 26:271–280.

Mills, J. G., and M. A. Zwarich. 1975. *Can J Soil Sci* 55:295–300.

Mishra, D. and M. Kar. 1974. *Bot Rev* 40:395–452.

Misra, S. G., and P. Pande. 1974. *Plant Soil* 40:679–684.

Mitchell, R. L. 1945. *Soil Sci* 60:63–70.

———— 1971. In *Minist Agric Fish Fed Tech Bull* 21:8–20. Her Majesty's Sta Office, London.

Mizuno, N. 1968. *Nature* 219:1271–1272.

Mortvedt, J. J., and P. M. Giordano. 1977. In H. Drucker and R. E. Wildung, eds. *Biological implications of metals in the environment.* CONF-750-929. NTIS, Springfield, VA.

Mulla, D. J., A. L. Page, and T. J. Ganje. 1980. *J Environ Qual* 9:408–412.

Murthy, G. K., U. S. Rhea, and J. T. Peeler. 1973. *Environ Sci Techol* 7: 1042–1045.

Nakos, G. 1982. *Plant Soil* 66:271–277.

National Academy of Sciences (NAS) 1975. *Nickel.* NAS, Washington, DC. 268 pp.

Nriagu, J. O., H. K. T. Wong, and R. D. Coker. 1982. *Environ Sci Technol* 16:551–560.

Ogoleva, V. P., and L. N. Cherdakova. 1980. *Sov Soil Sci* 12:554–558.

———— 1981. *Sov Soil Sci* 13:62–65.

Painter, L. I., S. J. Toth, and F. E. Bear. 1953. *Soil Sci* 76:421–429.

Peterson, P. J. 1971. *Sci Prog Oxf* 59:505–526.

Pierce, F. J., R. H. Dowdy, and D. F. Grigal. 1982. *J Environ Qual* 11:416–422.

Polacco, J. C. 1976. *Plant Physiol* 20:350–357.

———— 1977. *Plant Physiol* 59:827–830.

Polemio, M., S. A. Bufo, and N. Senesi. 1982. *Plant Soil* 69:57–66.

Pratt, P. F., F. L. Bair, and G. W. McLean. 1964a. *8th Intl Cong Soil Sci* 3:243–248, Bucharest, Romania.

———— 1964b. *Soil Sci Soc Am Proc* 28:363–365.

Proctor, J. 1971. *J Ecol* 59:827–842.

Quin, B. F., and J. K. Syers. 1978. *N Z J Agric Res* 21:435–442.

Richter, R. O., and T. L. Theis. 1980. In J. O. Nriagu, ed. *Nickel in the environment,* 189–202. Wiley, New York.

Roberts, B. A. 1980. *Can J Soil Sci* 60:231–240.

Russell, Jr. L. H. 1979. In F. W. Oehme, ed. *Toxicity of heavy metals in the environment.* Dekker Inc, New York.

Rutherford, G. K., and C. R. Bray. 1979. *J Environ Qual* 8:219–222.

Schmidt, J. A., and A. W. Andren. 1980. In J. O. Nriagu, ed. *Nickel in the environment,* 93–135. Wiley, New York.

Schroeder, H. A. 1971. *Environment* 13:18–32.

Severne, B. C. 1974. *Nature* 248:807–808.

Severne, B. C., and R. R. Brooks. 1972. *Planta* (Berlin) 103:91–94.

Sevin, I. F. 1980. In H. A. Waldron, ed. *Metals in the environment,* 263–292. Academic Press, New York.

Sheaffer, C. C., A. M. Decker, R. L. Chaney, and L. W. Douglass. 1979. *J Environ Qual* 8:455–459.

Shewry, P. R., and P. J. Peterson. 1976. *J Ecol* 64:195–212.

Soane, B. D., and D. H. Saunder. 1959. *Soil Sci* 88:322–330.

Sommers, L. E., D. W. Nelson, and K. J. Yost. 1976. *J Environ Qual* 5:303–306.

Sommers, L. E., and W. L. Lindsay. 1979. *Soil Sci Soc Am J* 43:39–46.

Sposito, G., and S. V. Mattigod. 1980. In *Geochem.* Dept Soil & Environ Sci, Univ of California, Riverside.

Sterrett, S. B., C. W. Reynolds, F. D. Schales, R. L. Chaney, and L. W. Douglas. 1983. *J Am Soc Hort Sci* 108:36–41.

Stokes, P. M., T. C. Hutchinson, and K. Krauter. 1973. *Can J Bot* 51:2155–2168.

Swaine, D. J. 1955. *The trace element content of soils.* Common Bur Soil Sci Tech Commun 48.

Swaine, D. J. 1980. In J. O. Nriagu, ed. *Nickel in the environment,* 67–92. Wiley, New York.

Swaine, D. J., and R. L. Mitchell. 1960. *J Soil Sci* 11:347–368.

Temple, P. J., and S. Bisessar. 1981. *J Plant Nutr* 3:473–482.

Terry, R. E., D. W. Nelson, and L. E. Sommers. 1979a. *Soil Sci Soc Am J* 43:494–499.

———— 1979b. *J Environ Qual* 8:343–347.

Traynor, M. F., and B. D. Knezek. 1974. *Trace Subs Environ Health* 7:75–81.

Trudinger, P. A., D. J. Swaine, and G. W. Skyring. 1979. In P. A. Trudinger and D. J. Swaine, eds. *Biogeochemical cycling of mineral-forming elements,* 1–27. Elsevier, Amsterdam.

Tyler, L. D., and M. B. McBride. 1982. *Soil Sci* 134:198–205.

US Environmental Protection Agency (US-EPA). 1976. *Quality criteria for water.* US-EPA, Washington, DC.

Valdares, J. M. A. S., M. Gal, U. Mingelgrin, and A. L. Page. 1983. *J Environ Qual* 12:49–57.

Vanselow, A. P. 1951. *Calif Citrog* 37(2):77–80.

———— 1965. In H. D. Chapman, ed. *Diagnostic criteria for plants and soils,* 302–309. Quality Printing Co, Abilene, TX.

Vinogradov, A. P. 1959. *The geochemistry of rare and dispersed chemical elements in soils.* Consultants Bureau, New York.

Walker, R. B. 1954. *Ecology* 35:259–266.

Wallace, A., E. M. Romney, J. W. Cha, S. M. Soufi, and F. M. Chaudry. 1977. *Commun Soil Sci Plant Anal* 8:757–764.

Welch, R. M. 1981. *J Plant Nutr* 3:345–356.

Welch, R. M., and E. E. Cary. 1975. *J Agric Food Chem* 23:479–482.

Whitby, L. M., and T. C. Hutchinson. 1974. *Environ Conserv* 1:191–200.

Whitby, L. M., J. Gaynor, and A. J. MacLean. 1978. *Can J Soil Sci* 58:325–330.

Williams, C. H. 1977. *J Aust Inst Agric Sci* (Sept–Dec):99–109.

Wither, E. D., and R. R. Brooks. 1977. *J Geochem Explor* 8:579–583.

12
Selenium

1. General Properties of Selenium

Selenium (atomic no. 34) is a member of Group VI, also known as the S family, in the periodic table. It has specific gravities of 4.79 g/cm³ for the metallic (gray) form, or 4.28 g/cm³ for the vitreous (black) form. It has an atomic weight of 78.96, melting point of (gray form) 217°C, and boiling point of (gray form) 684.9 ± 1°C. Because of its chemical similarity to S, it resembles S both in its forms and compounds and accounts for their many interrelations in biology. It has six stable isotopes in nature with the following percentage isotopic composition: ^{74}Se, 0.87; ^{76}Se, 9.02; ^{77}Se, 7.58; ^{78}Se, 23.52; ^{80}Se, 49.82; and ^{82}Se, 9.19. The most important oxidation states of Se are $-$II, II, IV, and VI. Selenium can be easily oxidized from Se(0), elemental Se, to Se(IV), (SeO_3^{2-}), and to Se(VI), (SeO_4^{2-}). Selenites (SeO_3^{2-}) are stable in alkaline to mildly acidic conditions and can be found in nature. Some of the known commercial Se compounds are H_2Se, metallic selenides Se($-$II), SeO_2, H_2SeO_3, SeF_4, Se_2Cl_2, and H_2SeO_4 (selenic acid). Selenium also forms a large number of organic compounds that are analogous to those of S.

2. Production and Uses of Selenium

Selenium is rather a rare element, there being no large deposits anywhere. It is mainly obtained commercially as a by-product of the electrolytic refining of Cu. Production in the United States and worldwide is shown in Figure 12.1 and indicates a steep rise in worldwide production from 1976 to 1979. The current leading producers of Se, in decreasing order are: Canada, Japan, the United States, Mexico, Sweden, and Belgium, with both Canada and Japan contributing about 450 tonnes per year. The USSR is known to be a major producer, but data are insufficient to estimate annual production.

In the United States, the following are estimates of Se consumption by end-use categories in 1978 and 1979 (US Bureau of Mines, 1980): electronic

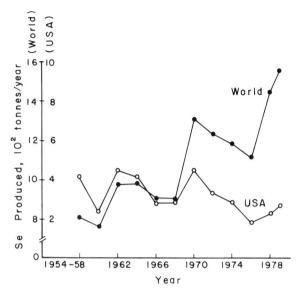

FIGURE 12.1 Trends in selenium production in the United States and worldwide, 1954–1979. *Source:* US Bureau of Mines, *Minerals yearbook,* several issues.

and photocopier components—35%; glass manufacturing—30%; chemicals and pigments—25%; and others—10%. Compounds of Se are used in photoelectric cells (e.g., in photographic exposure meters, photometers, counting devices, and light-controlled switches); as maroon and orange pigments, in combination with Cd sulfide, for plastics and ceramics; in increasing the resistance of rubber to heat, oxidation, and abrasion; and as lubricants to increase the machinability of stainless steel. Selenium sulfide is used as a therapeutic agent in shampoos for certain scalp conditions. Selenium is also used in glass manufacture and as an alloy of steel and Cu.

3. Natural Occurrence of Selenium

The abundance of Se in the earth's crust is reportedly about 0.05 to 0.09 ppm, which is ≈1/6,000 of the abundance of S and 1/50 of the abundance of As (Lakin, 1972; NAS, 1974). It is ranked 68th in crustal abundance among elements, ahead of Hg (Krauskopf, 1979). Although Se is inconsistently dispersed in geologic materials, it is detectable in most earth materials (Table 12.1) and is frequently enriched in black shales in concentrations ranging up to 675 ppm (Table 12.1). It tends to concentrate in carbonaceous debris in sandstones; is also enriched in phosphate rocks (1–300 ppm) and consequently may contaminate soil from applications of phosphatic fertilizers.

TABLE 12.1. Commonly observed selenium concentrations (ppm) in various environmental matrices.

Material	Average concentration	Range
Igneous rocks[a]	0.05	—
Limestone[b]	0.08	0.1–6
Sandstone[a,b]	0.05	—
Shale[a]	0.60	<1–675
Petroleum	—	0.01–1.4
Coal[c]	3	0.04–10
Fly ash:[d]		
Bituminous	5.7	—
Sub-bituminous	1.2	—
Lignite	4.4	—
Phosphate rocks[a,b]	mostly <20	1–300
Superphosphate[a]	mostly <4	up to 25
Soils[a]	<0.1–2	up to 5,000
Grasses[e]	0.26	<0.01–9.0
Plants[a]	0.1–15	some up to 1,200
Se accumulator plants[b]	—	100–10,000
Vegetables and fruits[e]	0.05	0.01–0.20
Woody angiosperm[c]	0.029	—
Woody gymnosperm[c]	0.026	—
Freshwater[a]	0.0002	0.0001–0.4
Sea water[c]	0.0002	0.0001–0.0002

Sources: Extracted from
[a] Swaine (1978).
[b] Gamboa-Lewis (1976).
[c] Bowen (1979).
[d] Adriano et al (1980).
[e] Cannon (1974).

The primary sources of Se in nature are volcanic emanations and metallic sulfides associated with igneous activity; secondary sources are biological pools in which Se has accumulated (NAS, 1974). Examples of primary sources are western shales of Cretaceous Age and the sandstone ores of the Colorado Plateau. In general, shales have the highest concentrations of Se and are the primary sources of high-Se soils in the Great Plains and Rocky Mountain foothills in the United States (Lakin, 1961). The higher contents in shales can be attributed partly to biological activity (Swaine, 1978). Although about 50 Se minerals are known, Se is commonly associated with heavy metal (Ag, Cu, Pb, Hg, Ni, etc.) sulfides, where it occurs as a selenide, or as a substitute ion for S in the crystal lattice. Neither Se nor S is an essential constituent of rock-forming silicate minerals. The highest Se contents of sulfides are associated with uranium ores in sandstone-type deposits in the western United States. Coleman and Delevaux (1957) reported the following maximum levels of Se: pyrite, 5%; marcasite, 0.65%; and chalcocite, 5%. The highest resources of Se are reported for chalcopyrite-pentlandite-pyrrhotite deposits geologically related to basic and ultrabasic rocks of the Precambrian Age.

Certain regions in some countries have soils that contain Se well above the normal levels. These soils, generally referred to as "seleniferous," are in the arid and semi-arid areas of China, Hawaii, Mexico, Colombia, and the western areas of the United States and Canada. In the arid and semi-arid areas of the United States having mean annual rainfall of <50 cm, soils have been reported to produce "seleniferous" vegetation (Trelease and Beath, 1949). In general, high-Se soils do not occur in humid regions.

4. Selenium in Soils

4.1 Total Soil Selenium

Selenium is present in detectable but highly variable amounts in soils. Swaine (1955) stated that most soils contain 0.1 to 2 ppm of total Se. Byers et al (1938) found, from several thousand soil analyses, that Se varied from trace amounts to a maximum of 82 ppm and concluded that Se levels of 1 to 6 ppm are about normal for seleniferous soils. Seleniferous soils of North America extend into the Canadian provinces of Alberta, Manitoba, and Saskatchewan, west to the Pacific coast, and south into Mexico. In the United States, the region extends primarily from North Dakota to Texas and west to the Pacific coast (excluding the Pacific northwest). Seleniferous soils in the three Canadian provinces showed Se values of 0.1 to 6 ppm (Byers and Lakin, 1939). Some seleniferous soils in the United Kingdom had Se contents of 0.9 to 91.4 ppm (Nye and Peterson, 1975). However, normal soils for the United Kingdom reportedly average 0.6 ppm, with values ranging from 0.2 to 1.8 ppm ($n = 114$) (Archer, 1980). Other Se values reported are 0.41 ppm for Indian soils (Misra and Tripathi, 1972), 0.60 ppm for New Zealand soils (Wells, 1967), and 0.70 ppm for the soils of Japan (Tsuge and Terada, 1949).

Selenium can be expected to be present in excessive amounts only in semi-arid and arid regions in soils derived from cretaceous shales (Rosenfeld and Beath, 1964) but in humid climates or in irrigated areas, most of the Se is leached from soils of this origin. In some cases, however, Se may be insufficient in soils, as far as animal nutrition is concerned, particularly those derived from igneous rocks; these amounts in the surface layers may be further depleted by intensive irrigation (Oldfield, 1972).

4.2 Extractable (Available) Selenium in Soils

Total Se in soil, as with many other soil-borne trace elements, is not a reliable index of plant-available Se (Johnson, 1975; Nye and Peterson, 1975). For this reason, several extracting solutions have been employed to predict potential bioaccumulation of Se in plants. Normal soils contain

<50 ppb of water-soluble Se (Workman and Soltanpour, 1980). Soils supporting vegetation containing toxic quantities of Se can be expected to have water-soluble Se in the tenths of a ppm range (Lakin, 1972). Byers et al (1938) observed water-soluble levels in more than 100 US soils of Se that ranged from 0.1 ppm to 38 ppm. However, in about 80% of the extracts, Se did not exceed 0.1 ppm. Such water-soluble Se has been identified as the selenate form in soils in the midwestern United States (Williams and Byers, 1936), in Wyoming soils (Rosenfeld and Beath, 1964), in soils of Ireland (Fleming and Walsh, 1957), and in soils of Canada (Levesque, 1974b). Water-soluble Se varied from 2.0% to 7.0% of the total Se (0.197 to 0.744 ppm) for some Canadian soils (Levesque, 1974b) and from 0.33% to 2.90% of total Se (20.4 to 850 ppm) for some Irish soils (Fleming and Walsh, 1957). Water-extractable Se can be of doubtful utility as a plant-available indicator in Se-deficient soils, but can be a useful criterion for assessing the behavior of Se applied to soil (Levesque, 1974b).

Other extractants tested that show promise in predicting plant-available Se include hot water and NH_4HCO_3-DTPA (Fig. 12.2) (Soltanpour and Workman, 1980), $CaCl_2$, $Ca(NO_3)_2$, or K_2SO_4 solutions (Cary and Allaway, 1969; Geering et al, 1968; Olson and Moxon, 1939). Sodium arsenite ($NaAsO_2$) has also been used to extract Se from soil as a potential test for sorbed or exchangeable selenite (Cary and Allaway, 1973). Soltanpour and Workman (1980) found that hot water and NH_4HCO_3-DTPA extracted approximately equal amounts of Se from soil (Fig. 12.2). Significantly high correlation ($r^2 = 0.96$) was found between amounts of Se extracted by NH_4HCO_3-DTPA and Se uptake by alfalfa plants grown in a growth chamber. DTPA-extractable soil Se levels of over 100 ppb produced alfalfa plants containing levels of Se of 5 ppm or higher. These levels of Se in forage may be harmful to animals, if the high-Se forage makes up a high percentage of the diet.

4.3 Selenium in the Soil Profile

The distribution of Se in the soil profile is influenced by several factors, such as the parent material, organic matter, pH, ferruginous material, and to some extent, rainfall. In a very extensive survey involving 54 profiles taken from areas in Canada with possible animal production problems due to the low Se contents in crops, Levesque (1974a) found that, in general, the highest Se values ($\bar{x} = 0.940$) were obtained with the surface organic layers. He indicated the importance of organic matter in retaining Se in the surface horizon. Selenium distribution somewhat varied with the soil order. For the podzols, the surface layer in the A horizon gave a total Se content of 0.084 ppm compared to a content of 0.073 ppm in the C horizon. Among the mineral horizons, the B horizon produced the highest Se content (0.521 ppm). In gleysolic soils, marked accumulations occurred in the A (0.479 ppm) and in the B horizon (0.246 ppm) as compared to only

4. Selenium in Soils   395

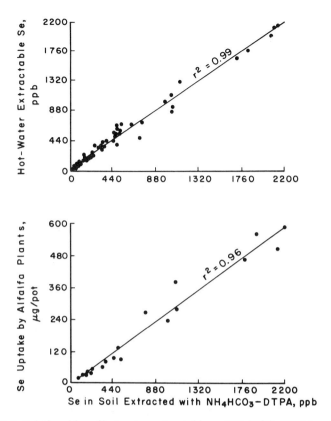

FIGURE 12.2 Relationships of hot water-extractable vs. NH₄HCO₃-DTPA-extractable selenium in soils before planting and after harvest of alfalfa (upper); selenium uptake by two cuttings of alfalfa as a function of NH₄HCO₃-DTPA extractable selenium in soils (below). *Source:* Soltanpour and Workman, 1980, with permission of the authors and Dekker, Inc.

0.100 ppm in the C horizon. In brunisolic soils, there were continuous decreases of Se with depth, 0.640 ppm Se in the surface layer vs. 0.110 ppm in the C horizon. Levesque (1974a) concluded that Se was closely associated with organic C in the upper horizons and with Fe compounds in the lower horizons.

In New Zealand, Wells (1967) indicated that, for soils with little profile development (A-C soils), the Se content of the topsoil depends upon the nature of the parent material; it is very low for granites and rhyolitic pumices, low for mica schist and graywacke (excluding volcanic graywacke), but is high for andesitic and basaltic ashes and for argillite. However, high amounts of Se occur in strongly weathered topsoils from which the clay and Fe compounds have not been eluviated. Wells (1967) found low-Se topsoils in the following soil groups: brown-gray earths, yellow-gray earths away from the andesitic ash zone, high country yellow-brown earths, podzols, pumice soils, and southern steepland soils. On the other hand, high-Se topsoils were found in: northern yellow-brown earths, rendzinas, and some andesitic and basaltic soils. His conclusions are somewhat similar to Levesque (1974a) in Canada in that the levels of Se are very high in well-developed B2 horizons ($\bar{x}$ = 1.4 ppm) where clay accumulation occurs and at a maximum in the concretion layer of ironstone soils and in the iron-pan layer of podzols. For comparison, the A horizons have an average Se content of 0.60 ppm.

Others correlating Se accumulation in the soil profile with organic matter were Fleming (1962) in seleniferous soils in Ireland and Kubota (1972) in the Ao horizons (0.22 ppm vs. 0.06 ppm in the C horizon) in highly weathered sandy soils in the state of New York. Slater et al (1937) also observed that the most ferruginous horizons of the soil are also the most highly seleniferous.

4.4 Forms of Selenium in Soils

The form of Se in soils largely determines its mobility and availability to plants and subsequently to animals. As depicted in Figure 12.3, Se is present in both organic and inorganic forms (Olson and Moxon, 1939; Byers et al, 1938; Cary et al, 1967; Cary and Allaway, 1969). The organic form originates from partially decayed seleniferous vegetation. Inorganic Se

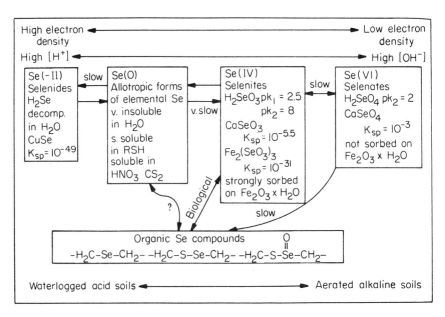

FIGURE 12.3 Generalized chemistry of selenium in soils and sediments as influenced by soil pH and redox potential. *Source:* Allaway (1968).

occurs as metal selenides, elemental Se, selenite [SeO_3^{2-}, (SeIV)], or as selenate [SeO_4^{2-}, (SeVI)]. The latter forms often occur as stable salts (e.g., ferric selenite, calcium selenate, etc.) or as adsorption complexes (Ganje, 1965). Selenite is the predominant mobile inorganic form of Se in soils of humid regions that is probably adsorbed on hydrous sesquioxides (Cary et al, 1967). Under ordinary alkaline conditions, Se is present as selenate, which does not form highly insoluble salts or stable complexes with the sesquioxides. The low solubility of Se in acid or neutral soils can be attributed to its presence as selenite in combination with ferric iron or as basic ferric selenite. Ferric selenite has a solubility product of 10^{-33}, whereas $Fe_2(OH)_4SeO_3$ has a solubility product of about 10^{-63} (Geering et al, 1968). Geering et al presented evidence that the solubility of selenite in soils that contain reactive iron oxides may be explained by the formation of ferric selenite-ferric hydroxide adsorption complexes. In some soils, Se reacts with active hydroxides in a manner similar to interactions with active ferric hydroxides. Geering et al also predicted the solubility and oxidation states of Se as influenced by pH, redox potential, and ions that interact with Se. Oxidation of selenite to selenate in alkaline soils is favored by a decrease in the stability of ferric hydroxide-selenite complex.

In general, it can be predicted that, under well-aerated conditions, Se is present in alkaline soils of semi-arid areas as selenate (Fig. 12.3). However, under acid and reducing conditions, as in humid areas, Se may exist dominantly in the selenite form. Peterson and his co-workers (1981) showed the predominance of selenate in an alkaline high-Se soil from South Dakota and the predominance of selenite in an acidic high-Se soil from Ireland. The rate of transformation of selenite to selenate and vice versa is slow, and the rate of transformation of selenite to elemental Se is even slower. Cary and Allaway (1969) found slow conversion of soil-added selenite to organic Se compounds and to elemental Se. They further indicated that when elemental Se is added to soils, part of it rapidly oxidizes to selenite. After the fast initial oxidation, the remaining elemental Se is quite inert and its oxidation then proceeds very slowly. Oxidation of elemental Se did not appear to be pH-dependent, but the rate varied among different soils. This selenite then reacts with the soil in a manner similar to that added as soluble selenite.

Chemical fractionation of residual ^{75}Se in cropped soil revealed that several forms of Se are present in soils that had acid to neutral pH after reaction with soluble [75]SeO_3^{2-} (Cary et al, 1967). The ^{75}Se was believed to be close to an equilibrium state in the soil since it was in the soil for over a year and since Se uptake by the indicator plants (alfalfa) has remained fairly constant. Soluble and exchangeable [75]Se (extracted by K_2SO_4 at pH 6.6) were present in six of eight soils treated, but did not exceed 15% of the residual [75]Se. Adsorbed ^{75}Se, probably present as adsorbed selenite, ranged from about 10% to 40% of the residual [75]Se. Organic ^{75}Se ranged from about 4% to 22% of the residual [75]Se. Roots and root residues are believed to have contributed these organic compounds. The presence of selenite $-^{75}Se$, extracted by 6 N HCl, either occluded in or precipitated on sesquioxides, was also found. They also found evidence of the reduction of Se to insoluble selenides, elemental Se, or both. Elemental Se could have been formed from selenite on the surface of plant roots. Thus, when Se is added to soils as selenite, especially those in the acid to neutral range, it is transformed to insoluble forms and becomes less available for plant uptake.

4.5 Sorption and Fixation of Selenium in Soils

The mobility and availability of the various forms of Se depend on such processes as adsorption, precipitation, etc. in the soil. Geering et al (1968) reported several forms of Se in soils (e.g., SeO_3^{2-}, SeO_4^{2-}, $HSeO_3^{2-}$, Se^0—see section 4.4) depending on the pH and redox potential of the soil, but SeO_3^{2-} and SeO_4^{2-} are the predominant forms in aerated soils. However, SeO_3^{2-} is the predominant mobile inorganic form of Se in soils of humid regions and is most likely adsorbed on hydrous sesquioxide (Cary et al, 1967). Hingston et al (1971) concluded that selenite was specifically adsorbed on gibbsite and goethite. John et al (1976) found that both amorphous and crystalline forms of Fe_2O_3 were related to Se adsorption by New Zealand soils. Lakin (1961) has indicated the role of ferric hydroxide in precipitating selenite in acidic to neutral soils and consequently decreasing Se uptake by plants. However, Geering et al (1968) presented evidence that selenite may form adsorption complexes with ferric oxides in soils rather than as crystalline ferric selenite. Plotnikov (1960, 1964) reported that the coprecipitation of selenite by $Fe(OH)_3$ is via adsorption and occlusion. Cary et al (1967), using a fractionation scheme of residual Se in soil, indicated possible occlusion of Se in sesquioxide particles or as solid phase of selenites in sesquioxides.

Adsorption of selenite- or selenate-Se by soil materials has been described by the Langmuir isotherm (Singh et al, 1981; Rajan and Watkinson, 1976). In contrast, Ames et al (1984) found that sorption of selenate-Se by basalt material obeyed the Freundlich isotherm. Several studies have demonstrated that selenite is sorbed on soil to a much greater extent than selenate (Gissel-Nielsen et al, 1984). Therefore, selenite is not as available as selenate is for plant uptake (also see section 6.3). Contrary to most findings, Singh et al (1981) found that selenate was always sorbed by soils in higher amounts than selenite and that for the two Se forms, sorption decreased in the order: high organic matter soil > calcareous soil > normal soil > saline soil > alkali soil. They found that certain soil properties, such as organic C, clay content, $CaCO_3$, and CEC were related to sorption capacity for Se. Similarly, John et al (1976) found that specific surface area, organic C, free forms of SiO_2, Al_2O_3, Fe_2O_3, and allophane content were closely related to Se adsorption by soils.

4.6 Transformations of Selenium in Soils

Methylations of certain trace elements, such as As, Hg, Pb, and Sn can yield organic compounds that are more toxic to higher organisms than the inorganic forms. Selenium is also known to be biomethylated, producing organic metabolites that are more volatile than the parent inorganic forms (Chau et al, 1976). Plants, fungi, bacteria, and microorganisms can produce methylated forms of Se when exposed to inorganic or certain organic forms. Non-accumulator plants, such as cabbage and alfalfa, give off $(CH_3)_2$ Se or other volatile Se compounds, and accumulator species, such as *Astragalus racemosus*, produce dimethyldiselenide, $(CH_3)_2Se$, when exposed to selenite (Lewis 1976; Lewis et al, 1966, 1974; Evans et al, 1968).

Several microorganisms have been isolated from the soil that are capable of methylating inorganic Se (Barkes and Fleming, 1974; Doran and Alexander, 1977). Doran and Alexander (1977) demonstrated that volatilization of Se from soils amended with Se^0, selenite, selenate, trimethylselenonium chloride $[(CH_3)_3SeCl]$, selenomethionine, and selenocysteine was, for all practical purposes, the result of microbial processes. They further observed that the conversion of selenite, selenate, and indigenous soil Se to volatile products was enhanced when soils were amended with organic matter. Under aerobic conditions, dimethylselenide was produced from all Se compounds tested; under anaerobic conditions, dimethylselenide was evolved from soil treated with three organic Se compounds and selenate; hydrogen selenide (H_2Se) was evolved from soil treated with Se^0, selenite, selenate, or selenocysteine. In lake sediments, conversion of inorganic and organic Se compounds by microorganisms to volatile Se compounds, such as dimethylselenide and dimethyldiselenide has also been observed (Chau et al, 1976). Biomethylation products were also detected from sewage sludge (Reamer and Zoller, 1980). Francis et al (1974) also found dimethylselenide evolution from soil amended with selenite. Similarly, Abu-Erreish et al (1968) found that addition of selenite, selenate, or organic matter to seleniferous soils resulted in Se volatilization. Major factors that determine the amounts of added Se absorbed by plants are the extent of fixation by soil particles and amounts volatilized.

5. Selenium in Plants

5.1 Essentiality of Selenium in Plants

Selenium first came to attention in the 1930s, when studies in the western and Great Plain regions of the United States showed many areas where levels of Se in plants were sufficiently high to be toxic to animals. Selenium was the first element found to occur in native vegetation at levels toxic to animals (Allaway, 1968). Livestock consuming plants containing excessive amounts of Se can be afflicted with two maladies, commonly known as "alkali disease" and "blind staggers." On the other hand, Se-deficient feeds can cause a disorder known as "white muscle disease." The safe range between "normal" and "toxic" levels in feedstuff is very narrow, one reason why Se levels in plants have received critical attention from soil, plant, and animal scientists. The minimal dietary level required to prevent "white muscle disease" is from 0.03 to 0.10 ppm Se, depending upon the level of vitamin E and possibly other substances in the diet (Allaway et al, 1967). However, the generally accepted critical Se level for preventing Se disorder in livestock is 0.10 ppm. Concentrations above 3 or 4 ppm are considered toxic (Coles, 1974; Underwood, 1977).

Worldwide surveys on Se concentrations in crops indicate that areas producing crops with Se contents too low (<0.1 ppm) to meet animal requirements are more common than areas producing toxic levels (>2 ppm) of Se in crops (Gissel-Nielsen et al, 1984).

While Se is an important element in animal nutrition, it is known to be nonessential for plant growth. However, there are plant genera, called "indicator" or "accumulator" plants, where Se may be required for normal growth (Shrift, 1969; Lewis, 1976; Johnson, 1975). But, since no beneficial effects were found in agronomic species when Se was added to purified cultures in which plants were grown (Broyer et al, 1966), Se does not meet the criteria for elemental essentiality for plants as established by Arnon and Stout (1939). In general, evidence that Se is essential to plants is not convincing.

5.2 Uptake and Accumulation of Selenium in Plants

Selenium is absorbed by plants in both the inorganic form, such as selenate or selenite, and the organic form (Johnson et al, 1967; Stadtman, 1974; Hamilton and Beath, 1963a, b, 1964). Probably volatile forms from soil are also absorbed by the foliage of plants. Uptake of added Se by plants is in the neighborhood of 1% of the total amount added (Levesque, 1974b; Gissel-Nielsen and Gissel-Nielsen, 1973). Peterson et al (1981) reported that selenate is taken up metabolically by plants, whereas selenite is largely taken up by passive processes, and at lower levels. Leggett and Epstein (1956) indicated that selenate is taken up through the same binding sites in the plant roots as sulfate. They concluded that the two ions are taken up by the same active absorption processes in competition with each other, while selenite is taken up through other sites. Trelease et al (1960) reported that Se partially replaced S in plant metabolism and occurred along with S in amino acids and proteins of several plant species. Furthermore, Stadtman (1974) reported that the mechanisms responsible for incorporating Se into living cells are accomplished by the same enzymes that metabolize sulfate.

In investigating about 20 range plants, Hamilton and Beath (1963a) observed that all species studied were able to absorb soil-added inorganic Se (selenate or selenite) and convert some or all of it into organic compounds. The majority of the plants were able to also convert soil-added organic Se into inorganic compounds. In addition, Hamilton and Beath (1964) found that, with the 19 vegetable species they studied, all the plants possessed the ability to convert a portion of the absorbed selenate-Se into an organic form. Likewise, these plants were also capable of absorbing some of Se added to the soil as organic compounds, converting at least a portion of it into other compounds and storing it in their tissues as both organic and inorganic compounds. Food crop plants apparently incorporate Se principally in their protein in a bound form when grown in soils high in Se. Those that normally contain high levels of S, such as cabbage, tend to accumulate more Se than those that are normally low in S. Stadtman (1974) suggested that marked substitution of Se for S in normal cell constituents of nonspecialized plants occurs, and this may account for their sensitivity to large amounts of Se in soil. The most likely chemical and biochemical changes in Se involved in its movement from different soils through various species and eventually to animals are depicted diagramatically in Figure 12.4.

Selenium content of crops receives popular attention because of its importance in the food chain. Selenium-responsive diseases, such as "white muscle disease" in livestock, are most likely to occur in areas producing crops low in Se. Consequently, attempts were made to delineate the continental United States into regions based on Se concentrations in forages (Allaway, 1973; Kubota et al, 1967), and then further to divide the regions into smaller areas (Carter et al, 1968). Kubota et al (1967) placed the Pacific Northwest, the Northeastern region, and the Southeastern seaboard states as low-Se areas (Fig. 12.5). More specifically, Carter et al (1968) ranked the western half of the states of Washington and Oregon and part of northern California as extremely low-Se areas. Whereas, in most of the west-central region, the Se contents of crops are predominantly in the protective, but nontoxic, range of Se concentrations. Other parts of the country are characterized by variable Se contents in plants.

Similarly, soils in the Atlantic coast provinces of Canada have been found to be very low in Se, consequently producing forage crops ($<<0.1$ ppm Se in dry matter) to be generally Se-deficient for poultry and livestock (Gupta and Winter, 1975; Winter et al, 1973; Winter and Gupta, 1979). Other countries that have reported low Se contents in plants are: Finland (Koljonen, 1974; Korkman, 1980), New Zealand (Davies and Watkinson, 1966a, b), and Scandinavian countries (Gissel-Nielsen and Gissel-Nielsen, 1973). Gissel-Nielsen et al (1984) reported that serious Se deficiency in livestock in Australia and New Zealand were recognized as early as about 1960. In Australia, the Se deficiency is most pronounced in the western and southern parts of the continent, while all of New Zealand is a low-Se area. Selenium deficiency in livestock in England and Scotland has been reported. The most severe Se-deficient area of the world is the Keshan region in southeastern China, where a great number of children have died from the Se-deficiency disease known as the "Keshan" disease.

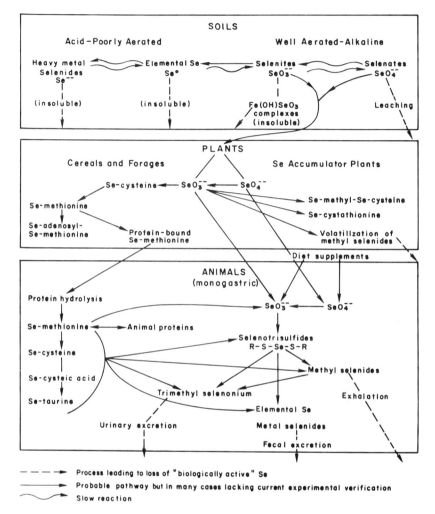

FIGURE 12.4 Chemical and biochemical changes of selenium in the soil-plant-animal pathway. *Source:* Allaway (1973).

Data on Se in crops from most European countries, especially northern Europe, are available (Fig. 12.6). In general, Se levels in plants in northern Europe are inadequate. Gissel-Nielsen et al (1984) reported that cereal crops in Sweden, Norway, Finland, and Denmark contain low Se levels, usually <40 ppb. Pasture crops, in general, contained more Se, about double that for the cereal crops. The low levels of Se in plants can be attributed to strongly leached soils that were developed on glacier-deposited material.

Toxic levels of Se (up to 500 ppm Se in crops) in Europe are limited to a few spots in Wales and some areas in Ireland (Fig. 12.6). Selenium in

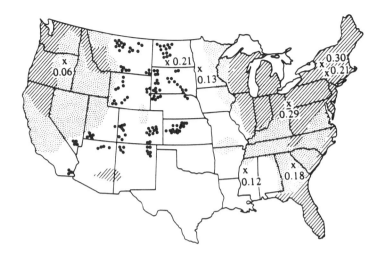

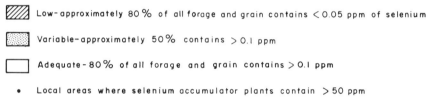

Low-approximately 80% of all forage and grain contains < 0.05 ppm of selenium

Variable-approximately 50% contains > 0.1 ppm

Adequate-80% of all forage and grain contains > 0.1 ppm

• Local areas where selenium accumulator plants contain > 50 ppm

X Selenium concentration in rainwater (μg/L) by sampling location

FIGURE 12.5 Selenium levels in forages, crops, and accumulator plants and their distribution in the United States. *Source:* Kubota et al (1975).

these soils, which ranges from 30 to >300 ppm, originates from Se leached from Avonian shales and deposited in low-lying valleys. Thus, the Se level of soils and plants is greatly influenced by soil parent material and climatic conditions, as demonstrated in the United States and Europe.

Certain plant species absorb soil Se very efficiently. These "indicator" or "accumulator" plants absorb Se from geological formations, virgin shale, or soil and play a very important role in the cycling of Se. The accumulator plants—which are always associated with semi-arid, seleniferous soils—biosynthesize organic Se compounds, such as Se-methylselenocysteine, and release these compounds to the soil upon decay (Shrift, 1964; Davis, 1972). Lewis (1976) defined Se "accumulators" as plants that can accumulate Se to levels 10^2 to 10^4 times greater than the levels in most other native and crop species. These plants have been reported to occur in at least 15 countries of the world, including Canada, Mexico, the United Kingdom, Ireland, Israel, Venezuela, Australia, and others (Lakin, 1972).

A range of seleno-amino acids has been isolated from various plants (Peterson et al, 1981). Plant accumulators synthesize selenocystathionine

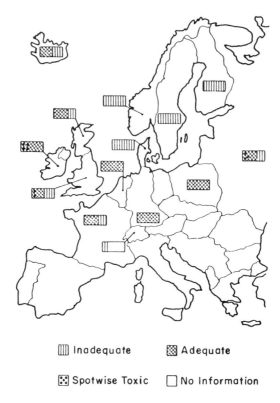

☐☐☐☐ Inadequate   ▨ Adequate

⊡ Spotwise Toxic   ☐ No Information

FIGURE 12.6 Selenium levels in fodder crops in Europe. *Source:* Gissel-Nielsen et al, 1984, with permission of the authors and Academic Press.

and Se-methylselenocysteine, whereas non-accumulators synthesize se-lenomethionine and Se-methylselenomethionine (Fig. 12.4). Peterson and Butler (1967) hypothesized that accumulator species may have evolved mechanisms to exclude selenocysteine and selenomethionine from their enzymes and other proteins as a detoxifying mechanism, whereas non-accumulator species incorporate these amino acids into their proteins.

Because of the extreme variations in plants' capacity to accumulate Se, several investigators (Moxon and Olson, 1974; Rosenfeld and Beath, 1964) divided plants into three categories (Table 12.2):

Group 1. Primary indicators—plants that contain large amount of Se (10^3 to 10^4 ppm Se dry wt) and seem to require Se for their growth. These include several species of *Astragalus* (Leguminosae), *Conopsis* (Compositae), *Stanleya* (Cruciferae), and *Xylorhiza* (Compositae). Selenium is present in most tissues primarily in inorganic form. Plants when consumed can cause "blind staggers," or acute Se poisoning, in livestock.

Group 2. Secondary indicators—plants that will accumulate Se, rarely over a few hundred ppm, when they grow on soils of high available Se

TABLE 12.2. Grouping of plants according to their capacity to accumulate selenium from soil.[a]

Primary group		
Astragalus		
albulus	*pattersonii*	*foliosa*
argillosus	*pectinatus*	*wardii*
asclepiadoides	*pectinatus platyphyllus*	*Stanleya*
beathii	*praelongus*	*albescens*
bisulcatus	*preussii*	*bipinnata*
cocalycis	*preussii latus*	*elata*
confertiflorus	*racemosus*	*glauca*
eastwoodiae	*rafaelensis*	*integrifolia*
ellisiae	*sabulosus*	*pinnata*
flaviflorus	*scobinatulus*	*tomentosa*
flavus	*sophoroides*	*viridiflora*
grayi	*toanus*	*Xylorhiza*
haydenianus	*urceolatus*	*glabriuscula*
limatus	*Conopsis*	*parryi*
moencoppensis	*argillacea*	*venusta*
osterhoutii	*condensata*	*villosa*
	engelmannii	
Secondary group		
Aster	*Atriplex*	*Mentzelia*
adscendens	*canescens*	*decapetala*
commutatus	*Castilleja*	*Sideranthus*
ericoides	*chromosa*	*grindelioides*
glaucus	*Machaeranthera*	
	ramosa	
Tertiary group		
Most cultivated agronomic and horticultural crops	Grain plants	Native grasses

[a] *Sources:* Ganje (1965) and Rosenfeld and Beath (1964).

contents, and do not appear to require Se for their growth. Most of the Se in these plants occurs as selenate with lesser amounts in the organic form. These plants may produce acute or chronic selenosis in livestock.
Group 3. Most cultivated crop plants, grains, and native grasses—generally accumulate low levels of Se (maximum, 30 ppm under field conditions). Selenium is associated primarily with plant proteins. "Alkali disease" in livestock may result from consumption of these plants.

Indicator plants are known otherwise as "converter" plants because of their unique ability to absorb Se from the soil in a form that is not available to other plants and recycling it in a form that is available (Moxon and Olson, 1974). Since most of these plants are deep-rooted, they can absorb available Se from the deep strata of the soil profile and deposit it on the surface where it is more readily accessible and probably more available to other plants.

5.3 Sensitivity to and Phytotoxicity of Selenium

Selenium is toxic to plants at low concentrations in nutrient cultures, and crops—including the forages—are sensitive to addition of small amounts of Se in soil. Broyer et al (1966) found that Se at 0.025 ppm in nutrient solutions decreased the yields of alfalfa. Mild toxicity was displayed in bush beans at 8 ppm Se, as Na_2SeO_3, in solution culture (Wallace et al, 1980); severe toxicity was obtained at 80 ppm Se concentration, with plants characterized by stunting and brown spots. Leaves, stems, and roots of these plants had 901, 1,058, and 1,842 ppm of Se, respectively. In sand culture, Davis et al (1978) found that the critical level of Se in solution was 5 ppm Se (as SeO_3^{2-}); the critical level in barley tissue was about 30 ppm, dry weight. In culture solutions void of sulfate, concentrations of selenate-Se as low as 0.1 ppm produced a slight detectable injury to wheat plants (Hurd-Karrer, 1933, 1935), but this was probably due to S deficiency since signs of injury did not occur in the presence of S. In soils, selenite-Se added at 20 ppm did not affect wheat germination but retarded plant growth (Beath et al, 1937). At the rate of 40 ppm, it reduced germination substantially and produced stunted and chlorotic plants. The characteristic toxic symptoms in wheat plants are chlorosis, stunting, and yellowing of the leaves. For most cereal crops, snow-white chlorosis is a typical symptom of Se toxicity. Selenate is a more toxic source of Se than selenite, but this depends on S concentrations in the medium; at S concentrations above 30 ppm, selenite-Se could become more toxic (Hurd-Karrer, 1935, 1937). Using sand culture, Mason and Phillis (1937) noticed that cotton growth was slightly affected at 20 ppm selenate-Se; at 50 ppm in the medium, growth was markedly depressed. Walsh and Fleming (1952) found that the Se content of plants grown in soils containing 40 to 88 ppm of Se and showing toxicity symptoms were 117 ppm in meadow fescue, 30 ppm in meadow sweet, and 7.2 ppm in barley grain.

In growth chamber experiment, soils treated with Na_2SeO_3 and yielding about 2 ppm of soil Se extracted with NH_4HCO_3-DTPA solution, were highly toxic to alfalfa plants and resulted in plant concentrations of over 1,000 ppm of Se (Soltanpour and Workman, 1980).

Under field conditions, visual symptoms of Se toxicity in cultivated crops have never been observed (Hemphill, 1972; NAS, 1976). It appears then that, under field conditions, readily available Se is not provided to plants at a level high enough to cause injury. Rosenfeld and Beath (1964) reported that crop plants would show no injury until they contain at least 300 ppm of the element, which is usually much higher than what plants would contain, even in seleniferous areas.

The toxicity of Se to plants is influenced by a number of factors. Plant type is very important, with the "accumulators" being very tolerant to high Se levels. The form of Se in the growth medium may also be important. Presence of competing ions, such as sulfate and possibly phosphate, should also be taken into account.

6. Factors Affecting the Mobility and Availability of Selenium

6.1 Parent Material

The Se content of plants has been shown to vary with the geological formation in which they grow (Moxon et al, 1950). Furthermore, close correlations have been found between the geological strata from which the soils have developed and the amount of Se found in the soil (Byers and Lakin, 1939). Igneous rocks, being inherently low in Se, give rise to a great many Se-deficient soils (Peterson et al, 1981). In many cases, parent materials of sedimentary origin produce soils especially high in Se, with the shales of particular importance. Shales, particularly the Cretaceous shales, generally have the highest concentrations of Se and are the major sources of high Se-soils of the Great Plains and Rocky Mountain states in the United States, and vast areas in the provinces of Alberta and Saskatchewan, Canada, South Africa, Australia, Ireland, and the United Kingdom (NAS, 1974; Peterson et al, 1981).

In addition, Peterson et al (1981) reported that high Se soils in Israel are derived from limestone alluvium, while those in South America are from black slate alluvium. The influence of soil parent material on Se concentration in plants has been demonstrated in the west-central area of Saskatchewan, Canada, where wheat plants had the highest Se concentrations when grown over lacustrine clay and glacial till; the Se concentrations were intermediate in plants grown on lacustrine silt; and the Se concentrations were lowest in plants grown on aeolian sand (Table 12.3) (also see sections 3 and 5.2).

TABLE 12.3. Selenium contents (median and range) of whole wheat plants and associated soil parent materials.[a]

| | | Se content, ppm dry wt | |
Soil parent material	Number of samples[b]	Wheat plants	C horizon soil
Aeolian sand	16:15	0.64	0.12
		0.42–4.00	0.07–0.26
Lacustrine silt	15:15	1.08	0.28
		0.38–3.60	0.18–0.63
Glacial till	15:15	1.50	0.26
		0.88–11.2	0.08–1.50
Lacustrine clay	16:16	2.18	0.37
		1.02–5.40	0.24–1.92

[a] *Source:* Doyle and Fletcher, 1977, with permission of the authors and the Agricultural Institute of Canada.
[b] Number of wheat samples:number of soil samples.

6.2 pH

It is known that Se is quite available to plants grown on well-aerated alkaline soils where Se tends to occur as selenates. In acid to neutral soils, however, the ferric oxide-selenite complex is formed, which is fairly insoluble and therefore, this Se is only slightly available to plants (see section 4.4). Geering et al (1968) determined the solubility of Se in soils as a function of pH and found that the lowest solubility occurred when the soil was slightly acid to neutral. Consequently, liming the soil could enhance Se uptake by plants. The effect of elevated soil pH on Se concentrations in barley and oats is demonstrated in Figure 12.7. Liming of soil to raise the pH from 6.0 to 6.9 has been shown to increase the uptake of added Se by alfalfa (Cary et al, 1967). Cary and Allaway (1969) further observed increased uptake of added Se, from several sources [selenite-ferric hydroxide coprecipitate [$Fe(OH)_3$-$HSeO_3$], sodium biselenite ($NaHSeO_3$), and elemental Se], by alfalfa when grown on higher pH treatment. Likewise, Gissel-Nielsen (1971b) found higher uptake of added selenite by Italian ryegrass in limed soils. Because of this trend, Allaway et al (1967) have indicated that applying Se to alkaline soils is likely to be more hazardous in producing feedstuff for livestock than applying the element in acid soils.

6.3 Carrier of Selenium

In general, increasing the level of soil Se increases Se uptake by plants. Selenate-Se is, in most instances, absorbed most efficiently, selenite-Se somewhat less efficiently, and organic-Se absorbed to a lesser extent (Hamilton and Beath, 1963a). Likewise, several investigators found that

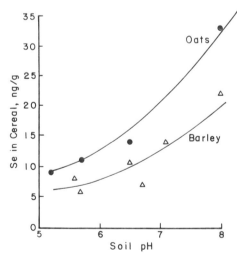

FIGURE 12.7 Relationships between selenium concentration in selected cereals (grain samples) and soil pH. *Source:* Modified from Lindberg and Bingefors, 1970, with permission of Acta Agriculture Scandinavica.

added selenate to soils was readily taken up by the plants and remained available in some soils (Gissel-Nielsen and Bisbjerg, 1970; Hurd-Karrer, 1935). Bisbjerg and Gissel-Nielsen (1969) found that uptake by plants of Se from Danish soils (pH 4.8 to 7.6) was, on an average, eight times more with selenate than with selenite. Others found that when selenite was added, it reverted to insoluble forms, and uptake by plants became limited (Cary et al, 1967; Davies and Watkinson, 1966a). Gile and Lakin (1941) observed that these two forms of Se are differentially translocated, with Se accumulating higher in the roots when supplied as selenite-Se; the reverse trend happened with selenate-Se as the source. However, some plants, such as the "indicator" species, absorbed comparatively larger amounts of organic Se from the soil. Hamilton and Beath (1964) showed that in some vegetable species, plant uptake of Se was greater when plants were grown on soils containing organic Se than when grown on selenate-Se.

The form of Se added to increase plant concentration to a useful level may well be tailored to the soil pH. Carter et al (1969) and Cary and Allaway (1969) suggested that selenites may be more useful than selenates as slowly available Se additives on alkaline soils, since selenates added to these soils can produce plants possessing Se concentrations toxic to livestock. Furthermore, Cary and Allaway (1969) indicated that soluble selenites may be the most practical form of Se to add to neutral or acid soils that produce crops very low in Se.

Gissel-Nielsen and Bisbjerg (1970) determined total uptake of Se, as a percentage of added Se, from a two-year field experiment using mustard and found values of 0.01% from Se^0, 4% from K_2SeO_3, and 30% from K_2SeO_4 and $BaSeO_4$. With lucerne, barley, and sugar beets, the uptakes were one-third of these values or less. Thus, the use of Se^0 is impractical as a Se source because of its ineffectiveness. With cowpeas as the indicator—a plant used as a fodder crop in India—Singh and Singh (1979), using a pot culture, found that inhibition of plant growth from soil-added Se was in the order $SeO_4^{2-} > H_2SeO_3 > SeO_3^{2-} > Se^0$. Addition of $BaCl_2$ to some alkaline seleniferous soils reduced Se uptake by lucerne plants. Presumably $BaSeO_4$ was formed, and the level of soluble Se in soil was reduced (Ravikovitch and Margolin, 1959). Because it is fairly insoluble, $BaSeO_4$ may serve as a slowly available Se source in alkaline soils low in Se.

6.4 Soil Texture and Organic Matter

In general, uptake of Se by plants from mineral soils can be expected to be inversely related to the clay content of the soil (Bisbjerg and Gissel-Nielsen, 1969; Gissel-Nielsen, 1971b; Cary and Allaway, 1969). Bisbjerg and Gissel-Nielsen (1969) observed that Se concentrations in plants (red clover, barley, and white mustard) they tested were sometimes twice as high on sandy than on loamy soils. In addition, Singh et al (1981) noted that in soil sorption studies of selenite- and selenate-Se, the degree of sorption was influenced positively by the clay content. Thus, if selenite were to be applied to farmlands low in Se, smaller amounts should be used for sandy than for loamy or clayey soils.

The effect of organic matter on Se availability is somewhat inconsistent. These inconsistencies might be due to the complex role of organic matter on Se stability in the soil environment. Selenium is associated with the organic matter in soils via the cycling of the element through decay of plant material. Thus, Se levels in the surface soil layers can be expected to be higher than in the subsoils due to higher organic matter content in the former (Levesque, 1974a). In soil sorption studies, Singh et al (1981) noticed that sorption of Se was influenced positively by the organic matter content, indicating its binding role on Se. Levesque (1972, 1974b) also indicated that Se was presumably bound to organo-metallic complexes. These findings, however, are not consistent with plant uptake studies since organic soils, such as peats, failed to inhibit Se uptake by plants (Davies and Watkinson, 1966b; Gile and Lakin, 1941).

In Se transformation in soil, organic matter apparently plays an important role. Conversion of inorganic forms of Se, both added and indigenous, to volatile products by soil microorganisms may be enhanced if soil is amended with organic materials (Doran and Alexander, 1977).

6.5 Sulfur and Phosphorus

The close chemical similarity between S and Se and the fact that almost all S compounds are known to have Se analogs led to the concept that Se functions in selenoamino acids in animals and plants alike (Frost, 1972). In biological systems, enzymes responsible for assimilation of the S atom would also accommodate the Se atom. Frost and Lish (1975) advocated studying Se cycle in relation to S cycle. Because of their similarity, Se and S interactions in plant metabolism could be expected. However, results are somewhat conflicting, although evidence is more overwhelming toward the antagonistic effects of S on Se nutrition (Ravikovitch and Margolin, 1959; Hurd-Karrer, 1933, 1935; Cary and Gissel-Nielsen, 1973; Martin, 1936; Martin and Trelease, 1938; Williams and Thornton, 1972). Gissel-Nielsen (1971a) noted that plant uptake of applied Se was highest where a fertilizer low in S was used. In some instances, however, additions of S to soil may fail to inhibit the uptake of Se by plants (Gupta and Winter, 1975; Carter et al, 1969), particularly in seleniferous soils (Franke and Painter, 1937). Carter et al (1969) found that uptake of Se by alfalfa from $BaSeO_4$ applied to an alkaline silt loam in the field was enhanced by the application of S, as $BaSO_4$, particularly when the S/Se ratio approached

10. They indicated that perhaps with low-Se, alkaline soil and small application of Se, S applications would have a positive effect on Se uptake. When S fertilizers are applied to farmlands, diminution of Se concentrations in plant tissues may actually be the result of dilution by plant biomass due to increased growth rather than direct competition between sulfate and selenate/selenite ions (Walker, 1971; Westermann and Robbins, 1974; Pratley and McFarlane, 1974). Shrift (1969) cautioned, however, that the many resemblances between S and Se assimilation should not be construed as universal, since there are a number of examples that show that the two elements are not always metabolically similar.

Selenium-S interactions have also been observed in soil systems. Sulfate ions, added as $BaSO_4$ to the soil, aided in leaching Se from soil (Brown and Carter, 1969). They further observed that when a gypsum ($CaSO_4 \cdot 2H_2O$) solution was used for leaching, more Se was removed from the soil columns than when water alone was used. They concluded that addition of SO_4^{2-} increased the solubility of $BaSeO_4$ in these alkaline soils ($pH \approx 7.8$). This is consistent with the observation of Carter et al (1969) that application of SO_4^{2-} to an alkaline field soil increased Se uptake by alfalfa. Singh et al (1981) found that sulfate and phosphate ions desorbed Se from soils. They explained this antagonism between Se versus S and also versus P through exchange mechanism, since it has been demonstrated (Rājan and Watkinson, 1976) that most of the SeO_3^{2-} and SeO_4^{2-} is retained in soils through one of the mechanisms by which phosphate and sulfate are retained.

Like the Se-S interaction, the interaction between Se and P in plants has also been shown to occur. The pattern has also been inconsistent, although more results showed that the effect of P tended to be more positive (Carter et al, 1972; Levesque, 1974b). Carter et al (1972) had two plausible explanations for the Se-P interaction: (1) evidently Se and P ions compete for the same reaction sites in soils; when P is added to the system, it replaces some Se in the sorption sites, rendering the free Se available to plants; (2) P applications may stimulate plant growth, including greater root proliferation, enabling the plants to absorb more Se. The greater the root system, the greater is the root-soil contact and the capacity to absorb ions.

6.6 Plant Species and Age

Considerable data are available indicating differences in Se accumulation among plant species, particularly on soils high in available Se (Byers and Lakin, 1939; Hamilton and Beath, 1963a, b; Miller and Byers, 1937; Olson et al, 1942; Rosenfeld and Beath, 1964). Under these conditions, differences tenfold or more are common. However, there are only a few studies conducted on plant species differences in soils indigenously low in available Se or in typical agricultural soils (Ehlig et al, 1968; Bisbjerg and Gissel-

TABLE 12.4. Range and average concentration of selenium in Danish-grown forage crops.[a]

Crop	Year	Number of samples	Average Se concentration, ppb	Range of Se concentration, ppb
Barley	1972	140	17	6–110
Barley	1973	178	18	2–110
Winter wheat	1973	40	21	7–87
Spring wheat	1973	41	20	4–67
Oats	1973	44	16	3–54
Rye	1973	25	16	6–72

[a] Source: Gissel-Nielsen et al, 1984, with permission of the authors and Academic Press.

Nielsen, 1969; Grant, 1965; Davies and Watkinson, 1966a). In general, differences among plant species in Se accumulation from soils low in Se were small.

A survey on Se concentrations in Danish forages (Table 12.4) indicates that Se levels in the same species vary considerably within a small area, such as Denmark. However, differences in Se levels among species are generally small.

Stage of growth and harvesting can also determine the Se concentration in plants. Selenium concentrations in plant tissues usually decline with maturity (Rosenfeld and Beath, 1964) or with subsequent cuttings of pasture or forage crops (Gissel-Nielsen, 1971b; Grant, 1965; Davies and Watkinson, 1966a). Exceptions for this trend can occur in low-Se soils.

7. Selenium in Drinking Water and Food

The US-EPA standard for Se in drinking water is 10 ppb Se (US-EPA, 1976). Likewise, the World Health Organization has a 10-ppb limit for Se in drinking water (WHO, 1984). In some cases, the drinking water standard may be exceeded. For example, Valentine et al (1978) found that water from house wells in Milan, New Mexico, had a mean content of 537 $\pm$ 650 ppb Se ($n = 19$). Presumably the house wells had become contaminated from seepage from a nearby uranium mill tailing pond. Some surface waters, such as those in the Colorado River Basin, were reported to have Se as high as 400 ppb (Fairhall, 1941). The high value was apparently caused by irrigation drainage effluent from seleniferous soils. Otherwise, extensive analyses of waters from the major watersheds of the United States showed only very rare occasions when Se levels exceeded 10 ppb (NAS, 1976). Even lower levels were found in public, finished water supplies, where a mean of 8 ppb was found for samples from 194 sources (Taylor, 1963). Much lower levels of Se were reported (NAS, 1976) for waters from some Australian villages ($\bar{x} < 1$ ppb) and tap and mineral

TABLE 12.5. Estimated dietary intakes and levels of selenium by food class.[a]

| Food class | μg Se/day | % of total | Concentration[b], ppm wet wt | |
			Range	Median
Dairy products	0.0	0.0	0.009–0.05	0.02
Meat, fish, and poultry	56.3	37.6	0.11–0.24	0.17
Grain and cereal	92.5	61.8	0.11–0.30	0.17
Potatoes	0.65	0.4	0.007–0.05	<0.02
Leafy vegetables	0.0	0.0	0.005–<0.06	<0.01
Legume vegetables	0.0	0.0	0.005–<0.06	0.016
Root vegetables	0.25	0.2	0.004–0.039	<0.02
Garden fruits	0.0	0.0	0.006–0.008	0.016
Fruits	0.0	0.0	—	—
Oils and fruits	0.0	0.0	—	—
Sugars and adjuncts	0.0	0.0	—	—
Beverages	0.0	0.0	—	—
Totals	149.7			

[a] Based on FY 1973 total diet survey data of the US Food and Drug Administration, Mahaffey et al (1975).
[b] Tanner and Friedman (1977).

waters from Stuttgart, Federal Republic of Germany ($\bar{x}$ = 1.6 and 5.3 ppb, respectively).

Because of the differences in dietary habits, human Se intake varies considerably. The greatest variation in Se intake occurs where the intake of plant food predominates. For example, Gissel-Nielsen et al (1984) reported that the daily intakes of Se in China range from <10 μg where soil Se is low to >500 μg where food plants are grown on seleniferous soils. In their review, Lo and Sandi (1980) reported that the estimated typical dietary intake of Se ranges from 4 to 35 μg/day in infants to 60 to 250 μg/day in adults. The per capita value reported for the United States was 150 μg Se/day (Table 12.5). However, a much lower estimate of 62 μg/day was reported for the northeastern United States (Schroeder et al, 1970). The reported values, in μg Se/day, for other countries are: Canada, 113–220 (Thompson et al, 1975); the Netherlands, 110; Italy, 141; France, 166 (Shamberger, 1981); New Zealand, 25 (Robinson, 1976); Japan, 208; the United Kingdom ≈ 200 (Hamilton and Minski, 1972); and Federal Republic of Germany, 60 (Pfannhauser and Woidich, 1980; Schelenz, 1977). Other values were reported by Thorn et al (1978) for the average British diet (60 μg/day) and by Clemente et al (1978) for the Italian diet (14 μg/day). Other values reported are presented in Table 12.6. The USDA has recently recommended that the daily intake of Se should range from 100–200 μg/day (Shamberger, 1981). However, the WHO/FAO of the United Nations were unable to make any recommendations about dietary intakes of Se (Russell, 1979).

Lo and Sandi (1980) indicated that, in adults, approximately 99% of

TABLE 12.6. Estimated dietary selenium intakes and blood selenium levels in some countries around the world.[a]

Country	Dietary Se intake (μg/day)	Blood Se level (μg/liter)
Belgium	55	123
Canada	98–224	182
		144 (plasma)
China		
Endemic selenosis area	4990	3180
Keshan disease area	11	8
Finland	30	
South of Finland Helsinki		87
		66 (serum)
Southeast of Finland Lappeenranta		56
		42 (serum)
North of Finland Utsjoki		108 (serum)
New Zealand		
North Island	56	83
South Island	28	59
United States	62–216	157–265
Venezuela	218	355

[a] *Source:* Gissel-Nielsen et al, 1984, with permission of the authors and Academic Press.

ingested Se comes from cereals, grains, fish, meat, and poultry (Table 12.5). In average British diets, about 50% of Se ingested comes from cereals and cereal products and another 40% from meat and fish (Thorn et al, 1978). Milk, table fats, fruits, and vegetables provided little or no Se. In Canadian diets (Thompson et al, 1975), cereals provided the most Se (62–112 μg/day), followed by meat, poultry, and fish (25–90 μg/day), and dairy products (5–25 μg/day).

8. Sources of Selenium in the Terrestrial Environment

Although low-Se soils may produce forage insufficient to that required by animals for normal nutrition (i.e., 0.03–0.10 ppm in the diet, depending on vitamin E and other factors), Se compounds are infrequently added to these soils or plants. There are three conventional means of increasing the concentration of Se in plants: soil application, foliar spray, and seed treatment. For soil applications, rates of application may vary from tens of grams to a few kilograms per hectare, depending on soil type, type of crops, and other factors. In New Zealand (Grant, 1965) top-dressing pasture with about 0.07 kg Se/ha, as Na_2SeO_3 or Na_2SeO_4, produced Se concentrations toxic to livestock for periods of more than a year after application. The high Se levels in plant tissues were attributed to foliar absorption and resulted in uptake that was greater than if the Se had been applied directly to the soil. Alfalfa grown in the field on acid soils treated with 1.12 kg Se/ha (as Na_2SeO_3) contained up to 2.7 ppm, within the considered safe limit for livestock (Allaway et al, 1966). In another field trial (Gupta and Winter, 1981), barley and timothy grown on an acid soil treated with up to 2.24 kg Se/ha (as Na_2SeO_3) were found to have adequate levels of Se for up to two years in barley grain and up to five years in timothy. Elemental Se applied at rates as low as 0.25 to 0.5 kg/ha can provide adequate Se for up to two crops of alfalfa (Carter et al, 1969). However, it may be an impractical carrier of Se because of its short and low recovery by plants. Field application of Se should be exercised with caution when Se is top-dressed or when the soil is limed because of the greater availability of Se with increasing pH. Foliar application of Se is potentially efficient and safe for increasing Se in forages and feeds (Cary and Rutzke, 1981; Gupta et al, 1983a). Seed treatment of Se has been shown to be effective also in increasing Se concentrations in plants (Gupta et al, 1983b). Application of Se, as Na_2SeO_3 solution to seeds, at 50 to 200 g/ha produced forages containing greater than 0.1 ppm Se for at least three harvests. A big advantage of foliar application over soil application is that it avoids the influence that soil conditions can exert on plant uptake.

Another source of Se in agriculture is the fertilizer materials, especially phosphatic fertilizers. Selenium occurs in phosphate rocks, ranging from trace amounts to several hundred ppm (Rader and Hill, 1936; Robbins and Carter, 1970; Williams, 1977). Selenium concentrations in seven Florida phosphate rocks ranged from 0.7 to 7.0 ppm (Robbins and Carter, 1970); the range was from 1.4 to 178 ppm in seven samples of phosphorus fertilizer produced from western United States phosphate mines. Robbins and Carter concluded that normal phosphate fertilization practices could provide the required Se amounts for livestock, provided the fertilizer contains sufficient Se. In general, however, the contribution of fertilizers to the total Se content of plants is negligible (Gissel-Nielsen, 1971a).

Because large amounts of coal residues are being produced yearly ($\approx$ 80×10^6 tonnes/year) by coal-fired power plants, this solid waste looms as a potential source of Se in the environment. Selenium from fly ash is quite available to plants and could place some constraints on land application of fly ash (Adriano et al, 1980; Straughan et al, 1978; Mbagwu, 1983). It can accumulate in plant tissues in the range of several hundred ppm and up to two orders of magnitude of that in the growth medium (Adriano et al, 1980; Gutenmann et al, 1976).

Because of the volatile nature of Se, this element is omnipresent in the atmosphere, usually in unmeasurable amounts. In addition to soil parent material (section 6.1), other natural sources of Se include volcanic activity, volatilization from plants, soil microorganisms, animals, and burning of seleniferous vegetation (Kubota et al, 1975; West, 1971). Burning of fossil fuels, including coal and fuel oils, can also be an important source of Se in the atmosphere (Hashimoto et al, 1970). Mining and smelting of sulfide ores also contribute Se to the environment (Ragaini et al, 1977) and in isolated cases could cause toxicity of plants in the vicinity of the refinery (Burton and Phillips, 1981).

In a regional survey of trace elements in humus layers of Norwegian forest soils, Låg and Steinnes (1978) found that Se concentration decreased with increasing distance from the ocean and was shown to be significantly correlated with the annual rainfall. Thus, they attributed the soil input of Se to be primarily via precipitation. They considered natural sources, such as marine and terrestrial sources, rather than anthropogenic sources, to be the major source of atmospheric Se input.

Organic residues, like animal manures and sewage sludges, applied on land at normal rates should pose practically no hazard in terms of harmful effects of Se (Allaway, 1977; Frank et al, 1979; Page, 1974).

References

Abu-Erreish, G. M., E. I. Whitehead, and O. E. Olson. 1968. *Soil Sci* 106: 415–420.

Adriano, D. C., A. L. Page, A. A. Elseewi, A. C. Chang, and I. Straughan. 1980. *J Environ Qual* 9:333–344.

Allaway, W. H. 1968. *Trace Subs Environ Health* 2:181–206.

——— 1973. *Cornell Vet* 63:151–170.

——— 1977. In L. F. Elliot and F. J. Stevenson, eds. *Soils for management of organic wastes and water wastes,* 283–298. Am Soc Agron, Madison, WI.

Allaway, W. H., D. P. Moore, J. E. Oldfield, and O. H. Muth. 1966. *J Nutr* 88:411–418.

Allaway, W. H., E. E. Cary, and C. F. Ehlig. 1967. In O. J. Muth, ed. *Selenium in biomedicine,* 273–296. AVI Publ Co Inc, Westport, CT.

Ames, L. L., P. F. Salter, J. E. McGarrah, and B. A. Walker. 1984. *Chem Geol* 43:287–302.

Archer, F. C. 1980. *Minist Agric Fish Food* (Great Britain) 326:184–190.

Arnon, D. I., and P. R. Stout. 1939. *Plant Physiol* 14:371–375.

Barkes, L., and R. W. Fleming. 1974. *Bull Environ Contam Toxicol* 12:308–311.

Beath, O. A., H. F. Eppson, and C. S. Gilbert. 1937. *J Am Phar Assoc* 26:394–405.

Bisbjerg, B., and G. Gissel-Nielsen. 1969. *Plant Soil* 31:287–298.

Bowen, H. J. M. 1979. *Environmental chemistry of the elements.* Academic Press, New York. 333 pp.

Brown, M. J., and D. L. Carter. 1969. *Soil Sci Soc Am Proc* 33:563–565.

Broyer, T. C., D. C. Lee, and C. J. Asher. 1966. *Plant Physiol* 41:1425–1428.

Burton, M. A. S., and M. L. Phillips. 1981. *J Plant Nutri* 3:503–508.

Byers, H. G., J. T. Miller, K. T. Williams, and H. W. Lakin. 1938. In USDA Tech Bull 601. USDA-ARS, Washington, DC. 74 pp.

Byers, H. G., and H. W. Lakin. 1939. *Can J Res* 17-B:364–369.

Cannon, H. L. 1974. In P. L. White and D. Robbins, eds. *Environmental quality and food supply,* 143–164. Futura Publ Co, Mt Kisco, NY.

Carter, D. L., M. J. Brown, W. H. Allaway, and E. E. Cary. 1968. *Agron J* 60:532–534.

Carter, D. L., M. J. Brown, and C. W. Robbins. 1969. *Soil Sci Soc Am Proc* 33:715–718.

Carter, D. L., C. W. Robbins, and M. J. Brown. 1972. *Soil Sci Soc Am Proc* 36:624–628.

Cary, E. E., G. A. Wieczorek, and W. H. Allaway. 1967. *Soil Sci Soc Am Proc* 31:21–26.

Cary, E. E., and W. H. Allaway. 1969. *Soil Sci Soc Am Proc* 33:571–574.

———— 1973. *Agron J* 65:922–925.

Cary, E. E., and G. Gissel-Nielsen. 1973. *Soil Sci Soc Am Proc* 37:590–593.

Cary, E. E., and M. Rutzke. 1981. *Agron J* 73:1083–1085.

Chau, Y. K., P. T. S. Wong, B. A. Silverberg, P. L. Luxon, and G. A. Bengert. 1976. *Science* 192:1130–1131.

Clemente, G. F., L. C. Rossi, and G. P. Santaroni. 1978. *Trace Subs Environ Health* 12:23–30.

Coleman, R. G., and M. H. Delevaux. 1957. *Econ Geol* 52:499–527.

Coles, L. E. 1974. *J Assoc Public Anal* 12(3):68–72.

Davies, E. B., and J. H. Watkinson. 1966a. *N Z J Agric Res* 9:317–327.

———— 1966b. *N Z J Agric Res* 9:641–652.

Davis, A. M. 1972. *Agron J* 64:823–824.

Davis, R. D., P. H. T. Beckett, and E. Wollan. 1978. *Plant Soil* 49:395–408.

Doran, J. W., and M. Alexander. 1977. *Soil Sci Soc Am J* 41:70–73.

Doyle, P. J., and W. K. Fletcher. 1977. *Can J Plant Sci* 57:859–864.

Ehlig, C. F., W. H. Allaway, E. E. Cary, and J. Kubota. 1968. *Agron J* 60:43–47.

Evans, C. S., C. J. Asher, and C. M. Johnson. 1968. *Aust J Biol Sci* 21:13–20.

Fairhall, L. T. 1941. *New England Water Works Assoc J* 55:400–410.

Fleming, G. A. 1962. *Soil Sci* 94:28–35.

Fleming, G. A., and T. Walsh. 1957. *Royal Irish Acad Proc* 58B:151–165.

Francis, A. J., J. M. Duxbury, and M. Alexander. 1974. *Appl Microbiol* 28:248–250.

Frank, R., K. I. Stonefield, and P. Suda. 1979. *Can J Soil Sci* 59:99–103.

Franke, K. W., and E. P. Painter. 1937. *Ind Eng Chem* 29:591–595.

Frost, D. V. 1972. *CRC Critical Rev Toxicol* 1:467–514.

Frost, D. V., and P. M. Lish. 1975. *Ann Rev Phar* 15:259–332.

Gamboa-Lewis, B. A. 1976. In J. O. Nriagu, ed. *Environmental biogeochemistry,* vol. 1: 389–409. Ann Arbor Science, Ann Arbor, MI.

Ganje, T. J. 1965. In H. D. Chapman, ed. *Diagnostic criteria for plants and soils,* 394–404. Quality Printing Co Inc, Abilene, TX.

Geering, H. R., E. E. Cary, L. H. P. Jones, and W. H. Allaway. 1968. *Soil Sci Soc Am Proc* 32:35–40.

Gile, P. L., and H. W. Lakin. 1941. *J Agric Res* 63:559–581.

Gissel-Nielsen, G. 1971a. *J Agric Food Chem* 19:564–566.

———— 1971b. *J Agric Food Chem* 19:1165–1167.

Gissel-Nielsen, G., and B. Bisbjerg. 1970. *Plant Soil* 32:382–396.

Gissel-Nielsen, G., and M. Gissel-Nielsen. 1973. *Ambio* 2:114–117.

Gissel-Nielsen, G., U. C. Gupta, M. Lamand, and T. Westermarck. 1984. *Adv Agron* 37:397–460.

Grant, A. B. 1965. *N Z J Agric Res* 8:681–690.
Gupta, U. C., and K. A. Winter. 1975. *Can J Soil Sci* 55:161–166.
—— 1981. *J Plant Nutr* 3:493–502.
Gupta, U. C., H. T. Kunelius, and K. A. Winter. 1983a. *Can J Soil Sci* 63:455–459.
—— 1983b. *Can J Soil Sci* 63:641–643.
Gutenmann, W. H., C. A. Bache, W. D. Youngs, and D. J. Lisk. 1976. *Science* 191:966–967.
Hamilton, J. W., and O. A. Beath. 1963a. *J Range Mgmt* 16:261–264.
—— 1963b. *Agron J* 55:528–531.
—— 1964. *J Agric Food Chem* 12:371–374.
Hamilton, E. I., and M. J. Minski. 1972. *Sci Total Environ* 1:375–394.
Hashimoto, Y., J. Y. Hwang, and S. Yanagisawa. 1970. *Environ Sci Technol* 4:157–158.
Hemphill, D. D. 1972. *Ann New York Acad Sci* 199:46–61.
Hingston, F. J., A. M. Posner, and J. P. Quirk. 1971. *Disc Farraday Soc* 52:334–342.
Hurd-Karrer, A. M. 1933. *Science* 78:660.
—— 1935. *J Agric Res* 50:413–427.
—— 1937. *Am J Bot* 24:720–728.
John, M. K., W. M. H. Saunders, and J. W. Watkinson. 1976. *N Z J Agric Res* 19:143–151.
Johnson, C. M., C. J. Asher, and T. C. Broyer. 1967. In O. J. Muth, ed. *Selenium in biomedicine*. AVI Publ Co, Westport, CT.
Johnson, C. M. 1975. In D. J. D. Nicholas and A. R. Egan, eds. *Trace elements in the soil-plant-animal systems,* 165–180. Academic Press, New York.
Koljonen, T. 1974. *Oikos* 25:353–355.
Korkman, J. 1980. *J Sci Agric Soc Finland* 52:495–504.
Krauskopf, K. B. 1979. *Introduction to geochemistry,* 2nd ed. McGraw-Hill, New York. 617 pp.
Kubota, J. 1972. *Ann New York Acad Sci* 199:105–117.
Kubota, J., W. H. Allaway, D. L. Carter, E. E. Cary, and V. A. Lazar. 1967. *J Agric Food Chem* 15:448–453.
Kubota, J., E. E. Cary, and G. Gissel-Nielsen. 1975. *Trace Subs Environ Health* 9:123–130.
Låg, J., and E. Steinnes. 1978. *Geoderma* 20:3–14.
Lakin, H. W. 1961. In *Selenium in agriculture*. Agric Handbook no 200. USDA-ARS, Washington, DC.
Lakin, H. W. 1972. *Geol Soc Am Bull* 83:181–190.
Leggett, J. E., and E. Epstein. 1956. *Plant Physiol* 31:222-226.
Levesque, M. 1972. *Soil Sci* 113:346–353.
—— 1974a. *Can J Soil Sci* 54:63–68.
—— 1974b. *Can J Soil Sci* 54:205–214.
Lewis, B. G., C. M. Johnson, and C. C. Delwiche. 1966. *J Agric Food Chem* 14:638–640.
Lewis, B. G., C. M. Johnson, and T. C. Broyer. 1974. *Plant Soil* 40:107–118.
Lindberg, P., and S. Bingefors. 1970. *Acta Agric Scand* 20:133–136.
Lo, M. T., and E. Sandi. 1980. *J Environ Pathol Toxicol* 4:193–218.
Mahaffey, K. R., P. E. Corneliussen, C. F. Jelinek, and A. Fiorino. 1975. *Environ, Health Perspec* 12:63–69.

Martin, A. L. 1936. *Am J Bot* 23:471–483.

Martin, A. L., and S. F. Trelease. 1938. *Am J Bot* 25:380–385.

Mason, T. G., and E. Phillis. 1937. *Emp Cotton Grow Rev* 14:308–309.

Mbagwu, J. S. C. 1983. *Plant Soil* 74:75–81.

Miller, J. T., and H. G. Byers. 1937. *J Agric Res* 55:59–68.

Misra, S. G., and N. Tripathi. 1972. *Indian J Agric Sci* 42:182–183.

Moxon, A. L., and O. E. Olson. 1974. *In* R. A. Zingaro and W. C. Cooper, eds. *Selenium*, 675–707. Reinhold, New York.

Moxon, A. L., O. E. Olson, and W. V. Searight. 1950. In South Dakota Agric Exp Sta Tech Bull no. 2. Brookings, SD. 93 pp.

National Academy of Sciences (NAS). 1974. *Geochem Environ* 1:57–63.

——— 1976. *Selenium*. NAS, Washington, DC. 203 pp.

Newland, L. W. 1982. In O. Hutzinger, ed. *The handbook of environmental chemistry*, vol 3, part B:27–67. *Anthropogenic compounds*. Springer-Verlag, Berlin.

Nye, S. M., and P. J. Peterson. 1975. *Trace Subs Environ Health* 9:113–121.

Oldfield, J. E. 1972. *Geol Soc Am Bull* 83:173–180.

Olson, O. E., and A. L. Moxon. 1939. *Soil Sci* 47:305–311.

Olson, O. E., D. F. Jornlin, and A. L. Moxon. 1942. *Agron J* 34:607–615.

Page, A. L. 1974. *Fate and effects of trace elements in sewage sludge when applied to agricultural lands.* EPA-670/2-74-005. US-EPA, Cincinnati, OH. 98 pp.

Peterson, P. J., L. M. Benson, and R. Zieve. 1981. In N. W. Lepp, ed. *Effect of heavy metal pollution on plants.* vol. 1, *Effect of trace metals on plant function*, 279–342. Applied Science Publ, London.

Peterson, P. J., and G. W. Butler. 1967. *Nature* 213:599–600.

Pfannhauser, W., and H. Woidich. 1980. *Toxicol Environ Chem Rev* 3:131–144.

Plotnikov, V. I. 1960. *Russ J Inorg Chem* 5:351–354.

——— 1964. *Russ J Inorg Chem* 9:245–247.

Pratley, J. E., and J. D. McFarlane. 1974. *Aust J Exp Agric Anim Husb* 14:533–538.

Rader, L. F., Jr., and W. L. Hill. 1936. *J Agric Res* 51:1071–1083.

Ragaini, R. C., H. R. Ralston, and N. Roberts. 1977. *Environ Sci Technol* 11:773–781.

Rājan, S. S. S., and J. H. Watkinson. 1976. *Soil Sci Soc Am J* 40:51–54.

Ravikovitch, S., and M. Margolin. 1959. *Emp J Exp Agric* 27:235–240.

Reamer, D. C., and W. H. Zoller. 1980. *Science* 208:500–502.

Robbins, C. W., and D. L. Carter. 1970. *Soil Sci Soc Am Proc* 34:506–509.

Robinson, M. F. 1976. *J Human Nutr* 30:79–91.

Rosenfeld, I., and O. A. Beath. 1964. *Selenium*. Academic Press, New York.

Russell, L. H., Jr. 1979. In F. W. Oehme, ed. *Toxicity of heavy metals in the environment,* part 1:3–23. Dekker Inc, New York.

Schelenz, R. 1977. *J Radioanal Chem* 37:539–548.

Schroeder, H. A., D. V. Frost, and J. J. Balassa. 1970. *J Chron Dis* 23: 227–243.

Shamberger, R. J. 1981. *Sci Total Environ* 17:59–74.

Shrift, A. 1964. *Nature* 201:1304–1305.

——— 1969. *Ann Rev Plant Physiol* 20:475–494.

Singh, M., and N. Singh. 1979. *Soil Sci* 127:264–269.

Singh, M., N. Singh, and P. S. Relan. 1981. *Soil Sci* 132:134–141.

Slater, C. S., R. S. Holmes, and H. G. Byers. 1937. USDA Tech Bull 552. USDA-ARS, Washington, DC.

Soltanpour, P. N., and S. M. Workman. 1980. *Commun Soil Sci Plant Anal* 11:1147–1156.

Stadtman, T. 1974. *Science* 183:915–922.

Straughan, I., A. A. Elseewi, and A. L. Page. 1978. *Trace Subs Environ Health* 12:389–402.

Swaine, D. J. 1955. *The trace element content of soils.* Common Bur Soil Sci (Great Britain) Tech Commun 48.

——— 1978. *Trace Subs Environ Health* 12:129–134.

Tanner, J. T., and M. H. Friedman. 1977. *J Radioanal Chem* 37:529–538.

Taylor, F. B. 1963. *J Am Water Works Assoc* 55:619–623.

Thompson, J. N., P. Erdody, and D. C. Smith. 1975. *J Nutr* 105:274–277.

Thorn, J., J. Robertson, D. H. Buss, and N. G. Bunton. 1978. *Brit J Nutr* 39:391–396.

Trelease, S. F., and O. A. Beath. 1949. *Selenium.* Champlain Printers, Burlington, VT.

Trelease, S. F., A. A. DiSomma, and A. L. Jacobs. 1960. *Science* 132:618.

Tsuge, F., and S. Terada. 1949. *J Agric Chem Soc Japan* 23:421–425.

Underwood, E. J. 1977. *Trace elements in human and animal nutrition.* Academic Press, New York. 545 pp.

US Bureau of Mines. 1980. *Minerals yearbook,* vol. 1. *Metals and minerals.* US Dept of Interior, Washington, DC.

US Environmental Protection Agency (US-EPA). 1976. *Quality criteria for water.* US-EPA, Washington, DC.

Valentine, J. L., H. K. Kang, and G. H. Spivey. 1978. *Environ Res* 17:347–355.

Walker, D. R. 1971. *Can J Soil Sci* 51:506–508.

Wallace, A., R. T. Mueller, and R. A. Wood. 1980. *J Plant Nutr* 2:107–109.

Walsh, T., and G. A. Fleming. 1952. *Intl Soc Soil Sci Trans* (Comm II and IV) 2:178–183.

Wells, N. 1966. *N Z J Geol Geophys* 10:198–208.

——— 1967. *N Z J Sci* 10:142–179.

West, P. W. 1971. In *Proc Intl Sympos Ident and Measure of Environ Pollut* (Ontario, Canada), 38–43. Campbell Print, Ottawa.

Westermann, D. T., and C. W. Robbins. 1974. *Agron J* 66:207–208.

Williams, C., and I. Thornton. 1972. *Plant Soil* 36:395–406.

Williams, C. H. 1977. *J Aust Inst Agric Sci* (Sept–Dec):99–109.

Williams, K. T., and H. G. Byers. 1936. *Ind Eng Chem* 28:912–914.

Winter, K. A., U. C. Gupta, H. G. Nass, and H. T. Kunelius. 1973. *Can J Anim Sci* 53:113–114.

Winter, K. A., and U. C. Gupta. 1979. *Can J Anim Sci* 59:107–111.

Workman, S. M., and P. N. Soltanpour. 1980. *Soil Sci Soc Am J* 44:1331–1333.

World Health Organization (WHO). 1984. *Guidelines for drinking-water quality.* Vol. 1—*Recommendations.* WHO, Geneva, Switzerland. 130 pp.

13
Zinc

1. General Properties of Zinc

Zinc (atomic no. 30) is a bluish-white, relatively soft metal with a density of 7.133 g/cm³. It belongs to Group IIb of the periodic table, has an atomic weight of 65.37, melting point of 419.6°C, and boiling point of 907°C. Zinc is divalent in all its compounds. It is a composite of five stable isotopes: ^{64}Zn, ^{66}Zn, ^{67}Zn, ^{68}Zn, and ^{70}Zn. Their relative abundances are: 48.89%, 27.81%, 4.11%, 18.56%, and 0.62%, respectively. Six radioactive isotopes have been identified: ^{62}Zn, ^{63}Zn, ^{65}Zn, ^{69}Zn, ^{72}Zn, and ^{73}Zn with ^{65}Zn (t½ = 245 days) and ^{69}Zn (t½ = 55 minutes) being the most commonly used.

The ion Zn^{2+} is colorless and exists in a hydrated form in acidic and neutral aqueous solutions; however, the hydroxide is precipitated in alkaline solution. With excess base, the hydroxide redissolves to form zincate ion, $Zn(OH)_4^{2-}$. The Zn ion forms many complex ions in aqueous solutions, such as $Zn(NH_3)_4^{2+}$ and $Zn(CN)_4^{2-}$, as well as, others (Shamberger, 1979).

The oxidation state of Zn in a natural environment, like soils, is exclusively II.

2. Production and Uses of Zinc

Most of the Zn produced in the world comes from ores containing Zn sulfide minerals. Although more than 80 Zn minerals are known, there are only a few important commercial ores. The principal ores are the sulfides—sphalerite and wurtzite and their weathering products, mainly smithsonite ($ZnCO_3$) and hemimorphite [$Zn_4Si_2O_7(OH)_2 \cdot H_2O$]. The more common ore minerals of Zn, their compositions, and their geographic locations are shown in Table 13.1.

In 1930, world mine production was about 1.6 million tonnes, with Australia, Canada, Germany, Mexico, and the United States providing about

TABLE 13.1. Composition and principal geographical location of the common ore minerals of zinc.[a]

Mineral	Composition	Zinc concentration, %	Location
Sphalerite (Zinc blende)	ZnS	67.0	Worldwide; major ore mineral
Zincite	ZnO	80.3	USA (NJ), Brazil
Franklinite	$(Fe, Zn, Mn)(Fe, Mn)_2O_4$	15–20	USA (NJ)
Smithsonite (Dry-bone)	$ZnCO_3$	52.0	Thailand, USA
Hydrozincite	$Zn_5(OH)_6(CO_3)_2$	56.0	Brazil
Willemite	Zn_2SiO_4	58.5	USA (NJ), Australia, Brazil
Hemimorphite (Calamine)	$Zn_4(OH)_2Si_2O_7 \cdot H_2O$	54.2	Brazil, Thailand

[a] *Source:* Adapted from Cammarota, 1980, with permission of John Wiley & Sons, Inc, New York.

75% of the total (Fig. 13.1). By 1950, production was 2.15 million tonnes, with Australia, Canada, Mexico, the United States, and the USSR producing 65% of the total. By 1979, world mine production was almost 6.0 million tonnes, with Australia, Canada, Peru, the United States, and the USSR as major producers. The United States was the largest Zn metal producer in the world from 1901 to 1971, but has substantially depended on imports for its smelter operation since 1941. On the other hand, from 1950 to 1977, Canada's growth dramatically increased to place Canada as the world's largest producer.

Zinc ranks fourth among metals of the world in annual consumption, behind Fe, Al, and Cu. The automobile industry accounts for almost one third of United States slab Zn consumption. Zinc is also extensively used as a protective coating on a number of metals to prevent corrosion and in alloys such as brass and bronze. Galvanized metals have a variety of applications in the building, transportation, and appliance industries. Gal-

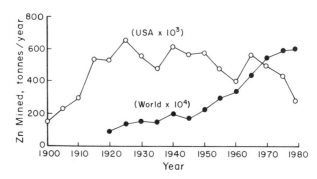

FIGURE 13.1 Trends in zinc production in the United States and worldwide 1900–1979. *Source:* Adapted from Cammarota et al (1980).

vanized pipes are commonly used in domestic water delivery systems, and Zn solubilized by corrosion is thought to make some contribution to the Zn concentrations observed in waste waters. Zinc and its compounds are ingredients of many household items, including utensils, cosmetics, powders, ointments, antiseptics and astringents, paints, varnishes, linoleum, rubber, and others. They are also used in the manufacture of parchment papers, glass, automobile tires, television screens, dry cell batteries, and electrical apparatus. Other uses include agricultural micronutrient fertilizers, insecticides, hardeners in cement and concrete, and in printing and drying of textiles. They are also used in the production of adhesives, as a flux in metallurgical operations, and as a wood preservative. Detailed uses of Zn and its compounds are described by Cammarota (1980) and the National Research Council Subcommittee on Zinc (1979).

3. Natural Occurrence of Zinc

Zinc is the 24th most abundant element in the earth's crust, with the average value quoted as 70 ppm (Krauskopf, 1979). Zinc is infrequently present in metamorphic and igneous rocks as the sulfide (sphalerite), but most of it is distributed as a minor constituent of rock-forming minerals, especially those rich in iron, such as magnetite (Fe_3O_4), the pyroxenes [$(Mg,Fe)_2Si_2O_6$ and $Ca(Mg,Fe)Si_2O_6$], the amphiboles such as $Ca_2(Mg,Fe)_5Si_8O_{22}(OH)_2$, biotite, spinel, garnet, and staurolite. Some of the rock-forming minerals have the following Zn concentrations, in ppm: magnetite (<25–2,500), olivine (50–82), garnet (<30–5,275), staurolite (2,000–6,000), pyroxene (<30–2,250), amphibole (34–8,900), and biotite (40–2,540). The common clay minerals have the following Zn concentrations, in ppm: muscovite (2–200), illite (120), chlorite (33–1,600), montmorilonite (73–156), and kaolinite (14–264). Zinc abundance in the different minerals is influenced by the zinc concentration of the magma, premetamorphic rock, etc., and the ability of the crystal lattice to incorporate this element (Wedepohl, 1978). Examples of the extreme variability in Zn concentrations of soil-forming rocks are presented in Appendix Table 1.2. The Zn concentrations of 15 groups of New Zealand samples studied by Whitton and Wells (1974) ranged from <5 to 310 ppm. Zinc is also very variable in other environmental matrices, as exemplified by the data in Table 13.2.

4. Zinc in Soils

4.1 Total Soil Zinc

To a large extent, the Zn content of soils depends on the nature of the parent rocks, organic matter, texture, and pH. As soils develop from the parent material of the earth's surface, they acquire, in varying degrees,

TABLE 13.2. Commonly observed zinc concentrations
(ppm) in various environmental matrices.

Material	Average concentration	Range
Igneous rocks[a]	65	5–1,070
Limestone[b]	20	<1–180
Sandstone[b]	30	5–170
Shale[b]	97	15–1,500
Petroleum[c]	30	—
Coal[c]	50	3–300
Fly ash:		
Bituminous[d]	20	—
Sub-bituminous[d]	15	—
Lignite[d]	14	—
Lime[e]	6	<5–8
Phosphate fertilizers[e]	305	40–600
Organic wastes[e]	390	8–1,600
Sewage sludges[f]	2,250	1,000–10,000
Soils[c]	90	1–900
Herbaceous vegetables[c]	—	1–160
Common crops[g]	6–200	—
Common fruit trees[g]	2–200	—
Ferns[c]	—	60–400
Fungi[c]	—	14–340
Lichens[c]	—	20–60
Woody angiosperm[c]	—	34–68
Woody gymnosperm[c]	—	20–90
Freshwater[c]	15*	<1–100*
Sea water[c]	5*	<1–48*

* µg/liter
Sources: Extracted from
[a] Natl. Res. Council (1979).
[b] Wedepohl (1978).
[c] Bowen (1979).
[d] Adriano et al (1980).
[e] Whitton and Wells (1974).
[f] Page (1974).
[g] Chapman (1966).

the elements present in the parent material. Soils formed from basic rocks are richer in Zn, whereas soils from granites, gneisses, etc., are poorer (Vinogradov, 1959).

The most quoted range for total Zn in normal soils is 10 to 300 ppm (Swaine, 1955), although the wider range of 1 to 900 ppm, with an average of 90 ppm, has been given (Table 13.2). For world soils, a mean of 50 ppm Zn was indicated by Vinogradov (1959), whereas Aubert and Pinta (1977) indicated means ranging from 50 to 100 ppm. More recently, Berrow and Reaves (1984) reported a mean content of 40 ppm for world soils. Other values reported for total soil Zn are given in Appendix Table 1.17.

For Canadian soils, a range of 10 to 200 ppm Zn, with a mean of 74 ppm, has been given, as contrasted with United States soils having a mean of 54 ppm (McKeague and Wolynetz, 1980). More specifically, in a survey of agricultural areas of Ontario, Canada which included 296 farms, the soils were found to have a mean of 54 ppm, with a range of 5 to 162 ppm (Frank et al, 1976). Zinc content of the Ontario soils was related to soil texture. Sandy soils were much lower in Zn (40 ppm) than loamy soils (64 ppm), clayey soils (62 ppm), and organic soils (66 ppm).

In England and Wales, from a survey that involved 225 randomly selected farms that yielded 748 samples, Archer (1980) gave the median value of 77 ppm Zn, with a range of 5 to 816 ppm. However, about 63% of the samples had Zn concentrations between 40 and 99 ppm. In the Poles'ye area of the Ukrainian region of the USSR, the total Zn content of soils averaged 40 ppm, with a range of 14 to 95 ppm (Golovina et al, 1980). In China, total Zn in soils was found to vary from 9 to 790 ppm, with an average of 100 ppm (Liu et al, 1983). The variation was primarily attributed to the influence of the underlying soil parent material. Geographically, the Zn concentration in soil decreases from the southern to the northern region of China, with calcareous soils having lower Zn than acidic soils. The mean total Zn content of tropical Asian paddy soils ranges from 35 to 88 ppm (Appendix Table 1.18).

4.2 Extractable (Available) Zinc in Soils

Zinc in soil that is water-soluble and adsorbed at exchange sites of colloidal materials is considered to be plant-available. The amount present in water-soluble form is virtually nonexistent, while the amount removed by an exchanger, such as NH_4OAc, is very small. It is a general consensus that plant-available Zn in soil can best be predicted by the use of extractants that remove only a fraction of the total amount.

There is vast literature on extractants used to measure plant-available Zn (e.g., water, salt solutions, acids, bases, and chelates). The most commonly used Zn extractants differ widely in their capacity to extract this element, from almost zero to several parts per million (Bauer, 1971). The amounts extracted are only a small fraction of the total Zn content. For example, Stewart and Berger (1965) found 4.5 ppm Zn with 0.1 N HCl vs. 0.81 ppm with 2 N $MgCl_2$ in soils having an average total Zn content of 55.2 ppm. Similarly, Trierweiler and Lindsay (1969) found approximately 5, 1.6, and 1.0 ppm Zn using 0.1 N HCl, EDTA-$(NH_4)_2CO_3$, and NH_4OAc and dithizone, respectively in soils averaging 59 ppm of total Zn. In soil tests, soils that have the highest total contents may not have the highest amount of plant-available Zn, since there are numerous factors, including inherent soil factors, that affect its extractability.

There is no universal Zn extractant for plant-availability measurement. The needs vary by country, crops, and soil type and even by region in the United States. The more popular extractable Zn tests are based on the amounts of Zn extracted by inorganic acids (Nelson et al, 1959; Wear and Evans, 1968) and by chelating agents (Brown et al, 1971; Trierweiler and Lindsay, 1969; Lindsay and Norvell, 1978; Minami et al, 1972). Acid-extractable Zn generally is a poor indicator of available Zn in near neutral and alkaline soils. Zinc extracted by EDTA (ethylenediamine tetraacetic acid), DTPA (diethylenetriamine pentaacetic acid), and dithizone (diphenylthiocarbazone) have separated near-neutral and alkaline soils into Zn-deficient and Zn-sufficient categories based on the growth of corn plants in the greenhouse (Trierweiler and Lindsay, 1969; Brown et al, 1971).

In "Cerrado" vegetation in Brazil, Lopez and Cox (1977) found the median level for Zn extracted with double acids (0.05 N HCl + 0.025 N H_2SO_4) to be 0.6 ppm, with a range of 0.2 to 2.2 ppm. The critical level for these soils is set at about 0.8 ppm. Although Zn deficiency does not occur in the Netherlands, the NH_4OAc extractant produced the best correlation between the soil and crop contents of Zn (van Eysinga et al, 1978). In Ontario, Canada, of the nine extractants used, NH_4OAc was also the best predictor of plant-available Zn from soils contaminated with metals at varying degrees (Haq et al, 1980).

Plants' need for supplemental Zn can be determined by soil testing or by plant tissue analysis. An example is the DTPA-extractable Zn test (Table 13.3) adopted by several states in the United States. Through this test, critical levels of extractable Zn have been established for certain crops and soils. For example, 0.5 ppm DTPA-extractable Zn is the "critical level" in soils for some midwestern states for growing corn, sorghum, soybeans, and pinto beans (Whitney et al, 1973). In California, the critical level for DTPA-extractable Zn was also set at 0.5 ppm (Brown et al, 1971). In Virginia, 0.8 ppm EDTA-extractable Zn was considered critical for corn (Alley et al, 1972). More recently, Lindsay and Norvell (1978) in Colorado found DTPA-extractable Zn to be critical for corn at below 0.8 ppm. In rice, the critical level using this extractant was higher at 1.65 ppm (Gangwar and Chandra, 1976). Sakal et al (1982) found that the DTPA-extractable Zn in 23 calcareous soils ranged from 0.34 to 3.42 ppm, which produced Zn levels in rice leaves of 15 to 50 ppm. The critical DTPA-extractable Zn in these soils was 0.78 ppm.

TABLE 13.3. Levels and interpretation of the DTPA-extractable test for soil zinc.[a,b]

DTPA-extractable zinc, ppm	Test ranking	Comments
0–0.50	Low	Likelihood of a response to Zn is good on corn, sorghum, soybeans, and pinto beans with good management.
0.51–1.00	Medium	Questionable range where response to Zn may be obtained under adverse conditions.
More than 1.00	High	Response to Zn is not likely to occur.

[a] Source: Whitney et al, 1980, with permission of the authors.
[b] The DTPA extraction procedure was adapted from Lindsay and Norvell (1978).

Havlin and Soltanpour (1981) have demonstrated that the NH_4HCO_3-DTPA extractant was as effective as the DTPA soil test of Lindsay and Norvell (1978) in predicting plant-available Zn in Colorado soils. The NH_4HCO_3-DTPA extractant developed by Soltanpour and Schwab (1977), simultaneously extracts both the macronutrients and micronutrients from soil. The critical Zn level for corn using this extractant was 0.9 ppm, slightly higher than the critical level established by the DTPA extractant for the same crop growing on Colorado soils. For soils in the Sub-Himalayan Peninsula, the following critical levels for the rice plant were found for various extractants: DTPA-$CaCl_2$, 0.76 ppm; DTPA-NH_4HCO_3, 0.86 ppm; EDTA-NH_4OAc, 0.85 ppm; and EDTA-$(NH_4)_2CO_3$, 1.18 ppm (Sakal et al, 1984).

4.3 Zinc in the Soil Profile

In more than 100 soil profiles in Scotland, Swaine and Mitchell (1960) found that total Zn was, in general, distributed quite uniformly from horizon to horizon. However, the distribution of extractable Zn showed greater variation with soil horizon than does total Zn content. In most well-drained agricultural soils, and in typical podzols, there was normally a decrease in extractable Zn with depth in the profile.

In Canada, John (1974) reported that total Zn and extractable Zn contents in horizons of seven soil profiles generally declined with increasing soil depth. Similarly, Roberts (1980) noted that total Zn somewhat declined with profile depth in two Canadian serpentine soils. The inconsistencies in profile distribution of Zn can perhaps be best exemplified by data for forest soils (Table 13.4).

From several profile studies done in the United States and abroad, it

TABLE 13.4. Mean total concentrations (ppm) of zinc in profiles of forest soils derived from differing parent materials.[a]

Soil horizon	Hard limestone	Mica schist	Calcareous tertiary deposit	Siliceous tertiary deposit	Argillaceous flysch	Sandy flysch	"Ultrabasic"[b] igneous
O	125	96	115	98	96	76	106
Ah	137	72	86	73	90	63	121
B)	136	83	69	79	102	72	125
C	—	83	67	75	105	77	101
Average[c]	136	79	74	76	99	71	116
R[d]	22	83	39	61	72	58	NA[e]

Source: Nakos, 1983, with permission of the author.
"Ultrabasic" includes basic and ultrabasic igneous rocks.
Average of all mineral horizons.
R indicates underlying rock.
NA = not analyzed.

can be generalized that extractable Zn decreased with depth, while total Zn was somewhat uniformly distributed throughout the profile, although there were exceptions. This was the case in Colorado, where the soils had 62, 60, and 52 ppm total Zn for the topsoil, subsoil, and parent material (C horizon), respectively (Follett and Lindsay, 1970). Values for the DTPA-extractable Zn for the respective horizons were 1.62, 0.59, and 0.29 ppm. In Hawaii, Kanehiro and Sherman (1967) reported that the concentration of total Zn appeared to be less dependent on depth than was the acid-extractable Zn. In some profiles, where the extractable Zn followed the general pattern of decreasing with soil depth, the total Zn remained essentially the same. In Louisiana, the majority of the soils examined had the highest amounts of total Zn in the subsurface horizons, whereas the highest concentration of extractable Zn was in the surface horizons (Karim and Sedberry, 1976). In 50 profiles in California, total Zn appeared to be uniformly distributed in the profile (Bradford et al, 1967). In Kansas, acid-extractable Zn from calcareous and noncalcareous soils was concentrated within the A horizon (Travis and Ellis, 1965). The B or C horizon usually contained insufficient amounts of extractable Zn for optimum crop production.

A decreasing trend of DTPA-extractable Zn in the soil profile has important implications relative to plant nutrition. If the surface horizon is removed, as in the case of leveling the land for irrigation purposes, the subsoils would probably be deficient in Zn and supplemental Zn fertilizer may be required.

The content and distribution of Zn in the profile have been linked to various soil processes and factors, such as soil formation, weathering, organic matter, cation exchange capacity, clay content, pH, and others. Although soils within a given soil group may have similar properties, they can differ widely in total and extractable Zn contents because of variations in local pedological and biological soil-forming factors. It is a general belief that the surface accumulation of Zn is caused by its acquisition by plant roots from deeper horizons, decay of organic matter, and subsequent immobilization at the surface (Fig. 13.2). Cases of increased extractable Zn in deeper horizons are indicative of parent materials having high Zn content, or of the overlying horizons being depleted because of Zn removal by plants.

Extractable Zn has been positively correlated with total Zn (John, 1972; 1974; Kanehiro and Sherman, 1967), organic matter (Follett and Lindsay, 1970; John, 1974; MacLean and Langille, 1976; Martens et al, 1966), clay content (John, 1972, 1974; Dankert and Drew, 1970), and cation exchange capacity (Follett and Lindsay, 1970; John, 1974) and inversely correlated with free $CaCO_3$ (Travis and Ellis, 1965), soil pH (John, 1972, 1974; MacLean and Langille, 1976), and base saturation (John, 1974). In soils with horizons of clay illuviation, Zn may be transferred downward in association with clay (Dankert and Drew, 1970).

FIGURE 13.2 Distribution of extractable zinc in the soil profile
and its association with the organic matter content. *Source:*
Follett et al, 1981, with permission of the authors and Prentice-
Hall.

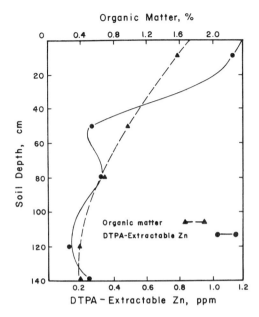

In soils treated with sewage sludge, Williams et al (1984) found that Zn
has moved by only about 10 cm below the area of sludge incorporation
done over a period of six years, whereas Chang et al (1984b) observed
that at least 90% of the heavy metals, including Zn, stayed within the 0-
to 15-cm soil depth in soils treated with sewage sludge for six years. Minimal movement of metals occurred below the 30-cm depth.

4.4 Forms and Speciation of Zinc in Soils

Trace elements, including Zn, can be partitioned into forms or pools
(sinks), based primarily on ways that elements are bound in soil or held
against extraction by various chemical agents. These forms (or fractions)
may include (1) water-soluble (both ionic and complexed with soluble organic compounds); (2) exchangeable, from soil surfaces; (3) extractable,
from organic and inorganic sites not released to extractants of exchangeable ions; (4) occluded by soil hydrous oxides; (5) precipitates; (6) immobilized in living organisms and biological residues; and (7) constituents
of the lattice structures of primary and secondary minerals. Factors controlling the distribution of trace elements between the soil matrix and the
soil solution largely govern both their dissolution and fate in biological
systems. Consequently, Zn in the different soil fractions varies in plant
availability. Zinc present in water-soluble and exchangeable forms is
readily available to plants; Zn in the other forms is either not available
or not as readily available to plants. Exchangeable Zn is adsorbed as a

cation by electrostatic attraction to negatively charged sites formed on colloidal mineral solids and organic materials. Zinc that is fixed within the crystal lattice becomes available only through mineral weathering.

The exchangeable forms can be estimated using extractants like either NH_4OAc, $CaCl_2$, HOAc, or $MgCl_2$ (Gibbs, 1973; Iyengar et al, 1981; Keefer and Estepp, 1971; Tiller and Hodgson, 1962). The organically-bound form can be estimated with H_2O_2 (DeRemer and Smith, 1964; Gupta and Chen, 1975), NaOCl (Gibbs, 1973), or $K_4P_2O_7$ (Sims and Patrick, 1978). Zinc associated with soil sesquioxides can be extracted with sodium-citrate-dithionite-bicarbonate (Jenne et al, 1974; White, 1957) or acid-ammonium oxalate (McKeague and Day, 1966; Sims and Patrick, 1978). Zinc remaining after other extractions (residual fraction) can be estimated by total digestion of residue, with either $HClO_4$-HF-H_2SO_4 mixture, HF-HNO_3 mixture (Sedberry and Reddy, 1976; Shuman, 1979), or Na_2CO_3 fusion (Pratt and Bradford, 1958).

The amounts of different Zn forms in soils vary considerably. Shuman (1979) found that, for southeastern United States soils, the exchangeable form was within the 1% to 7% range reported by White (1957) for Tennessee soils. The greater part of Zn was in the colloidal fractions—40.4% in clay and 11.5% in silt. The remainder was in the following fractions: 12% in organic, 20% in noncrystalline Fe-oxide, and about 12% in sand. The fine-textured soils had very high amounts of Zn in the clay, compared to the other fractions, but the coarse-textured soils had proportionately more in the organic matter fraction. In another study of southeastern United States soils, Iyengar et al (1981) found ≈0.4, 3.3, ≈2.5, 2.0, 25.4, and 69.6% of total Zn distributed among the following respective forms: exchangeable (nonspecifically adsorbed), exchangeable (specifically adsorbed), organically-bound, Mn-oxide bound, Al- and Fe-oxide bound, and residual (Table 13.5). The data in Table 13.5 clearly demonstrate that most of the Zn in these soils is in the Al- and Fe-oxide and residual forms, which are unavailable to plants. Those forms that are relatively available represent only a small fraction in the total soil. Sedberry and Reddy (1976) found values of 1.7%, 0.9%, 2.6%, 4.4%, and 86.4% of total Zn in Louisiana soils in the water-soluble, exchangeable, chelated, organic, and residual forms, respectively. Fractionation of Egyptian soils resulted in the following distribution: 0.01%—water soluble + exchangeable; 1.20%—weakly bound in inorganic sites; 28.6%—organically-bound; 21.5%—occluded as free oxide material; and 45.5%—residual, mainly in the mineral lattice. The very low amounts in the water-soluble + exchangeable forms were attributed to alkaline reactions (7.7–8.3 pH) of these soils that decreased Zn solubility. Sims and Patrick (1978) found that greater amounts of Zn were found in either the exchangeable or organic form at low pH and Eh values than at high pH or Eh. In contrast, amounts of Zn in the remaining forms (oxalate and residual) generally were greater at high than low pH or Eh, indicating that Zn precipitated and occluded as oxides and hydroxides was solubilized by low pH and soil saturation.

TABLE 13.5. Levels (ppm) and distribution (%) of the various forms of zinc for southeastern United States soils.[a]

	Zn fraction[b]											
	Exchangeable											
Soil type	Nonspecifically adsorbed		Specifically adsorbed		Organically bound		Mn-oxide bound		Al- and Fe-oxide bound		Residual	
	ppm	%	ppm	%	ppm	%	ppm	%	ppm	%	ppm	%
Appalachian region												
Dunmore sil[d]	0.16	0.6	0.50	1.8	0.44	2.6	0.14	0.5	4.4	15.8	22.6	81.7
Emory cl	0.09	0.1	2.71	2.6	2.20	2.2	1.99	1.9	26.0	25.4	64.7	63.2
Frederick sil	0.39	1.0	0.83	2.1	0.76	1.9	0.71	1.8	8.5	21.5	19.8	61.7
Groseclose sil	0.58	1.1	3.23	6.2	3.84	7.4	2.14	4.1	11.9	22.9	23.6	45.4
Hayter sil	0.04	<0.1	0.77	1.2	4.24	6.4	1.36	2.1	13.9	21.1	38.7	59.0
Litz sil	0.04	<0.1	0.54	0.5	0.92	0.8	1.14	1.0	64.7	53.5	55.4	45.8
Westmoreland sicl	0.04	<0.1	2.41	1.9	3.35	2.6	2.78	2.2	55.6	43.7	72.7	57.1
Coastal Plain region												
Altavista sl(I)	0.21	0.4	6.67	14.0	0.97	2.0	2.15	4.5	4.7	9.8	33.2	69.5
Altavista sl(II)	0.14	0.4	3.84	12.0	2.24	7.0	1.10	3.4	7.3	22.9	19.8	61.7
Dragston sl	0.02	0.1	0.20	1.0	0.76	4.0	0.18	0.9	2.0	10.4	15.9	83.7
Ruston s	0.44	2.2	2.06	10.6	0.37	1.9	0.40	2.0	6.9	35.6	12.1	61.8
Piedmont region												
Cecil l	0.03	<0.1	1.41	2.0	0.46	0.6	1.09	1.5	16.1	22.7	54.9	77.3
Congaree l	0.09	0.1	0.75	0.6	0.48	0.4	1.14	1.0	42.6	36.1	74.9	63.4
Cullen cl	0.05	<0.1	0.75	1.3	0.71	1.2	0.65	1.1	14.5	24.6	43.3	69.0
Davidson c	0.02	<0.1	2.71	1.7	1.12	0.7	4.20	2.6	23.5	14.7	143.4	89.4
Hiwassee sil	0.13	0.2	0.16	0.3	2.59	4.7	1.48	2.7	10.6	19.1	47.8	86.1
Penn sil	0.34	0.3	0.77	0.7	0.56	0.5	1.22	1.1	21.0	19.0	96.4	87.3
Starr sicl	0.10	<0.1	2.10	1.7	2.71	2.2	3.28	2.7	36.9	29.9	90.0	73.0
Tatum sil	0.06	<0.1	1.11	1.2	<0.01	<0.1	1.00	1.1	31.6	33.9	64.0	69.0
Average	0.16	~0.4	1.76	3.3	~1.51	~2.5	1.48	2.0	21.2	25.4	52.8	69.6

[a] *Source:* Iyengar et al, 1981, with permission of the authors and the Soil Sci Soc of America.
[b] The following extractants were used: Exchangeable (nonspecifically adsorbed)—0.05 M CaCl$_2$; exchangeable (specifically adsorbed)—2.5% HOAc; organically bound—0.1 M K$_4$P$_2$O$_7$; Mn-oxide bound—0.1 M NH$_2$OH · HCl in 0.01 M HNO$_3$; Al- and Fe-oxide bound—Tamm's solution; and residual—aqua regia–HF mixture.
[c] Percent of the total Zn.
[d] sil = silty loam; cl = clay loam; sicl = silty clay loam; sl = sandy loam; s = sand; l = loam; c = clay.

Hickey and Kittrick (1984) found that the majority (i.e., 39%) of Zn in heavily polluted samples was associated with the Fe and Mn oxide fraction. The next important fraction was the carbonate fraction, containing approximately 28% of the Zn. Samples having high levels of inorganic C also had high levels of carbonate-bound Zn. Zinc in the exchangeable and organic fractions was also appreciable, but the residual fraction was only slightly enriched. These polluted samples included three soils that received the metals from Cu smelting or sewage sludge and a harbor sediment. Chang et al (1984a) noted a shift in the chemical forms of Zn in soils after treatment with sewage sludge for seven consecutive years. In sludge-treated soils, the percentage of Zn in KNO_3 (exchangeable), NaOH (organic-bound), and EDTA-extracted (carbonate-bound) fractions increased significantly. Prior to sludge application, the HNO_3 (sulfide/residual fraction)-extractable fraction accounted for >80% of the total soil Zn, but was reduced to <50% after sludge treatment. Based on the same long-term study of Chang et al, in which soils were extracted sequentially with reagents (KNO_3, H_2O, NaOH, EDTA, HNO_3) and with DTPA, LeClaire et al (1984) concluded that: (1) Zn extracted by KNO_3 and H_2O in sequence is associated with highly labile, soluble pool dominated by Zn^{2+}, which determines immediate bioavailability to barley; (2) Zn extracted by KNO_3, H_2O, and NaOH in sequence has a strong positive correlation with DTPA-extractable Zn (Fig. 13.3), and therefore, is associated with the labile,

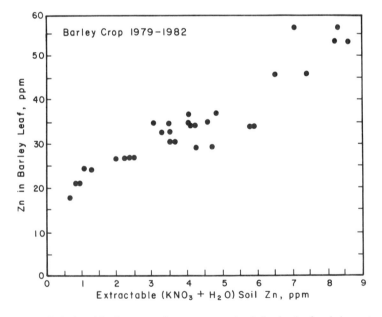

FIGURE 13.3 Relationships between zinc concentration in barley leaf and zinc extracted in sequence by KNO_3 and H_2O from a soil receiving composted municipal sewage sludge. *Source:* LeClaire et al, 1984, with permission of the authors and the Soil Sci Soc of America.

plant-available Zn pool; (3) EDTA-extractable Zn represents a reservoir of potentially bioavailable Zn; and (4) HNO_3-extractable Zn is associated with nonlabile pool that is not bioavailable.

In wetland rice culture, Murthy (1982) suggested that the $Cu(OAc)_2$ and acidified ammonium oxalate-extractable Zn are the prevalent forms of Zn controlling the availability of Zn to the rice plants. The $Cu(OAc)_2$ reagent reportedly extracted the exchangeable and complexed Zn fractions, whereas the ammonium oxalate extracted the Zn bound by amorphous sesquioxides.

The predominant Zn species in soil solution below pH 7.7 is Zn^{2+} (Lindsay, 1979). Above this pH, $ZnOH^+$ is more prevalent. The neutral species $Zn(OH)_2^0$ is the primary species above pH 9.1, while $Zn(OH)_3^-$ and $Zn(OH)_4^{2-}$ are never significant solution species in the pH range of soils (see also section 7.1).

4.5 Adsorption and Fixation of Zinc in Soils

Adsorption of Zn in soil is an important factor governing Zn concentration in the soil solution and Zn availability to plants. The adsorption capacity of soil, in general, far exceeds the few kg of Zn per hectare applied to soil to correct Zn deficiency. Apparently, the usual amounts of Zn applied are adsorbed almost completely within a short period of time and held against leaching and runoff. Fixed Zn is differentiated from adsorbed Zn in that adsorbed Zn can be removed by using an exchangeable extractant, such as 1 N NH_4OAc.

Adsorption isotherms have been used to determine the nature of various types of adsorption processes. They not only serve as useful models for physical adsorption, but sometimes can also describe chemisorption. The Langmuir adsorption isotherm was derived for the adsorption of gas molecules on solids (Langmuir, 1918), but since has been used to characterize the relationship between the adsorption of ions by solids and the concentration of ions in solutions (Boyd et al, 1947). The isotherm is simple to apply and gives parameters that can be interpreted in light of soil properties. It has been used to describe Zn adsorption in soils by several investigators (Singh and Sekhon, 1977; Shuman, 1975, 1976; Wada and Abd-Elfattah, 1978; Udo et al, 1970).

The other commonly used adsorption isotherm is the Freundlich (1926), which has been used for Zn adsorption by Shukla and Mittal (1979), El-rashidi and O'Connor (1982), and Mehta et al (1984).

The various adsorption mechanisms to soil surfaces for some trace elements, including Zn, have been discussed extensively by Ellis and Knezek (1972). They discussed cation exchange reactions, covalent bonding, and hydrolysis as important mechanisms. Some trace element cations, like Zn^{2+}, are known to enter into the crystal lattice of layer silicates through isomorphous substitution or solid-state diffusion into the crystal structure, as suggested by Hodgson (1963). Complexation with organic compounds through chelation or sequestration is also an important bonding mechanism in soils. Although precipitation of trace elements in soils is not considered an adsorption mechanism, it gives difficulties in adsorption studies since most trace elements form relatively insoluble hydroxides or carbonates.

Adsorption and fixation in soils are influenced by several factors, such as pH, clay mineral, CEC, organic matter, and soil texture. Clay minerals, hydrous oxides, carbonates, and soil organic matter are constituents known to be important in the adsorption, fixation, or precipitation of added Zn in soils.

Clay minerals vary in their capacity to adsorb or fix Zn (Farrah and Pickering, 1977; Reddy and Perkins, 1974; Tiller and Hodgson, 1962), which may be attributed to differences in CEC of clays, specific surface area, and their basic structural makeup. Elgabaly (1950) suggested that some of the applied Zn was irreversibly fixed by clay. However, Bingham et al (1964) found that Zn can be an exchangeable cation and that amounts in excess of the CEC were retained as $Zn(OH)_2$. Reddy and Perkins (1974) hypothesized that the greater fixing capacity of the 2:1 clays bentonite and illite over the 1:1 clay kaolinite was caused by Zn^{2+} entrapment in the interlattice wedge zones of the clay structure when the zones expanded due to hydration (wetting) and contracted upon drying. However, they did not rule out precipitation of $Zn(OH)_2$ at the exchange sites as a factor in Zn fixation. Tiller and Hodgson (1962) characterized clay-bound Zn as dominantly reversible associated with clay surface groups, the other existing in an irreversible nonexchangeable form associated with lattice entrapment. Similarly, Nelson and Melsted (1955) showed that a portion of adsorbed Zn on clay was not exchangeable but was acid-soluble.

Zinc adsorption by carbonates (Leeper, 1952; Jurinak and Bauer, 1956; Navrot et al, 1967; Udo et al, 1970), precipitation of Zn hydroxide or carbonates (Saeed and Fox, 1977; Sharpless et al, 1969; Singh and Sekhon, 1977; Udo et al, 1970), or formation of insoluble calcium zincate (Jurinak and Thorne, 1955; Sharpless et al, 1969) are believed partly responsible for Zn unavailability in calcareous and alkaline soils. Since soil reaction is the most important factor influencing Zn availability in soils, calcareous soils, which inherently have high pH values, have been noted for their Zn deficiencies. Thus, the overliming of acid soils, such as the highly leached and weathered ultisols in the tropics, can induce Zn deficiency in crops.

In experiments with various soils (Shuman, 1975), clays (Bingham et al, 1964; Reddy and Perkins, 1974), and organic matter (Randhawa and Broadbent, 1965; Tan et al, 1971) it was found that more Zn was adsorbed in basic than in acidic media, which was partly attributed to decreased competition from protons for adsorption sites (Farrah and Pickering, 1977). However, there may be instances where dispersion of organic matter in soils high in pH may actually reduce Zn adsorption due to the formation of metal-organic complexes (Kuo and Baker, 1980; Prokhorov and Gromova, 1971; Udo et al, 1970).

In general, there was a direct relationship between CEC and Zn adsorption, which may be even more important than organic matter (Kuo and Baker, 1980) and pH (Sharpless et al, 1969). The latter found that, in their alkaline soils, approximately 75% of the Zn initially retained was present in the exchangeable form, 15% to 20% was in the acid-extractable form, and the remainder was not extracted by either method.

Acid soils may contain large amounts of hydrous oxides of Fe, Al, or Mn, which coat the clays and may form clay-size particles themselves. These "cementing" compounds play an important role when considering Zn adsorption by soils (Gadde and Laitinen, 1974; Jenne, 1968; Kalbasi and Racz, 1978; Kalbasi et al, 1978b; Shuman, 1976, 1977). Jenne (1968) proposed that some trace element ions may be occluded and coprecipitated with hydrous oxides of Fe and Mn, a principal matrix in which metal ions are held. These amorphous and partially crystalline hydrous oxides have an amphoteric CEC, especially Al, and have the capacity to adsorb cations. Stanton and Burger (1967) found that Zn is adsorbed via two mechanisms in ferric and aluminum oxides. One mechanism involved OH^- and the other HPO_4^{2-}.

The highly pH-dependent retention of Zn in a nonexchangeable form in the soil perhaps was partly due to the existence of oxide surfaces in mineral soils whose clay fractions were dominated by layer silicates (McBride and Blasiak, 1979). The practical significance of these oxides was indicated by White (1957) who found that large fractions of Zn (30%–60%) in soils of the eastern United States were tied up with the hydrous ferric oxides. In addition, the increase in Zn content in the sand fraction in deeper soil profiles in South Africa was attributed to increased ferruginous concretions that had higher Zn than the surrounding soil from which they were taken (Stanton and Burger, 1967). Cavallaro and McBride (1984) concluded that the oxide constituent of clays could be more important than the organic constituent in Zn sorption. Their conclusion was based on their finding that pretreatment of these clays to remove organic matter by NaOCl did not decrease Zn sorption.

Elrashidi and O'Connor (1982) found that the presence of EDTA in the soil suspension decreased Zn sorption by soils, which indicates that Zn is strongly complexed with EDTA and that EDTA can compete effectively with soil sorption sites for Zn. Furthermore, they found that complex formation of Zn with the anions Cl^-, NO_3^-, and SO_4^{2-}, as well as the ionic strength of these anions (up to 0.1 M) did not have significant effects on Zn sorption. Thus, they concluded that the presence of these inorganic ligands would not affect Zn mobility in soils. However, the presence of synthetic chelates, such as EDTA and DTPA, could maintain most of the Zn in mobile form, thus increasing its potential of contaminating ground waters.

4.6 Complexation of Zinc in Soils

Soil organic matter, otherwise known as humus or soil humic matter, plays several important roles, including the increase in buffering capacity and exchange capacity of soils. It has been estimated that between 70% and 80% of the organic matter in predominantly mineral soils consists of humic materials, namely humic acid, fulvic acid, and humin (Schnitzer, 1979). The humified material is the most active fraction of humus due to its high content of oxygen-containing functional groups, including COOH, phenolic-OH, and $C=O$ structures of various types. These functional groups are responsible for the complexing ability of humic and fulvic acids for trace element cations, like Zn^{2+}. Recent comparative studies (Chen et al, 1978; Schnitzer, 1977) indicate that humic and fulvic acids extracted from soils formed under differing geographic and pedologic environments (Argentina, Canada, Israel, Italy, Japan, and West Indies) had basically similar analytical characteristics and chemical structures. Soil organic matter reacts with divalent metal cations in a manner similar to chelation reaction (Himes and Barber, 1957). Evidence for complexation of metals, like Zn, by humic acids is based on: (1) positive correlation between humus content and Zn retention in soils; (2) ability of known complexing agents to extract metals while solubilizing part of the humus; (3) selective retention of metal ions by humic and fulvic acids in the presence of cation exchange resins (Stevenson, 1972).

Organic matter has been implicated with Zn complex formation, since Zn retention by soil has been correlated with its organic matter content (Hibbard, 1940). At least three types of sites in Zn retention by humic acid have been identified: the less stable complex was associated with phenolic hydroxyl groups and weakly acidic carboxyls, while the more stable complex is associated with strongly acidic carboxyls (Randhawa and Broadbent, 1965). Tan et al (1971) found that Zn complexes with fulvic acid extracted from sewage sludge involved the formation of coordinate covalent bonds between OH groups and Zn and electrovalent linkages between COO^- and Zn. In comparing mineral and humic soils, Matsuda and Ikuta (1969) found that the percentage of added Zn in the exchangeable form was higher in mineral soil, while the percentage of organically-complexed Zn was higher in humic soil. They also found that complexation increased with humification of organic matter.

Commercial chelating agents have been available to supply micronutrient cations to plants for about 30 years. There is unquestionable evidence that chelated micronutrient cations in soils and nutrient solutions are soluble complexes (Cheng et al, 1972; Lindsay, 1974; Kalbasi et al, 1978a; Wallace, 1963). The most complete experimental data on the chemical reaction and stability of Zn chelates in soils have been obtained by Dr. Lindsay and co-workers at Colorado State University (Lindsay et al, 1967; Lindsay and Norvell, 1969; Norvell and Lindsay, 1969, 1972; Norvell, 1972). ZincEDTA and ZnDTPA are the most commonly used chelated Zn. Both are fairly stable at pH levels below 7, but above this pH,

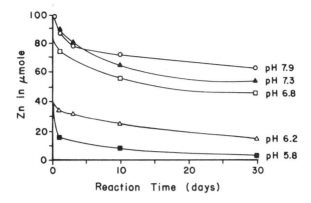

FIGURE 13.4 Concentrations of soluble zinc in soil during reaction of 100 μM Zn-DTPA with five soils of different pH. *Source:* Norvell and Lindsay, 1972, with permission of the authors and the Soil Sci Soc of America.

ZnDTPA is more stable (Lindsay and Norvell, 1969). At pH 7.5, about 12% of the EDTA and 82% of the DTPA were complexed with Zn. Similar differences between the two chelates were found by Elsokkary (1980). It can be generalized that ZnDTPA is unstable in acid soils, moderately stable in slightly acidic soils, and fairly stable in calcareous and alkaline soils (Lindsay and Norvell, 1969; Norvell, 1972). This increased ability of DTPA to chelate Zn in soils with increasing pH is shown in Figure 13.4. The low stability under acid conditions was apparently caused by the relatively high solubility of soil iron at low pH, while at high pH, the cation can be displaced by Ca^{2+} ions, followed by precipitation of Zn as sparingly soluble compounds.

Chelating agents, either natural or synthetic, are important in Zn mobility and plant nutrition. Because chelated Zn is generally soluble, it is subject to root surface contact or interception via diffusion and mass flow (Elgawhary et al, 1970; Hodgson, 1969; O'Connor et al, 1971). Thus, chelating agents play an important role in transporting insoluble nutrients to roots.

Zinc also forms complexes with Cl^-, PO_4^-, NO_3^-, and SO_4^{2-} (Lindsay, 1979). The complexes $ZnNO_3^+$, $Zn(NO_3)_2^0$, $ZnCl^+$, $ZnCl_2^0$, $ZnCl_3^-$, $ZnCl_4^{2-}$, and $ZnH_2PO_4^+$ do not contribute significantly to Zn in soil solutions. The species $ZnHPO_4^0$ may be important in neutral and calcareous soils, depending upon pH and activity of phosphate.

5. Zinc in Plants

5.1 Essentiality of Zinc in Plants

Zinc is one of the essential nutrients for plant growth. It is required in only minute amounts, unlike the major nutrients ordinarily supplied in mixed fertilizers. For example, a hectare of healthy oats contains only about 70 g of Zn in the dry matter. Yet without this small amount, no crop would grow at all. Because of this small amount needed by crops, Zn is classed as a micronutrient. Zinc is very important in plant nutrition because it is vitally involved in a number of metallo-enzymes, is essential to the stability of cytoplasmic ribosomes, catalyzes the process of oxidation, is concerned with the synthesis of auxin indole acetic acid, and the transformation of carbohydrates (Price et al, 1972; Sauchelli, 1969; Vallee, 1959).

5.2 Deficiency of Zinc in Plants

Worldwide, Zn deficiency is more common than deficiency of any other micronutrient. Treated areas amount to several million hectares a year, requiring thousands of tonnes of Zn compounds in the form of fertilizers and sprays. In the United States, Zn deficiency is known in 39 states (Sparr, 1970). Abroad, Zn deficiencies occur in Canada, Australia, New Zealand, Africa, Asia, South America, and Europe—and probably in all areas where commercial crops are grown.

Zinc deficiency is caused primarily by three factors: (1) low content of Zn in the soil; (2) unavailability of Zn present in the soil to the plant; and (3) management practices that depress the Zn availability or its uptake. Zinc is low in highly leached, acid, sandy soils such as those found in many coastal areas. In southern and western Australia, Zn deficiency has been found on large areas of sandy soils (Leeper, 1970). In most cases, the total quantity of Zn present is sufficient, but essentially all of that present is unavailable to the plant because of one or more of the factors listed in section 7. This is because most of the Zn is fixed in an unavailable form, and the extractable Zn present is insufficient to sustain normal plant growth. Several management practices could contribute to Zn deficiency.

Because plants vary in their optimum requirement for Zn, even among various species and cultivars, it is difficult to establish a single critical value. However, plants with Zn contents below 20 ppm in dry tissue can be suspected of Zn deficiency, with normal values ranging from 25 to 150 ppm Zn (Jones, 1972). Lucas (1967) gave the following sufficiency range (ppm dry wt) for Zn: corn—ear leaf at tasseling, 20–70; alfalfa—top 15 cm, 20–70; soybeans—top fully open leaves, 20–50; and vegetables—top fully open leaves, 30–100. A survey of 182 citrus groves showed that the average Zn content of the leaves was 34 ppm. However, about half of the groves had <30 ppm, which is considered low (Stewart et al, 1955). Wheat grown on deficient Darling Downs black earths in Australia had critical Zn concentrations in the tops of 20 ppm, which corresponded to 0.60 ppm EDTA-extractable Zn in the soil (Radjagukguk et al, 1980).

Zinc deficiency is widespread in China, especially in calcareous soils, and calcareous paddy soils (Liu et al, 1983). These soils, which are extensive on the loess plateau and North China Plain, have DTPA-extractable

Zn averaging 0.37 pm (range of trace to 3 ppm). The average value of 0.37 ppm is below the considered critical value of 0.5 ppm for Zn deficiency (Table 13.3). Zinc application on these soils has increased the yields of rice, corn, legumes, and fruit trees.

Zinc deficiency in lowland rice is widespread in many Asian countries, according to scientists at the International Rice Research Institute in the Philippines (Yoshida et al, 1973). They grew rice in the greenhouse on Zn-deficient soils collected from the Philippines, Korea, Thailand, Sri Lanka, Indonesia, Taiwan, and Pakistan and found that with 19 ppm or less of Zn in the shoot, rice responded to Zn application in the field. They further found that, when Zn content of the shoot was 9 to 12 ppm, growth was severely retarded; at 15 to 20 ppm Zn, visible symptoms showed up in the foliage. They set up the following criteria for diagnosing Zn deficiency of rice:

Zn content in the

Whole shoot	Diagnosis
<10 ppm	Definite Zn deficiency
10–15 ppm	Very likely Zn deficiency
15–20 ppm	Likely Zn deficiency
>20 ppm	Unlikely Zn deficiency

Only recently this physiological disorder in rice has been diagnosed as Zn deficiency. It was previously known in other countries as Akagare Type II, Khaira disease, Taya-Taya, or Hadda, caused by high fertilization of alkaline or calcareous soils (Tanaka and Yoshida, 1970; Yoshida and Tanaka, 1969).

5.3 Phosphorus-Zinc Interactions in Plants

Phosphorus, along with N and K, is usually applied at or before planting in quantities determined by soil analysis and crop needs. When large amounts of P are supplied, particularly when placed close to the seed, luxury uptake of P may occur. Often nutrients, such as N and K, may also be taken up in large quantities early in the growing cycle. The resulting high concentrations of P in plant tissues, roots and shoots, are sometimes associated with reduced uptake of other nutrients, particularly the micronutrients. Such an association of reduced micronutrient concentration with high P concentration has been termed "P-induced micronutrient deficiency." Phosphorus-Zn interaction has been the subject of intensive investigations in many countries. Wallace et al (1978) stated that a 1970 review in the USSR listed 110 references on P-Zn relationships alone. More than 100 scientific reports on that subject have appeared since then. Results have been quite variable and conflicting, reflecting the complexity of the subject.

Many plants (beans, corn, potatoes, soybeans, sorghum, flax, citrus, rice, wheat, tomatoes, and hops) have been reported to show P-Zn interaction with consequent detrimental effects on plant growth. Only a few species will be given as examples.

Beans have exhibited P-Zn interaction under both controlled and field conditions. Boawn et al (1954), Lessman and Ellis (1971), and Khan and Soltanpour (1978) reported significant effects on plant growth under field conditions. Effects on yield were conflicting. The most adverse effect appeared to occur when high levels of P were applied in the presence of low or deficient levels of available Zn. Reductions in yields were associated with high P/Zn ratios in plant tissues (Lessman and Ellis, 1971; Khan and Soltanpour, 1978). However, other field experiments failed to produce P-induced Zn deficiency, even when conditions of low soil-available Zn were present and no meaningful relationships between yield and P/Zn ratio could be deduced (Boawn et al, 1954).

Although most investigators reported either significant P-Zn interaction in terms of induced Zn deficiency and lower plant Zn concentrations (Ellis et al, 1964) other studies produced synergistic interaction in which P application enhanced Zn uptake (Chaudhry et al, 1977, Pauli et al, 1968; and Watanabe et al, 1965). Others (Peck et al, 1980; Orabi et al, 1982) also reported positive correlations between P and Zn. In a field experiment, Peck et al (1980) found that fertilizer P, when combined with Zn fertilizer, especially at the 30 kg P/ha rate, increased Zn concentrations in peas, snap beans, cabbage, and table beet tissues, as compared with no P fertilizer. However, they found that fertilizer P without Zn decreased the Zn concentration in plant tissues. In addition, fertilizer Zn with P increased Zn in plant tissues more than fertilizer Zn without P. Similarly, Orabi et al (1982) found that Zn uptake by the tomato plants, grown in potted soils, was enhanced by P application, particularly on a calcareous soil.

Corn is another major crop prone to P-Zn interaction. The most severe antagonistic interactions have been observed in instances where high application rates of P were superimposed on marginal or deficient Zn levels in the soil. Corn yield reductions or poor growth of plants have been attributed to disturbed Zn nutrition as a result of P fertilization (Adriano and Murphy, 1970; Burleson et al, 1961; Ganiron et al, 1969; Rudgers et al, 1970). Variations in results were noted in other studies where yields were unaffected despite depressed Zn concentrations associated with high tissue concentrations of P (Adriano et al, 1971; Edwards and Kamprath, 1974; Giordano and Mortvedt, 1978; Stukenholtz et al, 1966). Variations in grain and dry matter yields in field and greenhouse studies point out the dependence of P-Zn interaction on a number of other factors, including climatic condition.

A number of studies conducted with other crops with a wide range of soil and climatic conditions generally indicate that high P fertilization does induce or accentuate Zn deficiency, with concomitant decreases in yields when adequate supplemental Zn is not provided.

Under some conditions, high levels of available P in the growth medium have decreased the Zn concentration in the aerial portion of plants compared to the roots (Stukenholtz et al, 1966). Lower concentrations of Zn in the aerial portions of plants could be attributed to difficulties in translocating Zn from the roots to shoots. However, faster growth of plants receiving high rates of P could affect Zn concentrations by dilution.

Collectively, results suggest that P-induced Zn deficiency in plants cannot be ascribed to precipitation of Zn as insoluble P-Zn compounds in soils. Olsen (1972) reviewed the various possible causes of such induced Zn deficiency and concluded that dilution of an already low Zn concentration by growth response to applied P was the chief cause of such induced Zn deficiencies. This is not the entire explanation, however, in instances where applied P caused a net reduction in Zn uptake by the aerial portion of the plants. Evidence also exists for P inhibition of Zn absorption into the root or interferences with translocation of Zn from roots to metabolic sites in the leaves (Khan and Zende, 1977; Malavolta and Gorostiaga, 1974; Paulsen and Rotimi, 1968; Soltanpour, 1969; Youngdahl et al, 1977). Additional theories of physiological imbalance due to P-Zn interaction (Boawn and Brown, 1968; Watanabe et al, 1965) have not adequately described all aspects of the phenomenon and consequently, the explanation for the interaction remains vague.

5.4 Accumulation and Toxicity of Zinc in Plants

At the other end of spectrum, opposite deficiency, is toxicity, which arises when there is excessive uptake of Zn from the soil. Occurrence of Zn toxicity has been associated with Zn smelting (Singh and Lag, 1976), naturally high localized Zn concentrations (Staker and Cummings, 1941), or production practices that add extremely large quantities of Zn to the soil (Lee and Page, 1967). Zinc concentrations in acid southeastern United States soils that produced toxicity to corn ranged from 450 to 1,400 ppm extractable Zn, and to cowpeas, from 180 to 700 ppm (Gall and Barnette, 1940). Tissue Zn concentrations (ppm, dry wt) for plants showing visual toxicity symptoms were: cotton and orange—200; tung—485; tomato—526; and oats—1,700 (Chapman, 1966; Ohki, 1975). Once plant leaf levels of Zn exceed 400 ppm, toxicities can be expected (Jones, 1972). Using a fine sandy loam, Benson (1966) found stunted apple seedlings with tissue concentration of 100 ppm, and at 200 ppm, there was no top growth. In pot experiments, with different soil types and rates of Zn of up to 250 ppm Zn in soil, Zn concentrations of 792 ppm in corn, 523 in lettuce, and 702 ppm in alfalfa, were associated with yield depressions (MacLean, 1974). Some species, however, are sensitive to even low Zn level in the tissue. For example, 50 ppm in pea tissue was toxic and reduced growth (Melton et al, 1970). In a solution culture, the 5-day acute-toxicity threshhold for lettuce was 4.3 mg Zn/liter (Berry, 1978), above which root growth ceased.

TABLE 13.6. Zinc concentrations (ppm) of leafy vegetables field-grown with normal and excessive rates of Zn (as $ZnSO_4$) fertilization.[a]

Crop	Sample description	Zn treatment, (kg/ha)						
		0	11	56	112	224	448	896
Head lettuce	Market size heads	38	45	64	94	144	165	248
Leaf lettuce	Market size plants	38	46	64	125	157	239	269
Romaine lettuce	Whole plant prior to heading	32	40	56	78	108	146	179
Romaine lettuce	Market size heads	48	50	62	76	100	117	122
Endive	Market size plants	32	38	73	142	247	308	343
Parsley	Market size plants	58	50	86	107	188	296	438
Swiss chard	Market size plants	80	72	153	325	615	704	862[b]
Spinach	Market size plants	139	119	148	175	240	344	340[b]
Chinese cabbage	Whole plant prior to heading	54	48	68	84	89	112	114
Chinese cabbage	Market size heads	46	42	60	71	112	248	389
Mustard	Market size plants	32	32	36	43	58	131	364
Collard	Rosette of young leaves	33	34	38	42	63	104	366
Cabbage	Market size heads	22	20	23	28	34	54	73
Brussels sprouts	Market size heads	50	47	56	50	62	73	79

[a] *Source:* Adapted from Boawn, 1971, by courtesy of Marcel Dekker, Inc, New York.
[b] At the indicated rate of Zn fertilization plants showed normal color, but were stunted.

For rice, a 10% reduction in yield occurred when extractable (0.1 N HCl) Zn was 460 ppm in paddy soils in Hokkaido, Japan (Chino, 1981b). Ichikura et al (1970) reported that from 250 to 1,000 ppm of total Zn in soil was harmful to rice plants. Chino (1981b) summarized the Zn toxic levels for rice tops as 100 to 300 ppm, and rice roots as 500 to 1,000 ppm. Other crops can tolerate much higher application of Zn, as shown in Table 13.6. This research was conducted on a noncalcareous fine sandy loam in which $ZnSO_4$ was broadcast at rates indicated in the table. Of the 12 crops, only Swiss chard and spinach displayed stunted growth and any tendency to be Zn accumulators.

6. Zinc in Natural Ecosystems

Relatively little information is available on the background concentrations, distributions, or fluxes of trace elements in natural, wildland ecosystems. Long-term ecological behavior of trace elements, including pathways and rates of dispersion, residence times in various components of ecosystems, and chemical transformation, is largely unknown. Forest ecosystems can be viewed as nutrient-element conserving systems, controlled by nutrient availability and climatic constraints (O'Neill et al, 1975). Forest productivity depends upon the recycling of nutrient elements through higher plants, consumers, decomposers, predators, and others. Nutrient cycling mechanisms in forests collectively constitute a closed cycle in which all major processes are critical for the system to be maintained. Inhibition or disruption of one of these processes can adversely affect forest productivity, as when certain trace elements in ecosystem components become significantly enriched from anthropogenic activities. These metals become airborne from industrial operations, are transported over distant areas, and are deposited on the earth's surface, usually with precipitation. Cadmium, Zn, Cu, and Pb are some metals likely to be detected in somewhat elevated concentrations in natural ecosystems even in very remote areas.

Van Hook et al (1977) determined the distribution and cycling of Cd, Pb, and Zn in a mixed deciduous forest at the Walker Branch Watershed, located in eastern Tennessee. The major anthropogenic sources of atmospheric emissions in the vicinity (14-km radius) of this watershed are three coal-fired electric power generating stations with a combined coal consumption of 7×10^6 tonnes/year. Annual discharges of Zn from these plants were estimated at approximately 9.5 tonnes/year. But in spite of this input, this system is still considered a relatively unpolluted one. Concentrations of Zn in living vegetation there generally followed the pattern: roots > foliage > branch > bole (Table 13.7). Zinc concentrations generally were higher in *Carya* spp. and *Quercus velutina* than the other species. Generally, Zn concentrations in O2 litter were higher than those in the O1 litter (Table 13.8). The A1 soil horizon contained the highest concentration of Zn, which was associated with its high organic matter content. The standing pools of Zn for the soil profile (100 cm deep) averaged approximately 354 kg/ha. The inventory in Figure 13.5 indicates that the watershed soil is the major sink for trace elements in a given forested ecosystem. Similar results were obtained in the Solling Mountains in central Germany, where 262 to 315 kg/ha of Zn were found in the soil (0–50 cm), leaving only 1.5% to 4.5% of the total inventory tied up in the vegetation (Heinrichs and Mayer, 1980). In tropical moist forest, soil (0–30 cm) inventory for Zn is much lower (average of 134 kg/ha for two sites), but the amount contained in the vegetation ($\approx$5% of total) is still comparable with the other ecosystems (Golley et al, 1975).

In another relatively unpolluted forested ecosystem in New Hampshire (Reiners et al, 1975), it was found that the concentrations of Zn in organic horizons of soils distributed over an elevational gradient fell within a broad range of levels to represent natural conditions (Table 13.8). The Zn levels in these organic layers compare favorably with those from other ecosystems. Why Zn was sometimes lower in the mostly decomposed organic layer than the least decomposed layer in this particular ecosystem, is not well understood. In forest stands along an elevational gradient on Camels Hump Mountain, northern Vermont, Friedland et al (1984) found that Zn accumulation in the forest floor during the period 1966 to 1980 in the northern hardwood, transition, and boreal forests was 0.18, 0.41, and 0.09 kg Zn/ha per year, respectively. The accumulated amounts of Zn in the

TABLE 13.7. Zinc concentrations (ppm) of some tree species from relatively unpolluted areas.

Species	Needles or leaves	Twigs (wood & bark)	Branches (wood & bark)	Trunk (wood)	Trunk (bark)	Roots (bark & wood)
Hubbard Brook Exp. Forest,[a] New Hampshire, USA						
Sugar maple	52	62	19	8	29	46
Yellow birch	334	226	154	32	259	118
American beech	25	40	21	7	10	47
Mt. maple	45	63	45	12	97	75
Red spruce	21	40	59	12	103	119
Pine cherry	24	22	27	9	30	28
University Forest,[b] Maine, USA						
Red spruce	45	46	44	8	50	67
Balsam fir	50	67	48	11	45	40
Hemlock	10	29	28	2	15	13
White pine	52	68	55	11	65	22
White birch	77	91	100+	28	99	100+
Red maple	41	49	48	29	78	69
Aspen	100+	88	100+	17	97	47
Walker Branch Watershed,[c] Tennessee, USA						
Shortleaf pine	10	—	6	4	—	29
Chestnut oak	14	—	1	<1	—	17
Red maple	19	—	13	8	—	—
Hickory	36	—	11	14	—	—
Yellow poplar	15	—	5	3	—	29
White oak	18	—	4	3	—	—
Northern red oak	17	—	3	3	—	—
Tupelo	15	—	10	6	—	—
Dogwood	16	—	12	5	—	—
Black oak	28	—	10	5	—	—

Sources: Extracted from
[a] Likens and Bormann (1970).
[b] Young and Guinn (1966).
[c] Van Hook et al (1977).

TABLE 13.8. Zinc levels (ppm) in various ground
components of relatively unpolluted forest ecosystems.

Location[a]	ppm Zn	Source
New Hampshire, USA		Reiners et al (1975)
Hardwoods		
O1	113	
O2 (F)	169	
(H)	73	
Spruce fir		
O1	96	
O2 (F)	70	
(H)	39	
Fir		
O1	121	
O2 (F)	118	
(H)	72	
Fir-krummholz		
O1	96	
O2 (F)	73	
(H)	57	
Alpine tundra		
O1 + O2	29	
Tennessee, USA		Van Hook et al (1977)
Yellow poplar		
O1	58	
O2	130	
Chestnut oak		
O1	42	
O2	110	
Oak-hickory		
O1	48	
O2	125	
Pine		
O1	56	
O2	59	
Heath, Sweden		Tyler et al (1973)
Aboveground litter	53, 68	
Below-ground litter	72, 80	
Spruce needle litter, Sweden		Nilsson (1972)
Least decomposed	60, 65, 86	
Intermediate	72, 90, 91	
Most decomposed	78, 126	

[a] O1 horizon corresponds to L (litter) layer. O2 horizon corresponds to F (fermentation) and H (humic) layers.

forest floor prior to this period were estimated to be 6.08, 2.13, and 6.97 kg Zn/ha in the respective stands. Zinc concentration in the vertical profile of the forest floor of the boreal forest decreased significantly with depth, with the highest concentration occuring in the F-layer (103 ppm in 0–2 cm depth vs. 67 ppm in the 8–10 cm depth).

Zinc concentrations for various tree species in two forested ecosystems

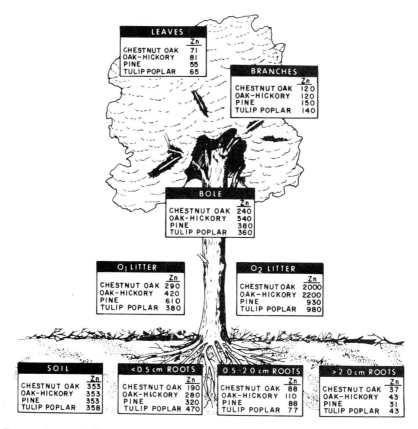

FIGURE 13.5 Distribution of zinc in the major vegetative components (g · ha⁻¹) and soils (kg · ha⁻¹) in a relatively unpolluted eastern Tennessee mixed deciduous forest. For the soil inventory, the 100-cm depth of the profile was considered. *Source:* Adapted from Van Hook et al, 1977, with permission of the authors.

in New England are shown in Table 13.7. The data indicate that birch is perhaps an accumulator of this element.

Forested ecosystems close to large urban or industrial areas have markedly higher concentrations of certain trace elements than ecosystems in rural areas. This was shown near the Chicago-northwestern Indiana area (Parker et al, 1978); in New Haven, Connecticut (Smith, 1973); and in the Belgian Kempen (Bosmans and Paenhuys, 1980). Zinc concentrations in the surface litter (O1 + O2 horizons) for the urban site near Chicago (Parker et al, 1978) were about 10 times greater than those reported by Reiners et al (1975) for the New Hampshire forests. The bulk of the metals in the urban site is retained in the soil layers that contain appreciable organic matter, the O1 + 2 and A1 horizons. Of the total metal load in the ecosystem, these two horizons contained 80% of the Zn. Smith (1973) found that Zn in six woody plant species in New Haven, Connecticut was present in "above normal" amounts based on "normal" values obtained for samples from New Hampshire and Vermont. In the Belgian Kempen, the affected forest surface soils had a mean Zn concentration of 38 ppm, compared to the normal content of 5 to 15 ppm. In white pine stands in central Massachusetts, Siccama et al (1980) found that the total Zn content of the forest floor increased by 0.71 kg/ha after 16 years. However, Zn concentrations in both the F and H layers decreased significantly, which was attributed primarily to significant increase in forest floor total dry weight.

Other means to increase Zn levels in trees in forested ecosystems include: smelter operation (Buchauer, 1973; Jackson and Watson, 1977; Whitby and Hutchinson, 1974), sludge application (Sidle and Kardos, 1977; Korcak et al, 1979; Lepp and Eardley, 1978), and fertilizer application (McKee, 1976).

7. Factors Affecting the Mobility and Availability of Zinc

7.1 pH

When the soil reaction is above pH 7.0, the availability of Zn becomes very low. Severe Zn deficiency is often associated with high soil pH. However, Zn is most soluble, and hence phytoavailable, under acidic conditions. In calcareous soils, characterized by the presence of free $CaCO_3$, Zn deficiency can also be prevalent. There are numerous reports elucidating the depression of Zn uptake by crops in alkaline or calcareous soils (Melton et al, 1970; Pauli et al, 1968; Saeed, 1977; Lutz et al, 1972; Jugsujinda and Patrick, 1977; Wallace et al, 1978; Chino, 1981b). Liming the soil is the most commonly followed practice in elevating the pH to the desired level. Liming of an acid sandy loam soil (pH 4.9) to about the neutral point markedly reduced the amounts of extractable Zn (MacLean, 1974). Zinc activity in soil solution declined sharply when soils were limed from pH 4.3 to pH above 5.0 (Friesen et al, 1980). In another case (Gupta et al, 1971), liming a strongly acid sandy loam soil to pH 5.6 and above reduced the availability of Zn to forage crops. They indicated that at pH 6.1, Zn supply may be severely restricted. Pepper et al (1983) found that liming plots amended with anaerobically digested sewage sludge reduced Zn levels in corn tissues. The lime raised soil pH from 4.6 to 6.5. In a greenhouse experiment, Chang et al (1982) found that uptake by barley of sludge-borne Zn applied to soils was much lower from a soil whose pH was 7.8 than from the other two soils whose pH were 7.1 and 6.0, especially at the 100 tonnes/ha rate compared to the 20 tonnes/ha rate. Reduction in yield of corn grown on a limed podzol in the greenhouse was attributed to reduction in Zn uptake (Grant et al, 1972). In potatoes, lime application increased the foliage and tuber content of Ca but concurrently decreased Zn uptake (Laughlin et al, 1974). In soybeans grown on strongly acid soils containing Zn from peach sprays, lime application decreased the Zn content of leaves from 229 ppm in the acid soil (pH 5.4) to 77 ppm in limed soil (pH 6.4) (Lee and Craddock, 1969). However, overliming soils, especially highly leached sandy soils, can drastically reduce Zn availability and result in severe Zn deficiency among sensitive crops. Apparently, lime, which supplies Ca^{2+} and Mg^{2+} in the case of dolomitic lime, inactivates Zn in the soil through its reduction in solubility and/or induces ion competition on the root surface between Ca^{2+} and Zn^{2+} ions (Rashid et al, 1976).

Zinc deficiency of rice is widespread throughout Asia on neutral to alkaline, calcareous soils, especially those containing 1% or more organic matter. The incidence of Zn deficiency has been correlated positively with high pH (Yoshida et al, 1973; Forno et al, 1975). When soil pH was low, a low content of extractable Zn was not necessarily associated with Zn deficiency. In laboratory experiments, Zn uptake by rice plants under both aerobic and anaerobic conditions progressively decreased as soil pH was raised from 5.0 to 8.0 (Jugsujinda and Patrick, 1977). There was a particularly sharp decrease in uptake under aerobic conditions at pH 7.0 to 8.0.

The importance of pH on the predicted environmental behavior of Zn is depicted in Figure 13.6 (see also section 4.4).

7.2 Redox Potential

Redox condition has also a pronounced effect on the solubility of Zn and hence its availability to rice plants. In constantly flooded soils, relatively insoluble zinc sulfide may be formed under strongly reducing conditions.

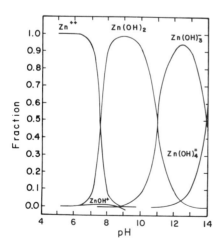

FIGURE 13.6 Simulated distribution of molecular and ionic species of zinc at different pH values. Note that the pH of most productive agricultural lands falls between pH 6.0 and 7.0. *Source:* Hahne and Kroontje, 1973, with permission of the authors and the Am Soc of Agronomy, Crop Sci Soc of America, Soil Sci Soc of America.

This occurs at about −150 mV redox potential (Connell and Patrick, 1968), in which case Zn may not be available to plants. However, the activity of the rice plant roots may be involved in metal absorption and translocation from flooded soils, averting nutritional problems (Chino, 1981a).

7.3 Organic Matter

Zinc availability to plants is generally low in organic soils. Thus, Zn deficiency oftenly occurs in muck or peat soils or even in old barnyards and corral sites. In comparing two Coastal Plain soils, Zn uptake by soybeans was lower from the higher organic matter Pocomoke soil than the lower organic matter Sassafras soil (White and Chaney, 1980). The former soil had 3.8% organic matter and CEC of 16 meq/100 g, compared with 1.2% organic matter and 5.4 meq/100 g for the latter. White and Chaney rated organic matter as more important than hydrous oxides of Fe and Mn in moderating the effects of excessive soil Zn to soybeans. Even paddy soil with high extractable Zn can be expected to produce Zn deficiency in rice if the soil contains high (>1%) organic matter (Yoshida et al, 1973). Addition of rice straw to Crowley silt loam at the Rice Branch Experiment Station in Arkansas caused a significant decrease in soil solution Zn concentration at 15 days after flooding, after which time no differences between treatments (1,500 and 6,000 kg/ha of rice straw) were observed (Yoon et al, 1975). Apparently Zn was temporarily immobilized by the bicarbonates, hydroxide, and organic ligands from decomposition of the straw. Another possibility is immobilization of Zn in microbial biomass during the process of decomposition or adsorption of Zn on exchange sites.

7.4 Management and Fertilization Practices

A. REMOVAL OF SURFACE SOIL

Removal of topsoil during the course of land leveling for irrigation purposes exposes the subsoils, which are often low in available Zn. Numerous instances are known where crops grown on newly leveled fields showed deficiency symptoms (Boawn and Leggett, 1963; Grunes et al, 1960; Adriano and Murphy, 1970).

B. ZINC CARRIER

Reports in the literature indicate that the source of Zn applied to the soil makes a difference in its availability to the crop. In general, Zn applied as a chelate is more readily utilized by the plants than Zn applied as an inorganic salt. Chelated sources of Zn are generally believed to be about five times as effective as inorganic sources in overcoming Zn deficiency (Judy et al, 1965; Keefer and Singh, 1968; Lucas, 1964; Wallace and Mueller, 1959). Boawn et al (1957) noted that when 2 ppm Zn was applied to an alkaline silt loam in pots, bean plants took up 3.5 times more Zn from ZnEDTA than from $ZnSO_4$. Based on Zn concentration of sweet corn, Wallace and Romney (1970) concluded that 0.90 kg of Zn/ha, as ZnEDTA, was as effective as 9.0 kg/ha as $ZnSO_4$. Field tests by Judy et al (1965) and Brinkerhoff et al (1966) in Michigan showed that 0.55 kg of Zn/ha as chelate was as effective as 2.7 kg applied as $ZnSO_4$. Both carriers were banded near the seed. The greater effectiveness of chelate sources was subsequently confirmed by later field and laboratory studies in Michigan (Mugwira and Knezek, 1971; Vinande et al, 1968). In field experiments with potatoes, Soltanpour et al (1970) found ZnEDTA to be a more effective Zn source than $ZnSO_4$, when 1.8 kg/ha Zn as EDTA produced a comparable yield to that of 7.1 kg/ha Zn as $ZnSO_4$.

The effectiveness of Zn sources may be influenced by the method of placement. Brown and Krantz (1966), using pot culture, noted that Zn uptake by corn from $ZnSO_4$ and from ZnEDTA was comparable when these materials were mixed into the soil, but when banded under the seed, ZnEDTA was more effective than $ZnSO_4$. Similarly, Chesnin (1963) reported that when the sources were banded under the corn seed in pot culture, Zn chelates produced higher dry matter than when $ZnSO_4$ was used. Shukla and Morris (1967) found no difference in Zn utilization by corn from ZnEDTA and $ZnSO_4$ when these sources were mixed into a loamy sand in pots. In rice plants grown in the greenhouse, application of equal amounts of $ZnSO_4$ and ZnEDTA showed leaf-burning symptoms with ZnEDTA but not with $ZnSO_4$, probably due to toxicity of excess EDTA or high accumulation of Zn in plant tissues (414 ppm) (Yoshida et al, 1973). In pot experiments, ZnEDTA gave the highest yield of paddy rice, followed by $ZnSO_4$ and ZnO when these materials were mixed with silty loam soil at rates of up to 10 ppm (Singh and Singh, 1980). It was pointed out in a previous section that chelates increase the diffusion of metals in soil to the root surface as a result of these compounds transforming solid phase forms of metal cations in soil into soluble metal complexes.

C. PLACEMENT OF PHOSPHORUS

Methods of application of P fertilizer have had significant effects on the magnitude of P-Zn interaction. When supplemental P is required, placement of the P close to the seed at planting has been recognized as a more highly effective method for inducing P absorption during early plant development than are preplant broadcast applications that are incorporated into the soil by tillage. Higher P concentrations are frequently noted in plants from banded P. This difference in early season P absorption is particularly pronounced when soil conditions are relatively cool and wet after seeding. Those same conditions tend to depress plant absorption of both P and Zn because of slow root development and slow metabolic activity.

Since Zn uptake is depressed under cool, wet soil conditions, banded placement of P tends to magnify the P-Zn interaction even further when Zn is deficient or near-deficient in the soil. Banded application of P near the seeds has been observed to depress Zn uptake by corn (Adriano and Murphy, 1970; Langin et al, 1962; and Stukenholtz et al, 1966). Adriano and Murphy (1970) provided strong evidence for the improved uptake of P from banded applications in a cool, wet spring, and leaf analyses for Zn demonstrated the subsequent depression of Zn concentrations in corn leaves when P was applied without Zn and when P was banded versus broadcast. When the Zn deficiency was eliminated, the P-Zn interaction became positive, producing higher grain yields.

7.5. Plant Factors

Different abilities to utilize nutrients are fairly common among cultivars of crops. It was noted that corn (Clark, 1978; Terman et al, 1975), barley (Chang et al, 1982), wheat (Shukla and Raj, 1974), cocoa (Chude and Obigbesan, 1983), and rice (Giordano and Mortvedt, 1974) inbreds differ in their ability to take up and use Zn. Similar differences have been reported in the literature in relation to the abilities of sorghum and soybeans to absorb Fe. The remarkable differences among cultivars of lettuce and corn in Zn nutrition is demonstrated in Table 13.9. Cultivars of soybeans and dwarf wheat reacted differently to P-Zn interaction (Paulsen and Rotimi, 1968; Sharma et al, 1968). Obviously, lines that have a poor ability to utilize micronutrients present in their root environment will be more subject to P-induced deficiencies. Brown and Tiffin (1962) reported that Zn deficiency is also dependent on plant species. Crops sensitive to Zn

TABLE 13.9. Zinc concentrations (ppm) in different cultivars of lettuce and corn.[a]

Lettuce		Corn		
Cultivar	Leaf	Inbred line	Leaf	Grain
Romaine	51	B77	61.8	44.6
Boston	63	R177	88.0	52.4
Bibb	68	H96	94.3	54.0
Valmaine	71	H99	102.9	33.8
Tania	72	H100	140.1	46.9
Dark green		R805	148.1	43.7
Boston	68	B37	164.2	64.3
Belmay	89	Oh545	170.6	51.6
Butterhead 1033	100	A619	193.3	70.0
Butterhead 1044	124	H98	217.2	54.5
Buttercrunch	94	Oh43	281.8	70.0
Butterhead 1034	104	—	—	—
Summer bibb	125	—	—	—

[a] Source: Adapted from CAST, 1980.

deficiency in the United States are: beans, potatoes, and peaches in the northwest; cotton and rice in the southwest; corn and soybeans in the central Great Plains; flax and corn in the northern Great Plains; corn and beans in the north central region; and corn and citrus in the southeastern region. In general, cereal crops have not been found to be sensitive to Zn deficiency (Viets et al, 1954).

8. Zinc in Drinking Water and Food

Zinc is an essential element in human and animal nutrition. The daily dietary allowances for Zn recently recommended by the National Academy of Sciences (1977) are as follows: adults, 15 mg/day; growing children over a year old, 10 mg/day; and additional supplements during pregnancy and lactation. Marginal or deficient Zn intake, rather than toxicity, is the major health concern in the general population. In general, the contribution of drinking water to the daily nutritional requirement of Zn is negligible. The US-EPA (1976) has set a limit of 5 mg/liter of Zn for domestic water supplies. The USSR has established a much lower limit of 1 mg/liter for other than health reasons.

Dietary intakes of Zn (mg/day) in several countries vary slightly: 14.3 in the United Kingdom; 16.1 in India; 14.0 in Japan; 18.0 in the United States; 4.7–11.3 in Italy; and 13.0 by the International Commission on Radiological Protection (Clemente et al, 1977; Hamilton, 1979; Soman et al, 1969; Russell, 1979). In India, drinking water contributed only 0.008 mg/day of Zn (Soman et al, 1969). The variation in intake can be partly attributed to the different eating habits and preferences of the people of the world. For example, foods of Asians may be dominated by fish and poultry, grain and cereal, leafy and root vegetables, and fruits in contrast to the diet of an average American adult as shown in Table 13.10. Foods vary not only by classes but also by individual items. Among the richest sources of Zn are: wheat germ and bran (40–120 ppm) and oysters (up to 1,000–1,500 ppm); white sugar, pome, and citrus fruits are among the lowest source, usually with <1 ppm of fresh edible portion (Underwood, 1973). Data extrapolated from Shacklette et al (1979) gave the Zn concentrations (ppm in dry matter) for the following vegetables: asparagus, 20–40; lima beans, 20–100; snap beans, 20–100; beet root, 30–50; blackeyed peas, 40–120; cabbage, 10–600; carrot, 10–30; corn, 50–280; sweet pepper, 20–40; soybeans, 70–150; and tomato, 10–60. Duke (1970) reported concentrations (ppm in dry matter) for pineapple as 20; for coconut, 16; for cassava, 16–27; for rice, 33–38; and for corn, 31–36. The tolerance level given for Zn in fresh vegetables in Canada to comply to the 1970 Food and Drug Act and Regulations is 50 ppm on a wet weight basis (MacLean, 1974). Assuming a moisture content of 94% for lettuce, this tolerance level is equal to 833 ppm Zn on a dry matter basis. Only in very rare instances can this high concentration be encountered in agricultural practices.

TABLE 13.10. Estimated dietary intakes and concentrations of zinc in various food classes.

Class	Food intake[a] g food/day	Food intake[a] % of total diet	Zn intake[b] µg Zn/day	Zn intake[b] % of total diet	Zn concentration, ppm wet wt[b] Average	Zn concentration, ppm wet wt[b] Range
Dairy products	769	22.6	3,837	21.4	5.1	3.4–7.3
Meat, fish, and poultry	273	8.0	6,600	37.2	30	20–58
Grain and cereal	417	12.3	3,370	18.8	7.4	5.1–8.2
Potatoes	200	5.9	1,198	6.7	4	1.6–7.8
Leafy vegetables	63	1.9	136	0.8	3	1.9–7.5
Legume vegetables	72	2.1	542	3.0	8.1	4.7–16.8
Root vegetables	34	1.0	80	0.5	2.6	1.6–4.2
Garden fruits	89	2.6	267	1.5	3.4	1.9–7.5
Fruits	222	6.5	194	1.1	—	—
Oil and fats	51	1.5	314	1.8	—	—
Sugars and adjuncts	82	2.4	254	1.4	—	—
Beverages	1,130	33.2	1,066	6.0	—	—
Total	3,402	100.0	17,918 (~18 mg)	100.0		

[a] Based on 1972–73 total diet survey by the US Food and Drug Administration for an adult person, Tanner and Friedman, 1977, and used with permission from Elsevier Sequoia SA, Lausanne, Switzerland.
[b] Based on 1973 total diet survey by the US-FDA (Russell, 1979).

Food processing can alter the Zn contents of foods. Results from 22 countries indicate that unpolished rice contains, on the average, 16.4 ppm Zn vs. polished rice, which contains 13.7 ppm Zn. Similar results were found for Indonesian rice: 19.0 ppm for unpolished vs. 17.3 ppm Zn for polished (Suzuki et al, 1980). Zinc may be lost during polishing since it is contained mostly in the germ and bran, which are eliminated during milling. The average per capita consumption of polished rice in countries like Bangladesh, Indonesia, Japan, Philippines, Singapore, and Taiwan is very high: 260, 300, 304, 235, 237, and 364 g/day. On the other hand, daily consumption in most European countries is very low, usually <10 g/person.

Toxicologically, Zn is relatively insignificant, since there is a wide range between the usual environmental levels and toxic levels. Therefore, neither WHO/FAO nor the US-FDA give guidelines for Zn other than as an essential micronutrient.

9. Sources of Zinc in the Terrestrial Environment

9.1 Zinc Fertilizers

Other than Zn in soils arising from the parent material, all other sources of Zn in soil are anthropogenic in nature. In agriculture, Zn-carrying fertilizers are by far the largest source of Zn. About 22,000 tonnes of Zn are used in fertilizer each year in the United States. There are two types of manufactured Zn fertilizers: inorganic and organic (Follet et al, 1981). Primary inorganic sources include zinc sulfate ($ZnSO_4$), the most commonly used Zn fertilizer; zinc oxide (ZnO); and zinc-ammonia ($Zn-NH_3$) complex. Primary organic sources include ZnEDTA, zinc lignin sulfonate, and zinc polyflavonoids. The latter two are wood by-products in paper production and sometimes are called *natural organic* complexes. Zinc applied for crop production can have residual effects for up to 2, 3, or more years (Follet and Lindsay, 1971; Brown et al, 1964; Boawn, 1974). In applying various rates of $ZnSO_4$ of up to 896 kg/ha Zn to a noncalcareous fine sandy loam soil, most of the extractable Zn was confined in the top 25 cm (Boawn, 1971).

Recommended rates of Zn for various crops ranged from <1 kg/ha Zn, as chelates, to as high as 22 kg/ha, as $ZnSO_4$ (Murphy and Walsh, 1972). The rate of application depends upon the crop species, soil, or leaf test recommendations, method of placement (foliar, banded, or broadcast), and carrier of Zn (chelates vs. inorganics). While foliar application and seed treatments can be satisfactory in supplying Zn to crops, it is generally accepted that there is no better substitute than soil application. Foliar application should be followed as a supplement only during critical periods. Very high rates of Zn applied to soil can be tolerated by crops, however. In a field experiment on a Plainfield loamy sand, rates up to 363 kg Zn/ha, as $ZnSO_4$, failed to affect the yields of snap beans, cucumber, or corn deleteriously, even if the Zn concentrations in the first two crops exceeded 350 ppm (Walsh et al, 1972).

9.2 Sewage Sludges

The other source of Zn in agriculture and other environments (e.g., drastically disturbed lands, forests, etc.) that is becoming more popular is municipal sewage sludge. Because most sludge studies conducted in the United States, Canada, and several European countries are experimental in nature, application rates are usually in tens or sometimes in hundreds of tonnes, giving much larger amounts of Zn than commercial fertilizers. In the resource-limited Far Eastern countries, raw sewage has been applied directly to soils as a source of fertilizer for many centuries. Although reports vary, there is a general agreement that most metals, including Zn, from sludge applied to soils tend to remain in the surface. Page (1974), using data obtained from Anderson and Nilsson (1972) reported that, after 12 years of adding 84 tonnes/ha of sludge to soil, practically all the Zn remained in the surface 20 cm. Boswell (1975) found that very little Zn had moved deeper than 15 cm in the clay loam soil after two years of sewage sludge application. Page and Chang (1975), in reviewing the literature on trace element movement, concluded that in general, Zn in soils treated with sludge is restricted to the depth of tillage (plow layer).

An example of Zn distribution in the soil profile from long-term application of sludge is presented in Appendix Table 1.8. However, others reported deeper penetration of Zn. Hinesly et al (1972) observed that metals had moved to the 45 cm-depth in the soil following a three-year furrow irrigation of corn plots with 65 cm of digested sludge (168 dry tonnes/ha). Touchton et al (1976), after only two years and eight additions of sludge, found Zn movement to the 30-cm depth. There are numerous cited evidences of increases in total or extractable Zn from sludge application (Gaynor and Halstead, 1976; LeRiche, 1968; Webber, 1972; Purves, 1972; Abdou and El-Nennah, 1980).

The review by Page (1974), citing numerous field and greenhouse studies, indicates that metal concentrations of plants grown on sludge-amended soils are dependent on loading rates, soil properties—particularly pH—and plant species. He also demonstrated that there is a wide range of metal concentrations in sewage sludges and that the chemical forms of these elements differ among sludges from different treatment plants. Sludges that contain large amounts of metals may adversely affect soil productivity and lead to food chain accumulation by some metals. Based on these expectations, the US Environmental Protection Agency (1977), recommended that the amount of sludge applied on privately owned agricultural lands be partly limited by the cumulative amounts of Zn added, which in turn should be based on the soil CEC (see Chapter 1, section 4.2). For example, for soils with pH maintained at pH 6.5 or above, the maximum cumulative application of sludge-borne Zn is limited to 560 kg/ha for soils having CEC ranging from 5 to 15 meq/100 g. Acceptable metal loading rate is reduced by one-half for soils having CEC of <5 meq/100 g, and is doubled for soils having CEC in excess of 15 meq/100 g.

To date, results on the effect of sewage sludge on crop growth, soil productivity, and food chain quality are conflicting, signalling the complexity of the issue. Increases in yield can be expected due to the N and P in the material (Page, 1974; Coker, 1966a, 1966b). However, utilization of sludge in land often results in significant increases of toxic metals in soils, with Zn as one of the metals likely to cause phytotoxicity, particularly in acid soils (Berrow and Webber, 1972; Webber, 1972). Lunt (1953) attributed the poor growth of oats and beans on the acid soil treated with sludge rates above 65 tonnes/ha to the toxicity of Zn and Cu. Similarly, Rhode (1962) concluded that excessive amounts of Zn and Cu are mainly responsible for the poor growth of crops on sewage farms at Berlin and Paris. Microelement toxicities have also been reported at sludge rates of 90 (Burd, 1968) and 125 (Millar, 1955) tonnes/ha. Merz (1959), however, reported that 225 tonnes/ha was not harmful to barley. In England, annual applications of 2.5 cm of liquid sludge for 17 years, combined with regular applications of lime, have not resulted in heavy metal toxicities (Hills, 1971). Yields of several vegetables grown on a Sango silt loam (pH 6.4) amended with an anaerobically digested sewage sludge at 112 tonnes/ha were unaffected, with tomato and squash even showing yield increases (Giordano and Mays, 1977).

Since the solubility of nearly all toxic metals is enhanced by soil acidity,

456   13. Zinc

there is a consensus that soils must be limed to about pH 6.5 to minimize
the possibilities of metal accumulation and phytotoxicity (CAST, 1980;
Lunt, 1959; Chumbley, 1971; Patterson, 1971; King and Morris, 1972;
Webber, 1972; Chaney, 1973). In addition, Jones et al (1975) noted that
corn, grown under field conditions on a Blount silt loam treated with liquid
anaerobically digested sludge at various rates of up to 296 tonnes/ha, had

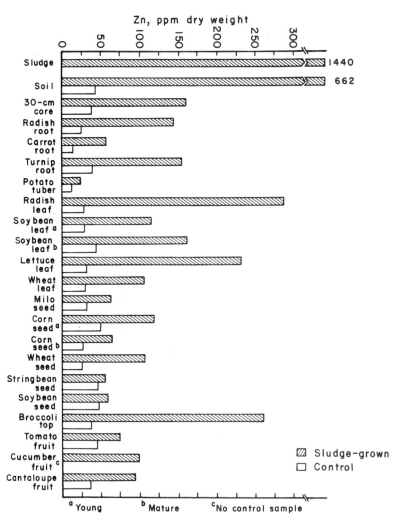

FIGURE 13.7 Zinc levels (ppm, dry wt basis) in various crops grown in the field amended
with 235 tonnes/ha of dried municipal sewage sludge (hatched bars) and in crops
grown in nontreated field (open bars). Concentrations in sludge, surface soil, and
soil core samples are indicated; a, b, and c indicate young tissue, mature tissue,
and no control, respectively. *Source:* McIntyre et al, 1977, with permission of the
authors and J. G. Press, Inc, Emmaus, PA.

leaf Zn contents substantially increased at soil pH below 6.3. However, above pH 7.0, increases in leaf content were not observed. Similarly, Bates et al (1975), in testing nine liquid digested sewage sludges applied to potted Grimsby sandy loam soil, found that the higher pH (pH 7 vs. pH 5) markedly reduced Zn concentrations in the ryegrass.

Numerous findings on the accumulation of trace metals in the tissues of edible vegetables grown on sludge-amended soils have been reported. There were increases in levels of Zn in beets and Swiss chard (Chaney et al, 1977; LeRiche, 1968; Ryan, 1977); small, but significant, increases in the Zn content of edible fruit, tuber, and root tissues of vegetable crops (Dowdy and Larson, 1975); and substantial increases in Zn levels of the edible portions of several vegetables (Giordano and Mays, 1977). Both of the latter investigators found that Zn concentrations were higher in the leaf or vegetative tissue than in the storage tissue. This type of uptake pattern between plant parts is shown in Figure 13.7 for radish, soybeans, and wheat. The figure also indicates that Zn accumulates in some species more than in others. In grain crops, some Zn may be translocated to the grain (Ham and Dowdy, 1978; Keeney and Walsh, 1975; Kelling et al, 1977; Soon et al, 1980) although in some instances (Kirkham, 1975) translocation is not significant in spite of large Zn accumulation in the soil.

The residual effectiveness of large amounts of sewage sludge applied to soil in elevating plant tissue Zn content may be long-lived (Hinesly et al, 1979; Touchton et al, 1976). The long-term impact of residual sludge-borne Zn on vegetative and edible portions of crops is exemplified in Table 13.11. This is expected, since Zn in soil accumulates primarily in the root zone of crops, giving easy access for plant roots, and this metal is not easily depleted by plant uptake, unlike the major nutrients.

9.3 Atmospheric Fallout

The release of Zn to the environment from mining and smelting activities is well documented in many parts of the world. The hazard from mining arises from dispersal of spoils and mine drainage water, while atmospheric discharge of metals and other pollutants is the main concern in smelting. In Halkyn Mountain, Clwyd, Wales, a limestone area containing Pb and Zn ores that were mined until recently, Davies and Roberts (1975) reported that Zn (total and extractable forms) for soil samples, collected from gardens and grazing fields in the area, averaged 2,923 and 330 ppm for total and extractable Zn, respectively, compared to only 94 ppm for total Zn and 4 ppm for extractable Zn from uncontaminated soils. Similar findings were reported by Thoresby and Thornton (1979) for soils in the Tamar Valley, southwest England, and again around Halkyn Mountain, Wales. With smelting, the impacts from aerial discharges appear to diminish substantially beyond the 10-km radius (Whitby and Hutchingson, 1974; John, 1975). Hogan and Wotton (1984) found that in a boreal forest in Manitoba,

TABLE 13.11. Concentrations of zinc (ppm) over several years in plants field-grown on soil that received only a single application of this metal in sewage sludge.[a,b]

Zn applied in 1971, kg/ha	Year[c]	Soil pH	Snap beans		Sweet corn	
			Leaves	Pods	Leaves	Grain
0 (Control)	1972	4.9	60	45	41	37
	1973	5.6	44	49	47	35
	1974	6.4	53	59	28	35
	1975	6.3	42	53	30	36
	1976	4.2	93	52	72	—
	1977	4.6	124	64	86	44
	1978	6.0	47	48	94	39
	1979	6.4	—	—	63	30
90	1972	5.3	158	61	94	43
	1973	4.9	171	72	153	47
	1974	5.2	282	86	98	48
	1975	5.5	128	71	130	45
	1976	4.2	253	66	209	—
	1977	4.5	191	88	126	54
	1978	5.9	68	60	164	47
	1979	6.4	—	—	107	43
180	1972	5.3	189	75	95	49
	1973	5.2	184	90	184	46
	1974	5.6	254	91	94	49
	1975	5.7	141	82	158	48
	1976	4.3	356	92	244	—
	1977	4.6	319	77	161	66
	1978	5.9	79	61	204	48
	1979	6.3	—	—	138	38
360	1972	5.6	164	83	97	44
	1973	5.4	187	90	207	54
	1974	5.9	296	87	130	63
	1975	6.1	128	79	172	51
	1976	4.4	499	92	234	—
	1977	5.0	408	77	159	65
	1978	6.1	137	83	273	60
	1979	6.6	—	—	137	44

[a] *Source:* Courtesy of Dr. P. M. Giordano, Tennessee Valley Authority, Muscle Shoals, Alabama.
[b] Decatur silt loam, CEC = 10 meq/100 g.
[c] Sulfur was applied in the fall of 1975 and limestone in the fall of 1977.

Canada, that had been exposed to smelter deposition for 50 years, significant deposition of metals has occurred at sites up to 35 km from the stacks (Fig. 13.8). The deposition of Zn from the source declined progressively with distance. The most heavily impacted sites were within the 6-km distance, where significant accumulation of metals at lower soil depths has occurred. Soils in the 10- to 15-cm depth at these sites averaged 260 ppm total Zn compared with 80 ppm total Zn in soils from this depth, 35 km further away. A typical example of the diminution in metal content in soils from smelter operations with distance is shown in Appendix Table 1.13 and Figure 13.8.

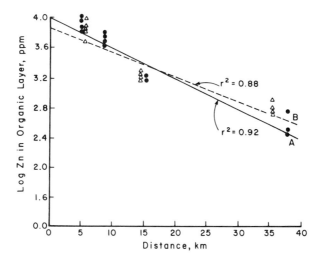

FIGURE 13.8 Relationships between zinc content of the forest floor in a boreal forest and the distance from the smelter. Triangles and circles represent measured points from the south (B) and southeast (A) transects, respectively. *Source:* Hogan and Wotton, 1984, with permission of the authors and the Am Soc of Agronomy, Crop Sci Soc of America, and Soil Sci Soc of America.

Aerial fallout from large industrial-urban complexes is also a major source of metal transfer to the soil. In the Chicago-northwestern Indiana area, it was estimated that industrial production released approximately 648,000 tonnes Zn to the atmosphere from 1900 to 1971 (Yost, 1973). Some metal enrichment has occurred in agricultural soils near Gary, Indiana, although extensive sampling in the region revealed no widespread contamination (Pietz et al, 1978). Similar findings were reported for other industrial-urban areas in Australia (Beavington, 1973) and Michigan (Klein, 1972).

Airborne Zn from automobile emissions is also another pathway of plant contamination with Zn. Values for Zn in plants and soil fall off rapidly with distance from the roadside and with decreasing traffic density (Motto et al, 1970). Apparently Zn originates from tire wear, fuel additives, brake linings, and motor oils. Significant contamination is due to surface deposition of Zn-bearing particulates (Flanagan et al, 1980).

9.4 Other Sources

Zinc from municipal refuse can also be markedly available to plants (King et al, 1974; Purves and MacKenzie, 1973; Terman et al, 1973; Webber and Beauchamp, 1975), although the use of this waste material is not as extensive as sewage sludge.

Another common source of Zn in agriculture is animal wastes in the form of dairy, feedlot, swine, or chicken manure. But like sludge, the area being affected is generally localized, usually in the vicinity of operators. Because animal feeds and rations are mostly of plant origin, the Zn contents of animal manures usually reflect those of the crop constituents in the ration, generally in the range of several ppm. In greenhouse pot culture, heavy application of farmyard manure (535 tonnes/ha) containing 100 ppm Zn, generally increased Zn uptake by ladino clover (Atkinson et al, 1958). Wallingford et al (1975) found that DTPA-extractable Zn from beef feedlot manure increased from the 10- to 20-cm soil depth.

Another source of Zn in agricultural operations is pesticides. It is used as zinc sulfate (20%–22% Zn) in insecticides for fruit trees and in fungicides Mancozeb (16% Mn, 2% Zn) and Zineb and Ziram (1%–18% Zn) for fruits and vegetables (Frank et al, 1976). In South Carolina, zinc oxysulfate is used in peach orchards to control the bacterial spot, *Xanthomonas pruni* (Lee and Page, 1967).

Coal combustion can contribute Zn input in the environment in two major ways: through atmospheric discharge or accumulation and disposal of coal residues (slag, fly ash, or scrubber sludge). Klein et al (1975) projected that, in the United States, the nationwide Zn flow due to coal combustion for power production was approximately 500 tonnes/year from atmospheric discharge and 26,000 tonnes/year from collected fly ash. Although Zn contents in fly ash are relatively high, the Zn does not appear to be readily available to crops, indicating that it is not a good source of this element for plant nutrition (Adriano et al, 1978; Martens, 1971).

References

Abdou, F. M., and M. El-Nennah. 1980. *Plant Soil* 56:53–57.

Adriano, D. C., and L. S. Murphy. 1970. *Agron J* 62:561–567.

Adriano, D. C., G. M. Paulsen, and L. S. Murphy. 1971. *Agron J* 63:36–39.

Adriano, D. C., T. A. Woodford, and T. G. Ciravolo. 1978. *J Environ Qual* 7:416–421.

Adriano, D. C., A. L. Page, A. A. Elseewi, A. C. Chang , and I. Straughan. 1980. *J Environ Qual* 9:333–344.

Alley, M. M., D. C. Martens, M. G. Schnappinger, and G. W. Hawkins. 1972. *Soil Sci Soc Am Proc* 36:621–624.

Anderson, A., and K. O. Nilsson. 1972. *Ambio* 1:176–179.

Archer, F. C. 1980. *Minist Agric Fish Food* (Great Britain) 326:184–190.

Atkinson, H. J., G. R Giles, and J. G. Desjardins. 1958. *Plant Soil* 10:32–36.

Aubert, H., and M. Pinta. 1977. *Trace elements in soils*. Elsevier, New York.

Bates, T. E., A. Haq, Y. K. Soon, and J. R. Moyer. 1975. In *Proc Intl Conf on Heavy Metals in the Environ*, 403–416. Toronto, Ontario, Canada.

Bauer, A. 1971. *Commun Soil Sci Plant Anal* 2:161–193.

Beavington, F. 1973. *Aust J Soil Res* 11:27–31.

Benson, N. R. 1966. *Soil Sci* 101:171–179.

Berrow, M. L., and J. Webber. 1972. *J Sci Food Agric* 23:93–100.

Berrow, M. L., and G. A. Reaves. 1984. In *Proc Intl Conf Environ Contamination*, 333–340. CEP Consultants Ltd, Edinburgh, UK.

Berry, W. L. 1978. In D. C. Adriano and I. L. Brisbin, eds. *Environmental chemistry and cycling processes*. CONF-760429 US Dept. of Energy, Washington, DC.

Bingham, F. T., A. L. Page, and J. R. Sims. 1964. *Soil Sci Soc Am Proc* 28:351–354.

Boawn, L. C. 1971. *Commun Soil Sci Plant Anal* 2:31–36.

——— 1974. *Soil Sci Soc Am Proc* 38:800–803.

Boawn, L. C., and G. E. Leggett. 1963. *Soil Sci* 95:137–141.

Boawn, L. C., and J. C. Brown. 1968. *Soil Sci Soc Am Proc* 32:94–97.

Boawn, L. C., F. G. Viets, Jr., and C. L. Crawford. 1954. *Soil Sci* 78:1–7.

——— 1957. *Soil Sci* 83:219–227.

Bosmans, H., and J. Paenhuys. 1980. *Pedologie* 30:191–223.

Boswell, F. C. 1975. *J Environ Qual* 4:267–272.

Bowen, H. J. M. 1979. *Environmental chemistry of the elements*. Academic Press, New York. 333 pp.

Boyd, G. E., J. Shubert, and A. W. Adamson. 1947. *J Am Chem Soc* 69:2818–2829.

Bradford, G. R., R. J. Arkley, P. F. Pratt, and F. L. Bair. 1967. *Hilgardia* 38:541–556.

Brinkerhoff, F., B. G. Ellis, J. Davis, and J. Melton. 1966. In *Quart Bull Mich Agric Exp Sta* 49:262–275.

Brown, A. L., and B. A. Krantz. 1960. *Calif Agric* 14:8–9.

——— 1966. *Soil Sci Soc Am Proc* 30:86–89.

Brown, A. L., B. A. Krantz, and P. E. Martin. 1964. *Soil Sci Soc Am Proc* 28:236–238.

Brown, A. L., J. Quick, and J. L. Eddings. 1971. *Soil Sci Soc Am Proc* 35:105–107.

Brown, J. C., and L. O. Tiffin. 1962. *Agron J* 54:356–358.

Buchauer, M. J. 1973. *Environ Sci Technol* 7:131–135.

Burd, R. S. 1968. In *Study of sludge handling and disposal*, 235–237. USDI, FWPCA Public WP-20-4.

Burleson, C. A., A. D. Pacus, and C. J. Gerard. 1961. *Soil Sci Soc Am Proc* 25:365–368.

Cammarota, V. A., Jr. 1980. In J. O. Nriagu, ed. *Zinc in the environment*, 1–38. Wiley, New York.

Cammarota, V. A., Jr., J. M. Lucas, and B. M. Gorby. 1980. In US Bureau of Mines, Mineral yearbook, 1978–79, 981–1019. US Dept of Interior, Washington, DC.

Cavallaro, N , and M. B. McBride. 1984. *Soil Sci Soc Am J* 48:1050–1054.

Chaney, R. L. 1973. In *Recycling municipal sludges and effluents on land*, 129–141. Natl Assoc of State Univ and Land Grant Colleges, Washington, DC.

Chaney, R. L., S. B. Hornick, and P. W. Simon. 1977. In R. C. Loehr, ed. *Land as a waste management alternative*, 283–314. Ann Arbor Science, Ann Arbor, MI.

Chang, A. C., A. L. Page, J. E. Warneke, and E. Grgurevic. 1984a. *J Environ Qual* 13:33–38.

Chang, A. C., J. E. Warnek , A. L. Page, and L. J. Lund. 1984b. *J Environ Qual* 13:87–91.

Chang, A. C., A. L. Page, K. W. Foster, and T. E. Jones. 1982. *J Environ Qual* 11:409–412.

Chapman, H. D., ed. 1966. *Diagnostic criteria for plants and soils.* Quality Printing Co, Abilene, TX.

Chaudhry, F. M., F. Hussain, and A. Rashid. 1977. *Plant Soil* 47:297–302.

Chen, Y., N. Senesi, and M. Schnitzer. 1978. *Geoderma* 20:87–104.

Cheng, S. M., R. L. Thomas, and D. E. Elrick. 1972. *Can J Soil Sci* 52:337–341.

Chesnin, L. 1963. *J Agric Food Chem* 11:118–122.

Chino, M. 1981a. In K. Kitagishi and I. Yamane, eds. *Heavy metal pollution in soils of Japan,* 81–94. Japan Sci Soc Press, Tokyo.

——— 1981b. In K. Kitagishi and I. Yamane, eds. *Heavy metal pollution in soils of Japan,* 65–80. Japan Sci Soc Press, Tokyo.

Chude, V. O., and G. O. Obigbesan. 1983. *Commun Soil Sci Plant Anal* 14:989–1004.

Chumbley, C. G. 1971. In *Permissible levels of toxic metals in sewage used on agricultural land.* Agric Dev and Advis Serv Advisory Paper no 10. M A F F, London. 12 pp.

Clark, R. B. 1978. *Agron J* 70:1057–1060.

Clemente, G. F., L. C. Rossi, and G. P. Santaroni. 1977. *J Radioanal Chem* 37:549–558.

Coker, E. G. 1966a. *J Agric Sci* 67:91–97.

——— 1966b. *J Agric Sci* 67:105–107.

Connell, W. E., and W. H. Patrick, Jr. 1968. *Science* 159:86–87.

Council for Agricultural Science and Technology (CAST). 1976. *Application of sewage sludge to cropland: appraisal of potential hazards of the heavy metals to plants and animals.* EPA-430/9-76-013. General Serv Admin, Denver. 63 pp.

——— 1980. In *Effects of sewage sludge on the cadmium and zinc content of crops.* CAST Report no. 83. Ames, Iowa. 77 pp.

Dankert, W. N., and J. V. Drew. 1970. *Soil Sci Soc Am Proc* 34:916–919.

Davies, B. E., and L. J. Roberts. 1975. *Sci Total Environ* 4:249–261.

DeRemer, D. E., and R. L. Smith. 1964. *Agron J* 56:67–70.

Dowdy, R. H., and W. E. Larson. 1975. *J Environ Qual* 4:278–282.

Duke, J. A. 1970. *Econ Botany* 24:344–366.

Edwards, J. H., and E. J. Kamprath. 1974. *Agron J* 66:479–482.

Elgabaly, M. M. 1950. *Soil Sci* 69:167–174.

Elgawhary, S. M., W. L. Lindsay, and W. D. Kemper. 1970. *Soil Sci Soc Am Proc* 34:66–70.

Ellis, B. G., and B. D. Knezek. 1972. In J. J. Mortvedt, P. M. Giordano, and W. L. Lindsay, eds. *Micronutrients in agriculture,* 59–78. Soil Sci Soc Am Inc, Madison, WI.

Ellis, R., Jr., J. F. Davis, and D. L. Thurlow. 1964. *Soil Sci Soc Am Proc* 35:935–938.

Elrashidi, M. A., and G. A. O'Connor. 1982. *Soil Sci Soc Am J* 46:1153–1158.

Elsokkary, I. H. 1980. *Plant Soil* 54:383–393.

Farrah, H., and W. F. Pickering. 1977. *Water, Air, Soil Pollut* 8:189–197.

Flanagan, J. T., K. J. Wade, A. Currie, and D. J. Curtis. 1980. *Environ Pollut* 1:71–78.

Follett, R. H., and W. L. Lindsay. 1970. In *Profile distribution of zinc, iron, manganese, and copper in Colorado soils.* Colorado Agric Exp Sta Tech Bull 110:1–78.

——— 1971. *Soil Sci Soc Am Proc* 35:600–603.

Follett, R. H., L. S. Murphy, and R. L. Donahue. 1981. *Fertilizers and soil amendments.* Prentice-Hall, Englewood Cliffs, NJ. 557 pp.

Forno, D. A., S. Yoshida, and C. J. Asher. 1975. *Plant Soil* 42:537–550.

Frank, R., K. Ishida, and P. Suda. 1976. *Can J Soil Sci* 56:181–196.

Freundlich, H. 1926. *Colloid and capillary chemistry*. Translated from the third German edition by H. S. Hatfield. Methven and Co., London.

Friedland, A. J., A. H. Johnson, and T. G. Siccama. 1984. *Water, Air, Soil Pollut* 21:161–170.

Friesen, D. K., A. S. R. Juo, and M. H. Miller, 1980. *Soil Sci Soc Am J* 44:1221–1226.

Gadde, R. R., and H. A. Laitinen. 1974. *Anal Chem* 46:2022–2026.

Gall, O. E., and R. M. Barnette. 1940. *J Am Soc Agron* 32:23–32.

Gangwar, M. S., and S. K. Chandra. 1976. *Commun Soil Sci Plant Anal* 7:295–310.

Ganiron, R. B., D. C. Adriano, G. M. Paulsen, and L. S. Murphy. 1969. *Soil Sci Soc Am Proc* 33:306–309.

Gaynor, J. D., and R. L. Halstead. 1976. *Can J Soil Sci* 56:1–8.

Gibbs, R. J. 1973. *Science* 180:71–72.

Giordano, P. M., and J. J. Mortvedt. 1974. *Agron J* 66:220–223.

Giordano, P. M., and D. A. Mays. 1977. In H. Drucker and R. Wildung, eds. *Biological implications of metals in the environment*. CONF-750929. NTIS, Springfield, VA.

Giordano, P. M., and J. J. Mortvedt. 1978. *Agron J* 70:531–534.

Golley, F. B., J. T. McGinnis, R. G. Clements, G. I. Child, and M. J. Duever. 1975. *Mineral cycling in a tropical moist forest ecosystem*. Univ Georgia Press, Athens. 248 pp.

Golovina, L. P., M. N. Lysenko, and T. I. Kisel. 1980. *Sov Soil Sci* 12:73–80.

Grant, E. A., A. A. MacLean, and U. C. Gupta. 1972. *Can J Plant Sci* 52:35–40.

Grunes, D. L., L. C. Boawn, C. W. Carlson, and F. G. Viets, Jr. 1960. *Agron J* 53:68–71.

Gupta, U. C., F. W. Calder, and L. B. MacLeod. 1971. *Plant Soil* 35:249–256.

Gupta, S. K., and K. Y. Chen. 1975. *Environ Letters* 10:129–158.

Hahne, H. C. H., and W. Kroontje. 1973. *J Environ Qual* 2:444–450.

Ham, G. E., and R. H. Dowdy. 1978. *Agron J* 70:326–330.

Hamilton, E. I. 1979. *Trace Subs Environ Health* 13:3–15.

Haq, A. U., T. E. Bates, and Y. K. Soon. 1980. *Soil Sci Soc Am J* 44:772–777.

Havlin, J. L., and P. N. Soltanpour. 1981. *Soil Sci Soc Am J* 45:70–75.

Heinrichs, H., and R. Mayer. 1980. *J Environ Qual* 9:111–118.

Hibbard, P. L. 1940. *Soil Sci* 49:63–72.

Hickey, M. G., and J. A. Kittrick. 1984. *J Environ Qual* 13:372–376.

Hills, L. D. 1971. *Compost Sci* 12(6):28–31.

Himes, F. L., and S. A. Barber. 1957. *Soil Sci Soc Am Proc* 21:368–373.

Hinesly, T. D., R. L. Jones, and E. L. Ziegler. 1972. *Compost Sci* 13:26–30.

Hinesly, T. D., E. L. Ziegler, and G. L. Barrett. 1979. *J Environ Qual* 8:35–38.

Hodgson, J. F. 1963. *Adv Agron* 15:119–159.

——— 1969. *Soil Sci Soc Am Proc* 33:68–75.

Hogan, G. D., and D. L. Wotton. 1984. *J Environ Qual* 13:377–382.

Holden, E. R., and J. W. Brown. 1965. *J Agric Food Chem* 13:180–184.

Ichikura, T., Y. Doyama, and M. Maeda. 1970. In *On the effects of various heavy metals on growth and yield of rice and onion. Bull Osaka Agric Res Center* 7:33–41.

Iyengar, S. S., D. C. Martens, and W. P. Miller, 1981. *Soil Sci Soc Am J* 45:735–739.

Jackson, D. R., and A. P. Watson. 1977. *J Environ Qual* 6:331–338.

Jenne, E. A. 1968. *Adv Chem Ser* 73:337–387.

Jenne, E. A., J. W. Ball, and C. Simpson. 1974. *J Environ Qual* 3:281–287.

John, M. K. 1972. *Soil Sci* 113:222–227.

―――― 1974. *Can J Soil Sci* 54:125–132.

―――― 1975. In *Proc Intl Conf on Heavy Metals in the Environ*, 365–377. Toronto, Ontario, Canada.

Jones, J. B., Jr. 1972. In J. J. Mortvedt, P. M. Giordano, and W. L. Lindsay, eds. *Micronutrients in agriculture*, 319–346. Soil Sci Soc Am Inc, Madison, WI.

Jones, R. L., T. D. Hinesly, E. L. Ziegler, and J. J. Tyler. 1975. *J Environ Qual* 4:509–514.

Judy, W., J. Melton, G. Lessman, B. G. Ellis, and J. Davis. 1965. *In* Mich State Univ Agric Exp Sta Rep no. 33:8.

Jugsujinda, A., and W. H. Patrick, Jr. 1977. *Agron J* 69:705–710.

Jurinak, J. J., and D. W. Thorne. 1955. *Soil Sci Soc Am Proc* 19:446–449.

Jurinak, J. J., and N. Bauer. 1956. *Soil Sci Soc Am Proc* 20:466–471.

Kalbasi, M., and G. J. Racz. 1978. *Can J Soil Sci* 58:61–68.

Kalbasi, M., G. J. Racz, and L. A. L. Rudgers. 1978a. *Soil Sci* 125:55–64.

―――― 1978b. *Soil Sci* 125:146–150.

Kanehiro, Y., and G. D. Sherman. 1967. *Soil Sci Soc Am Proc* 31:394–399.

Karim, H., and J. E. Sedberry, Jr. 1976. *Commun Soil Sci Plant Anal* 7:453–464.

Keefer, R. F., and R. N. Singh. 1968. In *Trans 9th Intl Cong Soil Sci* (Adelaide, Australia) 2:367–374.

Keefer, R. F., and R. Estepp. 1971. *Soil Sci* 112:325–329.

Keeney, D. R., and L. M. Walsh. 1975. In *Proc Intl Conf on Heavy Metals in the Environ*, 379–401. Toronto, Ontario, Canada.

Kelling, K. A., D. R. Keeney, L. M. Walsh, and J. A. Ryan. 1977. *J Environ Qual* 6:352–358.

Khan, A. A., and G. K. Zende. 1977. *Plant Soil* 46:259–262.

Khan, A., and P. N. Soltanpour. 1978. *Agron J* 70:1022–1026.

King, L. D., and H. D. Morris. 1972. *J Environ Qual* 1:425–429.

King. L. D., L. A. Rudgers, and L. R. Webber. 1974. *J Environ Qual* 3:361–366.

Kirkham, M. B. 1975. *Environ Sci Technol* 9:765–768.

Klein, D. H. 1972. *Environ Sci Technol* 6:560–562.

Klein, D. H., A. W. Andren, and N. E. Bolton. 1975. *Water, Air, Soil Pollut* 5:71–77.

Korcak, R. F., F. R. Gouin, and D. S. Fanning. 1979. *J Environ Qual* 8:63–68.

Krauskopf, K. B. 1979. *Introduction to geochemistry*. 2nd ed. McGraw-Hill, New York. 617 pp.

Kuo, S., and A. S. Baker. 1980. *Soil Sci Soc Am J* 44:969–974.

Langin, E. J., R. C. Ward, R. A. Olson, and H. F. Rhoades. 1962. *Soil Sci Soc Am Proc* 26:574–578.

Langmuir, I. 1918. *J Am Chem Soc* 40:1361–1403.

Laughlin, W. M., P. F. Martin, and G. R. Smith. 1974. *Am Potato J* 51:394–402.

LeClaire, J. P., A. C. Chang, C. S. Levesque, and G. Sposito. 1984. *Soil Sci Soc Am J* 48:509–513.

Lee, C. R., and N. R. Page. 1967. *Agron J* 59:237–240.

Lee, C. R., and G. R. Craddock. 1969. *Agron J* 61:565–567.

Leeper, G. W. 1952. *Ann Rev Plant Physiol* 3:1–6.

——— 1970. *Six trace elements in soils.* Melbourne Univ Press, Melbourne, Australia. 59 pp.

Lepp, N. W., and G. T. Eardley. 1978. *J Environ Qual* 7:413–416.

LeRiche, H. H. 1968. *J Agric Sci Camb* 71:205–208.

Lessman, G. M., and B. G. Ellis. 1971. *Soil Sci Soc Am Proc* 35:935–938.

Likens, G. E., and F. H. Bormann. 1970. In *Chemical analyses of plant tissue from the Hubbard Brook ecosystem in New Hampshire.* Yale School Forestry Bull no. 79.

Lindsay, W. L. 1974. In E. W. Carson, ed. *The plant root and its environment,* 507–524. Univ Virginia Press, Charlottesville.

——— 1979. *Chemical equilibria in soils.* Wiley, New York. 449 pp.

Lindsay, W. L., and W. A. Norvell. 1969. *Soil Sci Am Proc* 33:62–68.

——— 1978. *Soil Sci Soc Am J* 42:421–428.

Lindsay, W. L., J. F. Hodgson, and W. A. Norvell. 1967. In *Intl Soc Soil Sci Trans Comm II, IV* (Aberdeen, Scotland), 305–316.

Liu, Z., Q. Q. Zhu, and L. H. Tang. 1983. *Soil Sci* 135:40–46.

Lopez, A. S., and F. R. Cox. 1977. *Soil Sci Soc Am J* 41:742–747.

Lucas, R. E. 1967. In Mich State Univ Coop Ext Service, Ext Bull no. 486:2–12.

Lucas, R. E., and B. D. Knezek. 1972. In J. J. Mortvedt, P. M. Giordano, and W. L. Lindsay, eds. *Micronutrients in agriculture,* 265–288. Soil Sci Soc Am Inc, Madison, WI.

Lunt, H. A. 1953. *Water Sewage Works.* 100:295–301.

——— 1959. Conn (New Haven) Agric Exp Sta Bull no. 622.

Lutz, J. A., Jr., C. F. Genter, and G. W. Hawkins. 1972. *Agron J* 64:583–585.

MacLean, A. J. 1974. *Can J Soil Sci* 54:369–378.

MacLean, K. S., and W. M. Langille. 1976. *Commun Soil Sci Plant Anal* 7:777–785.

Malavolta, E., and P. L. Gorostiaga. 1974. In *Proc 7th Intl Coll Plant Anal Fert Probs* (Hanover, Germany):261–272.

Martens, D. C. 1971. *Compost Sci* 12:15–19.

Martens, D. C., G. Chesters, and L. A. Peterson. 1966. *Soil Sci Soc Am Proc* 30:67–69.

Matsuda, K., and M. Ikuta. 1969. *Soil Sci Plant Nutr* 15:169–174.

McBride, M. B., and J. J. Blasiak. 1979. *Soil Sci Soc Am J* 43:866–870.

McIntyre, D. R., W. J. Silver, and K. S. Griggs. 1977. *Compost Sci* May–June:22–29.

McKeague, J. A., and J. H. Day. 1966. *Can J Soil Sci* 46:13–22.

McKeague, J. A., and M. S. Wolynetz. 1980. *Geoderma* 24:299–307.

McKee, W. H., Jr. 1976. *Soil Sci Soc Am J* 40:586–588.

Mehta, S. C., S. R. Poonia, and R. Pal. 1984. *Soil Sci* 137:108–114.

Melton, J. R., B. G. Ellis, and E. C. Doll. 1970. *Soil Sci Soc Am Proc* 34:91–93.

Merz, R. C. 1959. *Water Sewage Works* 106:489–493.

Millar, C. E. 1955. *Soil fertility,* 324–326. Wiley, New York. 436 pp.

Minami, K., S. Shoji, and J. Masui. 1972. *Tohoku J Agric Res* 23:65–71.

Motto, H. L., R. H. Daines, D. M. Chilko, and C. K. Motto. 1970. *Environ Sci Technol* 4:231–237.

Mugwira, L. M., and B. D. Knezek. 1971. *Commun Soil Sci Plant Anal* 2:337–343.

Murphy, L. S., and L. M. Walsh. 1972. In J. J. Mortvedt, P. M. Giordano, and W. L. Lindsay, eds. *Micronutrients in agriculture,* 347–387. Soil Sci Soc Am Inc, Madison, WI.

Murthy, A. S. P. 1982. *Soil Sci* 133:150–154.

Nakos, G. 1983. *Plant Soil* 74:137–140.

National Academy of Sciences (NAS). 1977. *Drinking water and health*. NAS, Washington, DC. 939 pp.

National Research Council (NRC). 1979. In *Zinc*, 1–18. Univ Park Press, Baltimore, MD.

Navrot, J., B. Jacoby, and S. Rovikovitch. 1967. *Plant Soil* 27:141–147.

Nelson, J. L., and S. W. Melsted. 1955. *Soil Sci Soc Am Proc* 19:439–443.

Nelson, J. L., L. C. Boawn, and F. G. Viets, Jr. 1959. *Soil Sci* 88:275–283.

Nilsson, I. 1972. *Oikos* 23:132–136.

Norvell, W. A. 1972. In J. J. Mortvedt, P. M. Giordano, and W. L. Lindsay, eds. *Micronutrients in agriculture*, 115–138. Soil Sci Soc Am Inc, Madison, WI.

Norvell, W. A., and W. L. Lindsay. 1969. *Soil Sci Soc Am Proc* 33:86–91.

—— 1972. *Soil Sci Soc Am Proc* 36:778–783.

O'Connor, G. W., W. L. Lindsay, and S. R. Olsen. 1971. *Soil Sci Soc Am Proc* 35:407–410.

Ohki, K. 1975. *Physiol Plant* 35:96–100.

Olsen, S. R. 1972. In J. J. Mortvedt, P. M. Giordano, and W. L. Lindsay, eds. *Micronutrients in agriculture*, 243–261. Soil Sci Soc Am Inc, Madison, WI.

O'Neill, R. V., W. F. Harris, B. S. Ausmus, and D. E. Reichle. 1975. In F. G. Howell, J. B. Gentry, and M. H. Smith, eds. *Mineral cycling in southeastern ecosystems*, 28–40, ERDA CONF 740513.

Orabi, A. A., A. S. Ismail, and H. Mashadi. 1982. *Plant Soil* 69:67–72.

Page, A. L. 1974. *Fate and effects of trace elements in sewage sludge when applied to agricultural lands*. EPA-670/2-74-005. US-EPA, Cincinnati, OH. 98 pp.

Page, A. L., and A. C. Chang. 1975. In *Proc 2nd Natl Conf Water Resources: Water's Interface with Energy, Air, and Solids*. US-EPA, Chicago.

Parker, G. R., W. W. McFee, and J. M. Kelly. 1978. *J Environ Qual* 7:337–342.

Patterson, J. B. E. 1971. *Tech Bull Minist Agric Fish Food* 21:193–207.

Pauli, A. W., R. Ellis, Jr., and H. C. Moser. 1968. *Agron J* 60:394–396.

Paulsen, G. M., and O. A. Rotimi. 1968. *Soil Sci Soc Am Proc* 32:73–76.

Peck, N. H., D. L. Grunes, R. M. Welch, and G. E. MacDonald. 1980. *Agron J* 72:528–534.

Pepper, I. L., D. F. Bezdicek, A. S. Baker, and J. M. Sims. 1983. *J Environ Qual* 12:270–275.

Pietz, R. I., R. J. Vetter, D. Mosarik, and W. W. McFee. 1978. *J Environ Qual* 7:381–385.

Pratt, P. F., and G. R. Bradford. 1958. *Soil Sci Soc Am Proc* 22:399–402.

Price, C. A., H. E. Clark, and E. A. Funkhauser. 1972. In J. J. Mortvedt, P. M. Giordano, and W. L. Lindsay, eds. *Micronutrients in agriculture*, 231–242. Soil Sci Soc Amer Inc, Madison, WI.

Prokhorov, V. M., and Y. A. Gromova. 1971. *Sov Soil Sci* 11:693–699.

Purves, D. 1972. *Environ Pollut* 3:17–24.

Purves, D., and E. J. MacKenzie. 1973. *Plant Soil* 39:361–371.

Radjagukguk, B., D. G. Edwards, and L. C. Bell. 1980. *Aust J Agric Res* 31:1083–1096.

Randhawa, N. S., and F. E. Broadbent. 1965. *Soil Sci* 99:362–365.

Rashid, A., F. M. Chaudry, and M. Sharif. 1976. *Plant Soil* 45:613–623.

Reddy, M. R., and H. F. Perkins. 1974. *Soil Sci Soc Am Proc* 38:229–231.

Reiners, W. A., R. H. Marks, and P. M. Vitousek. 1975. *Oikos* 26:264–275.

Rhode, G. 1962. *J Inst Sewage Purif* Part 6:581–585.

Roberts, B. A. 1980. *Can J Soil Sci* 60:231–240.

Rudgers, L. A., J. L. Demeterio, G. M. Paulsen, and R. Ellis, Jr. 1970. *Soil Sci Soc Am Proc* 34:240–244.

Russell, L. H., Jr. 1979. In F. W. Oehme, ed. *Toxicity of heavy metals in the environment,* Part II, 3–23. Dekker Inc, New York.

Ryan, J. A. 1977. In *Proc Natl Conf Disposal of Residues on Land,* 89–105. Inform Transfer, Rockville, MD.

Saeed, M. 1977. *Plant Soil* 48:641–649.

Saeed, M., and R. L. Fox. 1977. *Soil Sci* 124:199–204.

Sakal, R., B. P. Singh, and A. P. Singh. 1982. *Plant Soil* 66:129–132.

Sakal, R., A. P. Singh, B. P. Singh, and R. B. Sinia. 1984. *Plant Soil* 79:417–428.

Sauchelli, V. 1969. *Trace elements in agriculture,* Van Nostrand, New York. 248 pp.

Schnitzer, M. 1977. In *Soil organic matter studies,* 117–132. Intl Atom Energy Agency, Vienna.

———— 1979. In M. K. Wali, ed. *Ecology and coal resource development,* 807–819. Pergamon Press, New York.

Sedberry, J. E., Jr., and C. N. Reddy. 1976. *Commun Soil Sci Plant Anal* 7:787–795.

Shacklette, H. T., J. A. Erdman, T. F. Harms, and C. S. E. Papp. 1979. In F. W. Oehme, ed. *Toxocity of heavy metals in the environment.* Part I, 25–68. Dekker Inc, New York.

Shamberger, R. J. 1979. In F. W. Oehme, ed. *Toxicity of heavy metals in the environment,* 689–796. Dekker Inc, New York.

Sharma, K. C., B. A. Krantz, A. L. Brown, and J. Quick. 1968. *Agron J* 60:453–456.

Sharpless, R. G., E. F. Wallihan, and F. F. Peterson. 1969. *Soil Sci Soc Am Proc* 33:901–904.

Shukla, U. C., and H. D. Morris. 1967. *Agron J* 59:200–202.

Shukla, U. C., and H. Raj. 1974. *Soil Sci Soc Am Proc* 38:477–479.

Shukla, U. C., and S. B. Mittal. 1979. *Soil Sci Soc Am J* 43:905–908.

Shuman, L. M. 1975. *Soil Sci Soc Am Proc* 39:454–458.

———— 1976. *Soil Sci Soc Am J* 40:349–352.

———— 1977. *Soil Sci Soc Am J* 41:703–706.

———— 1979. *Soil Sci* 127:10–17.

Siccama, T. G., W. H. Smith, and D. L. Mader. 1980. *Environ Sci Technol* 14:54–56.

Sidle, R. C., and L. T. Kardos. 1977. *J Environ Qual* 6:431–437.

Sims, J. L., and W. H. Patrick, Jr. 1978. *Soil Sci Soc Am J* 42:258–262.

Singh, B. R., and J. Lag. 1976. *Soil Sci* 121:32–37.

Singh, B., and G. S. Sekhon. 1977. *Soil Sci* 124:366–369.

Singh, M., and S. P. Singh. 1980. *Plant Soil* 56:81–92.

Smith, W. H. 1973. *Environ Sci Technol* 7:631–636.

Soltanpour, P. N. 1969. *Agron J* 61:288–289.

Soltanpour, P. N., J. O. Reuss, J. G. Walker, R. D. Heil, W. L. Lindsay, J. C. Hansen, and A. J. Relyea. 1970. *Am Potato J* 47:435–443.

Soltanpour, P. N., and A. P. Schwab. 1977. *Commun Soil Sci Plant Anal* 8:195–207.

Soman, S. D., V. K. Panday, K. T. Joseph, and S. J. Raut. 1969. *Health Phys* 17:35–40.

Soon, Y. K., T. E. Bates, and J. R. Moyer. 1980. *J Environ Qual* 9:497–504.

Sparr, M. C. 1970. *Commun Soil Sci Plant Anal* 1:241–262.

Staker, E. V., and R. W. Cummings. 1941. *Soil Sci Soc Am Proc* 6:207–214.

Stanton, D. A., and R. du T. Burger. 1967. *Geoderma* 1:13–17.

Stevenson, F. J. 1972. *BioScience* 22:643–650.

Stewart, I., C. O. Leonard, and G. Edwards. 1955. *Florida State Hort Soc* 1955:82–88.

Stewart, J. A., and K. C. Berger. 1965. *Soil Sci* 100:244–250.

Stukenholtz, D. D., R. J. Olsen, G. Gogan, and R. A. Olson. 1966. *Soil Sci Soc Am Proc* 30:759–763.

Suzuki, S., N. Djuangshi, K. Hyodo, and O. Soemarwoto. 1980. *Arch Environ Contam Toxicol* 9:437–449.

Swaine, D. J. 1955. *The trace element content of soils*. Common Bur of Soil Sci (Great Britain), Tech Commun 48.

Swaine, D. J., and R. L. Mitchell. 1960. *J Soil Sci* 11:347–368.

Tan, K. H., L. D. King, and H. D. Morris. 1971. *Soil Sci Soc Am Proc* 35:748–752.

Tanaka, A., and S. Yoshida. 1970. In *Nutritional disorders of the rice plant in Asia*. Intl Rice Res Inst Tech Bull 10.

Tanner, J. T., and M. H. Friedman. 1977. *J Radioanal Chem* 37:529–538.

Terman, G. L., J. M. Soileau, and S. E. Allen. 1973. *J Environ Qual* 2:84–89.

Terman, G. L., P. M. Giordano, and N. W. Christensen. 1975. *Agron J* 67:182–184.

Thoresby, P., and I. Thornton. 1979. *Trace Subs Environ Health* 13:93–103.

Tiller, K. G., and J. F. Hodgson. 1962. *Clays Clay Miner* 9:393–403.

Touchton, J. T., L. D. King, H. Bell, and H. D. Morris. 1976. *J Environ Qual* 5:161–164.

Travis, D. O., and R. Ellis, Jr. 1965. *Trans Kansas Acad Sci* 68:457–460.

Trierweiler, J. F., and W. L. Lindsay. 1969. *Soil Sci Soc Am Proc* 33:49–54.

Tyler, G., G. Christina, K. Holmquist, and A. Kjellstrand. 1973. *J Ecol* 61:251–268.

Udo, E. J., H. L. Bohn, and T. C. Tucker. 1970. *Soil Sci Soc Am Proc* 34:405–407.

Underwood, E. J. 1973. In *Toxicants occurring naturally in foods*. 2d ed., 43–87. NAS, Washington, DC.

US Environmental Protection Agency (US-EPA). 1976. *Quality criteria for water*. US-EPA, Washington, DC.

———— 1977. *Fed Register* 42(211):57420–57427.

Vallee, B. L. 1959. *Physiol Rev* 30:443–490.

van Eysinga, J. P. N. L., P. A. van Dijk, and S. S. deBes. 1978. *Commun Soil Sci Plant Anal* 9:153–167.

Van Hook, R. I., W. F. Harris, and G. S. Henderson. 1977. *Ambio* 6:281–286.

Viets, F. G., Jr., L. C. Boawn, and C. L. Crawford. 1954. *Soil Sci* 78:305–316.

Vinande, R., B. Knezek, J. Davis, E. Doll, and J. Melton. 1968. In *Quar Bull Mich Agric Exp Sta* 50:625–636.

Vinogradov, A. P. 1959. *The geochemistry of rare and dispersed chemical elements in soils*. Consultants Bureau Inc, New York. 209 pp.

Wada, K. and A. Abd-Elfattah. 1978. *Soil Sci Plant Nutr* 24:417–426.

Wallace, A. 1963. *J Agri Food Chem* 11:103–107.

Wallace, A., and R. T. Mueller. 1959. *Soil Sci Soc Am Proc* 23:79.

Wallace, A., and E. M. Romney. 1970. *Soil Sci* 109:66–67.

Wallace, A., R. T. Mueller, and G. V. Alexander. 1978. *Soil Sci* 126:336–341.

Wallingford, G. W., L. S. Murphy, W. L. Powers, and H. L. Manges. 1975. *Soil Sci Soc Am Proc* 39:482–487.

Walsh, L. M., D. R. Steevens, H. D. Seibel, and G. G. Weis. 1972. *Commun Soil Sci Plant Anal* 3:187–195.

Watanabe, F. S., W. L. Lindsay, and S. R. Olsen. 1965. *Soil Sci Soc Am Proc* 29:562–565.

Wear, J. I., and C. E. Evans. 1968. *Soil Sci Soc Am Proc* 32:543–546.

Webber, J. 1972. *Water Pollut Contr* 71:404–413.

Webber, L. R., and E. G. Beauchamp. 1975. In *Proc Intl Conf on Heavy Metals in the Environ*, 443–451. Toronto, Ontario, Canada.

Wedepohl, K. H., ed. 1978. *Handbook of geochemistry*. Springer-Verlag, New York.

Whitby, L. M., and T. C. Hutchinson. 1974. *Environ Conserv* 1:191–200.

White, M. L. 1957. *Econ Geol* 52:645–651.

White, M. C., and R. L. Chaney. 1980. *Soil Sci Soc Am J* 44:308–313.

Whitney, D. A., R. Ellis, Jr., L. S. Murphy, and G. Herron. 1973. Publ. L-360, Coop Ext Serv, Kansas State Univ, Manhattan, KS. 3 pp.

Whitton, J. S., and N. Wells. 1974. *N Z J Sci* 17:351–367.

Williams, D. E., J. Vlamis, A. H. Pukite, and J. E. Corey. 1984. *Soil Sci* 137:351–359.

Yoon, S. K., J. T. Gilmour, and B. R. Wells. 1975. *Soil Sci Soc Am Proc* 39:685–688.

Yoshida, S., and A. Tanaka. 1969. *Soil Sci Plant Nutr* 15:75–80.

Yoshida, S., J. S. Ahn, and D. A. Forno. 1973. *Soil Sci Plant Nutr* 19:83–93.

Yost, K. J. 1973. *The environmental flow of cadmium and other trace elements*. Vol. 1. Prog Rep 1 NSF-RANN GI-35106, Purdue Univ, West Lafayette, IN.

Young, H. E., and V. P. Guinn. 1966. *Tappi* 49:190–197.

Youngdahl, L. J., L. V. Svec, W. C. Liebhardt, and M. R. Teel. 1977. *Crop Sci* 17:66–69.

14
Other Trace Elements

Antimony

The bluish-white, lustrous, very brittle metal (atomic no. 51, atomic wt., 121.75; specific gravity, 6.691; melting pt., 631°C) is found in nature in more than 100 minerals. However, only about one dozen ores are commercially important, such as Sb oxides and sulfides, and complex Cu-, Pb-, and Hg-Sb sulfides, the most important of which is stibnite (Sb_2S_3). In nature, it occurs primarily in the III oxidation state.

Antimony is used about equally between nonmetal and metal products. For nonmetal products, Sb is used in paints and lacquers, rubber compounds, ceramic enamels, glass and pottery, and abrasives. For metallic products, it is alloyed with Pb and is used in the manufacture of bearings, battery parts, sheet and pipe, tubes and foil, and ammunition.

Antimony is 62nd in crustal abundance, slightly higher than Cd (Krauskopf, 1979). In soils, Bowen (1979) reported a median value of 1 ppm (range of 0.2–10 ppm). Surficial soils of the United States have a geometric mean of 0.48 ppm, with a range of <1 to 10 ppm (Fig. 14.1, Appendix Table 1.17). Data compiled by Kabata-Pendias and Pendias (1984) for world soils indicate a range of 0.05 to 2.3 ppm Sb. Thus, most of the world soils appear to have mean Sb values higher than the considered crustal mean of 0.2 ppm (Mitchell and Burridge, 1979). Surface soils from Ontario, Canada, were reported to contain a mean of 0.24 ppm Sb, with a range of 0.05 to 1.64 ppm (Frank et al, 1979). Norwegian surface soils have Sb content in the range of 0.17 to 2.20 ppm (Allen and Steinnes, 1979). The trend—e.g., concentrations in the southernmost areas of Norway were several times greater than in areas further north—for the various metals, particularly Pb, indicates that these metals were deposited via a long-range aerial transport.

Since Sb's geochemical behavior is similar to As, it is commonly associated with nonferrous deposits and therefore, is emitted to the environment during the smelting of these ores. For example, the largest anthropogenic source of As and Sb to Puget Sound is a large Cu smelter located near Tacoma, Washington (Crecelius et al, 1975). This particular smelter releases Sb, along with As, to the environment in three ways:

1. As stack dusts into the air—about 20×10^3 kg/year of Sb oxides;
2. As dissolved Sb species in liquid effluent discharged directly into Puget Sound—about 2×10^3 kg Sb/year; and
3. As crystalline slag particles dumped directly into the Sound—about 1.5×10^6 kg Sb/year.

On the other hand, only 250 kg/year of Sb is discharged into the Sound in the form of sewage treatment plant effluents. Surface soils (0–3 cm) within 5 km of a Cu smelter can be enriched up to 200 ppm with Sb (Crecelius et al, 1974).

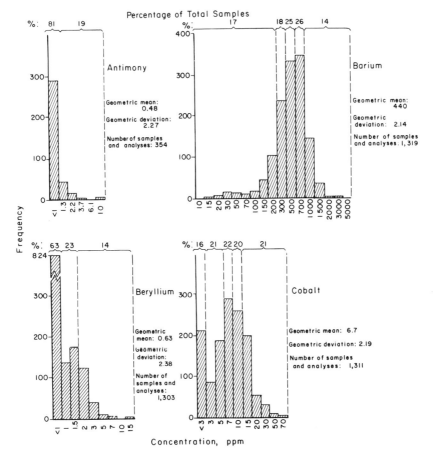

FIGURE 14.1 Trace element contents of soils and other surficial materials of the conterminous United States. *Source:* Shacklette and Boerngen (1984).

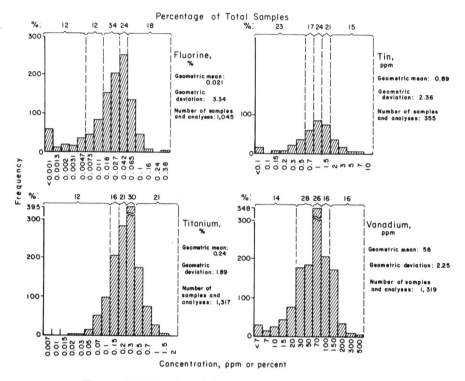

FIGURE 14.1 Continued (Shacklette and Boerngen, 1984).

Our knowledge of Sb behavior in soil is nil. If present in soil in a soluble form, it probably occurs as antimonate and would be adsorbed by the same soil constituents that bind phosphate and arsenate. Hence, Crecelius et al (1975) found that chemical extraction of sediments from Puget Sound showed that about 50% of the As and Sb was associated with extractable Fe and Al compounds.

In general, very little is known about plant uptake of Sb and its phytotoxicity; however, there are indications that normal levels in most terrestrial plants should be around <0.1 ppm.

Barium

The silvery-white metal (atomic no. 56; atomic wt., 137.36; specific gravity, 3.5; melting pt., 725°C) belongs to the alkaline earth group of elements, has a II oxidation state, and geochemically resembles Ca and Sr. The element is used in the form of Ba compounds, primarily as barite ($BaSO_4$), which in turn is produced from high-grade ore (75%–89%), often in association with granite and shale. The United States is a major producer of barite, with the states of Nevada, Missouri, and Georgia as leading producers.

Most of the ground barite is used extensively as a weighting agent in oil- and gas-drilling fluids. The other major uses for barite are in the production of Ba chemicals and in the glass, paint, and rubber industries.

At 500 ppm level in the earth's crusts, it ranks 14th among the elements in the order of abundance (Krauskopf, 1979). Bowen (1979) reported that in soils, Ba concentrations range from 100 to 3,000 ppm, the median being about 500 ppm (Appendix Table 1.17). Surficial soils in the United States have a geometric mean of 440 ppm and a range of 10 to 5,000 ppm (Fig. 14.1). Bertrand and Silberstein (1928) reported Ba contents of 80 to 1,700 ppm for some soils of France and Italy. For some Russian soils, Vinogradov (1959) reported a range of 100 to 1,500 ppm Ba. Soils near barite deposits may contain up to 37,000 ppm Ba.

Barium can accumulate in the Mn oxides in soil; thus, Ba content of sedimentary manganese deposits is usually high (Peterson and Girling, 1981). Similarly, binding of Ba by ferromanganese nodules in the ocean's bottom is considered one of the major sinks for this element (van der Weijden and Kruissink, 1977). In soil minerals, Ba can substitute for K, as in micas and feldspars.

Early interest in Ba in plant nutrition was spurred by the known toxicity of soluble Ba salts to animals. Barium was believed to be taken up by some pasture plants known as "locoweeds" *(Astragalus sp.)* which can be lethal to cattle (Vanselow, 1965a). As it turned out, it was probably the Se and certain alkaloids in these plants that caused the toxicity. Recently, Norrish (1975) pointed out that barite is soluble enough to be taken up by the plants growing on Ba-rich soils and accumulates at levels toxic to animals. Other than this special scenario, Ba toxicity to animals from ingestion of plant-enriched Ba should not be expected.

Results compiled by Vanselow (1965a) indicate that vegetative tissues of most agronomic crops contain Ba in the range of 10 to 100 ppm (dry wt), with the high range approaching 2,000 ppm. The pasture "locoweeds" contained 100 to 290 ppm Ba. Because of the chemical similarities between Ba, Sr, and Ca, competition among these ions for plant uptake is not unexpected. In solution culture, Ca decreased both Sr and Ba uptake by bush beans, with Ba being preferentially taken up over Sr (Wallace and Romney, 1971). Earlier study by Menzel (1954) showed that, indeed, uptake of Sr and Ba was inversely proportional to the exchangeable soil Ca.

Under experimental conditions, Ba toxicity to bush beans and barley was induced when the soil was spiked with 2,000 ppm Ba, as $Ba(NO_3)_2$, resulting in 1% to 2% Ba levels in leaf tissues (Chaudry et al, 1977). Applications of lime or sulfur to the soil tend to immobilize Ba due to the formation of sparingly soluble $BaCO_3$ and $BaSO_4$.

Beryllium

The silvery-gray metal (atomic no. 4; atomic wt., 9.013; specific gravity, 1.85; melting pt., 1,278 ± 5°C) is one of the lightest metals, but has one of the highest melting points among the lightest metals. In nature, it occurs in the II oxidation state. The most important commercial source of the element and its compounds is beryl (beryllium silicate, $3BeO \cdot Al_2O_3 \cdot 6SiO_2$). Other Be primary minerals are chrysoberyl ($BeAl_2O_4$) and phenacite (Be_2SiO_4). The major beryl-producing countries are Brazil, India, Zimbabwe, Argentina, and the United States. In 1979, about 3,000 tonnes were produced globally, with the United States consuming about 10% of the total produced. Beryllium and its salts are fairly toxic to humans, resulting in both acute and chronic effects.

Beryllium's major uses are as follows: as a hardening agent in alloys, primarily as Be-Cu, Be-Zn, and Be-Ni; as a neutron moderator in nuclear reactors and cladding material for U; in aircraft brakes; in inertial guidance systems in missiles and aircraft; and as structural components of space vehicles.

The average crustal abundance of Be (45th among elements) is estimated at around 3 ppm (Krauskopf, 1979); thus, it is ubiquitous in nature but in only trace amounts. However, values as high as 10 ppm may occur infrequently. Mitchell (1971) noted that for some typical Scottish surface soils, the Be levels are usually below 5 ppm for soils derived from sandstone, shale, andesite, olivine-gabbro, serpentine, and granite. Soils derived from trachyte, mica schist, and quartzite may have slightly higher Be contents. For United States surface soils, Be ranged from <1 to 15 ppm ($\bar{x}$ = 0.63 ppm) (Fig 14.1.); for USSR surface soils, it ranged from 1.2 to 13 ppm (Bakulin and Mokiyenko, 1966); and for Canadian surface soils, it ranged from 0.10 to 0.89 ppm (Frank et al, 1979). Bowen (1979) reported a median value of 0.3 ppm Be for soils (Appendix Table 1.17).

The primary environmental source of Be is coal combustion. Zubovic (1969) has estimated that combustion of 500×10^6 tonnes of coal containing an average Be concentration of 2.5 ppm would yield about 1,250 tonnes of Be. Beryllium in coal, where it was shown to be primarily bound to the organic matter of the coal, varies from <0.4 to 8 ppm ($\bar{x}$ = 1.5 ppm) in Australian bituminous coal (Swaine, 1979), and from 0.4 to 3 ppm ($\bar{x}$ = 1.8 ppm) in British coal (Lim, 1979). Coal ashes have been known to contain as much as 100 ppm Be.

In soil, Be occurs as a divalent cation and can be expected to also occur in most agricultural soils either as $(BeO_2)^{2-}$, $(Be_2O_3)^{2-}$, $(BeO_4)^{6-}$, or $(Be_2O)^{2+}$ and in calcareous soils as $Be(OH)CO_3^-$ and $Be(CO_3)_2^{2-}$ (Kabata-Pendias and Pendias, 1984).

Since Be readily complexes with organic compounds, it may accumulate in the surface layer of the soil profile and remain relatively unavailable for plant uptake. However, its inorganic complexes, such as $BeCl_2$ and $BeSO_4$, can be fairly mobile and, therefore, readily available for plant uptake.

Beryllium data for plants are scanty. Its typical concentrations in plants are at levels <0.1 ppm (dry wt) and <2 ppm in plant ashes of plant species collected over nonmineralized areas (Peterson and Girling, 1981). Accumulator species, such as *Vaccinium myrtillus, Vicia sylvatica, Aconitum excelsum,* and *Calamagrostis arundinacea,* however, may accumulate Be exceeding 10 ppm levels. Newland (1982) reported that hickory is an accumulator of Be, containing as much as 1 ppm (dry wt). Most results indicate that leaves contain more Be than either twigs or fruits, although some desert species contain more Be in the twigs.

Although very phytotoxic, availability of Be in low amounts may have beneficial effects on plants. These effects were manifested as the stimulation in growth of *Nicotiana tabacum* grown in nutrient culture containing 1 ppm Be, as $BeCl_2$ (Tso et al, 1973); as improved growth of timothy grown in sandy loam soil spiked with 10 ppm Be (Ruhland, 1958); and in activation of certain enzymes, such as acid phosphatase and inorganic pyrophosphatase (but Be seems to inhibit alkaline phosphatase) (Horovitz and Petrescu, 1964). The inhibition of the latter enzyme was shown to be due to Be substitution for Mg (Tepper, 1980).

Soluble Be has been shown to reduce the growth of various crops at concentrations greater than 2 ppm in nutrient solutions or at concentrations equivalent to 2% to 4% of the CEC of the soil (Romney and Childress, 1965). Beryllium in plants accumulates primarily in the roots and upon translocation has the tendency to accumulate more in leaves than in fruits and stems. This metal apparently interferes in plant nutrition, not only through inhibition of certain enzymes, but also through antagonism of Ca, Mg, and Mn nutrition. Since Be phytotoxicity can be prevalent only in acidic soils, liming the soil would render this element immobile (Williams and Riche, 1968). The US-EPA (1976) has issued several water quality criteria for Be, including limiting Be content to 100 ppb in water intended for continuous irrigation of soils, except that for neutral to alkaline fine-textured soils, the limit can be as high as 500 ppb.

Because of the known toxicological nature of Be and the likelihood of its increased releases into the environment with intensification of coal combustion, this metal, along with Tl and V, should be viewed as possible "problem" metals of the future.

Cobalt

Cobalt (atomic no. 27; atomic wt., 58.93; specific gravity, 8.9) is a silvery-white metal with a pink or blue tinge when polished. It is resistant to corrosion and to alkalis but is soluble in acids. It is harder but more brittle than Fe or Ni. It is geochemically similar to Ni and exhibits an oxidation state of II or III.

Cobalt occurs in the minerals cobaltite, smaltite, and erythrite, and is often associated with Ag, Ni, Pb, Cu, and Fe ores, from which it is commonly obtained as a by-product. The major ore deposits of Co occur as sulfides. Large deposits are found in Zaire, Morocco, and Canada.

Cobalt is used in the production of high-grade steels, alloys, superalloys, and magnetic alloys. More than 75% of the world's production of Co is used in the manufacture of alloys. It is also used in smaller quantities as a drying agent in paints, varnishes, enamels, and inks; as a pigment; and as a glass decolorizer. It is also an important source of catalysts in the petroleum industry.

Several values for the crustal abundance of Co have been quoted: 20 ppm (Bowen, 1979), 23 ppm (Mitchell, 1945), and 27 ppm (Carr and Turekian, 1961); Krauskopf (1979) gave 22 ppm, ranking Co 30th among the elements. Soils developed from ultrabasic rocks are usually enriched in Co, Ni, and Cr (Anderson et al, 1973). Mitchell (1945) stated that soils derived from basic igneous rocks or argillaceous sediments may contain 20 to 100 ppm total Co, whereas soils derived from sandstones, limestones, and acid igneous rocks usually have <20 ppm total Co.

Cobalt occurs in high amounts in ultrabasic rocks, where it is associated with olivine minerals. Serpentine soils in New Zealand can contain up to 1,000 ppm Co, with mean values around 400 ppm (Lyon et al, 1968, 1971). Kidson (1938) reported that the Co content of a wide variety of soils is related to the Mg contents of their parent rocks; e.g., serpentines that are enriched in Mg give soils high in Co, whereas soils derived from granite have low Co contents. Total Co concentrations of world soils are summarized by Young (1979) and Proctor and Woodell (1975), who show that serpentine soils generally contain 100 ppm or more of Co. Young (1979) stated that total Co in soils may range from about 0.3 ppm in severely Co-deficient areas to 1,000 ppm over mineralized areas; but most soils have Co levels within a range of 2 to 40 ppm.

In the conterminous United States, the mean total Co content of soils and other surficial materials was 6.7 ppm (Fig. 14.1), whereas for world soils, the median total content appears to be around 8 ppm (Appendix Table 1.17); the average for tropical Asian paddy soils was 56 ppm (Appendix Table 1.18).

There is a narrow range for extractable Co determined by common soil extractants. In Co-deficient farmlands of Scotland, 2.5% HOAc generally extracts <0.25 ppm Co, whereas for fertile soils, the values are above 0.30 ppm (Mitchell, 1945). Alban and Kubota (1960) found an increase in the Co content of black gum *(Nyssa sylvatica)* leaves with HOAc-extractable Co. In general, extractable Co, as a percentage of total Co, shows a very wide range, from 1% to 93%, but is usually between 3% and 20%

(Young, 1979). In terms of concentration, extractable Co varies from about 0.01 to 6.8 ppm but is usually in the range of 0.1 to 2 ppm.

Using ^{60}Co, Gille and Graham (1971) observed that 0.1 N HCl solution would give a good estimate of the quantity of Co in the soil and that 0.1 M CaCl$_2$ would give the best indication of the potential concentration of Co in the soil solution.

In typical soils of the Hawaiian Islands, the Co content is highest in the A horizon and appears to be correlated with the Fe oxide content of the soil (Fujimoto and Sherman, 1950). In the eastern United States, higher Co was dithionite-extractable from the A rather than from the B horizon of red-yellow podzols (Kubota, 1965).

Several factors are known to influence the sorption of Co by soils. Graham and Killion (1962) found that the uptake of Co by several plants from soil constituent was in the order: illite > kaolinite > Putnam clay > sedimentary peat > montmorillonite > fibrous peat. The peats were found to adsorb more Co than any of the clays. Using pot and field experiments, Adams et al (1969) found that the uptake of Co by clover was inversely related to the total Mn content of the soil. Tiller et al (1969) found that the Co sorption capacity of the soils they studied was highly correlated with Co content and surface area and to a lesser extent with Mn and clay content and pH. Taylor and McKenzie (1966) have shown that, in some cases, almost all of the Co in soil could be accounted for by that present in the Mn minerals, indicating that these particular minerals can be an important sink for Co in the soil. Hodgson (1960) identified two forms of bound Co in montmorillonite. One form, characterized as being slowly dissociable, appears to be bound in a monolayer by chemisorption and would exchange with Zn^{2+}, Cu^{2+}, or other Co^{2+} ions but not with a Ca^{2+}, Mg^{2+}, or NH_4^+ ions. A second form of Co is not dissociable and is believed to either enter the crystal lattice or be occluded in the precipitates of another phase.

Cobalt has not been demonstrated to be essential for the growth of higher plants, although it has been shown to be required by certain blue-green algae. Cobalt is essential for the N_2 fixation by free-living bacteria, blue-green algae, and symbiotic systems (Reisenauer, 1960). For example, Co is required by *Rhizobium,* the symbiotic bacterium that fixes N_2 in the root nodules of legumes. In higher plants, Co supplements have been reported to increase growth of rubber plants and tomatoes and to have increased the elongation of pea stem sections.

Although Co is not essential to plants, its level in plant tissues is of general concern because of its essentiality in animal nutrition. It is a constituent of vitamin B$_{12}$, which is required by all animals. Low levels of Co in feedstuff can cause nutritional diseases in ruminants, known as the "bush sickness" in cattle and sheep or "pining" in sheep. These maladies are also known as "coast disease" or "enzootic marasmus" in Australia,

"Morton Mains disease" in New Zealand, and "wasting disease" in Denmark. Young (1979) indicated that a level of about 0.07 ppm of Co in feedstuff is essential to maintain the normal health of animals, although the critical level varies slightly among countries and among animals. Tissues of higher plants are generally reported to contain <1 ppm Co, on dry weight basis. Leafy plants such as lettuce, cabbage, and spinach have a relatively high Co content, whereas Co is low in grasses and cereals (Kipling, 1980). Compared with grasses, legumes take up more Co and are more definitive of the Co status of the soil (Kubota, 1980).

As a general rule, plants growing on soils having low Co concentrations would have low Co contents. For example, in the eastern United States, coarse-textured podzols have been associated with Co deficiency of ruminant animals (Kubota, 1965). Peterson and Girling (1981) stated that Co deficiency in ruminants most commonly occurs on acid, highly leached sandy soils; on soils derived from granites; and on some highly calcareous soils and some peaty soils. Plants growing on such soils generally contain <0.07 to 0.08 ppm Co, and serious deficiency conditions commonly occur below 0.04 ppm.

In the United States, Kubota (1980) has delineated a geographical pattern of low and adequate Co areas. The lower Atlantic Coastal Plain has fairly low Co contents. Problem areas abound in sandy podzols that have 1 ppm or less total Co, with grasses and legumes grown on these soils having not quite enough Co to meet animal needs. The glaciated region of the northeast, including Michigan, is also characteristically low in total Co. In the United Kingdom, low Co contents can be found in soils derived from sandstones, sands, limestones, certain shales, and acid igneous rocks in the western and northern areas of England and Wales (Thornton and Webb, 1980).

Some plants appear to accumulate Co, and several "hyperaccumulator" species have been reported to inhabit mineralized areas (Malaisse et al 1979). The "hyperaccumulators" could contain Co in their tissues in excess of 1,000 ppm. Lambert and Blincoe (1971) reported Co concentrations over 20 ppm in range wheat grasses (*Agropyron* sp.) in northern Nevada, although they are not known accumulators of Co.

As with most metals, Co in plants is mostly associated with roots, but can be translocated to the foliage. Patel et al (1976) noticed that Co concentrations in leaves and stems of chrysanthemum significantly increased with increasing levels of Co in the nutrient solution and caused reduction in the levels of Fe, Mn, and Zn in the plant tissues. Toxicity symptoms due to excess Co exhibit the typically induced Fe-deficiency chlorosis and necrosis. Vanselow (1965b) reported that, in solution culture, small amounts of Co, sometimes as low as 0.1 ppm, can produce adverse effects on many crops.

Several factors can influence the availability of Co to plants. Vanselow (1965b) listed the following types of soil in which Co deficiency can be expected to occur.

1. Highly leached, acid, sandy soils
2. Soils derived from granites
3. Some highly calcareous soils
4. Atlantic Coastal Plain soils
5. Some peaty soils

In addition, he listed the following soil management and other practices that can affect the Co status of the soil:

1. Cobalt content of soils, in general, is related to Mg content of parent rocks. Common fertilizer treatments over many years should have no effect on Co status.
2. Soil acidification increases the availability of Co.
3. Liming of the soil can reduce the uptake of Co by plants, whereas application of gypsum to sandy loam could increase Co uptake.
4. Leaching of soils reduces available Co.

Several investigators (Filipovic et al, 1961; Price et al, 1955; Wright and Lawton, 1954) have reported decreased uptake of Co by plants with increasing soil pH. For example, lime appears to have a depressing effect on plant uptake of Co (Adriano et al, 1977; Askew and Dixon, 1937). Prokhorov et al (1979) noted that, of all the properties they studied, soil pH had the greatest effect on Co sorption by soil.

Various investigators found that Co accumulates in the hydrous oxides of Fe and Mn in soil. Kubota (1965) indicated that Co was associated with the Fe oxides. More recent findings by McKenzie (1975, 1978) indicates that Mn oxides play a key role in the chemistry of soil Co. However, Hodgson et al (1969) found that the adsorption of Co by certain soils was increased by the removal of Fe from the soil. In that case, Fe removal exposed clay mineral surfaces that were more reactive than the previously exposed Fe oxide surfaces.

Peterson and Girling (1981) summarized the following factors that can affect the Co status of plants: soil composition, including levels of elements such as Mn and Fe; soil pH and moisture; and plant factors such as stage of growth, plant species, and type of organ.

This element is relatively nontoxic to animals and man. A deficiency of Co is of far greater concern than potential toxic levels in plants.

Fluorine

Fluorine (atomic no. 9; atomic wt., 18.998) is the most electronegative and reactive of all elements. It is a pale yellow, corrosive gas that reacts with practically all organic and inorganic substances. In nature, it does not exist in the free state but always forms compounds in an oxidized state, $-I$.

Fluorine occurs mainly in the silicate minerals of the earth's crust, at a concentration of about 650 ppm, and ranks 13th among the elements. The most important F-bearing minerals are fluorite (CaF_2), fluorapatite [$Ca_5(PO_4)_3F$], and cryolite (Na_3AlF_6). The largest known deposits of fluorite (fluorspar or calcium fluoride) are located in the United States, England, and Germany.

The largest use of calcium (or sodium) fluoride is in steelmaking, in which it prolongs the fluidity of the ingot, thus facilitating the escape of undesirable gaseous products. Fluorides are also used as a flux in the smelting of Ni, Cu, Au, and Ag, as rodenticides, and as catalysts for certain organic reactions.

Fluorapatites are used industrially for the preparation of phosphoric acid, fertilizers, and mineral feeds. Cryolite is used as an insecticide, and in molten form, as an electrolyte in the production of Al from alumina. Fluorine and its compounds are used in producing U (from the hexafluoride form, UF_6) for the nuclear industry, high-temperature plastics, air conditioning, and refrigeration. Hydrofluoric acid is extensively used for etching the glass of light bulbs, etc.

Fluorine is ubiquitous in the environment and is always present in plants, soils, and phosphatic fertilizers. The F concentrations in these materials are on the order of 3×10^0, 3×10^2, and 3×10^4 ppm for plants, soils, and phosphatic fertilizers, respectively (Larsen and Widdowson, 1971). Fluorine is a common constituent of rocks and soils. The continental rocks of the earth's crust contained, on the average, 650 ppm of F (Fleischer and Robinson, 1963). The following mean values, in ppm of F, are reported for various geological materials: basalts, 360; andesites, 210; rhyolites, 480; phonolites, 930; gabbros and diabases, 420; granites and granodiorites, 810; alkalic rocks, 1,000; limestones, 220; dolomites, 260; sandstones and graywackes, 180; shales, 800; oceanic sediments, 730; and soils, 285 (Fleischer and Robinson, 1963). It is an essential element in the following minerals: fluorite, apatite, cryolite, topaz, phlogophite, lepidolite, and other less important minerals.

Very common soil minerals, such as biotite, muscovite, and hornblende, may contain as much as several percent of F and, therefore, would seem to be the main source of F in soils. However, Steinkoenig (1919) believed that micas, apatite, and tourmaline in the parent materials were the original source of F in soils. It appears, therefore, that the F content of soils is largely dependent on the mineralogical composition of the soil's inorganic fraction.

Total F concentrations in United States soils range from <100 ppm to over 1,000 ppm (NAS, 1971). For world soils, the median value reported ranged from 200 to 300 ppm, whereas an average of 430 ppm was reported for United States surface soils (Appendix Table 1.17). The F contents of 30 soil profiles ($n = 137$) of representative soils of varied texture, parent material, and geographic distribution varied from trace amounts to 7,070 ppm for an unusual Tennessee soil containing phosphate rocks (Robinson and Edgington, 1946). The average for surface samples of these soils was

292 ppm. In general, F was concentrated in colloidal material and increased with the depth of the soil.

Nömmik (1953) observed that the natural F content of soil increases with increasing depth, and that only 5% to 10% of the total F content of soil is water-soluble. Under natural conditions, the concentrations of F in soil solutions are usually below 1 ppm, but under severely F-polluted soils they may reach levels of about 10 ppm (Polomski et al, 1982). In soils in the humid temperate climate, F could be readily lost from minerals in the acid horizons (Omueti and Jones, 1980). A substantial amount of this F is retained in subsoil horizons, where it is complexed with Al that is most likely associated with phyllosilicates. The extent of F retention depends on the amount of clay and pH; therefore, the pattern of F distribution with depth follows clay pattern closely. Fluorine occurs only in trivial amounts in the organic matter fraction of these soils examined by Omueti and Jones (1980).

Several investigators observed that the solubility of F in soils is highly variable and has the tendency to be higher at pH below 5 and above 6 (Fig. 14.2). Fluorine solubility in soil is complex and may be controlled by solid phases even more insoluble than CaF_2. In addition, F solubility may be related to the solubility of Al or other ionic species with which it form complexes. Soils having high pHs and low levels of amorphous

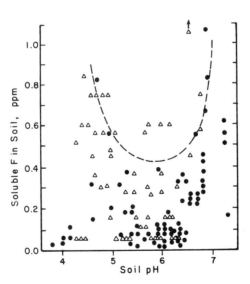

FIGURE 14.2 Possible relationships of soluble fluoride and pH. Data are fluoride measured in a 0.01 M $CaCl_2$ extract after 16 hours of shaking. Dots represent the data of Larsen and Widdowson (1971); triangles are of Gilpin and Johnson (1980). The dashed line represents the upper limit of solubility proposed by Larsen and Widdowson. *Source:* Gilpin and Johnson, 1980, with permission of the authors and the Soil Sci Soc of America.

Al species, clay, and organic matter generally sorb little F (Omueti and Jones, 1977a). Thus, it appears that the predominant retention mechanism is that of F exchange with the OH group of amorphous materials, such as Al-hydroxides (Flühler et al, 1982). In this case, the crystal lattice OH of clay minerals is replaced by F, resulting in a simultaneous release of Al and Fe. Other F retention mechanisms include the binding of F to soil cations (e.g., Ca^{2+}, Al^{3+}), or F precipitation as CaF_2, as in calcareous soils (Slavek et al, 1984).

Fluorine occurs in soils as the singly charged fluoride ion, F^-, or occasionally as a component of such complex anions as $(BF_4)^-$, $(AlF_6)^{3-}$, and $(SiF_6)^{2-}$ (Hopkins, 1977).

Fluorine is not an essential plant element but is essential for animals. However, continuous ingestion by animals of excessive amounts of F can lead to the disorder fluorosis, and sub-optimal levels in the diet can have an equally damaging effect. Therefore, plant content of F is of interest to livestock producers. In endemic fluorosis areas, the main F source is either the water or the dust-contaminated feedstuff (including pastures) or both (Underwood, 1977). In other areas, the main source of F is the naturally fluoridated drinking water derived from deep wells, which commonly contain 3 to 5 ppm F and sometimes as high as 10 to 15 ppm F. The chronic fluorosis of sheep, horses, and cattle, locally known as "darmous," that occurs in parts of North Africa is caused by contamination of water supplies and feedstuff with dusts originating from rock phosphate deposits and mines (Underwood, 1977).

Surface waters, such as are used for drinking and cooking in most areas, generally contain <1 ppm F, or even 0.1 ppm or less (Underwood, 1977). In regions where water comes from deep wells or artesian bores that are inherently high in F, fluorosis in humans has occurred, as in southern India and South Africa, where concentrations of F as high as 20 to 40 ppm have been reported.

Plants growing in areas free of F air pollution usually contain only trace amounts of F, usually <10 ppm in the foliage (Hansen et al, 1958; Kumpulainen and Koivistoinen, 1977). Kumpulainen and Koivistoinen (1977) reported that cereals usually contain <1 ppm F, where F tends to accumulate in the outer layer of the grain and in the embryo. The F contents of both leaf and root vegetables do not differ appreciably from those of cereals, with the exception of spinach, which is unusually enriched in F. Potato peelings can contain as much as 75% of the total F in the whole tuber. Tea is one of the most F-enriched foods, with about two thirds of the F in tea leaves being soluble in the beverage.

Brewer (1965) reported that the availability of soil F to plants is controlled by the following factors: pH, soil type and amount of clay, Ca, and phosphate. Liming acid soils to around pH 6 to 6.5 would decrease the levels of soluble F compounds present in or added to the soil. The use of phosphate fertilizers having low F concentrations can also be beneficial, as it can reduce F toxicity, presumably through ion competition, at low pH values as long as the F level does not exceed 180 ppm.

Several anthropogenic sources of F can enrich the soil with this element. Phosphatic fertilizers, especially the superphosphates, are perhaps the single-most important source of F in agricultural lands. Repeated applications of rock phosphates containing several percent of F was shown to have significantly elevated the F contents of soils (Omueti and Jones, 1977b). Typical additions of phosphate fertilizers (50–100 kg P_2O_5 per ha/ yr) could elevate soil F content by 5 to 10 ppm/yr (Gilpin and Johnson, 1980). Other sources that may affect soil status of F are precipitation, fallout of combustion products of coal and other industrially polluted air, and F-bearing insecticides (Adriano and Doner, 1982).

The effect of F-bearing fertilizers on the F content of plants is usually insignificant. However, plants growing in the vicinity of industries such as aluminum smelting can have substantially increased F contents.

The atmospheric release of F by industries is of ecological importance. The steel industry is the major source of atmospheric F in the United States and the third largest in Canada (NRCC, 1977). The primary aluminum production industry is the third largest source of atmospheric F in the United States and the largest source in Canada.

Silver

The lustrous, brilliant white metal (atomic no. 47; atomic wt., 107.87; melting pt., 962°C; specific gravity, 10.5) is very ductile and malleable. Metallic Ag occurs naturally and in ores such as argentite (Ag_2S) and hornsilver (AgCl); Pb, Pb-Zn, Cu, Ag, and Cu-Ni ores are also important commercial sources. Silver is also recovered during the electrolytic refining of Cu and smelting of Ni ores.

Some of the major uses of Ag are: silver alloys, such as sterling silver, used in silverware, jewelry, etc.; in the form of $AgNO_3$, it is used in photography, which accounts for about 30% of the amount used industrially in the United States; it is used for dental alloys, and in making high-capacity Ag-Zn and Ag-Cd batteries; it is used in mirror production; and silver iodide is used as a nucleating agent for weather modification.

In nature, Ag exists in the I oxidation state, although it could occur also in the II or III state. Silver is 66th in order of abundance in the earth's crust and is present at about 0.07 ppm level (Krauskopf, 1979). Smith and Carson (1977b) reported that the usual range of Ag in normal soils is <0.1 ppm to about 1 ppm, with some unusual values exceeding 5 ppm. For the conterminous United States, Shacklette and Boerngen (1984) found that surficial mineral soils averaged 0.7 ppm Ag, while soils high in organic matter had Ag content in the range of 2 to 5 ppm. Mitchell (1944) reported up to 2 ppm Ag in some Scottish soils. The Ag content of the surface soils in the agricultural area of Ontario, Canada, had a mean of 0.44 ppm Ag (range of 0.04–1.81 ppm) (Frank et al, 1979). In some southern California soils, Vanselow (1965c) found Ag in the range of 0.2 to 0.7 ppm.

In Table 14.1, the findings of Jones et al (1984) on concentrated HNO_3-extractable Ag in diverse United Kingdom surface soils are presented. Data are shown for noncontaminated sites and for areas contaminated by fluvial and aerial processes from derelict 19th century mine operations. Soils derived from black shales (usually 0.5 ppm Ag), or high in organic matter (0.5 ppm) are indigenously richer in Ag than the sandstones (0.05 ppm) or limestone-derived (0.07 ppm) soils. Contaminated soils in these areas may contain up to 10 ppm Ag, while mine spoils may contain Ag in the range of 4 to 65 ppm.

Silver can form complexes with nitrate, sulfate, and sulfite, but would not be sufficiently stable in soils to be significant; however, the complexes with I^-, Cl^-, and Br^- may be sufficiently stable to be significant (Lindsay and Sadiq, 1978). Under strongly reducing conditions, $AgHS^0$ could become an important Ag species in soil, which may prevent the formation of Ag sulfides unless enough Ag is present (Lindsay, 1979).

The typical level of Ag in plant tissues is <1 ppm (Vanselow, 1965c; Smith and Carson, 1977b). Silver accumulates in plant roots and is poorly translocated to the shoot (Figs. 14.3 and 14.4). When added to soil, as $AgNO_3$, at 75 ppm level, only 0.5 ppm Ag accumulated in the leaves of citrus plants (Vanselow, 1965c). Apparently, soil pH does not influence plant uptake of Ag.

Pettersson (1976) found that Ag was not appreciably translocated to the shoots of cucumber, tomato, and wheat plants grown in a nutrient solution containing 10 μM $AgNO_3$. Suzuki (1958) found only trace amounts of Ag taken up by the mint plant. Weaver and Klarich (1973), based on their experiment with wheat, maize, and soybeans grown on sandy and loamy soils spiked with AgI and $AgNO_3$, concluded that even large amounts of Ag, as AgI in soils, had practically no effect on plant growth. Silver from $AgNO_3$ is apparently more available to plants than Ag from AgI (Fig. 14.3).

Under field conditions, Horovitz et al (1974) found a range of 0.01 to 16 ppm Ag in the dry matter of 35 plant species, with the lower plant species having higher Ag contents. Klein (1978) concluded that Ag tends to accumulate mainly in the roots and that this element is excluded in the plant foliage even when present in soil at high concentrations. Ward et al (1979) found that as much as 90% of the Ag added to a nutrient solution can be immobilized in the roots of *Lolium perenne* and *Trifolium repens* (Fig. 14.4).

Data on the phytotoxicity of Ag are scarce. Wallace et al (1977) observed that Ag was very lethal to bush beans grown in solution culture containing 11 ppm Ag, as $AgNO_3$; even at 1.1 ppm level, yield was still significantly reduced, but the plants exhibited no symptoms and produced Ag levels

TABLE 14.1. Concentrated HNO_3-extractable silver levels (ppm dry weight) in diverse soils of the United Kingdom.[a]

Location	Rock/soil type	Number of samples	Uncontaminated (± S.E.)	Number of samples	Contaminated
Cardiganshire (Midwest Wales)	Lower Palaeozoic sandstones	19	0.05 ± 0.02 (Aeron Valley)	20	1–7 Ystwyth Valley
	Peaty upland soil	5	0.21	10	1–3.5 Rheidol Valley
Flintshire (Northeast Wales)	Carboniferous limestone	15	0.074 ± 0.03	15	0.5–3.5
Snowdonia (North Wales)	Black shale	5	0.15		
Co. Meath (Ireland)	Black shale	5	0.55		
Staffordshire (England)	Peaty moorland soil	5	0.47		

[a] Source: Jones et al, 1984, with permission of the authors and Elsevier Science Publishers, New York.

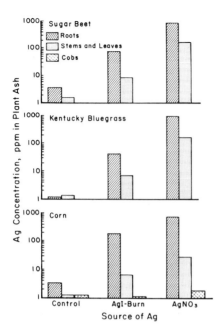

FIGURE 14.3 Concentrations of silver in corn, sugar beets, and Kentucky bluegrass from AgI burn complex and $AgNO_3$. Plants were grown in washed river sand in the greenhouse. *Source:* Teller and Klein, 1974, with permission of the authors.

of 5.8, 5.1, and 1,760 ppm (dry wt) in the leaves, stems, and roots, respectively. Davis et al (1978) noted that 0.5 ppm of Ag, as $AgNO_3$, in nutrient culture produced a critical level of about 4 ppm in foliage tissues of barley.

Because of the considerable unavailability of Ag to plants, the environmental concern about Ag arises from anthropogenic sources that could enrich the habitats of animals. Although there is no evidence that terrestrial animals biomagnify Ag from anthropogenic sources, aquatic species such as fish, mollusks, and crustaceans could be adversely affected. Silver ions are considered one of the most toxic heavy metal ions to microorganisms, particularly the heterotrophic bacteria, and to fishes (Cooper and Jolly, 1970; Charley and Bull, 1979). However, since Ag^+ easily forms insoluble compounds, it is considered harmless in the natural terrestrial environment.

The main potential environmental sources of Ag for plants are AgI used for weather modifications (estimated 2.6 tonnes of Ag, as AgI, used annually in the United States) and from mining, milling, smelting, and refining of Ag ores and other Ag-rich ores. Smith and Carson (1977b) estimated that of about 2,000 tonnes of Ag lost to the United States environment annually, about 12% would be dispersed to the air. None of these sources is likely to affect the terrestrial food chain, however.

FIGURE 14.4 Silver concentrations (ppm, dry wt) in whole roots and leaves of two pasture species grown for 90 days in a silica sand rooting medium treated with solutions containing 0–1000 μg/ml Ag (corresponding to 0–167 μg/g in the substrate). *Source:* Ward et al, 1979, with permission of the authors.

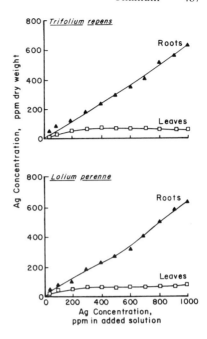

Thallium

Thallium (atomic no. 81; atomic wt., 204.39; specific gravity, 11.85; melting pt., 303°C) is a gray-white, soft, ductile metal.

Although Tl does not occur in free state in nature, several minerals contain Tl as a major constituent (16%–60%): crookesite $(Cu, Tl, Ag)_2Se$, lorandite $(TlAsS_2)$, and hutchinsonite $(Pb, Tl)_2(Cu, Ag)As_5S_{10}$. However, there are no known commercial deposits (Howe and Smith, 1970). Most of the ores containing relatively high levels of Tl occur in Switzerland, Yugoslavia, and Soviet Central Asia (Smith and Carson, 1977a).

In nature, the primary oxidation state of Tl is I, and Tl occurs in soil via its isomorphous substitution for K^+ in feldspars and silicates. Its geochemical behavior is similar to Rb and other alkali metal cations. It is concentrated in certain sulfides, and deposits characteristically high in As are usually also high in Tl.

Thallium is produced mainly as a by-product of Zn and Pb smelting and refining. Due to the volatile nature of most Tl compounds, the element becomes enriched on the flue dusts during the processing of ores. Practically all of the Tl produced in the world today at an annual rate of 10 to 12 tonnes is through Tl extraction of such dusts (Schoer, 1984).

When Tl use in rodenticides and insecticides was banned in the early 1970s in the United States, its commercial application became rather limited. Today, Tl is used in Pb, Ag, or Au-based bearing alloys which have very low friction coefficients, high endurance limits, and higher resistance to acids. The metal is also employed in the electrical and electronic industry, Hg vapor lamps, and deep-temperature thermometers. It is also used as a catalyst in the synthesis of olefins and hydrocarbons and other organic compound related reactions.

The crustal abundance of Tl has been estimated at about 1 ppm; Shaw (1952) gave it at 1.3 ppm, and Schoer (1984), at 0.3 to 0.6 ppm, probably a more realistic value. The element is widely distributed in the environment. Krauskopf (1979) reported a crustal concentration of 0.8 ppm, 59th among the elements. Interest in its environmental significance has been spurred by its known toxicity to living organisms in small amounts, potential biomethylation of Tl in the environment, and its release into the environment from the combustion of coal and from the cement industry (Peterson and Girling, 1981).

Combustion of coal is probably a larger environmental source of Tl than smelting. In coal, Tl occurs in the form of sulfides rather than in complexed forms with organic matter. This explains its low concentrations in anthracite coal. Bituminous coal was reported to contain 0.7 ppm Tl (Voskresenskaya, 1968) and as high as 76 ppm Tl could be present in coal fly ash (Davidson et al, 1976).

The mass flow of Tl in a United States coal-fired power plant is depicted in Figure 14.5. From such mass flow of Tl, various estimates on annual Tl emissions from coal combustion have been projected: United States,

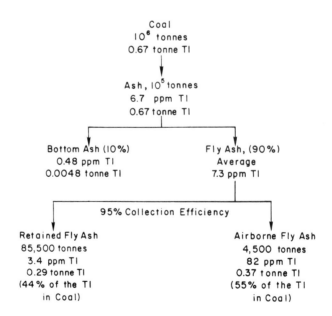

FIGURE 14.5 Estimated thallium flow in a coal-fired power plant. *Source:* Smith and Carson, 1977a, with permission of the authors.

140 tonnes; western Europe, 80 tonnes; and the rest of the world, 380 tonnes (Schoer, 1984). Smith and Carson (1977a) estimated that in the United States alone, nearly a total of 1,000 tonnes of Tl are released into the environment annually in the form of vapors and dusts (300 tonnes), nonferrous metals (60 tonnes), and aqueous and solid wastes (>500 tonnes).

Soil data are minimal, even on polluted soils. Bowen (1979) reported a median soil content of 0.2 ppm (0.1–0.8 ppm range). The highest value reported for United States soils was 5 ppm (Smith and Carson, 1977a). Soils sampled from the vicinity of a cement plant near Leimen, south-western Federal Republic of Germany had an average Tl contents of 3.6, 0.7, and 0.1 ppm in the 0 to 10, 40 to 50, and 60 to 70 cm depths, respectively (Schoer, 1984). This indicates that although Tl is somewhat mobile in soil, its mobility is limited.

Schoer (1984) concluded that the solubility and availability of Tl to organisms depends to a large extent on its source. Thallium present in flue dusts emanating from the cement industry is 75% to 100% soluble in water but can be rather rapidly immobilized once incorporated into the soil. Soils contaminated by the flue dusts could yield as much as 40% of their total Tl content when extracted with 1 N NH$_4$OAc in 0.01 M EDTA but could yield only 0.5% to 5% of the total Tl once the metal was incorporated in tailing material.

Some plant uptake data indicate that the amount of Tl that may be taken up by the plants is proportional to the Tl levels in the soil (Fig. 14.6). These results further indicate that the concentration ratio (i.e., calculated as the ratio of element concentration in plant tissue and element concentration in soil, all on dry weight basis) via root uptake may reach as high as 10, which is unusually high for metals. Because the geochemical be-

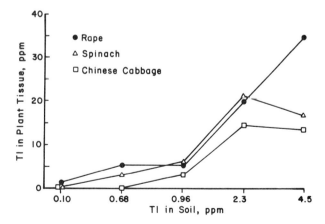

FIGURE 14.6 Thallium contents of selected plants as influenced by the soil thallium level. *Source:* Schoer, 1984, with permission of Springer-Verlag, New York.

havior of Tl is analogous to that of the essential element K, its uptake and distribution in plants maybe expected to be similar to that of K.

In general, Tl values for plant material are sparse. Smith and Carson (1977a) reported that terrestrial plants having detectable Tl contain from 0.01 to 3,800 ppm, on plant ash basis. However, they indicated that an ash value of around 0.5 ppm is probably more typical for a wide range of plants. Assuming that ash constitutes approximately 10% of the plant's dry matter, Tl concentrations then in the dry matter would be around 0.05 ppm. This extrapolated value is probably realistic in view of the finding of Geilmann et al (1960) that most crops grown in noncontaminated environment had Tl contents ranging from 0.02 to 0.04 ppm (dry wt) with some crops, such as leek, endive, and kale exhibiting higher values (0.075–0.125 ppm). Native plants in the USSR growing on soils with an average Tl content of 0.023 ppm had a range of 0.01 to 1.0 ppm, ash weight basis (Dvornikov et al, 1976). King (1977) considers values of 2 to 7 ppm Tl (ash wt basis) as anomalous and values >10 ppm, as highly anomalous. These anomalous levels could be observed in plants growing in soils containing 3 to 5 ppm total Tl (Table 14.2).

Highest levels in plants have been obtained in areas of unusual Tl enrichment, such as in the Alsar region of Yugoslavia, where animals grazing on the native vegetation had been intoxicated (Zyka, 1970). A case study on environmental pollution by Tl from a cement plant in Lengerich, Germany, indicates that foodstuff grown within an area having a radius of 2 km (the center being about 500 m east of the emission stack) can contain

TABLE 14.2. Thallium contents of various plant parts and their corresponding rhizosphere soils.[a]

Plant	Tl content of rhizosphere soil, ppm	Tl content of various plant parts, ppm dry wt	
Kohlrabi	3.7	Leaves	35.0
		Tuber	0.10
Turnip	9.4	Leaves	5.90
		Roots	0.40
Zucchini squash	5.2	Leaves	0.90
		Stalk	0.02
Cucumber	5.4	Leaves	0.70
		Fruits	0.10
Beet	5.2	Leaves	2.40
		Roots	0.60
Carrot	3.5	Leaves	0.30
		Roots	0.10
Onion	0.9	Stalk	0.10
		Tuber	0.01

[a] Source: Hoffman et al, 1982, with permission of the authors.

Tl levels exceeding the ingestion limit in the Federal Republic of Germany of 0.5 ppm (fresh wt) (Schoer, 1984). The Tl contents of some vegetables (fresh wt) were excessively high: kale–45 ppm; kohlrabi, savoy, and white cabbage–1 to 10 ppm.

Most plants showed phytotoxic symptoms when grown in media having relatively low levels of Tl. Carson and Smith (1977) indicated that, at about 7 ppm total Tl in soils, many crops may exhibit injury symptoms. Tobacco plants are especially sensitive, showing toxic effects at 1 ppm level in soil or 0.04 ppm in nutrient culture. Similarly, wheat growing in a sandy loam soil was injured with a concentration of 1.4 ppm in soil, and at 28 ppm level, death occurred. Significant reductions in photosynthesis of corn and sunflower were noticed when these plants were cultured in a nutrient solution containing 2 ppm Tl (Carlson et al, 1975). Excess Tl in growth medium has produced the following abnormalities in plants (Schoer, 1984): impairment of chlorophyll synthesis and seed germination, reduced transpiration due to interference of Tl in stomatal processes, growth reduction, stunting of roots, and leaf chlorosis. Indications are that Tl may be more toxic to plants than are Cd and Ni.

In general, our knowledge of the environmental biogeochemistry of Tl is very limited, particularly Tl in the food chain pathway.

Tin

The white-silvery, soft, pliable metal (atomic no. 50; atomic wt., 118.70; specific gravity, 5.75; melting pt., 232°C) exhibits in its geochemical occurrence two oxidation states, II and IV.

At 2.5 ppm crustal abundance, Sn is ranked 49th in its abundance among elements (Krauskopf, 1979). There are several known Sn-bearing minerals, but only cassiterite (SnO_2) is commercially important. The main region where Sn is produced is Southeast Asia, with Malaysia producing about 50% of the total amount of Sn produced by the free world. About 50% of that quantity is consumed by the United States.

Major uses of Sn are as a protective coating agent and as an alloying metal. About 80% of the total Sn produced is consumed in tinplate, bronze (1 part Sn: 9 parts Cu) and solder. Of the total amount used for tinplate, about 90% is used for cans, of which about 60% is used for food packaging, the rest for nonfood products. The Zn-Sn and Cd-Sn alloys are used as coatings in hydraulic brake parts, aircraft landing gear equipment, and engine parts in the automotive and aerospace industries. In agriculture, organotin compounds were used as fungicides for fungus control in potatoes and sugar beets (Barnes et al, 1971).

Tin is present in soils and plants in trace amounts. A common range of Sn in soil is 1 to 10 ppm (Peterson and Girling, 1981). Extensive analyses of surficial soil samples in the United States indicate a mean of 0.89 ppm, with a range of <0.1 to 10 ppm (Fig. 14.1). Laycock (1954) reported 1.3 to 3 ppm Sn in tea soils of Nyasaland, while Prince (1957) found 1 to 11 ppm Sn in several New Jersey soils.

Wallihan (1965) reported a narrow representative range of <0.1 to 3.0 ppm Sn in plants. A nonessential element in plant nutrition, it is fairly immobile in typical agricultural soil, especially in soils with near neutral pH. Plants tend to accumulate Sn in the roots, and it is poorly translocated to the foliage. Bean plants grown in soils spiked with up to 500 ppm Sn, as $SnCl_2$, showed only about 1 ppm Sn (dry wt) in leaves (Romney et al, 1975). In acidic soils, however, it can be fairly available and phytotoxic.

A survey of several vegetables and cereal grains indicate a usual range of 0.1 to 2 ppm in these foodstuffs (Schroeder et al, 1964). Kick et al (1971) reported a range of 0.8 to 6.8 ppm Sn in plant materials. Connor and Shacklette (1975) reported the Sn contents (ppm, ash wt) of the following vegetables: carrot—20; corn grain—30; and beet tuber—20. Duke (1970) reported 0.10 ppm Sn in dry avocado tissues. A fern, *Gleichenia linearis,* growing in Malaysia contained about 32 ppm Sn (dry wt), believed to be the highest recorded value for a terrestrial plant not contaminated by smelter operation (Peterson et al, 1976).

Because of the known toxicological importance of organotin compounds (Zingaro, 1979; Fishbein, 1974) and the possibility for Sn to be methylated (Summers and Silver, 1978), their biogeochemical behavior in the environment should be further pursued.

Titanium

A lustrous, whitish metal, Ti (atomic no. 22; atomic wt., 47.90; melting pt., 1,660 ± 10°C; specific gravity, 4.54) is the 9th most abundant element in the earth's crust (Krauskopf, 1979). It is almost always present in igneous rocks and igneous-derived sediments. It occurs in the following commercially useful minerals: ilmenite ($FeOTiO_2$), rutile (TiO_2), and sphene, and is present in titanates and in many iron ores. In igneous rocks, Ti occurs as rutile and can substitute for Al in pyroxene, hornblende, and biotite and as sphene in more siliceous parent rocks. It has oxidation states of II, III, or IV, with the latter being the most prevalent.

Titanium metal is used principally in aircrafts and missiles. Ilmenite is used primarily for manufacturing pigments, while rutile is used for making Ti metal, welding-rod coatings, and to a lesser extent, for alloys, carbides, ceramics, and fiberglass. Titanium dioxide pigment accounts for the largest use of the element; the pigment is extensively used for both house and artists' paints.

Titanium is present in the earth's crust at about 3,600 ppm and in normal soils at the range of 1,000 to 10,000 ppm (Trudinger et al, 1979). In United

States soils, the mean value is 2,400 ppm, with most of the values fluctuating from 1,500 to 5,000 ppm (Fig. 14.1). Kaminskaya (1941) reported that chernozem soils in the USSR had 4,500 to 5,300 ppm Ti, while podzols had a wider range of Ti content. Much higher values were reported for lateritic soils of Hawaii (as much as 15% Ti, as TiO_2), showing decreasing Ti content with depth (Sherman et al, 1949; Tamura et al, 1955). Soils in northern China typically contain about 3,000 ppm Ti (Lee, 1943). Occasionally, some Chinese soils contain 0.6 to 1.8% Ti and sporadically, as high as 3.6% Ti in the subsoil. In general, Ti content of soils in the tropical regions, especially lateritic soils, is higher than that in soils of the temperate regions. Because of the high levels of Ti in soil, plant tissues can be surficially contaminated with Ti and often may give misleading values. Berrow and Mitchell (1980) indicated that the extractable amounts of Ti in the mineral horizons of certain Scottish soils would not exceed 30 ppm and are usually below 3 ppm. In general, Ti is fairly insoluble in soil in the pH range of 4 to 8.

Titanium is not considered to be essential for the growth of higher plants. However, it does seem to play a role in increasing the yield of some crops (Pais et al, 1977, 1979) and in the N_2 fixation of legumes (Pratt, 1965b).

Since Ti salts become rapidly insoluble once added to soils, Pais et al (1977) used a chelated Ti solution for foliar spray. From such foliar application of Ti, they obtained from 16% increase in the chlorophyll content of tomato to 65% increase in apple. Likewise, the sugar content was increased from 9% to 17% over that of unsprayed plants. More recently, Ram et al (1983) concluded that foliar applications of Ti at low concentrations (1 ppm Ti or less) are beneficial for bean plants, since it promoted photosynthesis through an increase in the chlorophyll content of leaves, thereby increasing the plant dry matter yield by as much as 20% over the control plants. They also found that inorganic and organic sources of Ti were equally effective.

In a comprehensive review of the relationship between Ti and plant nutrition, Pais (1983) summarized the possible beneficial effects of Ti as follows: increased yields, in the range of 10% to 20%, of some of the most important agricultural crops in Hungary; increased photosynthetic activity, carbohydrate and protein synthesis; promoted enzyme activity such as catalase, peroxydase, nitrate reductase, and nitrogenase; promoted the Hill reaction; and exerted an auxin-type effect on plant growth. In spite of these findings, Ti is still far from being accepted as an essential element to plants.

Since Ti is only sparingly soluble in soils, plant toxicity under field conditions should not be encountered. In nutrient solutions, however, about 4.5 ppm Ti proved to be toxic to bush bean plants (Wallace et al, 1977), with most of the element accumulating in the roots. Toxicity symptoms included chlorosis, necrosis, and stunted growth. Leaf tissues of this chlorotic plant contained about 200 ppm Ti (dry wt).

Vanadium

Vanadium (atomic no. 23; atomic wt., 50.94; specific gravity, 6.11) compounds in nature have oxidation states of II, III, IV, or V. Its tendency to form various oxyanions is a property it shares with Mo, As, W, and P. In nature, it is found in about 65 different minerals, the most important commercial sources of the metal being carnotite, roscoelite, vanadinite, and patronite. It is also found in phosphate rocks and in some Fe ores, as well as in crude oils as organic complexes. The metal is bright white and is soft and ductile. Deposits are being mined mainly in the United States, Zimbabwe, and West Africa.

Vanadium is used in manufacturing steel for high-speed tools that require rust resistance. About 80% of the V now produced is used as ferrovanadium or as a steel additive. Vanadium pentoxide is used in ceramics and as a catalyst in the production of certain chemicals. The metal is a major constituent in high-strength Ti alloys. Its compounds are used as mordants in the dyeing and printing of cotton and for fixing aniline black on silk.

Vanadium is ubiquitous in nature, ranking 20th in abundance (110 ppm) among the elements in the earth's crust (Krauskopf, 1979). Its compounds commonly occur in the trivalent III oxidation state. Zenz (1980) reported that the average concentration of V in the earth's crust is about 150 ppm. Natural V occurs in igneous rocks, in titaniferrous magnetites, in certain deposits of phosphate rock, in shales, in some uranium ores, and in asphaltic deposits (Schroeder et al, 1963). The V contents of igneous rock, shale, sandstone, and limestone, in ppm, are: 135, 130, 20, and 20, respectively (NAS, 1974). In general, the total V contents of soils derived from sands and sandstones are low compared with soils developed from shales and clays.

Pratt (1965a) cited the V content of some soils ranged from 3 to 230 ppm, while Bowen (1979) reported a range of 3 to 500 ppm ($\bar{x}$ = 90 ppm). Typical Scottish soils derived from shale and olivine gabbro contained around 200 ppm, while soils derived from serpentine, sandstone, and granite had 100, 60, and 20 ppm, respectively (Thornton and Webb, 1980). Surface soils in the United States contain about 58 ppm V, on the average (Fig. 14.1). The mean concentration of V for tropical Asian paddy soils is 166 ppm (Appendix Table 1.18).

In nature, V occurs as vanadates and in various organic complexes (Peterson and Girling, 1981). In sediments and soils, V is known to occur primarily as oxovanadium (IV) in organic chelates (Yen, 1972).

In soil, V could substitute for Fe and can be sorbed onto Fe oxides, which explains the usually high correlations between the Fe and V contents of soil (Norrish, 1975). Although V tends to be more enriched in the A horizon, it is more or less uniformly distributed throughout the soil profile (Vinogradov, 1959). However, in cases where the B horizon contains higher amounts of clay and Fe oxides, V can be expected to accumulate more in this horizon (LeRiche and Weir, 1963).

Vanadium is not considered to be an essential element for higher plants. Welch and Huffman (1973) did not obtain any growth response of lettuce and tomato plants to added V. The plants were grown in purified nutrient solution with and without 50 ηg V/ml. All plants were healthy with similar growth and development patterns. However, some evidence exists to suggest that it is essential for the growth of certain bacteria and algae. The green algae *Scenedesmus obliquus* apparently requires it for their growth (Arnon and Wessel, 1953; Arnon, 1954). Bertrand (1950) found that trace amounts of V in the growth medium served as a growth stimulant for *Aspergillus niger*. Recently, it has been shown that during the fixation of N_2 by *Azotobacter*, V can replace Mo in the nitrogenase enzyme, but was not as effective as Mo-containing enzyme in fixing N_2 (Nicholas, 1975).

Vanadium could be beneficial in some plants since it has been demonstrated to enhance chlorophyll formation and Fe metabolism in tomato plants and to have increased the Hill reaction of isolated chloroplasts (Basiouny, 1984).

Vanadium is widely distributed in most biological materials. Bengtsson and Tyler (1976) indicated that it is not unreasonable to expect V concentrations to fall in the range of 0.5 to 2 ppm (dry wt) in plants growing in noncontaminated areas. Bertrand (1950) reported a range of 0.27 to 4.2 ppm V in 62 plant materials ($\bar{x}$ = 1 ppm) and found that plants growing on soils having higher V contents accumulated greater amounts of V than those growing on soils having less V. The roots of these plants contained higher V than the aerial parts. Cannon (1963) gave V values (in ppm, dry wt) for the following plants: grasses—1.4; legumes—0.84; forbs—1.2; deciduous trees—1.65; deciduous shrubs—2.7; ferns—1.28; fungi—0.22; and lichens—8.6. Vegetable crops generally contain <0.5 ppm V (Bertrand, 1950; Schroeder et al, 1963; Bengtsson and Tyler, 1976). Among the vegetables, radish and parsley could accumulate relatively more V (Söremark, 1967).

Thirty-five plant species, collected from 36 different locations in the Rocky Mountain region, had V contents ranging from nondetectable to 6.6 ppm (dry wt) whereas soil contents ranged from 0.65 to 98 ppm (Parker et al, 1978). In this region, increased plant and soil residues were obtained at locations near phosphate rock, ore smelting, and coal burning industrial facilities. In general, there was no relationship between the V contents of plants and soils and the geological deposits of V. It was concluded that soils possessing high V levels should not pose any hazard for V bioaccumulation in the food chain.

Plant uptake of V can depend largely on soil type (Peterson and Girling, 1981). Plant accumulation of V may occur in seleniferrous soils or river alluvium while calcareous soils may restrict V movement from the roots to the foliage.

Under field conditions, V toxicity is virtually nonexistent. However, when added to nutrient solution in concentrations of 0.5 ppm or greater, V could be phytotoxic (Pratt, 1965a; Davis et al, 1978). Since V is poorly translocated, analysis of vegetative tissues may not be a useful diagnostic tool for V toxicity. Some plant species suffering from excess V can contain only 1 to 2 ppm in their vegetative tissues while exhibiting several-fold concentrations in their roots (Pratt, 1965a; Warington, 1955, 1956; Davis et al, 1978). Berry (1978) concluded that V, along with Cd, is one of the most toxic trace elements in the nutrition of higher plants grown in solution culture.

Recent interest in the V content of plant material has been spurred by the essentiality of V for animal nutrition and its releases in large amounts to the atmosphere from the combustion of fossil fuels. In the United States, practically all of the coal burned contains some V, with the average content being about 30 ppm (Zenz, 1980). Australian bituminous coals have V contents ranging from 4 to 90 ppm ($\bar{x} = 20$ ppm) (Swaine, 1977).

In 1969, about 1,750 tonnes of V were emitted to the atmosphere in the United States as a result of combusting bituminous coal, with perhaps an additional 375 tonnes emitted from the combustion of anthracite coal (Zenz, 1980). In addition to releases of V from coal combustion, ash from petroleum-fired facilities may contain up to 38% V, as V pentoxide, but the concentration varies considerably, depending upon the source of the crude oil. Crude oils were found to contain as much as 1,400 ppm V with the Venezuelan oils characteristically enriched in this metal (Bengtsson and Tyler, 1976). Adoption of pollution abatement measures by electric power utilities has reduced, by as much as 85%, the release of V to the environment from fly ash. Since the critical level (25 ppm of total diet) of V for livestock would be unlikely to be exceeded under the usual field conditions, the environmental concerns about V would be primarily arising from the air pollution aspect.

Information on the environmental chemistry of V in terrestrial systems is generally sparse.

References

Adams, S. N., J. L. Honeysett, K. G. Tiller, and K. Norrish. 1969. *Aust J Soil Res* 7:29–42.

Adriano, D. C., M. Delaney, and D. Paine. 1977. *Common Soil Sci Plant Anal* 8:615–628.

Adriano, D. C., and H. E. Doner. 1982. In A. L. Page, ed. *Methods of soil analysis*. Part 2, 449–483. Am Soc Agron Inc, Madison, WI.

Alban, L. A., and J. Kubota. 1960. *Soil Sci Soc Am Proc* 24:183–185.

Allen, R. O., and E. Steinnes. 1979. In *Proc Intl Conf on Heavy Metals in the Environ*, 271–274. CEP Consultants Ltd, Edinburgh, UK.

Anderson, A. J., D. R. Meyer, and F. K. Mayer. 1973. *Aust J Agric Res* 24:557–571.

Arnon, D. I., and G. Wessel. 1953. *Nature*. 172:1039–1040.

Arnon, D. I. 1954. *Intl Cong Bot (8th) Proc* (Paris) 11:73–80.

Askew, H. O., and J. K. Dixon, 1937. *N Z J Sci Technol* 18:688–693.

Bakulin, A. A., and V. F. Mokiyenko. 1966. *Pochrovedenie* 4:66 (as cited by Kabata-Pendias and Pendias, 1984).

Barnes, R. D., A. T. Bull, and R. C. Poller. 1971. *Chem Ind* (London) 7:204.

Basiouny, F. M. 1984. *J Plant Nutr* 7:1059–1073.

Bengtsson, S., and G. Tyler. 1976. In *Vanadium in the environment*. MARC Tech Rep no 2. Monit Assess Res Center, Univ London.

Berrow, M. L., and R. L. Mitchell. 1980. *Trans Royal Soc* (Edinburgh) *Earth Sci* 71:103–121.

Berry, W. L. 1978. In D. C. Adriano and I. L. Brisbin, eds. *Environmental chemistry and cycling processes*, 582–589. CONF. 760429. NTIS, Springfield, VA.

Bertrand, D. 1950. *Bull Am Mus Nat Hist* 94:405–455.

Bertrand, D., and L. Silberstein. 1928. *Compt Rend Acad Sci* (Paris) 186:335–338.

Bowen, H. J. M. 1979. *Environmental chemistry of the elements*. Academic Press, New York. 333 pp.

Brewer, R. F. 1965. In H. D. Chapman, ed. *Diagnostic criteria for plants and soils*, 180–196. Quality Printing Co Inc., Abilene, TX.

Cannon, H. L. 1963. *Soil Sci* 96:196–204.

Carlson, R. W., F. A. Bazzaz, and G. L. Rolfe. 1975. *Environ Res* 10:113–120.

Carr, M. H., and K. K. Turekian. 1961. *Geochim Cosmochim Acta* 23:9–60.

Carson, B. L., and I. C. Smith. 1977. In Midwest Res Inst Tech Rep 5. Kansas City, MO. 386 pp.

Charley, R. C., and A. T. Bull. 1979. *Arch Microbiol* 123:239–244.

Chaudry, F. M., A. Wallace, and R. T. Mueller. 1977. *Commun Soil Sci Plant Anal* 8.795–797.

Connor, J. J., and H. T. Shacklette. 1975. In USGS Prof Paper 574-F. Denver, CO.

Cooper, C. F., and W. C. Jolly. 1970. *Water Resour Res* 6:88–98.

Crecelius, E. A., C. J. Johnson, and G. C. Hofer. 1974. *Water, Air, Soil Pollut* 3:337–342.

Crecelius, E. A., M. H. Bothner, and R. Carpenter. 1975. *Environ Sci Technol* 9:325–333.

Davidson, R. L., D. F. S. Natusch, C. A. Evans, and P. Williams. 1976. *Science* 191:852–854.

Davis, R. D., P.H.T. Beckett, and E. Wollan. 1978. *Plant Soil* 49:395–408.

Duke, J. A. 1970. *Econ Bot* 24:344–366.

Dvornikov, A. G., L. B. Ovsyannikova, and O. G. Sidenko. 1976. *Geokhimiya* 4:626–633 (as cited by Gough et al, 1979).

Filipovic, Z., B. Stankovic, and Z. Dusic. 1961. *Soil Sci* 91:147–150.

Fishbein, L. 1974. *Sci Total Environ* 2:341–371.

Fleischer, M., and W. O. Robinson. 1963. *Roy Soc Can Spec Publ* 6:58–75.

Flühler, H., J. Polomski, and P. Blaser. 1982. *J Environ Qual* 11:461–468.

Frank, R., K. I. Stonefield, and P. Suda. 1979. *Can J Soil Sci* 59:99–103.

Fujimoto, G., and G. D. Sherman. 1950. *Agron J* 42:577–581.

Geilmann, W., K. Beyermann, K. H. Neeb, and R. Neeb. 1960. *Biochem Z* 333:62–70 (as cited by Schoer, 1984).

Gille, G. L., and E. R. Graham. 1971. *Soil Sci Soc Am Proc* 35:414–416.

Gilpin, L., and A. H. Johnson. 1980. *Soil Sci Soc Am J* 44:255–258.

Graham, E. R., and D. D. Killion. 1962. *Soil Sci Soc Am Proc* 26:545–547.

Hansen, E. D., H. H. Wiebe, and W. Thorne. 1958. *Agron J* 50:565–568.

Hodgson, J. F. 1960. *Soil Sci Soc Am Proc* 24:165–168.

Hodgson, J. F., K. G. Tiller, and M. Fellows. 1969. *Soil Sci* 108:391–396.

Hoffman, G., P. Schweiger, and W. Scholl. 1982. *Landwirtsch Forschung*. 35:45–54 (as cited by Schoer, 1984).

Hopkins, D. M. 1977. *J Res USGS*. 5:589–593.

Horovitz, C. T., and O. Petrescu. 1964. In *Trans 8th Intl Cong Soil Sci* 4:1205–1213.

Horovitz, C. T., H. H. Schock, and L. A. Horovitz-Kisimova. 1974. *Plant Soil* 40:397–403.

Howe, H. E., and A. A. Smith. 1970. *J Electrochem Soc* 97:167C–170C.

Jones, K. C., P. J. Peterson, and B. E. Davies. 1984. *Geoderma* 33:157–168.

Kabata-Pendias, A., and H. Pendias. 1984. *Trace elements in soils and plants.* CRC Press Inc, Boca Raton, FL. 315 pp.

Kaminskaya, S. E. 1941. *Compt Rend Acad Sci* (USSR) 33:50–53.

Kick, H., R. Nosbers, and J. Warnusz. 1971. In *Proc Intl Symp on Soil Fertility Evaluation* (New Delhi) 1:1039–1045.

Kidson, E. B. 1938. *J Soc Chem Ind* (London) 57:95–96.

King, H. 1977. US Geol Survey Bull 1466:80.

Kipling, M. D. 1980. In H. A. Waldron, ed. *Metals in the environment,* 133–153. Academic Press, New York.

Klein, D. A. 1978. In D.A. Klein, ed. *Environmental impacts of artificial ice nucleating agents,* 109–126. Dowden, Hutchinson, and Ross Inc, Stroudsburg, PA.

Krauskopf, K. B. 1979. *Introduction to geochemistry.* 2nd edition. McGraw-Hill, New York. 617 pp.

Kubota, J. 1965. *Soil Sci* 99:166–174.

——— 1980. In B. E. Davies, ed. *Applied soil trace elements,* 441–466. Wiley, New York.

Kumpulainen, J., and P. Koivistoinen. 1977. *Residue Rev* 68:37–57.

Lambert, T. I., and C. Blincoe. 1971. *J Sci Fed Agric* 22:8–9.

Larsen, S., and A. E. Widdowson. 1971. *J Soil Sci* 22:210–221.

Laycock, D. H. 1954. *J Sci Food Agric* 5:266–269.

Lee, C. K. 1943. *Soil Sci* 55:343–349.

LeRiche, H. H., and A. H. Weir. 1963. *J Soil Sci* 14:225–235.

Lim, M. Y. 1979. In *Trace elements from coal combustion atmospheric emissions.* Intl Energy Agency, Coal Res, London.

Lindsay, W. L. 1979. *Chemical equilibria in soils.* Wiley, New York. 449 pp.

Lindsay, W. L., and M. Sadiq. 1978. In D. A. Klein, ed. *Environmental impacts of artificial ice nucleating agents,* 25–40. Dowden, Hutchinson, and Ross Inc, Stroudsburg, PA.

Lyon, G. L., R. R. Brooks, P. J. Peterson, and G. W. Butler. 1968. *Plant Soil* 29:225–240.

——— 1971. *J Ecol* 49:421–429.

Malaisse, F., J. Gregoire, R. S. Morrison, R. R. Brooks, and R. D. Reeves. 1979. *Oikos* 33:472–478.

McKenzie, R. M. 1975. In D. J. D. Nicholas and A. R. Egan, eds. *Trace elements in soil-plant-animal systems,* 83–93. Academic Press, New York.

——— 1978. *Aust J Soil Res* 16:209–214.

Menzel, R. G. 1954. *Soil Sci* 77:419–425.

Mitchell, R. L. 1944. *Proc Nutr Soc* 1:183–189.

——— 1945. Soil Sci. 60:63–70.

——— 1971. In Tech Bull no. 21. Her Majesty's Stat Off, London.

Mitchell, R. L., and J. C. Burridge. 1979. *Phil Trans R Soc* (London) B288:15–24.

National Academy of Sciences (NAS). 1971. *Fluorides.* In *Comm Biol Effects of Air Pollut.* NAS, Washington, DC. 295 pp.

——— 1974. In *Vanadium.* Washington, DC. 117 pp.

National Research Council of Canada (NRCC). 1977. In *Environmental fluoride.* NRCC no. 16081. Ottawa, Ontario. 151 pp.

Newland, L. W. 1982. In O. Hutzinger, ed. *The handbook of environmental chemistry.* Vol. 3(B): *Anthropogenic compounds,* 27–67. Springer-Verlag, Berlin.

Nicholas, D. J. D. 1975. In D. J. D. Nicholas and A. R. Egan, eds. Trace *elements in soil-plant-animal systems,* 181–198. Academic Press, New York.

Nömmik, H. 1953. *Acta Polytech* 127:1–121.

Norrish, K. 1975. In D. J. D. Nicholas and A. R. Egan, eds. *Trace elements in soil-plant-animal systems,* 55–81. Academic Press, New York.

Omueti, J. A. I., and R. L. Jones. 1977a. *J Soil Sci* 28:564–572.

——— 1977b. *Soil Sci Soc Am J* 41:1023–1024.

——— 1980. *Soil Sci Soc Am J* 44:247–249.

Pais, I. 1983. *J Plant Nutr* 6:1–131.

Pais, I., M. Feher, E. Farkas, Z. Szabo, and I. Cornides. 1977. *Commun Soil Sci Plant Anal* 8:407–410.

——— 1979. *Acta Agron Hung* 28:378–383.

Parker, R. D. R., R. P. Sharma, and G. W. Miller. 1978. *Trace Subs Environ Health.* 12:340–350.

Patel, P. M., A. Wallace, and R. T. Mueller. 1976. *J Am Soc Hort Sci* 101(5):553–556.

Peterson, P. J., and C. A. Girling. 1981. In N. W. Lepp, ed. *Effect of heavy metal pollution on plants.* Vol. 1., pp. 213–278. Applied Science Publ, London.

Peterson, P. J., M. A. S. Burton, M. Gregson, S. M. Nye, and E. K. Porter. 1976. *Trace Subs Environ Health* 10:123–132.

Pettersson, O. 1976. *Plant Soil* 45:445–459.

Polomski, J., H. Flühler, and P. Blaser. 1982. *J Environ Qual* 11:457–461.

Pratt, P. F. 1965a. In H. D. Chapman, ed. *Diagnostic criteria for plants and soils,* 480–483. Quality Printing Co Inc, Abilene, TX.

——— 1965b. In H. D. Chapman, ed. *Diagnostic criteria for plants and soils,* 478–479. Quality Printing Co Inc, Abilene, TX.

Price, N. O., W. N. Linkous, and R. W. Engel. 1955. *J Agr Food Chem* 3:226–229.

Prince, A. L. 1957. *Soil Sci* 84:413–418.

Proctor, J., and S. R. J. Woodell. 1975. *Adv Ecol Res* 9:255–366.

Prokhorov, V. M., L. P. Moskevich, and V. A. Kudyashov. 1979. *Sov Soil Sci* 11:161–167.

Ram, N., M. Verloo, and A. Cottenie. 1983. *Plant Soil* 73:285–290.

Reisenauer, H. M. 1960. *Nature* 186:375–376.

Robinson, W. O., and G. Edgington. 1946. *Soil Sci* 61:341–353.

Romney, E. M., and J. D. Childress. 1965. *Soil Sci* 100:210–217.

Romney, E. M., A. Wallace, and G. V. Alexander. 1975. *Plant Soil* 42:585–589.

Ruhland, W. 1958. In *Mineral nutrition of plants.* Springer-Verlag, Berlin. 1210 pp.

Schoer, J. 1984. In O. Hutzinger, ed. *The handbook of environmental chemistry.* Vol. 3(C):143–214. Springer-Verlag, Berlin.

Schroeder, H. A., J. J. Balassa, and I. H. Tipton. 1963. *J Chron Dis* 16:1047–1071.

—— 1964. *J Chron Dis* 17:483–502.

Shacklette, H. T., and J. G. Boerngen. 1984. *Element concentrations in soils and other surficial materials of the conterminous United States.* USGS Prof Paper 1270. US Govt Printing Office, Washington, DC.

Slavek, J., H. Farrah, and W. F. Pickering. 1984. *Water, Air, Soil Pollut* 23:209–220.

Shaw, D. M. 1952. *Geochim Cosmochim Acta* 2:118–154.

Sherman, G. D., Z. C. Foster, and C. K. Fujimoto. 1949. *Soil Sci Soc Am Proc* 1948. 13:471–476.

Smith, I. C., and B. L. Carson. 1977a. *Trace metals in the environment.* Vol. 1: *Thallium.* Ann Arbor Science, Ann Arbor, MI. 394 pp.

—— 1977b. *Trace metals in the environment.* Vol. 2: *Silver.* Ann Arbor Science, Ann Arbor, MI. 469 pp.

Söremark, R. 1967. *J Nutr* 92:183–190.

Steinkoenig, L. A. 1919. *J Indus Eng Chem* 11:463–465.

Summers, A. O., and S. Silver. 1978. *Ann Rev Microbiol* 32:637–672.

Suzuki, N. 1958. In *Sci Rep Res Inst* (Tohoku Univ) 42:149–157.

Swaine, D. J. 1977. *Trace Subs Environ Health* 11:107–116.

—— 1979. In *Trace elements in Australian bituminous coals and fly ashes.* Proc of Conf on Combustion of Pulverized Coal-Effect of Mineral Matter. Univ Newcastle, Newcastle, England.

Tamura, T., M. L. Jackson, and G. D. Sherman. 1955. *Soil Sci Soc Am Proc* 19:435–439.

Taylor, R. M., and R. M. McKenzie. 1966. *Aust J Soil Res* 4:29–39.

Teller, H. L., and D. A. Klein. 1974. In Oper Rep no 3, Colorado State Univ, Fort Collins. 105 pp.

Tepper, L. B. 1980. In H. A. Waldron, ed. *Metals in the environment,* 25–60. Academic Press, New York.

Thornton, I., and J. S. Webb. 1980. In B. E. Davies, ed. *Applied soil trace elements,* 381–439. Wiley, New York.

Tiller, K. G., J. L. Honeysett, and E. G. Hallsworth. 1969. *Aust J Soil Res* 7:43–56.

Trudinger, P. A., D. J. Swaine, and G. W. Skyring. 1979. In P. A. Trudinger and D. J. Swaine, eds. *Biogeochemical cycling of mineral-forming elements,* 1–27. Elsevier, Amsterdam.

Tso, T. C., T. P. Sorokin, and M. E. Engelhaupt. 1973. *Plant Physiol* 51:805–806.

Underwood, E. J. 1977. *Trace elements in human and animal nutrition*. Academic Press, New York. 545 pp.

US Environmental Protection Agency (US-EPA) 1976. In *Quality criteria for water*. US-EPA, Washington, DC.

van der Weijden, C. H., and E. C. Kruissink. 1977. *Marine Chem* 5:93–112.

Vanselow, A. P. 1965a. Barium. In H. O. Chapman, ed. *Diagnostic criteria for plants and soils,* 24–32. Quality Printing Co Inc, Abilene, TX.

────── 1965b. Cobalt. In H. D. Chapman, ed. *Diagnostic criteria for plants and soils,* 142–156. Quality Printing Co Inc, Abilene, TX.

────── 1965c. Silver. In H. D. Chapman, ed. *Diagnostic criteria for plants and soils,* 405–408. Quality Printing Co Inc, Abilene, TX.

Vinogradov, A. P. 1959. *The geochemistry of rare and dispersed chemical elements in soils*. Consultants Bureau Inc, New York. 209 pp.

Voskresenskaya, N. T. 1968. *Geochem Intern* 5:158–168.

Wallace, A., and E. M. Romney. 1971. *Agron J* 63:245–248.

Wallace, A., G. V. Alexander, and F. M. Chaudry. 1977. *Commun Soil Sci Plant Anal* 8:751–756.

Wallihan, E. F. 1965. In H. D. Chapman, ed. *Diagnostic criteria for plants and soils,* 476–477. Quality Printing Co Inc, Abilene, TX.

Ward, N. I., E. Roberts, and R. R. Brooks. 1979. *N Z J Sci* 22:129–132.

Warington, K. 1955. *Ann Appl Biol* 41:1–22.

────── 1956. *Ann Appl Biol* 44:535–546.

Weaver, T. W., and D. Klarich. 1973. In Final Res Rep no. 42. Mont Agric Exp Sta, Montana State Univ, Bozeman. 18 pp.

Welch, R. M., and E. W. D. Huffman, Jr. 1973. *Plant Physiol* 52:183–185.

Williams, R. J. B., and H. H. Riche. 1968. *Plant Soil* 29:317–326.

Wright, J. R., and K. Lawton. 1954. *Soil Sci* 77:95–105.

Yen, T. F. 1972. *Trace Subs Environ Health* 6:347–353.

Young, R. S. 1979. *Cobalt in biology and biochemistry*. Academic Press, London. 147 pp.

Zenz, C. 1980. In H. A. Waldron, ed. *Metals in the environment,* 293–327. Academic Press, New York.

Zingaro, R. A. 1979. *Environ Sci Technol* 13:282–287.

Zubovic, P. 1969. In Proc of Beryllium Conference, MIT (as cited by Tepper, 1980).

Zyka, V. 1970. *Sbornick Geol Ved Technol Geochem* 10:91–96 (as cited by Peterson and Girling, 1981).

Appendix

Scientific Names for Plants and Other Organisms Cited

PLANTS

Alfalfa	*Medicago sativa*
Algae (green)	*Scenedesmus obliquus*
Apple	*Malus sylvestris*
Apricot	*Prunus armeniaca*
Artichoke (Jerusalem)	*Helianthus tuberosa*
Ash	*Fraxinus* spp.
Asparagus	*Asparagus officinalis*
Aspen	*Populus* spp.
Aster (heath)	*Aster ericoides*
Avocado	*Persea* spp.
Banana	*Musa* spp.
Barley	*Hordeum vulgare*
Bean (bush)	*Phaseolus vulgaris*
Bean (common)	*Phaseolus vulgaris*
Bean (kidney)	*Phaseolus vulgaris*
Bean (navy)	*Phaseolus vulgaris*
Bean (lima)	*Phaseolus limensis*
Beech	*Fagus* spp.
Beech (American)	*Fagus grandifolia*
Beet (common and table)	*Beta vulgaris*
Beet (sugar)	*Beta saccharifera*
Bentgrass (creeping)	*Agrostis alba*
Bentgrass (velvet)	*Agrostis canina*
Bermudagrass	*Cynodon dactylon*
Bilberry	*Vaccinium* spp.
Birch	*Betula* spp.
Birch (white)	*Betula alba*
Birch (yellow)	*Betula alleghaniensis*

Blackberry	*Rubus* spp.
Blueberry	*Vaccinium* spp.
Bluegrass	*Poa* spp.
Bluegrass (Kentucky)	*Poa pratensis*
Breadfruit	*Artocarpus altilis*
Broadbean	*Vicia faba*
Broccoli	*Brassica oleracea* var. *italica*
Bromegrass	*Bromus* spp.
Bromegrass (Japanese)	*Bromus japonicus*
Bromegrass (smooth)	*Bromus secalinus*
Brussels sprouts	*Brassica oleracea* var. *gemmifera*
Cabbage (common, savoy)	*Brassica oleracea* var. *capitata*
Cabbage (Chinese)	*Brassica pekinensis*
Cantaloupe	*Cucumis melo*
Carrot	*Daucus carota*
Cassava	*Manihot esculenta*
Cauliflower	*Brassica oleracea* var. *botrytis*
Celery	*Apium graveolens*
Chard	*Beta vulgaris* var. *cicla*
Cherry	*Prunus* spp.
Cherry (choke)	*Prunus virginiana*
Chrysanthemum	*Chrysanthemum* vars.
Clover (alsike)	*Trifolium hybridum*
Clover (crimson)	*Trifolium incarnatum*
Clover (ladino)	*Trifolium repens forma lodigense*
Clover (red)	*Trifolium pratense*
Clover (sweet)	*Melilotus* spp.
Clover (white)	*Trifolium repens*
Clover (white sweet)	*Melilotus alba*
Clover (yellow sweet)	*Melilotus officinalis*
Clover (zigzag)	*Trifolium medium*
Cocklebur	*Xanthium* spp.
Cocoa	*Theobroma cacao*
Coconut	*Cocos nucifera*
Coffee	*Caffea* spp.
Collard	*Brassica oleracea* var. *acephala*
Corn (including sweet)	*Zea mays*
Cotton	*Gossypium hirsutum*
Cowberry	*Vaccinium* spp.
Cowpea	*Vigna sinensis*
Crabgrass	*Digitaria* spp.
Cress (curly)	*Lepidium sativum*
Cucumber	*Cucumis sativus*
Dallisgrass	*Paspalum dilatatum*
Deertongue	*Panicum clandestinum*

Dewberry	*Rubus caesius*
Dill	*Anethum* spp.
Dogwood	*Cornus* spp.
Elm	*Ulmus* spp.
Elm (American)	*Ulmus americana*
Endive	*Cichorium endivia*
Fescue	*Festuca* spp.
Fescue (meadow)	*Festuca elatior*
Fescue (red)	*Festuca rubra*
Fescue (tall)	*Festuca elatior*
Fig	*Ficus carica*
Fir (balsam)	*Abies balsamea*
Fir (Douglas)	*Pseudotsuga menziesii*
Grass (barnyard)	*Echinochloa crusgalli*
Flax	*Linum usitatissimum*
Foxtail	*Alopecurus* spp.
Grapes	*Vitis* spp.
Gum (black)	*Nyssa sylvatica*
Hairgrass	*Deschampsia* spp.
Hairgrass (tufted)	*Deschampsia cespitosa*
Hemlock (eastern)	*Tsuga canadensis*
Hickory	*Carya* spp.
Hops	*Humulus* spp.
Johnsongrass	*Sorghum halepense*
Kale	*Brassica oleracea* var. *acephala*
Kohlrabi	*Brassica oleracea* var. *gongylodes*
Larch	*Larix* spp.
Larkspur	*Delphinium* spp.
Lemon	*Citrus limon*
Lespedeza	*Lespedeza* spp.
Lespedeza (Korean)	*Lespedeza stipulacea*
Lespedeza (sericea)	*Lespedeza cuneata*
Lettuce (Romaine)	*Lactuca sativa*
Locoweeds	*Astragalus* spp.
Lovegrass (weeping)	*Eragrostis curvula*
Lucerne	*Medicago sativa*
Lupine	*Lupinus* spp.
Maize	*Zea mays*
Mangel	*Beta vulgaris* var. *macrorhiza*
Maple (mountain)	*Acer spicatum*
Maple (red)	*Acer rubrum*
Maple (silver)	*Acer saccharinum*
Maple (sugar)	*Acer saccharum*
Marigold	*Tagetes* spp.
Meadow sweet	*Spirea alba*

Millet (proso)	*Panicum miliaceum*
Mint	*Mentha* spp.
Mulberry	*Morus* spp.
Muskmelon	*Cucumis melo*
Mustard	*Brassica juncea*
Mustard (white)	*Brassica hirta*
Needle and thread	*Stipa comata*
Nutsedge	*Cyperus* spp.
Oak	*Quercus* spp.
Oak (black)	*Quercus velutina*
Oak (chestnut)	*Quercus prinus*
Oak (northern red)	*Quercus rubra*
Oak (white)	*Quercus alba*
Oats	*Avena sativa*
Oats (wild)	*Uniola latifolia*
Olive	*Olea* spp.
Onion	*Allium cepa*
Orange	*Citrus sinensis*
Orchardgrass	*Dactylis glomerata*
Palm	*Elaeis* spp.
Pansy	*Viola* spp.
Parsley (common)	*Petroselinum crispum*
Parsley (curly)	*Petroselinum crispum* var. *latifolium*
Parsnips	*Pastinaca sativa*
Pea	*Pisum sativum*
Pea (black-eyed)	*Vigna sinensis*
Pea (field)	*Pisum arvense*
Pea (sweet)	*Lathyrus odoratus*
Peach	*Prunus persica*
Peanut	*Arachis hypogaea*
Pear	*Pyrus communis*
Pecan	*Carya illinoensis*
Pepper (green)	*Capsicum annuum*
Peppermint	*Mentha piperita*
Persimmon (Japanese)	*Diospyros kaki*
Pigweed	*Amaranthus* spp.
Pine	*Pinus* spp.
Pine (eastern white)	*Pinus strobus*
Pine (loblolly)	*Pinus taeda*
Pine (Monterey)	*Pinus radiata*
Pine (red)	*Pinus resinosa*
Pine (Scotch)	*Pinus sylvestris*
Pine (shortleaf)	*Pinus echinata*
Pine (slash)	*Pinus elliottii*

Pine (western white)	*Pinus monticola*
Pineapple	*Ananas comosus*
Plantain	*Musa x paradisiaca*
Plum	*Prunus domestica*
Poplar (yellow)	*Liriodendron tulipifera*
Potato (Irish)	*Solanum tuberosum*
Potato (sweet)	*Ipomoea batatas*
Pumpkin	*Cucurbita pepo*
Quackgrass	*Agropyron repens*
Queen-of-the-meadow	*Filipendula ulmaria*
Radish	*Raphanus sativus*
Ragweed	*Ambrosia* spp.
Rape	*Brassica napus*
Raspberry	*Rubus* spp.
Red top	*Agrostis alba*
Reed (common)	*Phragmites communis*
Rice	*Oryza sativa*
Rose	*Rosa* spp.
Rutabaga	*Brassica napobrassica*
Rye	*Secale cereale*
Ryegrass (perennial)	*Lolium perenne*
Ryegrass (Italian)	*Lolium multiflorum*
Sandbur	*Cenchrus* spp.
Sesame	*Sesamum indicum*
Sorghum	*Sorghum vulgare*
Sorrel	*Rumex* spp.
Soybean	*Glycine max*
Spearmint	*Mentha spicata*
Spinach	*Spinacia oleracea*
Spruce	*Picea* spp.
Spruce (black)	*Picea mariana*
Spruce (Norway)	*Picea abies*
Spruce (red)	*Picea rubens*
Spruce (white)	*Picea glauca*
Squash	*Cucurbita pepo*
Strawberry	*Fragaria* spp.
Sudangrass	*Sorghum sudanense* stapf.
Sugarcane	*Saccharum officinarum*
Sunflower	*Helianthus annuus*
Switchgrass	*Panicum virgatum*
Sycamore (American)	*Platanus occidentalis*
Taro	*Colocasia* spp.
Tea	*Camellia* spp.
Tea (Labrador)	*Ledum* spp.
Thyme	*Thymus* spp.

Timothy	*Phleum pratense*
Tobacco	*Nicotiana tabacum*
Tomato	*Lycopersicon esculentum*
Trefoil (birdsfoot)	*Lotus corniculatus*
Tulip poplar	*Liriodendron tulipifera*
Tung	*Aleurites fordi*
Tupelo	*Nyssa* spp.
Turnip	*Brassica rapa*
Vetch	*Vicia* spp.
Vetch (crown)	*Coronilla varia*
Vetch (hairy)	*Vicia villosa*
Vetch (milk)	*Astragalus* spp.
Violet	*Viola* spp.
Walnut	*Juglans* spp.
Water-fern	*Azolla* spp.
Watergrass	*Echinochloa crusgalli*
Watermilfoil	*Myriophyllum* spp.
Wheat (common)	*Triticum aestivum*
Wheatgrass (range)	*Agropyron* spp.
Wheatgrass (crested)	*Agropyron cristatum*
Woodsorrel	*Oxalis* spp.
Yam	*Dioscorea* spp.
Zucchini squash	*Cucurbita pepo* var. *medullosa* Alef.

OTHER SPECIES

Oyster (eastern)	*Crassostrea virginica*
Oyster (Pacific)	*Crassostrea gigas*
Oyster (western, Olympia)	*Ostrea lurida*
Moth (codling)	*Carpocapsa pomonella*
Weevil (boll)	*Anthonomus grandis*
Beetle (Colorado potato)	*Leptinotarsa decemlineata*

Glossary of Certain Soil Terms Used in the Book

Alluvial soils - See Alluvium.

Alluvium - A general term for all detrital sediments deposited or in transit by streams, including gravel, sand, silt, clay, and all variations and mixture of these. Unless otherwise noted, alluvium is unconsolidated.

Black soils - A term used in Canada to describe soils with dark-colored surface horizons of the black (Chernozem) zone; includes Black Earth or Chernozem, Wiesenboden, Solonetz, etc.

Brown earth - See Brown gray earths or Yellow gray earths.

Brown gray earths - equivalent to the Alfisols and Inceptisols (See Soil classification).

Brunisolic soils - equivalent to the Inceptisols (See Soil classification).

Chernozem - equivalent to a Mollisol (See Soil classification).

Desert soils - equivalent to the Aridisols (See Soil classification).

Eluviated soils - See Eluviation.

Eluviation - The removal of soil material in suspension (or in solution) from a layer or layers of a soil. Usually, the loss of material in solution is described by the term "leaching." See illuviation and leaching.

Fluvioglacial - See Glaciofluvial deposits.

Glaciofluvial deposits - Material moved by glaciers and subsequently sorted and deposited by streams flowing from the melting ice. The deposits are stratified and may occur in the form of outwash plains, deltas, kames, eskers, and kame terraces.

Gley - indicates iron has been reduced or removed during soil formation as a result of saturation (See Soil classification).

Gray soils - may be equivalent to Alfisols (See Soil classification).

Gray forest soil - may be equivalent to an Alfisol (See Soil classification).

Illuviation - the process of deposition of soil material removed from one horizon to another in the soil, usually from an upper to a lower horizon in the soil profile. See Eluviation.

Lacustrine alluvium - See Lacustrine deposit.

Lacustrine deposit - Material deposited in lake water and later exposed either by lowering of the water level or by the elevation of the land.

Lateritic soils - equivalent to the Oxisols (See Soil classification).

Lateritic podzols - may be equivalent to Ultisols (See Soil classification).

Latosols - equivalent to the Oxisols (See Soil classification).

Leaching - the removal of materials in solution from the soil. See Eluviation.

Loess - material transported and deposited by wind and consisting of predominantly silt-sized particles.

Luvisolic soils - equivalent to Alfisols (See Soil classification).

Muck - highly decomposed organic material in which the original plant parts are not recognizable. Contains more mineral matter and is usually darker in color than peat. See Peat and Peat soil.

Peat - unconsolidated soil material consisting largely of undecomposed, or only slightly decomposed, organic matter accumulated under conditions of excessive moisture.

Peat (bog) soil - equivalent to a Histosol (See Soil classification).

Podzols - equivalent to the Spodosols (See Soil classification).

Pumice soils - soils developed from pumice rock.

Rendzinas - equivalent to the Mollisols (See Soil classification).

Soil classification - the systematic arrangement of soils into classes in one or more categories or levels of classification for a specific objective. Broad groupings are made on the basis of general characteristics; subdivisions on the basis of more detailed differences in specific properties.

The name and number of categories of the system of soil classification used by the National Cooperative Soil Survey since 1965 are: Soil Orders – 10; Soil Suborders – 47; Soil Great Groups – 185; Soil Subgroups – 970; Soil Families – 4,500; Soil Series – 10,466. The relationship between the orders of the present system and approximate equivalents of the previous system used in the United States are shown in Table A.1.

TABLE A.1. A comparison of the present United States soil classification system adopted in 1965 with the approximate equivalents in use before 1965.[a]

Soil order	Approximate equivalents
Alfisols	Gray Brown Podzolic, Gray Wooded soils, Noncalcic Brown soils, Degraded Chernozem, and associated Planosols and some Half-Bog soils
Aridisols	Desert, Reddish Desert, Sierozem, Solonchak, some Brown and Reddish Brown soils, and associated Solonetz soils
Entisols	Azonal soils and some Low-Humic Gley soils
Histosols	Bog soils
Inceptisols	Ando, Sol Brun Acide, some Brown Forest, Low-Humic Gley, and Humic Gley soils
Mollisols	Chestnut, Chernozem, Brunizem (Prairie), Rendzina, some Brown, Brown Forest, and associated Solonetz and Humic Gley soils
Oxisols	Laterite soils, Latosols
Spodosols	Podzols, Brown Podzolic soils, and Ground-Water Podzols
Ultisols	Red-Yellow Podzolic soils, Reddish Brown Lateritic soils of the U.S., and associated Planosols and Half-Bog soils
Vertisols	Grumusols

[a] *Source:* Soil Survey Staff, *Soil Taxonomy: A Basic System of Soil Classification for Making and Interpreting Soil Surveys,* USDA–Soil Conservation Service Agriculture Handbook 436 (December 1975), pp. 433–35.

Soil Classification

Order Order is the highest general category used in the soil classification system. The properties selected to distinguish the orders are reflections of the soil forming processes and the kinds of horizons present. The 10 soil orders are:

Alfisols – Soils with gray to brown surface horizons, medium to high supply of bases, and B horizons of illuvial clay accumulation. These soils form mostly under forest or savanna vegetation in climates with slight to pronounced seasonal moisture deficit.

Aridisols – Soils with pedogenic horizons, low in organic matter, that are never moist as long as three consecutive months. They have an ochric epipedon that is normally soft when dry or that has distinct structure. In addition, they have one or more of the following diagnostic horizons: argillic, natric, cambic, calcic, petrocalcic, gypsic or salic, or a duripan.

Entisols – Soils that have no diagnostic pedogenic horizons. They may be found in virtually any climate on very recent geomorphic surfaces, either on steep slopes that are undergoing active erosion or on fans and floodplains where the recently eroded materials are deposited. They may also be on older geomorphic surfaces if the soils have been recently disturbed to such depths that the horizons have been destroyed or if the parent materials are resistant to alteration, as is quartz.

Histosols – Soils formed from organic soil materials.

Inceptisols – Soils that are usually moist with pedogenic horizons of alteration of parent materials but not of illuviation. Generally, the direction of soil development is not yet evident from the marks left by the various soil-forming processes or the marks are too weak to classify in another soil order.

Mollisols – Soils with nearly black, organic-rich surface horizons and high supply of bases. These are soils that have decomposition and accumulation of relatively large amounts of organic matter in the presence of calcium. They have mollic epipedons and base saturation greater than 50% (NH_4OAc) in any cambic or argillic horizon. They lack the characteristics of Vertisols and must not have oxic or spodic horizons.

Oxisols – Soils with residual accumulations of inactive clays, free oxides, kaolin, and quartz. They occur mostly in tropical climates on old land surfaces.

Spodosols – Soils with illuvial accumulations of amorphous materials in subsurface horizons. The amorphous material is organic matter and compounds of aluminum and usually iron. These soils are formed in acid mainly coarse-textured materials in humid and mostly cool or temperate climates.

Ultisols – Soils that are low in supply of bases and have subsurface horizons of illuvial clay accumulation. They are usually moist, but during the warm season of the year some are dry part of the time. The balance between liberation of bases by weathering and removal by leaching is normally such that a permanent agriculture is impossible without fertilizers or shifting cultivation.

Vertisols – Clayey soils with high shrink-swell potential that have wide, deep cracks when dry. Most of these soils have distinct wet and dry periods throughout the year.

Suborder This category narrows the ranges in soil moisture and temperature regimes, kinds of horizons, and composition, according to which of these is most important. Moisture and/or temperature or soil properties associated with them are used to define suborders of Alfisols, Mollisols, Oxisols, Ultisols, and Vertisols. Kinds of horizons are used for Aridisols, composition for Histosols and Spodosols, and combinations for Entisols and Inceptisols.

Great Group The classes in this category contain soils that have the same kinds of horizons in the same sequence and have similar moisture and temperature regimes. Exceptions to the horizon sequences are made for horizons near the surface that may get mixed or lost by erosion if plowed.

Subgroup The great groups are subdivided into subgroups that show the central properties of the great group, intergrade subgroups that show properties of more than one great group, and other subgroups for soils with atypical properties that are not characteristic of any great group.

Family Families are defined largely on the basis of physical, chemical, and mineralogic properties of importance to plant growth.

Series The soil series is a group of soils having horizons similar in differentiating characteristics and arrangement in the soil profile, except for texture of the surface, slope, and erosion.

Soil horizon - A layer of soil or soil material approximately parallel to the land surface and differing from adjacent genetically related layers in physical, chemical, and biological properties or characteristics such as color, structure, texture, consistency, kinds and number of organisms present, degree of acidity or alkalinity, etc. See Tables A.2 and A.3.

TABLE A.2. Designation and description of master soil horizons and layers. [a]

Horizon designation	Description
O	Organic horizons of mineral soils. Horizons: (i) formed or forming in the upper part of mineral soils above the mineral part; (ii) dominated by fresh or partly decomposed organic material; and (iii) containing >30% organic matter if the mineral fraction is >50% clay, or >20% organic matter if the mineral fraction has no clay. Intermediate clay content requires proportional organic matter content.
O1	Organic horizons in which essentially the original form of most vegetative matter is visible to the naked eye. The O1 corresponds to the L (litter) and some F (fermentation) layers in forest soils designations, and to the horizon formerly called Aoo.
O2	Organic horizons in which the original form of most plant or animal matter cannot be recognized with the naked eye. The O2 corresponds to the H (humus) and some F (fermentation) layers in forest soils designations, and to the horizon formerly called Ao.
A	Mineral horizons consisting of: (i) horizons of organic matter accumulation formed or forming at or adjacent to the surface; (ii) horizons that have lost clay, iron, or aluminum with resultant concentration of quartz or other resistant minerals of sand or silt size; or (iii) horizons dominated by (i) or (ii) above but transitional to an underlying B or C.
A1	Mineral horizons, formed or forming at or adjacent to the surface, in which the feature emphasized is an accumulation of humified organic matter intimately associated with the mineral fraction.
A2	Mineral horizons in which the feature emphasized is loss of clay, iron, or aluminum, with resultant concentration of quartz or other resistant minerals in sand and silt sizes.
A3	A transitional horizon between A and B, dominated by properties characteristic of an overlying A1 or A2 but having some subordinate properties of an underlying B.
AB	A horizon transitional between A and B, having an upper part dominated by properties of A and a lower part dominated by properties of B, and the two parts cannot be conveniently separated into A3 and B1.
A & B	Horizons that would qualify for A2 except for included parts constituting <50% of the volume that would qualify as B.
AC	A horizon transitional between A and C, having subordinate properties of both A and C, but not dominated by properties characteristic of either A or C.
B & A	Any horizon qualifying as B in >50% of its volume, including parts that qualify as A2.
B	Horizons in which the dominant feature or features is one or more of the following: (i) an illuvial concentration of silicate clay, iron, aluminum, or humus, alone or in combination; (ii) a residual concentration of sesquioxides or silicate clays, alone or mixed, that has formed by means other than solution and removal of carbonates or more soluble salts; (iii) coatings of sesquioxides adequate to give conspicuously darker, stronger, or redder colors than overlying and underlying horizons in the same sequum but without apparent illuviation of iron and not genetically related to B horizons that meet requirements of (i) or (ii) in the same sequum; or (iv) an alteration of material from its original condition in sequums lacking conditions defined in (i), (ii), and (iii) that obliterates original rock structure, that forms silicate

TABLE A.2. *Continued*

	clays, liberates oxides, or both, and that forms granular, blocky, or prismatic structure if textures are such that volume changes accompanying changes in moisture.
B1	A transitional horizon between B and A1 or between B and A2 in which the horizon is dominated by properties of an underlying B2 but has some subordinate properties of an overlying A1 or A2.
B2	That part of the B horizon where the properties on which the B is based are clearly expressed characteristics, indicating that the horizon is transitional to an adjacent overlying A or an adjacent underlying C or R.
B3	A transitional horizon between B and C or R in which the properties diagnostic of an overlying B2 are clearly expressed but are associated with clearly expressed properties characteristic of C or R.
C	A mineral horizon or layer, excluding bedrock, that is either like or unlike the material from which the solum is presumed to have formed, relatively little affected by pedogenic processes, and lacking properties diagnostic of A or B but including materials modified by: (i) weathering outside the zone of major biological activity; (ii) reversible cementation, development of brittleness, development of high bulk density, and other properties characteristic of fragipans; (iii) gleying; (iv) accumulation of calcium or magnesium carbonate or more soluble salts; (v) cementation by accumulations such as calcium or magnesium carbonate or more soluble salts; or (vi) cementation by alkali-soluble siliceous material or by iron and silica.
R	Underlying consolidated bedrock; such as granite, sandstone, or limestone. If presumed to be like the parent rock from which the adjacent overlying layer or horizon was formed, the symbol R is used alone. If presumed to be unlike the overlying material, the R is preceded by a Roman numeral denoting lithologic discontinuity, such as II R.

a Source: In *Glossary of Soil Science Terms,* Soil Science Society of America, Madison, WI, 1984, 38 pp. With permission of Soil Sci Soc of America.

TABLE A.3. New designations for soil horizons and layers. *a*

Master Horizons and Layers

O horizons - Layers dominated by organic material, except limnic layers that are organic.

A horizons - Mineral horizons that formed at the surface or below an O horizon and (i) are characterized by an accumulation of humified organic matter intimately mixed with the mineral fraction and not dominated by properties characteristic of E or B horizons; or (ii) have properties resulting from cultivation, pasturing, or similar kinds of disturbance.

E horizons - Mineral horizons in which the main feature is loss of silicate clay, iron, aluminum, or some combination of these, leaving a concentration of sand and silt particles of quartz or other resistant materials.

B horizons - Horizons that formed below an A, E, or O horizon and are dominated by (i) carbonates, gypsum, or silica, alone or in combination; (ii) evidence of removal of carbonates; (iii) concentrations of sesquioxides; (iv) alterations that form silicate clay; (v) formation of granular, blocky, or prismatic structure; or (vi) combination of these.

C horizons - Horizons or layers, excluding hard bedrock, that are little affected by pedogenic processes and lack properties of O, A, E, or B horizons. Most are mineral layers, but limnic layers, whether organic or inorganic, are included.

R layers - Hard bedrock including granite, basalt, quartzite and indurated limestone or sandstone that is sufficiently coherent to make hand digging impractical.

TABLE A.3. *Continued*

The following is a synoptic comparison between the old (1962) and new (1981) designations for soil horizons and layers.[b]

1. The purpose of using designations remains unchanged. They reflect the describer's interpretations of the genetic relationships between the layers in a soil.
2. Capital letters, lowercase letters, and Arabic numerals are used to form the horizon designators.
   a. Capital letters are used to designate master horizons. This convention is unchanged.
   b. Lowercase letters are used as suffixes to indicate specific characteristics of the master horizon. This convention is unchanged.
   c. Arabic numerals are used as suffixes to indicate vertical subdivisions within a horizon and as prefixes to indicate discontinuities. This is a change. Previously, Arabic numerals were used as suffixes to indicate a kind of O, A, or B horizon and to indicate vertical subdivisions of a horizon, and Roman numerals were used as prefixes to indicate lithologic discontinuities.
3. The symbols used for many horizon characteristics have been changed. The comparison of symbols used to designate master horizons and layers and subordinate distinctions within master horizons and layers can be only approximate. Some designations in the old system can best be equated with a combination of master horizon symbol and subordinate symbol in the new system.

Master Horizons and Layers

Old (1962)	New (1981)
O	O
O1	Oi, Oe
O2	Oa, Oe
A	A
A1	A
A2	E
A3	AB or EB
AB	—
A & B	E/B
AC	AC
B	B
B1	BA or BE
B & A	B/E
B2	B or Bw
B3	BC or CB
C	C
R	R

TABLE A.3. *Continued*

Subordinate Distinctions Within Master Horizons

Old	New	
—	a	highly decomposed organic matter
b	b	buried soil horizon
cn	c	concretions or nodules
—	e	intermediately decomposed organic matter
f	f	frozen soil
g	g	strong gleying
h	h	illuvial accumulation of organic matter
—	i	slightly decomposed organic matter
ca	k	accumulation of carbonates
m	m	strong cementation
sa	n	accumulation of sodium
—	o	residual accumulation of sesquioxides
p	p	plowing or other disturbance
si	q	accumulation of silica
r	r	weathered or soft bedrock
ir	s	illuvial accumulation of sesquioxides
t	t	accumulation of clay
—	v	plinthite
—	w	color or structural B
x	x	fragipan character
cs	y	accumulation of gypsum
sa	z	accumulation of salts

4. The prime symbol is used in both the old and new systems, but the conventions for using it are different. In the old system, the prime was used to identify the lower sequum of a soil having two sequa, although not for a buried soil. In the new system, it may be appropriate to give the same designation to two or more horizons in a pedon if the horizons are separated by a horizon of a different kind. The prime is used on the lower of the two horizons having identical letter designations. If three horizons have identical designations, a double prime is used on the lowest.

a Source: In *Glossary of Soil Science Terms,* Soil Science Society of America, Madison, WI, 1984, 38 pp.
b Source: Guthrie, R. L. and J. E. Witty. 1982. *Soil Sci Soc Am J* 46:443-444, with permission of the authors and the Soil Sci. Soc. of America.

Tundra soils - The treeless land in arctic and alpine regions, varying from bare areas to various types of vegetation consisting of grasses, sedges, forbs, dwarf shrubs, mosses, and lichens.
Yellow-brown earths - equivalent to Inceptisols (See Soil classification).
Yellow-gray earths - equivalent to Alfisols (See Soil classification).

Index